Illness and the Environment

Illness and the Environment

A Reader in Contested Medicine

EDITED BY

*Steve Kroll-Smith, Phil Brown, and
Valerie J. Gunter*

New York University Press

NEW YORK AND LONDON

MW

NEW YORK UNIVERSITY PRESS
New York and London

Library of Congress Cataloging-in-Publication Data
Illness and the environment : a reader in contested medicine / edited by Steve
Kroll-Smith, Phil Brown, and Valerie J. Gunter.
p. cm.
Includes bibliographical references and index.
ISBN 0-8147-4728-0 (cloth : alk. paper) — ISBN 0-8147-4729-9 (pbk. : alk. paper)
1. Environmental health — Public opinion.   2. Environmental health — Political aspects.   3.
Environmental health — Government policy.   4. Environmentally induced diseases — Public
opinion.   I. Kroll-Smith, J. Stephen, 1947–   II. Brown, Phil.   III. Gunter, Valerie J. (Valerie
Jan), 1961–

RA566.I45 2000
615.9'02 — dc21   00-036125

New York University Press books are printed on acid-free paper, and their
binding materials are chosen for strength and durability.

Manufactured in the United States of America

10 9 8 7 6 5 4 3 2 1

07/24/03

This book is dedicated to
Amanda, Emma, Michael, Liza,
and Ella

Contents

Preface   *xi*
Acknowledgments   *xiii*

Introduction: Environments and Diseases in a Postnatural
World   1
Steve Kroll-Smith, Phil Brown, and Valerie J. Gunter

PART 1. Setting the Stage

1. Knowledge, Citizens, and Organizations: An Overview of
    Environments, Diseases, and Social Conflict   9
    Phil Brown, Steve Kroll-Smith, and Valerie J. Gunter

PART 2. Environments and Diseases: Professional Boundaries and the
Problem of Knowing

2. Limits of Epidemiology   29
    Steve Wing

3. Physicians' Knowledge, Attitudes, and Practice Regarding
    Environmental Health Hazards   46
    Phil Brown and Judith Kirwan Kelley

4. Environmental Illness as a Practical Epistemology and a
    Source of Professional Confusion   72
    Steve Kroll-Smith and H. Hugh Floyd

PART 3. Measurement Disputes and Health Policy

5. Environmental Endocrine Hypothesis and Public Policy   95
    Sheldon Krimsky

6. Who Cares If the Rat Dies? Rodents, Risks, and Humans in
    the Science of Food Safety   108
    Lawrence Busch, Keiko Tanaka, and Valerie J. Gunter

7. Threshold Limit Values: Historical Perspectives and Current
    Practice   120
    Grace E. Ziem and Barry I. Castleman

8. Threshold Limit Values in the 1990s and Beyond: A Follow-Up   135
    David Allen

PART 4. Toxins in the Workplace

9. An Axe to Grind: Class Relations and Silicosis in a 19th-Century Factory   145
    Janet Siskind

10. From Dust to Dust: The Birth and Re-Birth of National Concern about Silicosis   162
    David Rosner and Gerald E. Markowitz

11. Farmworker and Farmer Perceptions of Farmworker Agricultural Chemical Exposure in North Carolina   175
    Sara A. Quandt, Thomas A. Arcury, Colin K. Austin, and Rosa M. Saavedra

12. Competing Conceptions of Safety: High-Risk Workers or High-Risk Work?   192
    Elaine Draper

PART 5. Toxins in the Community

13. Pollution, Politics, and Uncertainty: Environmental Epidemiology in North-East England   217
    Peter Phillimore, Suzanne Moffatt, Eve Hudson, and Dawn Downey

14. Round and Round It Goes: The Epidemiology of Childhood Lead Poisoning, 1950–1990   235
    Barbara Berney

15. Lead Contamination in the 1990s and Beyond: A Follow-Up   258
    Patricia Widener

16. Suffering, Legitimacy, and Healing: The Bhopal Case, Critical Events   270
    Veena Das

PART 6. Living with Environments and Contested Diseases

17. Time   289
    Sandra Steingraber

18. A Cancer Death   308
    Martha Balshem

19. Notes from a Human Canary   333
    Lynn Lawson

PART 7. Citizen Responses to Contested Medicine

20. Reframing Endometriosis: From "Career Woman's Disease"
     to Environment/Body Connections   345
     Stella M. Čapek

21. Popular Epidemiology and Toxic Waste Contamination: Lay
     and Professional Ways of Knowing   364
     Phil Brown

22. Environmental Movements and Expert Knowledge: Evidence
     for a New Populism   384
     Stephen R. Couch and Steve Kroll-Smith

PART 8. Setting the Environmental Health Agenda

23. Competing Paradigms in the Assessment of Latent Disorders:
     The Case of Agent Orange   409
     Wilbur J. Scott

24. Environmental Politics and Science: The Case of PBB
     Contamination in Michigan   430
     Michael R. Reich

25. Environmental Health Research: Setting an Agenda by
     Spinning Our Wheels or Climbing the Mountain?   453
     John Eyles

     Contributors   473
     Index   475

Preface

Today, the words "environment" and "illness" are closely linked. It is hard to ignore the sometimes loud, sometimes measured, but almost always discernible public debate over the contamination of environments and their effects on human health and well-being. Sociologists have not ignored these debates, though until quite recently they have pursued them in two distinct specialities: the sociology of medicine and the sociology of the environment. Medical sociologists tended to examine the problem of environments and illnesses from the perspectives of doctor-patient relationships, lay explanatory models, and lay-professional differences in disease recognition and causality. Environmental sociologists, on the other hand, emphasized the sociology of risk, social movements, and community conflict, typically without reference to health as a social variable. Thus one group of sociologists talked about environments in the language of illness and knowledge while the other used the language of risks, movements, and community. The complex relationship between the two was not seriously examined.

The fact that these two specialities are now moving, however tenuously, toward some still ill-defined collaboration is an interesting commentary on the development of academic disciplines. The impetus to joining medical sociology and the sociology of the environment did not issue from respected sociologists or from federal or foundation research initiatives. Rather, the call to link the two originated with local and national citizen action groups who recognized the need to combine questions of illness, health, and environment in a context of grassroots political action. The toxic waste movement and the environmental justice movement focus considerable attention on the intersection of medicine, techniques of environmental measurement, and social action. This interdisciplinary focus is the starting point for thousands of citizens' groups which argue for public recognition of their miseries, lobby legislators, sue polluters, and challenge the sciences of medicine and epidemiology. Their sustained labors are drawing attention to the need to join the good insights of both medical and environmental sociology.

This book is a scholarly recognition of a fact long recognized by citizen activists: the modern relationships of environments to illnesses are almost always contested and are the sources of considerable political and technical debates. It is the first reader of its kind to bring together previously published and original essays on the complicated relationships of bodies, biospheres, science, and politics. Students of environmental and medical sociology and their colleagues in social movements and public policy will, hopefully, find in these essays ideas and arguments worth pursuing and challenging.

The authors would like to thank the editor-in-chief of New York University Press, Niko Pfund, who first encouraged this project and responded patiently and quickly to our steady stream of e-mail questions and concerns. We also want to thank Patricia Widener, a graduate student at the University of New Orleans, for her work in securing permission letters, many hours of library labor chasing down articles and book chapters, and for writing a follow-up to the discussion of children and lead poisoning. Amanda Smith also deserves mention for her patient work in the frenzied last days of editing and arranging. Steve Kroll-Smith would like to thank the University of New Orleans for research money to reduce his teaching load and purchase a computer capable of chatting with his colleagues, both near and far.

Acknowledgments

We would like to thank the following for kindly granting permission to reprint essays for this volume:

"The Limits of Epidemiology" by Steve Wing from *Medicine & Global Survival* 1, no. 2 (June 1994): 74–86. Reprinted by permission.

"Environmental Endocrine Hypothesis and Public Policy" by Sheldon Krimsky from *Comments Toxicology* 5, nos. 4–5 (1996): 487–502. Reprinted by permission of Gordon and Breach Publishers.

"Threshold Limit Values: Historical Perspectives and Current Practice" by Grace E. Ziem and Barry I. Castleman from *Journal of Occupational and Environmental Medicine* 31, no. 11 (November 1989): 910–18. Reprinted by permission.

"An Axe to Grind: Class Relations and Silicosis in a 19th Century Factory" by Janet Siskind from *Medical Anthropology Quarterly* 2, no. 3 (September 1988): 199–214. Reproduced by permission of the American Anthropological Association. Not for further reproduction.

"Farmworker and Farmer Perceptions of Farmworker Agricultural Chemical Exposure in North Carolina" by Sara A. Quandt from *Human Organization* 57, no. 3 (fall 1998): 359–68. Reproduced by permission of the Society for Applied Anthropology.

"Competing Conceptions of Safety: High-Risk Workers or High-Risk Work?" by Elaine Draper from her book *Risky Business: Genetic Testing and Exclusionary Practices in the Hazardous Workplace.* (New York: Cambridge University Press, 1991), pp. 37–58, 209–96. Copyright © 1991 by Cambridge University Press. Reprinted by permission.

"Round and Round It Goes: The Epidemiology of Childhood Lead Poisoning, 1950–1990" by Barbara Berney from *The Milbank Quarterly* 71, no. 1 (1993): 3–39. Copyright © 1993 by Milbank Memorial Fund. Reprinted by permission of Blackwell Publishers.

"Suffering, Legitimacy, and Healing: The Bhopal Case, Critical Events" by Veena Das from her book *Critical Events: An Anthropological Perspective on Contemporary India* (New Delhi: Oxford University Press, 1995), pp. 140–64. Copyright © 1995 by Oxford University Press. Reprinted by permission of Oxford University Press, New Delhi.

"Time" by Sandra Steingraber from her book *Living Downstream: An Ecologist Looks at Cancer and the Environment* (New York: Perseus Books, 1997), pp. 31–56. Copyright © 1997 by Sandra Steingraber. Reprinted by permission of Perseus Books Publishers, a member of Perseus Books, L.L.C.

"A Cancer Death" by Martha Balshem from *Cancer in the Community: Class and Medical Authority* (Washington, D.C.: Smithsonian Institution Press, 1993), pp. 91–124. Copyright © 1993 by Smithsonian Institution Press. Reprinted by permission of Smithsonian Institution Press.

"Popular Epidemiology and Toxic Waste Contamination: Lay and Professional Ways of Knowing" by Phil Brown from *Journal of Health and Social Behavior* 33 (September 1992): 267–81. Reprinted by permission.

"Environmental Movements and Expert Knowledge: Evidence for a New Populism" by Stephen R. Couch and Steve Kroll-Smith from *International Journal of Contemporary Sociology* 34, no. 2 (October 1997): 185–210. Reprinted by permission.

"Competing Paradigms in the Assessment of Latent Disorders: The Case of Agent Orange" by Wilbur J. Scott from *Social Problems* 35, no. 2 (April 1988): 145–61. Copyright © 1988 by The Society for the Study of Social Problems. Reprinted by permission.

"Environmental Politics and Science: The Case of PBB Contamination in Michigan" by Michael R. Reich from *American Journal of Public Health* 73, no. 3 (March 1983): 302–13. Copyright © 1983 by the American Public Health Association. Reprinted by permission.

"Environmental Health Research: Setting an Agenda by Spinning Our Wheels or Climbing the Mountain?" by John Eyles from *Health and Place* 3, no. 1 (1997): 1–13. Copyright © 1997 by Elsevier Science. Reprinted by permission of Elsevier Science.

Introduction

Environments and Diseases in a Postnatural World

Steve Kroll-Smith, Phil Brown, and Valerie J. Gunter

> Since World War II there has been an unprecedented growth of biological research; yet we remain astonishingly ignorant of the profound changes that, during that same period, have occurred in our own biological surroundings.
>
> —Barry Commoner (1972, 188)

Barry Commoner was not referring to facts about nature when he wrote about the profound ignorance of biology. He knew quite well that science was always flush with facts about air, water, soil, climates, vertebrae, invertebrates, and so on; the problem was interpretation. What do all these certainties mean? Granted, there are no facts independent of some theory, but the level of theorizing falls miserably short of providing any coherent picture of nature. In the almost three decades since Commoner pointed to the absence of a systemic understanding of nature and its interdependent relationship with humans, little has changed. There are, of course, more facts.

A 1996 report by the International Union for Conservation of Nature (IUCN) lists one hundred and fifty-six amphibian species as extinct, critical, endangered, or vulnerable to extinction. This represents 25 percent of all the amphibians on earth (RACHEL #590). The Nature Conservancy, a U.S. organization, in 1996 surveyed the status of 20,481 species of plants and animals in the United States and reported that 37.9 percent of amphibians are in danger of becoming extinct (RACHEL #590). Amphibians are nature's bellwether creatures. Closely associated to water and earth, their health is a key predictor of the health of the environment.

At the close of this century there is 25 percent more carbon dioxide in the atmosphere than there was at its beginning. Almost twenty years ago, an immense hole appeared in the ozone layer over the South Pole. If atmospheric concentrations of greenhouse gases increase as predicted, the average global temperature will increase by 1 to 3.5 degrees Celsius by the year 2100 (Shindell and Raso 1998).

To keep their houses smelling clean and free of bugs, Americans use an estimated

62.7 million pounds (28.5 million kilograms) of pesticides and 278.5 million pounds (126.6 million kilograms] of antimicrobials (disinfectants) each year (RACHEL #588). A study of indoor air in homes in Jacksonville, Florida, detected pesticides in the air in every home (Pogoda and Preston-Martin 1997). The EPA, however, can ensure the safety of only six out of six hundred active pesticide ingredients under its control (Duehring and Wilson 1994, 10). Moreover, the EPA has identified over nine hundred volatile organic chemicals in ordinary indoor environments, including offices and houses (Kroll-Smith and Floyd 1997, 9).

Every year approximately one thousand new industrial chemicals are added to the environment. In 1995, 70,000 chemicals were used in commercial manufacturing. Only 2 percent of them were tested for human health effects, and 70 percent had not been tested for health effects of any type. The chemical industry continues to grow at a rate of 3.5 percent each year, thus doubling in size every twenty years (RACHEL #197, #199).

In 1986, an EPA Executive Summary on chemicals in human tissue found measurable levels of styrene and ethyl phenol in 100 percent of adults living in the United States. The Summary also found 96 percent of adults with clinical levels of chlorobenzene, benzene, and ethyl benzene; 91 percent with toluene; and 83 percent with polychlorinated byphenols (Stanley 1986). Each of these chemicals is classified either as a human carcinogen or a probable human carcinogen (Harte et al. 1991).

Finally, a longitudinal *Index of Environmental Trends* published in April 1995 by the National Center for Economic and Security Alternatives in Washington, DC, and based on reliable data gathered by national governments, measured twenty-five-year trends in a wide range of serious environmental problems facing industrial societies. Twenty-one indicators measured the amount of environmental deterioration caused by human intervention into environments.

Ranking from least to most environmental deterioration by percentage of biosphere affected in 1970–1995, we find:

Denmark: –10.6%	Japan: –19.4%
Netherlands: –11.4%	United States: –22.1%
Britain: –14.3%	Canada: –38.1%
Sweden: –15.5%	France: –41.2%
West Germany: –16.5%	

(National Center for Economic and Security Alternatives 1995)

Amidst this bewildering array of facts we sit, as Commoner reminds us, ignorant about nature. But there is one conclusion we are invited to draw from these truths: there is little doubt we live in what McKibben calls a "postnatural world" (1990, 60). Nature as a natural phenomenon will survive in our imaginations, but it is no longer the nature that nurtured us and assured our survival as a species. It is a manufactured nature. We made it, though unintentionally and without a clear understanding of its effects on human well-being. The contemporary idea of disasters is a useful way to characterize this shift from a natural to a postnatural world.

When nature was just nature, the word "disaster" almost always referred to natural calamities, floods, hurricanes, tornadoes, droughts, and so on. These "acts of

God" were not welcomed events, but they were expected, if not often well planned for. In a postnatural world, however, where nature is increasingly mediated by human actions, disasters are less likely to be natural and more likely to be the results of human error, malfeasance, or simply bad planning. On the one hand, the rapid increase in technological disasters forces us to acknowledge that it is not God or nature that always puts us in harm's way; on the other hand, the idea that humans bear some responsibility for what were typically called natural disasters implicates culture and politics in explaining what were once considered the unfortunate consequences of fate, bad luck, the divine, or simply nature's will.

Not long ago, we thought nature was on the side of human life, in spite of the periodic disaster. In a postnatural world, however, a manufactured nature appears to have shifted loyalties and is now a known, or, more often, suspected cause of sickness and death. In his state of the world survey sponsored by the Gallup Institute, Riley Dunlap and his collaborators (Dunlap, Gallup, and Gallup 1992) found a significant increase in concern throughout the world for the deleterious effects of manufactured environments on health. In 1982, for example, 27 percent of a random sample of British citizens believed their own health was affected a "great deal" or a "fair amount" by local and global environments. By 1992, on the other hand, 53 percent of British citizens reported their health was adversely affected by environmental causes, an increase of almost 100 percent. Moreover, 79 percent of these same respondents believed environments would affect the health of their children over the next twenty-five years. The numbers in the United States also show a substantial increase in concern. In 1982, 45 percent of a random sample of Americans report the environment adversely affecting their health. By 1992 that number had increased to 67 percent. Moreover, 83 percent of the U.S. sample believed the health of their children would suffer over the next twenty-five years because of environmental problems.

Indeed, all of the twenty-two countries surveyed reported substantial increases from 1982 to 1992 in the number of people who believed their bodies were at risk from environmentally induced illnesses. Likewise, an average of 73 percent of respondents from all the countries surveyed believed environments would pose health risks for their children over the next twenty-five years (Dunlap, Gallup, and Gallup 1992). There appears to be a close relationship between an increasingly postnatural world and an increasing anxiety over the role of environments in the causes of sickness and death. Further evidence of this relationship can be found in the increased media coverage of diseases and environments. A search of the *New York Times* for 1998 using the key words "toxic waste and disease," "environmental asthma," "sick building syndrome," and "water pollution and disease," found one hundred and fifteen articles. Indeed, there is likely an interactive effect between increased media coverage and increased worry about environment and health.

The point is a simple one: the problem of diseases and environments is endemic to a postnatural world. Its resolution, however, is far from simple. While the belief that the "American public is being swept by a medical epidemic characterized by doubt of certitude, recognition of error, and discovery of hazard" (Cournand 1977, 700), seems excessive, the relationship between environment and disease in a postnatural world is a clearly complex and increasingly serious public problem. The relationship of a postnatural environment to disease is often a source of conflict and controversy between citizens, experts, social movements, industries, and government agencies. The relationship of diseases to environments is typically contested, open to several competing resolutions, and an important area of sociological inquiry.

This book uses original and previously published material to examine the controversies surrounding the relationships of environments to disease in a postnatural world. It assumes that a key sociological issue in these controversies is the problem of knowing or knowledge. A controversy like pesticide application and sickness among farmworkers, for example, is expressed in the social distribution of knowledge between farmworkers who experience the pesticides in one fashion, their employers who know them in another, the pesticide manufacturers, and the EPA. The problems of identifying cancer clusters and their relationships to low-level exposures to groundwater contamination issue in part from the absence of certain knowledge about the relationship between cancers and low-level exposures. A problem with many relatively new disorders, like endometriosis, chronic fatigue syndrome, and multiple chemical sensitivity, is the likelihood that their etiologies are, at least in part, environmental, though the clinical connection between environments and disorders remains contentious and open to multiple interpretations.

Contested in these medical disputes is what will count as valid knowledge of the relationships between environments and diseases. Validity, as we will see, depends more often on social movements, lawsuits, and the court of public opinion than on clinical and epidemiological research. While the historical voice of science as the final arbiter of important mysteries is by no means over, it plainly does not utter the uncontested last word in matters of human health and environments.

The book is organized into eight parts, each representing a key theme in the sociological study of environments and illnesses. "Setting the Stage," an original essay prepared by the editors, is a general introduction to the sources of conflict among medical professionals, citizens, legislators, and other stakeholders over how environments specifically cause diseases, who is responsible, what constitutes a victim, what is a proper course of treatment, and compensation. It pays particular attention to the complicated problem of uncertain knowledge in practically all environment and health controversies.

The essays in Part 2, "Environments and Disease: Professional Boundaries and the Problem of Knowing," direct attention to the problems medical researchers and clinicians experience in trying to make sense of the body and its well-being in postnatural environments. Part 3, "Measurement Disputes and Health Policy," focuses on the difficulties in measuring the adverse effects of the environment on health and the politicized contexts of shaping environmental health policies. The next two parts, "Toxins in the Workplace" and "Toxins in the Community," discuss the contested and onerous issues in identifying and responding to problems of environments and illnesses at work sites and neighborhoods. Part 6, "Living with Environments and Contested Diseases," provides a close-up look at lives fundamentally changed by contested diseases whose causes others may dispute, but are believed by the afflicted to be environmental. "Citizen Responses to Contested Medicine" is the focus of Part 7, which examines aspects of citizen health movements organized to make the relationships of environments to bodies and well-being a political and public-health issue. Finally, Part 8 on "Setting the Environmental Health Agenda" is an extended discussion of several of the key issues that must be addressed if the complicated relationships of bodies, illnesses, diseases, and postnatural environments are to be addressed in a judicious and humane manner.

Each section is prefaced by an introduction that provides an overview of the discussions represented in the essays to follow. At the end of each chapter a list of study questions invites further reflection on key issues. Two of the unpublished essays are "follow-ups" that bring two important topics in illnesses and environments— lead and threshold limit values—up to date. These follow-up pieces insure the topic is current, while providing both students and researchers with a long view of the development of key controversies.

Published essays were found in a diverse array of journals, representing multiple disciplines such as sociology, public health, anthropology, medicine and toxicology. Two criteria guided our selection. First, the chapter comments on key aspects of the social, medical, and political controversies that predictably accompany environment and illness issues. Second, each piece is clearly written and accessible to upper-division undergraduate and graduate students.

This combination of provocative research—new and old—is intended to provide a foundation for study and pedagogy and, as the first collection of its kind, promote further work on the deeply contentious debates swirling around the problems of bodies, environments, and the politics of disease.

REFERENCES

Commoner, Barry. 1972. *The Closing Circle.* New York: Bantam Books.
Cournand, A. 1977. "The Code of the Scientist and Its Relationship to Ethics." *Science* 198: 699–705.
Duehring, Cindy, and Cynthia Wilson. 1994. *The Human Consequences of the Chemical Problem.* White Sulphur Springs, MT: Chemical Injury Information Network.
Dunlap, Riley E., George H. Gallup, Jr., and Alec M. Gallup. 1992. *The Health of the Planet Survey.* Princeton, NJ: George H. Gallup International Institute.
Harte, John, Cheryl Holdren, Richard Schneider, and Christine Shirley. 1991. *Toxics A to Z.* Berkeley: University of California Press.
Kroll-Smith, Steve, and H. Hugh Floyd. 1997. *Bodies in Protest: Environmental Illness and the Struggle over Medical Knowledge.* New York: New York University Press.
McKibben, Bill. 1990. *The End of Nature.* New York: Anchor Books.
National Center for Economic and Security Alternatives. *Index of Environmental Trends.* Washington, DC: National Center for Economic and Security Alternatives, 1995.
Pogoda, Janice M., and Susan Preston-Martin. 1997. "Household Pesticides and Risk of Pediatric Brain Tumors." *Environmental Health Perspectives* 105, no. 11: 1214–1220.
RACHEL: Environment and Health Weekly. Environment and Health Research. 5036 Annapolis, MD 21403 (cited issues 590, 588, 197, 199).
Shindell, Sidney, and Jack Raso. 1998. *Global Climate Change and Human Health: A Position*

Paper of the American Council on Science and Health. http://www.acsh.org/publications/
reports/global.html
Stanley, J. S. 1986. "Broad Scan Analysis of Human Adipose Tissue, Executive Summary, vol.
1." EPA Contract—560/86/035. Springfield, VA: National Technical Information Service.

Part 1

Setting the Stage

Chapter One

Knowledge, Citizens, and Organizations
An Overview of Environments, Diseases, and Social Conflict

Phil Brown, Steve Kroll-Smith, and Valerie J. Gunter

Early analysts of disease and society viewed the conditions of the surrounding environment as central to health and commented forthrightly on the politics of well-being in industrial society. Friedrich Engels in *The Condition of the Working Class in England*, Rudolf Virchow in *Poverty in Spessart: Discussion of the Ongoing Typhus Epidemic in Oberschlesien*, 1848, and George Orwell in *The Road to Wigan Pier* were among those who considered poor working conditions, crowded slums, polluted water, and contamination of the air as central to health status. It is now well accepted that the largest part of mortality decline from infectious diseases in the 19th century stemmed from improved housing and sanitation, modern sewage systems, cleaner water, and other public health measures, rather than from advances in medical technology and science (McKeown 1976).

When epidemiologists study infectious diseases, they typically take into account a range of social, economic, and political factors, yet when they study toxic waste or chemical changes in the biosphere that induce diseases, they usually disclaim those same social factors. Likewise, government officials who would be embarrassed to ignore such social causation of infectious diseases, are quite ready to dismiss such claims when it comes to environmental contaminants from commercial, industrial, or military practices.

It is, of course, difficult to connect the seemingly endless human-made effluents in the biosphere with human disease, but we see key obstacles to such study in the adherence to the status quo by government officials, regulatory agencies, health professionals, and scientific organizations. Virtually all diseases and conditions that can be attributed to environmental causes are highly contested and the source of considerable confusion, anger, and resentment. Precisely because environmental diseases are often linked to the production and consumption practices of modern societies, acknowledging these diseases and taking actions to reduce them is more often the result of political action than routine medical intervention.

People who reject the claim that there are many environmentally induced diseases often argue that it is not always easy to pinpoint the environmental causation of diseases. This, of course, is true. But in some diseases the environmental causation is clear, and we still have social conflict. Consider black lung disease. Even so clear a connection between particulate matter and lung disease was the source of great contention, with coal miners and their union pitted against mining corporations and the federal government (Smith 1981). Mesothelioma (cancer of the pleura) caused by asbestos and asbestosis (a pulmonary disease) share a similar history with black lung. The hazards to health posed by asbestos were recognized early in its application, though both diseases generated considerable conflict both inside and outside the courts (Brodeur 1985). The sources of contest and conflict when environments are implicated in diseases are complex, likely to combine in different ways on a case-by-case basis, and protean—changing as new legislation is passed, experiments are conducted, and populist movements arise. In spite of the complications, however, it is worth identifying a few of the key sources of enmity and discord in this arena.

The Problem of Uncertain Science

A key source of the acrimony and discord accompanying environmental health controversies is the problem of uncertainty in both toxicology and epidemiology. Consider the following abbreviated list of uncertainties in the interactions between modern environments and human bodies. One uncertainty is the *body's past exposures* to possibly adverse environments. With the possible exception of employees who work with dangerous chemicals, routine medical care does not include histories of exposure to adverse environments (Harte et al. 1991). The medical history of a mother who disinfects her house with monoethanolamine, commonly found in "Easy Off," kills weeds in her garden with diazonon, found in most herbicides, and lives down the road from a petroleum cracking facility is likely to be written as if these environments are not health related.

Another uncertainty is the *dose-response relationship* (Harte et al. 1991). Clinical medicine is quite good at determining the body's response to acute poisoning. However, it is only just beginning to appreciate the need to understand the effects of low-level exposures on human health. Valid dose-response curves that predict health effects at lower doses are not available for most environmental substances. A third uncertainty are the *synergistic effects* of adverse environmental toxins on the body (Harte et al. 1991). Tobacco smoke, radon, and asbestos, for example, all cause lung cancer. Determining how, in fact, these carcinogens interact to cause a cancer is currently beyond the science of toxicology.

A fourth source of confusion is aptly called *"etiological uncertainty"* (Vyner 1988, 61). It is almost impossible to document conclusively that a specific disease is caused by exposure to specific environmental effluents. There is difficulty in understanding the relationships between toxins in the biosphere and morbidity and mortality. Is a woman's cancer caused by irresponsible personal choices, smoking for example, or living downstream from a paper mill and chlorine-contaminated water? Is a man's infertility caused by exposure to the insecticide dibromochloropropane, or is it a genetic disorder? Do eight spontaneous abortions in a four square block residential area constitute a true miscarriage cluster, or just a statistical anomaly? Answers to these and similar questions are anything but straightforward. The relationships between modern environments and diseases, it would appear, are often more compli-cated than our normal clinical and epidemiological models can, with assurance, detect.

Finally, consider the critical problem of *"diagnostic uncertainty"* (Vyner 1988, 62). Physicians typically do not possess the requisite technology or knowledge to make the link between exposure to adverse environments and a specific disease. The account of a former nuclear engineer severely disabled by what she assumes to be multiple chemical sensitivity illustrates this dilemma.

> I went to three neurologists, two cardiologists, one rheumatologist, two internists, two hospital clinics for full evaluation including the Mayo clinic. . . . [I had] medical proce-dures like MRI, EEG, ultrasound, X-rays repeatedly, and [was] hospitalized three times. Among the diagnoses [I] received were: temporal arteritis, fibromyalgia, conversion reaction disorder, and scleroderma. (Kroll-Smith and Floyd 1997, 113)

Without the benefit of exposure histories, accurate dose-response predictions, knowledge of synergistic effects, valid etiology models, and diagnostic capabilities there is a considerable amount of guessing, speculation, and editorializing among both medical professionals and those whose lives are turned inside out by fear of environmental diseases.

The Problem of Class and Race

A second and related source of conflict in illnesses and environments is the confound-ing problems of class and race. While no one is completely protected from adverse environments, there is a rough but identifiable association between wealth and race on the one side and exposure to dangerous environments on the other (Austin and Schill 1994; Szasz 1994). The poorer, and/or less white a person is, the higher the risk of environmentally induced illness. This is not surprising. It is widely recognized in medical sociology that the less control people have over their lives the more likely they are to experience poor health (Black 1980; Freund and McGuire 1991). In 1988, for example, 68 percent of African American children living in families with annual incomes of less than $6000 had lead blood levels that exceeded baseline standards considered safe (ATSDR 1988). In 1990, Greenpeace reported that the average income of families who live in communities with hazardous waste incinerators is 15 percent below the national average (Bullard 1994, 18). Finally, consider the unsettling fact that three out of five African Americans live on top of, adjacent to, or down the road from an abandoned hazardous waste facility (Bullard 1994, 17).

The point here is both simple and important: environment and health problems are likely to remain outstanding or, at best, receive grudging acknowledgment from powerful corporate or government interests who find it comparatively easy to ignore the powerless. But we overstate this point. The powerless, as we will see in later sections, are beginning to mobilize around environment and health issues. Social movements, even those made up of poor people, are rarely powerless.

The Problem of Organizational Deceit

A third source of discord and enmity in environment and health controversies is the resistance of powerful corporate and government organizations whose members place the well-being of the organization before the well-being of employees and citizens. Clarke refers to this as the tendency for organizations to be deceitful (1989, 176). "Organizational deceit" is a likely outcome of the need to make complex administrative operations appear rational, at least to those on the outside of the organization. In a classic study of the PCB contamination of an office building in Binghamton, New York, medical and government officials muddled "through ambiguous situations, constructing solutions before goals were known, and without any assurances they would solve problems" (Clarke 1989, 177). The point is for organizations to appear to others to be in control of dangerous situations. Organizational deceit is more likely to occur in those cases where risks are ambiguous and there is thus a greater need to appear in charge (see also Clarke 1997).

The case of bovine spongiform encephalopathy (BSE), popularly known as Mad Cow Disease, illustrates nicely Clarke's idea of organizational deceit. BSE is a degenerative disease of the brain that affects cows who eat food prepared with the carcasses of dead ruminants. It was widely believed when the disease appeared in 1986 that people who ate beef from cows who died with BSE would be at risk of contracting Creutzfeldt-Jacob disease, a degenerative human brain disorder.

BSE first emerged in Britain. The British Ministry of Health attempted to act as if it was in control of the problem by first attempting to deny its existence. When that strategy failed, the government acted slowly and deliberately at every stage, in a doomed attempt to give the impression that it was considering every possibility. Publicity gimmicks were used to downplay the issue in lieu of addressing it. Finally, officials refused to speak openly about the potential health problem in press conferences, arguing that it was still under review. In response to the attempts by the Ministry of Health to appear in control of the BSE controversy, "the Royal Society and the Association of British Science Writers called their own press conference on the grounds that: 'the public remains confused about its (BSE's) dangers' " (Irwin 1995, 23). The Ministry's organizational approach to the problem, perhaps best called "limiting the damage and maintaining public confidence," failed, of course, on both counts. Examples of organizational deceit are found in many of the following chapters.

We have briefly covered three of the more prominent causes of social conflict in cases of environments and illnesses: the problem of uncertainty, the statistical probability that the relatively powerless are the more likely victims of environmentally induced illnesses, and the likelihood that corporations and governments will engage in some organizational deceit to manage the impression that they are in control of a crisis. Obviously, these three factors will not act in the same fashion nor necessarily appear together in every health and environment dispute. The BSE controversy, for example, does not appear to be class or race biased. With this caveat in mind, however, these three sources of discord are often found in health crises that appear to be caused by human intervention into the environment.

Two contemporary health controversies illustrate the contentious political climate created by illnesses and environments. The first is a debate on measurement issues involving air particles under 10 microns in diameter. Researchers estimate that these cause 50,000–60,000 deaths per year, yet regulatory practices deal only with larger particles (Dockery and Pope 1996). Industry lobbying groups and their scientists are vigorously resisting possible Environmental Protection Agency (EPA) regulatory changes that might result from recent research on small particles. The second controversy is over the complicated data of exposure to toxins and, perhaps, radioactivity in the Persian Gulf War. Among the contested issues are: Who was exposed? How did exposure occur? What specifically were people exposed to? What is the pathophysiology of exposures? First, the case of unregulated air particles.

Asthma and Other Diseases Linked to Small, Unregulated Air Particles

Evidence is mounting that small, unregulated air particles are causing increased asthma and other diseases. Asthma incidence in the United States increased from 1.73 per 1000 in 1979 to 2.57 per 1000 in 1987, mostly in children 0–4 (Gergen and Weiss 1990). National Health Interview Survey (NIHS) data for 1988 found that 4.3 percent of children under 18 had asthma, that black children were 20 percent and poor children 10 percent more likely to have asthma. Poor children were hospitalized 40 percent more often than others (Halfon and Newacheck 1993). In poor and minority areas, the rates are higher. A 1994 survey of 662 households in the Bronx, New York, found that 14.3 percent had asthma at some time, and 8.6 percent had asthma in the previous year. Hispanic and poor children had significantly higher cumulative prevalence. These rates are twice as high as those in the 1988 NHIS. Hospitalization rates are five times the national average, which might indicate that not only are there more cases, but that these cases may be more severe (Crain et al. 1994).

A four-state longitudinal study of hospitalization rates from 1985 to 1994 found that Maine and Vermont rates did not change significantly, New Hampshire rates decreased 5.8 percent per year, and New York rates increased 3.8 percent per year. The largest increase occurred in children living in low-income zip codes (4.1 percent per year). Hispanics had annual increases of 6.8 percent, black non-Hispanics 4.5 percent, and white non-Hispanics a decrease of 0.9 percent. Children in urban areas had annual increases of 3.7 percent compared to children in nonurban areas with a 1.6 percent decrease. These findings suggest that the current explanations for the increase in asthma may not be useful (in particular, changes in disease severity, diagnostic substitution of bronchitis and other respiratory illnesses by asthma, or differences in supply and character of medical care), since these would all point to a reasonably uniform national trend (Goodman et al. 1998).

While some argue that increased asthma is largely an artifact of diagnosis, children may be underdiagnosed. A study of two Detroit schools determined which students previously had a diagnosis, and then provided a physical examination to all children. Undiagnosed asthma was found in 14.2 percent of all the students (Joseph et al. 1996). Asthmatic children have more learning disabilities and grade failure (Fowler et al. 1992). Asthma is financially burdensome—children with asthma in one HMO incurred 88 percent more costs per year than those without asthma, filled 2.77 times as many prescriptions, made 65 percent more nonurgent outpatient visits, and had twice as many inpatient days (Lozano et al. 1997). Despite growing medical concern over asthma, and the environmental concern with the impact of small air particles, there is scant public health recognition of the crisis. In an April 1996 survey of state health departments, the Centers for Disease Control (CDC) found that only eight had implemented asthma control programs in the previous ten years. While twenty-six states reported making interventions at some point, only three were currently doing any intervention, and only one had a system to monitor success. While twenty-two states had environmental control programs, only eight included active control measures, such as dust or allergen reduction (C. Brown et al. 1997).

Some researchers attribute asthma increases and inequalities to poor health care, but there is growing support for a linkage to air pollution. In spite of the fact that 50,000 to 60,000 deaths a year are caused by exposure to air particles under 10 microns, regulatory practices deal only with larger particles. Dockery and Pope (1996) found positive relationships between hospital visits for asthma and increased levels of PM_{10} particles in the four days prior to the visit. There was a dose-dependent increase with no apparent threshold. In other words, there was an adverse reaction at any dose, however small. Mean PM_{10} concentration was responsible for about 12 percent of asthma emergency visits (Schwartz et al. 1993). A clear dose-response curve was also found in the operation of a steel mill; when operating, more PM_{10} was measured and asthma hospitalization increased (Pope 1991).

Dockery and Pope's work on PM_{10} and mortality led to an EPA-sponsored conference, and to major federal debates on more stringent regulation. In March 1998, the National Research Council issued a report strongly urging EPA to expand research in this area, since an estimated 15,000 deaths per year could be prevented by more stringent standards (Chandler 1998). With the health of poor, minority groups at stake in the politics of particulate matter, environmental justice groups are beginning to mobilize around this issue. The importance of citizen action in defining and shaping environmental health disputes is a key theme in this Reader. Consider the second case.

Gulf War Syndrome

Many Gulf War veterans have claimed that toxic exposure during their service in Iraq and Kuwait caused a variety of ailments. Approximately 70,000 have sought treatment for service-related illnesses (Haley et al. 1998), and an estimated 2.7 million people (veterans, family members, and civilians) are eligible for some form of Gulf War Era VA benefits. Research has been plagued by poor data and by disputes over how to study health effects. There are glaring inconsistencies in the lists kept by the Veterans Administration (VA) and Department of Defense (DOD) regarding who served where in Iraq and Kuwait, who received immunizations and vaccinations, and even what dates constitute possible exposure. Veterans faced multiple potential sources of disease, including organophosphate pesticides, nerve gas, pyridostygmine bromide (nerve gas antidote), oil fires, and depleted uranium used in ordnance (Clauw 1998).

The National Academy of Sciences 1994 study, sponsored by the President's Advisory Committee, claimed that Gulf War veterans exhibited only a form of shell shock, common to all combat. In contrast, Haley and his colleagues (Haley and Kurt 1997; Haley et al. 1998) have found evidence for three specific neurological disorders. They argue that the cause is not a single factor, but rather a synergistic effect of many sources. The DOD still insists there was no toxic exposure, despite DOD's repeated embarrassment over missing records and its selective amnesia regarding the U.S. demolition of the Khamisiya arms dump, a likely nerve gas location. Concern over such actions, coupled with frustration over scientists' slowness in collecting and analyzing data, led other governmental actors to seek answers by going beyond traditional research approaches. Representative Christopher Shays ordered the General Accounting Office (GAO) to investigate the quality of federal research, which led to criticism of several government studies. Congress mandated VA-university collaboration on Gulf War research, and three centers were set up, at the State University of New Jersey Medical School, the University of Oregon, and Boston University.

The VA has maintained a strict medical approach, by demanding a specific diagnosis. This is a difficult demand, since disagreement rages over whether there is a single syndrome or various diseases. Caught in the politics of medical research and diagnostic uncertainty, Gulf War veterans are unable to legitimate their illness (Landrigan 1997).

Dissatisfied with VA research and treatment, Representative Lane Evans introduced a bill in 1998 to have the Institute of Medicine compare Gulf War veterans with other GIs; any conditions found to be different would lead to a "presumption of disability," and to treatment and compensation without concern over finding a cause. Readiness to assume something is there when in fact it is not comes much closer to the public health model of prevention. The public health model contrasts with a strict epidemiological one; its application could have major ramifications for many other diseases considered environmentally related. Representative Joseph Kennedy has also written a bill asking for an NIH study focused more on causes of Gulf War Syndrome (Monteiro 1998).

The absence of certain knowledge and the politics of disease that follow this uncertainty is evident in both the air quality and the Gulf War cases. Knowledge necessary to define the problems and fashion solutions was fragmented, contradictory, or perhaps simply not available. Moreover, the role of race and class is highly visible in the environmental asthma case and strongly implied in the Gulf War case. The majority of ground troops fighting in wars are young, relatively uneducated, and often poor. Finally, in both cases, but most visibly with Gulf War Syndrome, the role of organizational deceit is easily discernible.

Highlighting the Problem of Uncertainty

In sorting through the various relationships between knowledge and uncertainty, race and class, and the problem of organizational deceit represented in the literature on environments and illnesses, we were struck with the endemic presence of uncertain knowledge in practically all of the discussions. From the clinical literature on environments and diseases to the social science literature, the problem of knowing is never very far from the center of the discussion. Logically and empirically this makes sense. If knowledge regarding environments and diseases was clinically unambiguous, like the relationship between sodium intake and blood pressure, it is unlikely that serious, long-term conflicts would follow the recognition of a problem. Certainty could be expected to reduce discord. But certainty is typically absent when illnesses are related to environments, replaced by confusion, doubt, and differing appraisals of risk.

Moreover, the endemic problem of uncertainty is directly linked to the emergence of citizen groups organized to answer questions that are eluding the experts while holding appropriate industry or government parties responsible for their miseries. The problem of uncertainty and citizen action requires further discussion.

The Problem of Uncertainty in a Postnatural World

A postnatural environment is dense with "manufactured risks" that are both created by corporations and governments and, not surprisingly, difficult for them to measure, predict, and control (Giddens 1990). Ironically, the more culturally mediated the environment, the more volatile and mysterious it becomes. Enlightenment thinkers, of course, promised just the opposite. Human knowledge and intervention would tame nature, harness it to social ends. Environments do increasingly serve social ends, but they appear anything but tame. A key problem in a postnatural world is the considerable lag between the rapid changes taking place in environments and the limited capacity of experts and their systems for making coherent sense of the changes. The problem of coherence, Giddens explains, is experienced by ordinary people as an issue of trust. Can the experts be trusted? When experts disagree, which one can be trusted? Can I trust experts whose knowledge does not make sense of my experiences? Should I trust my experiences? Trust, or importantly, its fragile nature, is at the heart of the contemporary conflict between people, environments, and expert systems.

Trust, of course, is foremost a personal phenomenon. It is a person who trusts or does not trust. Giddens expands the personal significance of trust by linking it closely to feelings of security in contemporary society. A person who has confidence in her personal relationships, her material surroundings, and her body and its relative well-being enjoys a measure of "ontological security" (1990, 92). She approaches her environment and personal and social health as if they are routine, predictable, and require little if any conscious attention. A key problem in contemporary society, Giddens argues, is the increasing distrust people harbor towards the potentially adverse affects of manufactured environments. Ontological insecurity is a particular issue in the conflicts over knowing that shape the social realities of people, diseases, and environments.

Complementing and adding to the work of Giddens are the ideas of Ulrich Beck (1986, 1995). Coining the term "risk society" (1986), Beck argues that contemporary societies are increasingly organized around the distribution of ecological dangers. Modern societies produce both wealth and environmental hazards and disasters. Indeed, wealth production is tied inextricably to hazard production. The more wealth, the more ecological danger. Surplus capital, in other words, is always complimented by a surplus of dangers to human health and well-being. While wealth may be a source of security for some, it is also a source of insecurity and uncertainty. "In the risk society," he writes, "the unknown and unintended consequences come to be a dominant force in history and society" (1986, 22).

Beck pays particular attention to the increasing role of uncertainty in the organization of human affairs and the emergence of citizen groups who challenge the legitimacy of official accounts of human, environment relationships. The more we know about the effects of human activities on natural processes, the more questions we have about the effects of postnatural processes on humans. In other words, as we tinker with and change the environment, we know, or suspect, that it is tinkering with and changing us, too often from moderately healthy people to sick ones. But, as we will see, the links between environments and diseases are often vague and open to opposing interpretations.

In addition to encouraging a focus on the social production of environmental dangers and the indeterminate, unpredictable, and indefinite associations they are likely to create between bodies, ecosystems, and medicine, Beck adds a popular focus by calling attention to citizen responses to the inchoate voices of experts. Citizens cannot ignore experts and their expertise. Indeed, modern protest, with few exceptions, "must speak the language of science" (1995, 80). Beck continues:

> Of necessity, people have passed a kind of crash course in the contradictions of threat . . . in the risk society: on the arbitrariness of acceptable levels and calculation procedures, on the unimaginableness of long-term consequences, and on the ability of statistics to make them anonymous. They have acquired more information, more vividly and more clearly, than even the most hostile critique could have taught them. (1995, 32)

Successfully passing a "crash course" on how and what to know about the science of environmental dangers and their possible effects on human health is often done in collaboration with a few critical experts who abandon their allegiance to corporate or governmental sponsors and join citizens in their pursuit of an alternative rational account. What follows is a short account of the emergence of citizen activists searching for knowledge of environments and diseases, followed by a brief account of professionals who side with the activists. Readers will encounter both of these themes throughout this book.

The Problem of Uncertainty and Grassroots Mobilization

Important knowledge about environmental health effects has come from people actually or potentially affected by toxic contamination. In some instances, citizens are responding to an acute major contamination episode with immediate health effects, such as the spreading of dioxin-laced fuel oil to contain road dust in Times Beach, Missouri. Elsewhere a chronic contamination episode with continuing health effects, the Love Canal, for example, prompts them to observe their bodies and environment while systematically questioning themselves and their neighbors. In yet another variation, a contamination episode without immediate health effects, such as the accident at Three Mile Island, leads to long-term lay health monitoring in anticipation of late onset cancers.

In other cases, citizen groups discover prior contamination episodes, such as both accidental and deliberate releases of radioactivity from the Hanford nuclear facility in southeast Washington. Sometimes they find unexplained clusters of diseases like the leukemia cluster in Woburn, Massachusetts. In short, ordinary, nonprofessional people are playing pioneering roles in such areas as environmental causes of breast cancer, the effect of endocrine disrupters on sexual development and reproduction, and the effects of air pollution on respiratory diseases. Indeed, in many cases, scientific knowledge lags behind lay identification of diseases and their causes.

Laypeople's sustained self-education, self-organization, and scientific success in identifying environmentally induced diseases and their likely causes is one of the most significant phenomena in the field of environmental health. A growing literature, especially ethnographic case study research, demonstrates the capacities of nonspecialists with little or no higher education to bring to public light the extensive problems of toxic substances on human health (Balshem 1993; Brown and Mikkelsen 1990/1997; Edelstein 1988; Fox 1991; Irwin 1995; Kaplan 1997; Kroll-Smith and Floyd 1997).

Phil Brown started the discussion of lay medical expertise in his essays on environmental health controversies and citizen research (1992, 1997; Brown and Mikkelsen 1990/1997). His case study of Woburn, Massachusetts, charted the activities of several citizens who sought some credible explanations for the unusually high incidences of childhood leukemia in their neighborhoods (1992; Brown and Mikkelsen 1990/1997; see chapter 21 in this volume). Coining the term "popular epidemiology," he described ordinary people who were not satisfied with professional accounts of their troubles and sought to construct their own rational, epidemiological explanations for the untoward cancer clusters in their community. Knowing that the only way to refute expert judgments is through more accurate expertise, citizens sought to counter the official account by sponsoring an alternative study. Aided by sympathetic public health researchers at Harvard University, a study was conducted that found correlations between water contamination and leukemia in the community (Brown and Mikkelsen 1990/1997).

A key element of popular epidemiology is the possibility that citizens who sponsor or conduct their own medical research are likely to challenge some of the traditional assumptions of research that organize normal scientific investigation. Citizens and their experts are not just replicating conventional science. Their science, rather, is often a claim to know something rational and valid about the world based on novel research practices and assumptions. One simple but powerful assumption that citizens make in designing their research is that human experience is a valid mode of knowing the world. Normal science begins by discounting the importance of the individual's sensory understanding of the world. Subjectivity, in other words, must be rigorously excluded from the research design. Citizen science, on the other hand, privileges the human senses, according them an important place in the overall work to understand the problem. Noticing what appears to be an unusual number of local people with upper respiratory infections, or asking neighbors to relate their experiences with a water source suspected of carrying toxins, is not only a reason to begin inquiry; it is an authoritative source of data in its own right.

In addition to placing human experience in the center of their research design, citizens are also challenging normal science's claim to value neutrality. Citizens do not claim their science is value-free. Indeed, what they are scrambling to learn often has a direct bearing on their health and well-being. And, importantly, few citizens believe that corporate or state-sponsored science is value-free. Acknowledging that their research is value-driven encourages citizens to see that any research is, like all human enterprises, loaded with biases and weighted in favor of someone or some group's agenda. Citizen or lay science is at once a claim to know something rational and valid about the world and a potent form of protest against state- and corporate-sponsored research.

Nancy Hartsock, among others, suggests that the validity of beliefs is closely related to the social status of believers (1983; Smith 1981). An accurate perspective on social life, she argues, can only stem from members of beleaguered or oppressed groups. For our purposes, this argument suggests that people faced with the prospects of losing their health to contaminated environments are more likely to possess useful, valid knowledge of their unwelcome predicament than corporate or medical professionals are prepared to acknowledge. For Hartsock, knowledge from adversely affected groups is not merely another voice in a relativistic world. Its source of authority is moral rather than epistemological. Community power, coupled with social justice, makes their claim to know something important about their medical troubles difficult to ignore (see Couch and Kroll-Smith, chapter 22 in this volume). But, as Beck noted, lay action alone is not the whole of this story.

Oppositional Professionals

Local and regional environmental health movements are likely to draw on the expertise of oppositional professionals, experts who are willing to support communities in their efforts to seek recognition and assistance from industry or government. Oppositional experts, of course, are not new.

Marine biologist Rachel Carson worked for the U.S. Fish and Wildlife Service during the 1940s. Her book *Silent Spring* (1962), an exhaustive inquiry into the dangers of DDT, was written for non-experts and is widely recognized as the first shot across the bows of the U.S. chemical industry. The industry responded to the book with a vigorous campaign to discredit the author and persuade publisher Houghton Mifflin to pull the book from circulation. *Time Magazine* accused her of using "emotion fanning words" (Gore 1994). Despite considerable resistance, *Silent Spring* remains in print and Rachel Carson is considered, by many, to be the inspiration for the modern toxics movement (Gibbs 1998; Gore 1994).

Beverly Paigen was a conventional epidemiologist working for the state of New York when she encountered the citizens of the Love Canal:

> Before Love Canal, I also needed a 95 percent certainty before I was convinced of a result. But seeing this rigorously applied in a situation where the consequences of an error meant that pregnancies were resulting in miscarriages, stillbirths, and children with medical problems, I realized I was making a value judgment . . . whether to make errors on the side of protecting human health or on the side of conserving state resources. (1982)

John Till, who heads the Dose Reconstruction Study at Hanford, which is reconstructing historical dosages of radioactive iodine releases, recounted how he had been transformed by the process of being part of what he terms a "public study." He came to grasp the importance of openness to the public, and of access to classified information. He found a great empathy for the Hanford downwinders and a special understanding of the concerns of Native Americans in the area who viewed Hanford as an assault on their entire heritage (Till 1994).

On March 23, 1999, four EPA employees and the mother of a young man who died after being exposed to EPA-approved "sludge" held a press conference at the National Press Club in Washington DC. The group called on the EPA to stop harassing employees who bring to public attention important environmental health information. The mother recounted her son's death and thanked the four EPA scientists for their willingness to assist her and others in New Hampshire who are campaigning against the government-approved toxic "sludge" (NWC 1999).

The complicated role of lay and oppositional professional knowledge in the contested character of environment and disease is nicely illustrated in the case of the environmental causes of breast cancer.

Is Breast Cancer an Environmentally Induced Disease?

Breast cancer continues to escalate, with U.S. incidence among the world's highest. Debates concerning environmental causes of breast cancer have led to a burgeoning movement of activists who have played highly visible research, advocacy, and policy roles.

Genetic predispositions play a minor role; despite its much-heralded discovery, the BRCA1 gene accounts for only approximately 5 percent of cases (Davis and Bradlow 1995). Much attention is now on environmental carcinogens ingested through air, water, and food (Davis et al. 1997; Steingraber 1996). Perhaps the first modern oppositional professional, Rachel Carson (1962) suspected organochlorines in the etiology of breast cancer early in the development of the problem. Colborn et al.'s *Our Stolen Future* (1996) brought much attention to the vast research in many disciplines that linked xeno-estrogens to a wide range of reproductive disorders and abnormalities. These estrogen mimics are found in many chemicals, fertilizers, pesticides, and foods. Synthetic estrogens are potentially able to disrupt the body's endocrine system, interfering with its ability to deliver hormones to the blood stream. The critics argue that humans ingest many natural estrogens, and thus it is wrong to blame chemical producers and users (Safe and Ramamoorthy 1998). Cancer activists, environmentalists, and sympathetic scientists counter by saying that routine exposure to natural sources is not a reason to allow unchecked exposure to synthetic estrogens (Steingraber 1996). This debate is important because it raises questions of economic, political, and social responsibility, which are the sources of dispute.

Breast cancer activists convinced the Massachusetts legislature to appropriate $1.4 million for the Cape Cod Breast Cancer and the Environment Study. Their research, conducted by the Silent Spring Institute, has thus far failed to find environmental estrogenic results in the Cape Cod breast cancer excess, but they are continuing to do research on specific toxins that may prove to be carcinogens. A single study with scientific and political clout can put much pressure on a developing hypothesis. For example, the Nurses Health Study (Hunter et al. 1997) failed to find a relationship between organochlorines and breast cancer. The Long Island Study of a major breast cancer excess is completed but not yet public, and can be expected to play a similarly significant role. Breast cancer victims and their scientific allies remain convinced that exposures, such as massive DDT spraying, are related to breast cancer.

The endocrine disrupter debate represents more than just another scientific disagreement. It has led to what Krimsky (1998) terms a "quiet revolution in science and policy" concerning chemical risks. This "precautionary approach" sees an obligation to control the dangerous substances even before there is a definitive causal link to health effects. This departure from traditional risk assessment has influenced the 1996 Food Quality Protection Act and the Amendment to the Safe Drinking Water Act, which mandated the EPA to develop a screening and testing program for endocrine disrupting chemicals not just on a case-by-case basis, but taking into account cumulative, additive, or synergistic effects. The endocrine disruption hypothesis has also affected toxicity theory, since animal studies show that there may be harmful effects at small doses that are not observed at larger doses. Last, the hypothesis seeks to replace increasing specialization with more integrative biological studies (Krimsky 1998; Vom Saal 1999).

In 1998, EPA announced the specifics of a testing program to evaluate a large number of chemicals for their estrogenicity. This is a milestone with significance beyond just the endocrine disruption hypothesis. Many thousands of chemicals have never been tested, despite many regulations on, and even banning of, specific chemicals. Estrogenicity screening marks the first time that any government agency has decided to examine a major proportion of chemicals now in use. This offers the potential for broader screening of those chemicals' effects.

Conclusion

Offered in this chapter is a broad overview of the key ideas and themes found throughout this book. Three broad factors were identified as probable causes of the enmity and conflict that frequently accompany questions about illnesses and environments: uncertainty and the problem of knowledge, the race and class bias of environmental contamination, and corporate or government resistance to acknowledging responsibility for environment and illness problems. While helpful as conceptual tools, these three factors are not exhaustive of the range of variables that affect the social, cultural, and political responses to environmentally induced illnesses. We offer them as guideposts only and invite others to speculate on additional sources of animosity and hostility.

There does appear to be one factor that remains constant in this genre of contested illnesses: uncertainty and the problem of knowledge. This theme appears in one guise or another throughout the chapters of this book. The dispute over valid knowledge of bodies and environments remains a troubling issue in setting environmental health policy, often serves as a veil for business or government interests, and, importantly, is a major catalyst in the formation of citizen movements.

Lay involvement is essential in the development of significant clinical and epidemiological knowledge of environment and health problems. Citizen action is necessary because the active discovery of environmental health effects is not a major part of established professional groupings, government health and environment agencies, and universities and research centers. Indeed, government agencies and professionals often side with business in actively opposing connections between health and the environment.

Citizens approach environmental health effects from the standpoint of actual or potential suffering, and they are unwilling to wait for "perfect" studies to show some effect. Laypeople's concerted action has had deep impacts on scientific discovery, social policy, and environmental regulation. Further, because of the way that laypeople have influenced these arenas, they have expanded grassroots democratic participation in general.

It is informative to watch these environmental disputes unfold around us, often appearing on the front pages of our newspapers. By studying the conflicting perspectives in these and other environmental health issues, readers of this book can improve their ability to analyze other ongoing disputes that will very likely develop. By doing so, they will be better able to protect the health of themselves, their families, and their neighbors.

REFERENCES

Agency for Toxic Substances and Disease Registry [ATSDR]. 1988. *The Nature and Extent of Lead Poisoning in Children in the United States: A Report to Congress.* Atlanta: U.S. Department of Health and Human Services.

Austin, Regina, and Michael Schill. 1994. "Black, Brown, Red, and Poisoned." In Robert D. Bullard (ed.), *Unequal Protection.* San Francisco: Sierra Club Books, pp. 53–74.

Balshem, Martha. 1993. *Cancer in the Community: Class and Medical Authority*. Washington, DC: Smithsonian Institution Press.

Beck, Ulrich. 1986. *Risk Society*. London: Sage.

———. 1995. *Ecological Enlightenment*. Atlantic Highland 5, NJ: Humanities Press.

Black, Douglas. 1980. *Inequalities in Health: Report of a Research Working Group*. London: Department of Health and Social Services.

Brodeur, Paul. 1985. *Outrageous Misconduct: The Asbestos Industry on Trial*. New York: Pantheon Books.

Brown, Clive M., Henry A. Anderson, and Ruth A. Etzel. 1997. "Asthma: The States' Challenge." *Public Health Reports* 112: 198–205.

Brown, Phil. 1992. "Toxic Waste Contamination and Popular Epidemiology: Lay and Professional Ways of Knowing." *Journal of Health and Social Behavior* 33: 267–81.

———. 1997. "Popular Epidemiology Revisited." *Current Sociology* 45: 137–56.

Brown, Phil, and Edwin J. Mikkelsen. 1990/1997. *No Safe Place: Toxic Waste Leukemia, and Community Action*. Berkeley: University of California Press.

Bullard, Robert D. 1994. "Environmental Justice for All." In Robert D. Bullard (ed.), *Unequal Protection*. San Francisco: Sierra Club Books, pp. 3–22.

Carson, Rachel. 1962. *Silent Spring*. New York: Houghton Mifflin.

Chandler, David. 1998. "EPA Urged to Refocus on Health Risk." *Boston Globe*, April 1, 1998.

Clarke, Lee. 1989. *Acceptable Risk*. Berkeley: University of California Press.

———. 1997. "Supertanker Politics and Rhetorics of Risk: The Wreck of the *Exxon Valdez*." In J. Steven Picou, Duane A. Gill, and Maurie J. Cohen (eds.), *The Exxon Valdez Disaster*. Dubuque, IA: Kendall/Hunt Publishing Co., pp. 55–70.

Clauw, Daniel. 1998. Testimony before House Committee on Government Reform and Oversight, Subcommittee on Human Resources, February 23, 1998.

Colborn, Theo, Diane Dumanoski, and John Meyer. 1996. *Our Stolen Future: Are We Threatening Our Fertility, Intelligence, and Survival? A Scientific Detective Story*. New York: Dutton.

Crain, Ellen F., Kevin B. Weiss, Polly E. Bijur et al. 1994. "An Estimate of the Prevalence of Asthma and Wheezing among Inner-City Children." *Pediatrics* 94:356–62.

Davis, Devra Lee, and H. Leon Bradlow. 1995. "Can Environmental Estrogens Cause Breast Cancer?" *Scientific American* October 1995:167–72.

Davis, Devra Lee, Nitin T. Telang, Michael P. Osborne, and H. Leon Bradlow. 1997. "Medical Hypothesis: Bifunctional Genetic-Hormonal Pathways to Breast Cancer." *Environmental Health Perspectives* 105 (Supplement 3): 571–76.

Dockery, Douglas, and Arden Pope. 1996. "Epidemiology of Acute Health Effects: Summary of Times Series Studies." In Richard Wilson and John Spengler, (eds.), *Particles in Our Air: Concentrations and Health Effects*. Cambridge: Harvard University Press, pp. 123–48.

Edelstein, Michael. 1988. *Contaminated Communities: The Social and Psychological Impacts of Residential Toxic Exposure*. Boulder: Westview.

Fowler, Mary Glenn, Marsha G. Davenport, and Rekha Garg. 1992. "School Functioning of US Children with Asthma." *Pediatrics* 90: 939–44.

Fox, Steve. 1991. *Toxic Work: Women Workers at GTE Lenkurt*. Philadelphia: Temple University Press.

Freund, Peter, and Meredith McGuire. 1991. *Health, Illness, and the Social Body*. Englewood Cliffs, NJ: Prentice-Hall.

Gergen, Peter J., and Kevin B. Weiss. 1990. "Changing Patterns of Asthma—Hospitalization among Children: 1979–1987." *Journal of the American Medical Association* 264:1688–92.

Gibbs, Lois. 1998. *The Story Continues*. New York: New Society Publications.

Giddens, Anthony. 1990. *The Consequences of Modernity*. Stanford, CA: Stanford University Press.

Goodman, David C., Therese A. Stukel, and Chiang-hua Chang. 1998. "Trends in Pediatric Asthma Hospitalization Rates: Regional and Socioeconomic Differences." *Pediatrics* 101: 208–13.

Gore, Albert. 1994. "Introduction." In Rachel Carson, *Silent Spring*. New York: Houghton Mifflin.

Haley, Robert W., Jim Hom, Peter S. Roldna et al. 1998. "Evaluations of Neurologic Functions in Gulf War Veterans: A Blinded Case-Control Study." *Journal of the American Medical Association* 227: 223–30.

Haley, Robert W., and Thomas C. Kurt. 1997. "Self-Reported Exposure to Neurotoxic Chemical Combinations in the Gulf War: A Cross-Sectional Epidemiologic Study." *Journal of the American Medical Association* 227: 231–37.

Halfon, Neal, and Paul W. Newacheck. 1993. "Childhood Asthma and Poverty: Differential Impacts and Utilization of Health Services." *Pediatrics* 91: 56–61.

Harte, John, Cheryl Holdren, Richard Schneider, and Christine Shirley. 1991. *Toxics A to Z*. Berkeley: University of California Press.

Hartsock, Nancy. 1983. "The Feminist Standpoint: Developing the Ground for a Specifically Feminist Historical Materialism." In Sandra Harding and Merrill Hintikka, (eds.), *Discovering Reality: Feminist Perspectives on Epistemology, Metaphysics, Methodology, and Philosophy of Science*. Dordrecht: Reidel/Kluwer, pp. 283–310.

Hunter, D. J., S. E. Hankinson, and F. Lader et al. 1997. "Plasma Organochlorine Levels and the Risk of Breast Cancer." *New England Journal of Medicine* 337 (18): 1253–58.

Irwin, Alan. 1995. *Citizen Science*. London: Routledge.

Joseph, Christine L. M., Betsy Foxman, Frederick E. Leickly, Edward Peterson, and Dennis Ownby. 1996. "Prevalence of Possible Undiagnosed Asthma and Associated Morbidity among Urban Schoolchildren." *Journal of Pediatrics* 129: 735–42.

Kaplan, Louise. 1997. "The Hanford Education Action League: An Informed Citizenry and Radiation Health Effects." *International Journal of Contemporary Sociology* 34: 255–66.

Krimsky, Sheldon. 1998. "The Precautionary Approach to Endocrine Disrupting Chemicals." In press, *Forum for Applied Research and Public Policy*.

Kroll-Smith, Steve, and H. Hugh Floyd. 1997. *Bodies in Protest: Environmental Illness and the Struggle over Medical Knowledge*. New York: New York University Press.

Landrigan, Philip. 1997. "Illnesses in Gulf War Veterans: Causes and Consequences." *Journal of the American Medical Association* 277: 259–61.

Lozano, Paula, Paul Fishman, Michael Von Korff, and Julia Hecht. 1997. "Health Care Utilization and Cost among Children with Asthma Who Were Enrolled in a Health Maintenance Organization." *Pediatrics* 99: 757–64.

McKeown, Thomas. 1976. *The Modern Rise of Population*. New York: Academic Press.

Monteiro, Lois. 1998. "Ill-Defined Illnesses and Medically Unexplained Symptoms Syndrome." *Footnotes* 26, no. 2: 3, 6.

National Whistleblower Center [NWC]. 1999. Press Release March 23. (www.whistleblowers.org/PCrelease.htm)

Paigen, Beverly. 1982. "Controversy at Love Canal." *Hastings Center Reports* 12(3): 29–37.

Pope, Arden III. 1991. "Respiratory Hospital Admissions Associated with PM10 Pollution in Utah, Salt Lake, and Cache Valleys." *Archives of Environmental Health* 46: 90–97.

Safe, Stephen, and Kavit Ramamoorthy. 1998. "Disruptive Behavior: Endocrine Disruptors, Sperm Counts, and Breast Cancer." In press, *Forum for Applied Research and Public Policy*.

Schwartz, Joel, Daniel Slater, Timothy V. Larson, William E. Pierson, and Jane Q. Koenig.

1993. "Particulate Air Pollution and Hospital Emergency Room Visits for Asthma in Seattle." *American Review of Respiratory Disease* 147: 826–31.

Smith, Barbara Ellen. 1981. "Black Lung: The Social Production of Disease." *International Journal of Health Services* 11: 343–59.

Steingraber, Sandra. 1996. *Living Downstream: An Ecologist Looks at Cancer and the Environment.* Reading, MA: Addison-Wesley.

Szasz, Andrew. 1994. Ecopopulism. Minneapolis: University of Minnesota Press.

Till, John. 1994. "Update from the Hanford Dose Reconstruction Project." Presentation at Radiation Health Effects and Hanford Conference, Saturday September 10, 1994, Spokane, WA.

Vom Saal, Frederick. 1999. "A Challenge to Risk Assessment: Low-dose Effects of Endocrine Disruptors." In press, *Forum for Applied Research and Public Policy.*

Vyner, Henry. 1988. *Invisible Trauma.* Lexington, MA: Lexington Books.

Part 2

Environments and Diseases
Professional Boundaries and the Problem of Knowing

Disputes over the existence of environmentally induced diseases are clearly disputes about what should be done, that is, what *actions* should be taken. Victims, their families and friends, and sympathetic others want to get acknowledgment of responsibility, medical care, and future medical surveillance for sufferers, clean-up of sites that caused the problem, new government policies and regulations to prevent similar problems in the future, and even financial settlements from responsible parties.

These actions also presuppose a conflict over *knowledge*. If contending parties agreed that a certain problem existed, they might be more likely to work together to resolve it. But there is a knowledge gap between parties. Struggles over environmentally induced diseases are struggles over the very nature of *what exists* and *how we know the nature of the phenomenon.* When people seek the actions mentioned above, they are forced to present what they know. Often, they do not know enough because they have not been privy to relevant knowledge. Other times, they do know enough, but their knowledge is discounted by those in power.

In this section, we present essays that deal with the knowledge issues of epidemiologists and physicians, two key disciplines involved in the recognition and treatment of environmentally induced diseases. These disciplines face conflicts about professional boundaries in several ways. They are called upon to deal with issues that may be unfamiliar to them, due to the newness of much of the field of environmental health. They are often asked to play an advocate role for sufferers, rather than a detached scientific role. They are requested to provide fairly definitive knowledge in testimony to courts and regulatory bodies, even though this type of knowledge cannot always be so definitive. They are asked to contradict other members of their profession, and often to take stances that go against the dominant belief systems of their professions. And they are often asked to break down boundaries between lay and professional knowledge, as toxic contamination victims request joint efforts.

These requests are often very conflictual. Yet they also contain the kernels of positive new steps. These engagements can yield new knowledge not otherwise available. They can help professionals to overcome retrograde limitations of their fields. They can humanize the relationship between lay and professional, helping to produce a more democratic society.

Steve Wing, an epidemiologist, writes in chapter 2 on "Limits of Epidemiology" of the many conservative traditions in his field that make it hard to work with toxic victims. He offers a new perspective on the kinds of questions epidemiologists should ask, the kinds of social perspectives they should take, and the personal engagement they should have with these issues. As one of the growing number of epidemiologists who work with community groups, Wing represents a major shift in his discipline, a discipline that so many victims turn to for help.

Phil Brown and Judith Kirwan Kelley, in chapter 3 on "Physicians' Knowledge, Attitudes, and Practice Regarding Environmental Health Hazards," begin with the awareness that physicians have generally not been in the forefront of local environmental health crises. They show that physicians have been poorly educated in environmental health and do not have adequate resources to learn about it once they are practicing medicine. Even when faced with highly publicized environmental contamination episodes in their towns, physicians do not appear to be more aware of environmental health effects than physicians in non-affected areas.

Steve Kroll-Smith and H. Hugh Floyd's "Environmental Illness as a Practical Epistemology and a Source of Professional Confusion" comes from their book, *Bodies in Protest: Environmental Illness and the Struggle over Medical Knowledge*. In this excerpt, they point to how most of the symptoms and illness experience of people with multiple chemical sensitivity appears as anomalies to doctors. In this newly recognized symptom constellation, sufferers are at a great disadvantage because their experiences are quite simply *not believed*. Through studying this phenomenon, we can see the unfolding knowledge contests between lay and professional worldviews.

Chapter Two

Limits of Epidemiology

Steve Wing

Epidemiology is the field of study of health and disease in populations. As such it is considered to be the basic science of public health, the activity of preventing disease and promoting health in populations [1]. This focus on primary prevention of disease in large groups has distinguished public health from medicine, which focuses on treatment and prevention of disease in individual patients. To provide the basic science for public health, as the clinical disciplines do for medicine, epidemiology must try to explain, and thereby suggest ways to improve, the experience of disease and health in populations as distinct from individual patients.

Treatment for a patient is affected by the explanations that the basic medical sciences provide for a particular condition. Thus one kind of basic science suggests a pharmacological treatment, while another suggests a nutritional intervention. Furthermore, the understanding of pathological processes changes over time with basic scientific knowledge and the experience of clinical application, so that medical approaches for treating a given condition change.

Similarly, public health policy depends on epidemiological explanations of health and disease in populations. These explanations, as in any science, depend on the concepts and methods used to understand a given problem. Epidemiology is a fairly young discipline, yet it has undergone a variety of changes, including those brought on by the hygiene movement, the bacteriological revolution, the increasing application of statistical survey methods, molecular biology, and the environmental movement.

Despite a history of changing ideas about the causes and treatments of disease in populations, modern quantitative epidemiology is increasingly presented as the objective technique for uncovering universal laws about causes of disease in populations [2]. It is increasingly applied in clinical medicine, and some claim that it is a generic tool that can be used for the investigation of any human condition [3]. All sciences change in character over time, however, and the current dominant epidemiological practice is not, in historical terms, the end of the road, but rather a particular form of inquiry and explanation that will continue to be transformed in the context of changes in public health, medicine, science and society.

This chapter first describes the character of explanation in modern epidemiology, a practice that has coalesced in English-speaking countries during the latter half of the twentieth century [4]. Conceptual problems with the object of inquiry in modern epidemiology are raised. They suggest the field should adopt a less reductionistic approach. Some examples of alternative epidemiological perspectives are given in the third section. The limits of modern epidemiology in public health practice are considered in the fourth section, and the final section addresses issues of scientific objectivity and social responsibility. Epidemiological investigation of health effects of ionizing radiation is used as a primary example.

Explanation in Modern Epidemiology

The dominant mode of epidemiological explanation takes place fully within the limits of a scientific practice that has been termed Cartesian reductionism, an analytical approach characterized by a focus on factors considered in isolation from their context [5]. Disease in populations, the stated focus of the discipline and the central feature distinguishing epidemiology from the clinical sciences, is to be explained in terms of a series of agents, exposures, or risk factors. Generically these include microbes, chemicals, or nutrients, and also anthropometric, physiological, and genetic features as well as behaviors, mental states, race, and socioeconomic status. The method of the discipline is to observe whether disease occurs more (or less) commonly among individuals who have the exposure or factor than those who do not. The broader goal of explanation of the occurrence of disease in populations is pursued by enumerating all the risk and protective factors (the "independent variables") and the form of their relationships with a list of disease outcomes derived primarily from clinical practice (the "dependent variables"). The results of this research can be recognized in the lists of carcinogens, cardiovascular risk factors, and health risk behaviors that are targeted for modification by hygienic, behavioral, or pharmacological intervention.

Studying Disease in Populations

Before going further it may be helpful to clarify what is meant when it is said that epidemiology is the study of disease in populations. The populations of modern epidemiology are counts—numbers of individuals—who are grouped according to their exposure and disease status. Populations, in this context, are vehicles for making comparisons of rates or averages; they are not inherently defined as organized groups with unique histories involving economic, social and ecological relationships. Their features of organization, in the epidemiological context, are not considered to have etiological consequences. Epidemiological studies that do address factors such as economic position or occupation generally treat them only as individual attributes or exposure markers rather than as aspects of social and economic organization that provide the context for biopsychosocial development. Thus race, as a feature of individuals, can be studied without recognizing racism as a historical feature of the organization of populations. Epidemiological studies labeled as "ecological" examine associations between average levels of exposure and disease in groups [6, 7], not population organization and ecological relationships [8, 9].

The view that disease in populations is a function of essential exposure-disease relationships is mirrored in the model for modern epidemiological study designs, the randomized experiment. In this approach, subjects with specific characteristics, including absence of a disease or outcome of interest, can be chosen for study. Next they can be allocated to be exposed or unexposed to a factor according to rules that, as in the well-shuffled decks of repeated card games, tend towards an even distribution of the heterogeneous study subjects between exposure groups over the course of many trials. During the period of application of controlled amounts of exposure (or non-exposure), all other conditions affecting the subjects can be held constant. Finally, the subjects are available to the researcher for determination of the outcome characteristics in members of each group, using a standardized protocol. The analysis of such a study amounts to a comparison of the frequency of the outcomes of interest in the groups. Differences in frequency that persist over many trials, or that are obtained in a small number of large trials, are attributed to the action of the experimental agent. It should be clear that any questions of context, such as where the exposures have come from, why some individuals but not others were exposed, or what other changes occurred in order to produce the exposures, have been eliminated from the realm of scientific interest.

Observational studies attempt to imitate the controlled experiment in various ways. An occupational study of the effects of gamma radiation on cancer might be restricted to workers of a certain type (e.g., males employed at a specific facility who worked longer than six months) to avoid some initial differences between exposure groups, and might compare workers that had received different cumulative radiation exposures within strata of age, other occupational exposures, and behavioral attributes of interest, to provide a summary estimate of the exposure-disease relationship "adjusted" for those other factors. In attempting to yield results that would have been obtained in an experiment, the observational study controls "extraneous" factors (and the context) in search of the separate, independent effect of the exposure. The assumption is that well-designed studies will provide estimates of the radiation-cancer dose response relationship that converge around the underlying value which characterizes the change in cancer rates for each unit change in radiation.

Problems in Epidemiological Explanation

Problems in epidemiological explanation can be approached from two related perspectives. This section addresses logical problems with the object of investigation of modern epidemiology, exposure-disease associations, and with interpretation of evidence about associations. Later sections address problems with the nature of the public health impact of the application of this knowledge.

Problems in interpretation of exposure-disease associations receive great attention in the epidemiological literature. This attention is primarily focused on issues of measurement error and uncontrolled differences between groups being compared, issues that create difficulties because the magnitude and dose response form of observed associations may reflect these sources of bias rather than the association of interest. Despite these ambiguities, much refinement of method has occurred, and modern epidemiology has contributed to the identification of many pathogenic agents. Some, like cigarette smoke, produce large effects that are hard to miss. Others, like asbestos, have specific effects (mesothelioma) that rarely occur in the absence of the exposure. Much of epidemiology today, however, is focused on a search for evidence about weaker relationships and low exposure levels, where poor measurement and the presence of unmeasured differences between exposure groups become major potential problems [10]. Relatively small differences in disease occurrence, such as those that are suspected in the case of low level radiation and many other environmental contaminants, are difficult to detect because very accurate measurements are necessary for quantifying exposures and disease excesses that are not far above the "background" levels. But small increments in disease incidence can have a great population impact when many people are exposed [11].

What Should be Measured?

A more fundamental problem in epidemiological explanation is the development of theory about which aspects of exposure and outcome to measure. *Outcomes generally reflect experience with diagnosis and treatment more than with etiological inquiry.* Exposures are often measured due to convenience, availability of data, or convention, rather than based on biological models of disease process. Among workers exposed to penetrating ionizing radiation over long periods, for example, the total cumulative dose (derived by applying assumptions about relative biological effectiveness to measures of biologically absorbed radiation of different types) is typically studied. Sometimes only the doses received up to a certain number of years in the past are counted, on the assumption that cancers take time to develop. Alternatively, doses received in the distant past might not be relevant, and the doses that should be counted might only be those accumulated in some window of time around the emergence of the hypothetical mutation leading to a particular cancer death [12,13]. Then again, it might not be the simple cumulative dose that is critical, but whether the dose is delivered in one or a few short time periods, or is drawn out slowly. Other possible aspects of radiation exposure that might be important to measure are the peak dose, or the dose accrued in the context of other cocarcinogenic exposures or susceptibility states. *Lack of understanding about mechanisms of radiocarcinogenesis means that problems of measurement are secondary to the more fundamental problem of knowing what to measure* [14].

Like problems of measurement and comparability, questions about the biomedical basis for modeling exposure-disease associations are of concern within modern epidemiology [e.g., 2, 15]. The former technical concerns lead investigation in the field to emphasize increasing control over measurement and extraneous factors that can distort the exposure-disease association, while concerns over the latter conceptual issues focus the discipline's theoretical attention on pathological processes of individual organisms. Because control over measurement and extraneous factors is hindered when investigations are embedded in complex social and historical situations, this combination of influences supports the movement of the discipline away from engagement with issues of social theory, population biology and human ecology, and towards a more fundamental commitment to biomedical approaches.

The Broader Context of Exposure-Disease Associations

This direction is justified on the assumption that it will lead to the more accurate description of the underlying exposure-disease relationships that account for health and disease in populations. These relationships are seen as being self-contained, homogeneous, and independent phenomena. They are therefore appropriately studied in isolation, one at a time, according to the approach formalized in the experiment, in order to move "from time- and place-specific observations to an abstract universal statement" [2, p. 96]. In this approach, the experiences of particular populations "are only exploited to learn about the relation at issue in the abstract (in general), that is, without any spatiotemporal referent" [3, p. 16]. Thus, modern epidemiology is oriented towards identifying the fundamental laws, not of the universe, but of exposure-disease associations.

The idea that epidemiology is about universal exposure-disease associations may create some discomfort, at least because epidemiologists presumably believe that biological organisms evolve and therefore are not historically constant vehicles for such processes. From a more contemporary perspective, it is already recognized in epidemiology that exposure-disease associations vary between different groups depending on host characteristics or other exposures. This variation, often called interaction or effect modification, raises important questions about the underlying phenomenon being investigated. For strong exposure-disease relationships, such as smoking and lung cancer, effect modification may be of minor interest. In some cases, however, and especially for low-level exposures, variations in exposure-disease associations under different conditions may make the difference between no relationship and an important one. For ionizing radiation, it is recognized that associations may be modified according to age of exposure, sex, and perhaps other factors including diet, presence of other chemicals, and genetic characteristics. On the other side of the exposure-disease equation, disease categories are inevitably heterogeneous, with the association of exposure and disease showing more or less variability for subgroups of broad disease categories used in epidemiological studies. For example, among A-bomb survivors, radiation is related to neoplasms, to solid tumors, and to leukemias in general, but to some types of each more than others [16]. We are always studying average effects.

In epidemiological practice, while effect modification is recognized, it is treated as a special case, a sub-class of universal associations that is "an inherent characteristic . . . an unalterable fact of nature" [17, p. 586]. This status preserves the logical commitment to methods and theories predicated on the search for self-contained independent relationships that are, if not completely universal, at least homogeneous within subgroups. The belief in independent exposure-disease relationships leads to the centrality of the issue of controlling confounding bias through design and analysis. But there remains an uneasy contradiction within the field between, on the one hand, expanding epidemiological research (traditionally limited to middle-aged white men) to include women, nonwhites or the elderly, because of potential variation in exposure-disease relationships, and, on the other hand, the interest in techniques for quantitatively combining results from different studies [18] in order to produce a more reliable estimate of an exposure-disease association based on the assumption that different studies of the same exposure-disease relationship are providing estimates of a universal underlying phenomenon.

Contextual Complications

Although some relationships may be fairly stable over a large range of contemporary contextual variation (this is argued by some, for example, in the case of serum cholesterol and coronary disease in adult males), change in biological response to exposure should be expected on longer evolutionary time scales. The view that exposure-disease associations generally vary with context suggests that there is no underlying universal dose response relationship to be uncovered [19]. Rather, for any given exposure and outcome, there is some range of contexts in which the change in outcome per unit change in exposure exhibits more or less stability. From this perspective the fundamental object of inquiry in modern epidemiology, dose response, should be recognized as essentially contextual (developmental or historical) rather than universal, vastly complicating the reductionist program and, indeed, challenging the very de-contextualization on which it is based.

Description of the quantitative impact of disease agents is important, has practical implications, and should continue to be a part of epidemiology. The practical importance of the activity, however, does not mean that the study of exposure-disease associations constitutes a sufficient or rigorous object of inquiry for a basic science of public health. This is so because the object of inquiry itself, the exposure-disease association, is not a self-contained, homogeneous, or independent phenomenon as presumed by the approach that treats the experiment as a "paradigm" for research [20]. This approach has been called modern epidemiology [2], analytical epidemiology [21], occurrence research [3], and risk factor epidemiology [22, 23]. It is essentially a kind of human toxicology, an approach limited to the identification of risk factors using an analytical approach in which the historical context is a nuisance to be avoided by design or controlled by analysis. Such an approach justifies conclusions of leading methodologists in the field that war and epidemics are not problems for epidemiology [3, pp. 4–5], and that "social class is presumably causally related to few if any diseases" [2, p. 90] because it is only agents or risk factors, not characteristics of the organization of populations, that are eligible to be causes of disease. This limitation of causal explanation to the pathogenic action of risk factors in individual organisms is codified in the lists of "criteria for establishing causality" rehearsed in textbooks and journals [e.g., 2, 21, 24].

Expanding Epidemiology

Modern epidemiology can be contrasted with perspectives that recognize the roles of specific exposures but that place these exposures in a context that is itself of interest in scientific explanation and public health intervention, an approach closer to human ecology than to human toxicology [25]. Examples of broader views of the scope and goals of epidemiology can be found throughout its history.

The mid-nineteenth-century work of the young Rudolf Virchow has been revived as an early example of a quantitative approach to understanding disease in populations that, while recognizing the importance of specific agents or exposures, did not reduce the explanation of disease to a matter of these isolated factors themselves [26–28]. Virchow, in investigating an epidemic of typhus in Silesia, was deeply moved by the suffering of the people, and his explanations stressed the conditions that fostered the epidemic: lack of agricultural land, malnutrition, poor housing, low wages, and language barriers for the large Polish minority. His report to the government advocated land reform, progressive taxation, establishment of agricultural communes, local political autonomy, and, lastly, creation of a system of public hospitals. Virchow's conclusion: "Medicine is a social science, and politics nothing but medicine on a grand scale" [26, 27].

Pioneers of a New Epidemiology

Other nineteenth-century research that did not exclude the context of exposures was done in France [29] and England. The public health research that Friedrich Engels conducted in England was particularly insightful [26]. He documented the health problems that arose from crowding, lack of sanitation, malnutrition, and abuse of alcohol to alleviate chronic pain. Like Virchow, he quantified excesses of disease and death through statistical study without naming the analyzed factors, be they occupational, nutritional or behavioral, as the autonomous causes of disease. Unlike Virchow, however, Engels did not believe that reform of the political system would ever create the underlying conditions for adequate public health. Rather, he identified the capitalist economic system itself as the source of ill health.

Increasingly reductionist epidemiological thinking emerged in the context of Pasteur's discovery of microbes, the development of a more effective (and less damaging) allopathic medical practice, and the increasing dominance of statistics in quantitative investigation. Not all epidemiology, however, was converted to the theory that so successfully reduced the cause of disease to a germ, simultaneously distracting attention from the material living and working conditions in which disease arose. During the 1920s Joseph Goldberger showed that pellagra was not an infectious disease, as many at the time thought, that it was related to nutritional deficiency, and that its occurrence in the U.S. South depended on a share-cropping system that locked large numbers of people in poverty [30]. Unfortunately, his more global analysis of the economic arrangements in which pellagra and other public health problems proliferated was ignored by policy makers in favor of supplementing flour with niacin, a solution that probably contributed (along with access to electricity, refrigeration and dietary improvements) to the reduction of pellagra, but left in place the rural South's underdeveloped economic circumstances that continue to make the region the location of some of the worst public health conditions in the U.S. Yet Goldberger's well-documented research strategies can help contribute to today's critical evaluation of methodology in the field.

More holistic approaches to epidemiology continued in some third world countries. In Chile, the physician Salvador Allende came to believe that he would make the greatest contribution to the health of his people not by treating patients one at a time, but by working against the devastating effects of underdevelopment [26]. As president of Chile, he realized Virchow's vision of politics as medicine on a grand scale. Latin America—where it is clear that the immediate public health problems of disease in populations have less to do with specific exposures than with a position in the international economic system that sustains a lack of decent jobs, housing, clean water, food, and democratic control of institutions—is now home to a number of alternative currents of development in epidemiology. In these circumstances it is difficult to sustain the first world mirage that substantial public health advances can be achieved through the enumeration and regulation of unhealthy exposures on a case-by-case basis.

Pressure for change in epidemiology also comes from groups that have been exploited on the basis of race, gender, and class, and from environmental and peace activists. One aspect of an expanded public health agenda that is drawing attention from official agencies and academics in the U.S. is environmental racism [31, 32]. Increasingly, studies are documenting the systematic preferential location of toxic waste sites and polluting industries in areas that are predominantly inhabited by people of color and the poor. The primary issues here are not the identification of specific chemicals associated with particular diseases, threshold exposures for health effects, or dose response estimates; rather, the generation of exposures or potential exposures, environmental equity, and democratic decision-making are of primary concern. The specific effects of contaminants as well as medical means for treating problems once they occur are of great interest and importance, but as those issues are viewed in a larger context, they can potentially be connected to other health problems and the development of coordinated solutions. Thus, contamination of Native American lands in the southwestern U.S. by uranium mine tailings, nuclear weapons testing, and radioactive waste disposal can be connected to the long history of expropriation and destruction of Native lands, ecosystems, and means of subsistence, which have devastated Native public health for centuries.

Application of Epidemiology to Public Health

Epidemiology as a basic science of public health affects the nature of public health interventions. Because of its declared object of inquiry, modern epidemiology generally leads to interventions directed at specific individual exposures, and it is therefore important to consider the consequences of trying to intervene on specific exposure-disease associations in isolation from their context. Before addressing radiation and health, let us consider smoking and health, a textbook example of the early application of modern epidemiology for which public health consequences can be evaluated over three decades.

Despite widespread clinical observations of smokers' symptoms and the physical and biological plausibility of smoke as a lung pathogen and carcinogen, it was evidence from epidemiological studies of the 1940s and 1950s that led, in the United States, to the Surgeon General's 1964 report on the health hazards of smoking [33]. Modern non-infectious disease epidemiology had its first major success in identifying associations between cigarette smoking and lung cancer and other prominent diseases. Subsequently, public health efforts were initiated to reduce the prevalence of smoking through education about smoking hazards and through control of cigarette advertising. Over the last three decades there has been a remarkable shift in the prevalence of smoking and the burden of smoking-related diseases, a shift that can be attributed in part to the epidemiological explanation of the cause of smoking-related diseases. Smoking prevalence has declined substantially among better educated, higher income people in North America, parts of Western Europe, Australia and New Zealand [34]. Smoking prevalence has declined little or none among lower educated and lower income people in those countries [35], and is increasing rapidly in many of the most populous parts of the world [36]. Thus in the three decades following epidemiology's major success, more people are exposed to and made sick from the disease agent than ever before [37].

Tobacco and Epidemiology

The populations studied by modern epidemiologists were the exposed and the unexposed defined by the model of the experiment. They were not highly organized groups with economic systems and social relations. Thus, the cause of the lung cancer epidemic was identified as cigarette smoking, an individual behavior, while tobacco agribusiness, the commercial sale of cigarettes, and the social circumstances that make smoking a rewarding habit, could not be recognized by epidemiological studies as targets for intervention. When educational efforts and social options led some groups to reduce smoking levels, tobacco companies redirected advertising to replace those markets with others, often assisted by governments with their own financial stakes in tax revenues and contributions to trade balances [38].

Not only has the identification of the association between smoking and disease failed to stem the world-wide epidemic of smoking-related diseases, it has redistributed smoking prevalence so that it increasingly adds to inequalities in health between the poor and the rich within countries of the North [39], and between countries of the North and the South [40]. Such redistributions may be expected when interventions are directed only towards consumption rather than towards both production and consumption [41], and it is only the behavior of consuming cigarettes, not the organized production and promotion of tobacco, that has been considered by epidemiological studies. Health inequalities are further exacerbated through replacement of local food crops by tobacco for commercial markets and by toxic exposure of industrial and agricultural workers during manufacture and application of agricultural chemicals.

Three Sources of Radiation Exposure

A related situation can be described with radiation and health. The problem as now defined in the terms of modern epidemiology is one of quantifying the extent of the increase in cancer (or birth defects or some other physiological or disease outcome) produced by every unit increase in an individual's dose of ionizing radiation. The accurate quantification of this relationship will supposedly provide the basis for the rational determination of how much radiation exposure would be permissible, and regulations would be designed accordingly. As in the case of the evaluation of smoking and health, the situation of an individual's exposure is separated from the context of the production of the exposure and the other effects of that production on health and society.

Most manufactured exposures to radiation take place within the context of the energy, military, and medical industries. The military industry, which has produced many of the exposures studied by epidemiologists, developed in support of the creation of tens of thousands of nuclear weapons. Nuclear proliferation and the threat of catastrophic nuclear war continue. Attempts to clean up the most toxic remains of this production are just beginning, will go on for decades if not centuries, and cannot possibly restore many areas to safe states. The history of environmental contamination and human suffering in areas used for production and testing is horrible, and the stories of the affected people and places are indeed chilling [42, 43]. The public health consequences of direct contamination, however, are only a small part of the much larger health impact of the exposure context, the industry without which the exposures would not have occurred. This military enterprise and the research infrastructure that supports it have used huge proportions of national budgets and engaged hundreds of thousands of people in activities that have as their main purpose not social welfare, but development of means of destruction, diverting human and economic resources from potential projects to improve living conditions throughout the world. The expense and sophistication of nuclear technology itself has generated a scientific and bureaucratic elite that has perpetuated itself through secrecy, concentration of power, and elaboration of a xenophobic and divisive brand of patriotism [44]. This social context is highly undemocratic and contributes to inequalities of wealth and power justified on the basis of special knowledge of an elite group that is supposedly uniquely qualified to make decisions, promoting disregard for protection and rights of workers, indigenous peoples and others that are excluded from participation in decisions affecting health conditions.

The Health Effects of Industries and Social Policies

The use of nuclear technology for power production could not have occurred without the support of the infrastructure and research base created for military purposes. Once the commitment had been made, however, the commercial industry contributed to the hopes for ever-increasing energy consumption with the promise of a clean and cheap source of power. The industry proliferated widely in the absence of political and technical solutions to the problems of waste disposal. Energy would be produced without limit in centralized locations controlled by a technical elite rather than with a technology that could be widely distributed and controlled more democratically. This vision of electric power generation has in part prevented the development of a policy alternative to unlimited growth that is ecologically sound, sustainable, equitable, and consistent with reduced inequalities in health between the majority of the world's people that use little energy and the minority that consumes most of the energy.

Medical uses of ionizing radiation preceded the military and energy uses by decades, and while there have been important diagnostic and therapeutic applications, many disasters and victims were created in the process [44]. These include patients subjected to x-ray and radium treatments, children exposed to x-rays during pregnancy, and many of the clinicians that treated them. In medicine, consideration of the context of radiation exposures could encourage greater attention to preventive rather than diagnostic and curative measures, and, within the context of the latter, could draw attention to overuse of tests and treatments with marginal benefits.

In all three radiation industries, the existence of a technology that proponents pushed as a quick fix for complex social problems—international conflict, energy policy, and medical care—encouraged the attitude that it would not be necessary to address the global issues of political relationships, ecological sustainability, disease prevention, and humane healing, but only to put faith in a technology that would provide the power to solve what was perceived to be a specific problem by itself. The health effects of these policies occur not only through agents like ionizing radiation, but more importantly through their effects on society, on social inequalities, and on living conditions that are essential to public health.

Smoking and health and radiation and health are only two examples of how the nature and object of epidemiological explanation limits its scientific scope and public health application. Similarly, a focus on high-fat diet as individual behavior fails to address consequences of the animal-oriented agriculture systems that support mass high-fat diets. These consequences include production of export crops in the context of local malnutrition, use of vast quantities of non-renewable energy resources, occupational exposure to pesticides and herbicides, topsoil loss, and generation of methane greenhouse gases.

Two responses to this situation are suggested by approaches reviewed above. One is that epidemiology is simply not about these broader issues, rather, they are the responsibility of other disciplines. Another is that epidemiology has addressed the broader issues of context in the past, and that it has only lost that ability and interest because of the economic and political context of its dominant modern practice. Epidemiology is not alone among disciplines in its reductionism, and it cannot carry the burden, by itself, of adopting an ecological perspective. Rather, all health-related disciplines should adopt a broader perspective within their particular practice, encompassing and transforming the concepts and tools that have developed in each area.

Scientific Objectivity and Social Responsibility

Many scientifically trained people, as well as non-scientists who have felt comfortable putting their faith in the experts, have been attracted to reductionist science because of the very characteristics criticized here. According to this logic, it is only by excluding the context and focusing on particular factors considered independently of historical conditions that science can produce objective knowledge that has a greater claim to authority than other forms of knowledge. Although this perspective continues to thrive in the biomedical sciences, including epidemiology, it has been thoroughly critiqued by historians, philosophers, and practitioners of science of the latter half of the twentieth century [5, 45–52].

Science and Conceptual Frameworks

The basis of the critique of value-free objectivity is simple: it is impossible to know the world without intellectual tools, including languages and socially produced concepts. Whether explicit or not, all scientific investigations depend on conceptual frameworks. There can be no unmediated experience. These ideas have been extensively developed in various directions by many authors, especially since the seminal work of Thomas Kuhn [45], and they provide a basis for the most fundamental challenge to reductionist epidemiology by removing its justification as a unique means to provide objective analyses of health and disease problems. The choice is not between objective science and a science that is contaminated by social and political values. Risk factor epidemiology does not achieve objectivity by systematically examining exposure-disease associations separated from contexts of military, energy, or agriculture policy, and issues of economic inequalities and democracy. Rather, it makes a political commitment to the status quo by excluding these issues from public health consideration.

Shattering the myth that scientific inquiry can be independent of society amounts to recognizing the "distinction between the claim that the world is out there and the claim that truth is out there" [53]. Belief that truth is something which is found "out there" rather than something that is made from observations of the world using socially and culturally produced languages and concepts is partly a reaction to the mistaken perception that the only alternative to value-free scientific objectivity is relativism and the abandonment of any basis for making comparative evaluations of scientific explanations. But recognition that all knowledge, including scientific knowledge, is rooted in social constructs does not negate the idea of objectivity in the sense of fairness, justice, and intellectual honesty. This has been called "strong objectivity" by Sandra Harding [46], a concept that she has counterposed to the "weak objectivity" perpetuated by practitioners of Cartesian reductionism. Strong objectivity depends on indentifying assumptions and goals through self-critique rather than denying them by constructing a myth of socially unmediated scientific experience. Consideration of social responsibility as part of an analysis of the construction of scientific knowledge is therefore not something to be tacked on to scientific practice after the fact, but is a necessary part of being objective in the sense of being explicit about assumptions. The development of an alternative to reductionist epidemiology depends on such a commitment.

Recognizing social responsibility as an integral part of scientific inquiry, however, forces us to to make judgments about social responsibility, an ethical dilemma that is denied by those who believe that objectivity is obtained by separating science from human values. What, then, is to be the basis of moral judgment, and how are conflicts between different cultures and classes to be resolved? Most philosophical approaches consider this question against the standard "that there must be necessary, universal grounds for all moral principles" [54, p. 9], either searching to establish the basis for these universal grounds or accepting a defensive posture of moral relativism. Philosopher Cornel West argues that the search for universal moral certainty is doomed, but that various brands of moral relativism are unacceptable. He dissolves the problem of choosing between universal and relative moral principles by substituting a historical basis of thinking about morality which holds "that there are moral truths or facts, but that they are always subject to revision . . . relative to specific aims, goals, or objectives of particular groups, communities, cultures, or societies" [55, p. 10]. West calls for "a historical assessment and political reading of our morality and morale, in order to shed light on how we can make them more contagious to others captive to the prevailing cynicism and nihilism [of our culture]" [55, p. xiii]. Such an ethical position can support a socially responsible scientific practice but requires full engagement with, rather than denial of, the ethical aspects of the social construction of scientific knowledge.

An Alternative Epidemiology

If we accept a broad critique of the dominant practice of epidemiology, as opposed to the view that the discipline is essentially on track but needs fine tuning, the first question that arises is, "If this isn't the right way to do it, then what do you propose?" The answer to this question must emerge through the developing work of diverse groups of researchers and practitioners who are struggling to make the field more relevant to improving public health conditions throughout the world. There are many examples in the current literature of attempts to work out more or less contextual explanations of health and disease phenomena [5, 8, 9, 11, 25, 47, 51, 56–58], although they are not yet connected in a coherent set of theories, assumptions, and techniques that could constitute a real new paradigm [see 59 for a discussion of competing paradigms in epidemiology]. An alternative practice that is shared by large numbers of scientists will no doubt require much time and larger scientific, social and political change. Such change will be affected by relationships between professional practices including medicine and public health, and broader based efforts to improve public health conditions, including civil rights, economic justice, environmental, anti-war, and other activities. Epidemiology is already being affected by involvement of nonprofessionals with new ideas about appropriate research questions, evidence, and interpretations of findings [60].

How would such an epidemiology be practiced, and what would its products look like? First, it would not just ask questions about what is good or bad for health in general, but it would analyze differential effects—good or bad for whom? Second, it would look for connections between many diseases and exposures, what is common about them, instead of always isolating exposure-disease pairs. Third, it would look for side effects of interventions and exposures, the unintended consequences that may be more important than the intended ones. Fourth, it would develop ways to utilize historical information, the developmental narratives of particular populations and even individual people, with the aim of connecting the particular and the general. Fifth, it would address the conceptual framework of the research, including analysis of assumptions and the social construction of scientific knowledge, as a central part of any research reporting, as central to a research manuscript as consideration of measurement error or selection bias [49]. Sixth, such an epidemiology would recognize that the problem of controlling confounding factors comes from the search for independent relationships, not from the world we study, and that the "nuisance factors" of the reductionist perspective can become the essential context of exposure and disease of an ecological perspective. Finally, it would entail a humility about the scientific research process and an unrelenting commitment to playing a supportive role in larger efforts to improve society and public health.

Efforts to articulate an alternative epidemiology are connected by attention to the historical contexts of public health phenomena as well as to the science that addresses them. These positions are contrary to the dominant assumptions that there can be an ahistorical, value-free epidemiology that is about ahistorical, independent exposure-disease associations. Research into the health effects of ionizing radiation is a stereotypical example of the application of the reductionist program, preoccupied as it has been with the search for the shape and magnitude of a presumably universal dose-response relationship and its consistency across populations, and the quest for an ahistorical constant that will be analogous to the fundamental laws of nature supposedly uncovered by the discipline that produced the exposures under study, physics. Meanwhile, the public health disasters of runaway military spending, uncontrolled energy consumption, and dominance of high-technology curative medicine over preventive environmental and medical practices, go unseen by the "basic science of public health."

Our dominant epidemiology begins with the assumption that things work separately and independently, that exposures can be separated from the practices that produce them. An epidemiology oriented towards massive and equitable public health improvement requires reconstructing the connections between disease agents and their contexts. This is necessary for the successful application of scientific knowledge to public health practice, for the resolution of intractable technical and conceptual problems inherent in current exposure-disease studies, and for the development of a socially responsible epidemiology. The practical, technical, theoretical, and ethical goals are complementary. Such a direction will allow us to recognize the health consequences of industries and social and economic arrangements as well as the roles of specific disease agents. As social, economic, and political arrangements that provide the conditions for public health and human development become an explicit part of the epidemiological explanation of health and disease in populations, efforts to oppose injustice and inhumanity can be recognized as an integral part of a comprehensive public health agenda. Current global public health crises demand more than a piecemeal approach.

ACKNOWLEDGMENTS

Institutional and intellectual support was provided by the Department of Preventive Medicine, Federal University of Bahia, Salvador, Brazil, and the Brazilian National Research Council. I am grateful to the many colleagues and friends who provided input on earlier drafts. Rudi Nussbaum was instrumental in encouraging me to write this paper.

REFERENCES

1. Institute of Medicine Committee for the Study of the Future of Public Health. The Future of Public Health. Washington, DC: National Academy Press, 1988.
2. Rothman K J. Modern Epidemiology. Boston: Little, Brown and Co., 1986.
3. Miettinen O S. Theoretical Epidemiology: Principles of Occurrence Research in Medicine. NY: John Wiley and Sons, 1985.
4. Susser M. Epidemiology in the United States after World War II: The evolution of technique. Epidemiol Rev 1985:7:147–77.
5. Levins R and Lewontin R. The Dialectical Biologist, Conclusion: Dialectics, pp. 267–288. Cambridge: Harvard University Press, 1985.
6. Morganstern H. Uses of ecologic analysis in epidemiologic research. Am J Public Health 1982:72:1336–1344.
7. Greenland S. Divergent biases in ecologic and individual-level studies. Stat Med 1992;11: 1209–1223.
8. Loomis D and Wing S. Is molecular epidemiology a germ theory for the end of the twentieth century?. Int J Epidemiol 1990;19:1–3.
9. Wing S. Barnett E, Casper M, Tyroler H A. Geographic and socioeconomic variation in the onset of decline of coronary heart disease mortality in white women. Am J of Public Health 82:204–209, 1992.
10. McMichael A J. Setting environmental exposure standards: the role of the epidemiologist. Int J Epidemiol 1989;18:10–16.
11. Rose G. The Strategy of Preventive Medicine, NY: Oxford University Press, 1992.
12. Pearce N. Multistage modelling of lung cancer mortality in asbestos textile workers. Int J Epidemiol 1988;17:747–752.
13. Stewart A. Kneale G W. The Hanford data: issues of age at exposure and dose recording. PSR Quarterly 1993;3:101–111.
14. Wing S. A review of recent findings on radiation and mortality at Oak Ridge National Laboratory. In Lengfelder E, Wendhausen H (eds.), Neue Bewertung des Strahlenrisikos, Proceedings of the International Society of Radiological Protection, pp 217–228, Munich: Medizin Verlag, 1993.

15. Greenland S. Modeling and variable selection in epidemiologic analysis. Am J Pub Health 1989;79:340–349.
16. Committee on the Biological Effects of Ionizing Radiation. Health Effects of Exposure to Low Levels of Ionizing Radiation (BEIR V). Washington, DC: National Academy Press, 1990.
17. Rothman K. Causes. Am J Epidemiol 1976;104:587–592.
18. Dickersin K. Berlin J A. Meta-analysis: State-of-the-science. Epidemiol Rev 1992;14:154–176.
19. Lewontin R. Analysis of variance and analysis of causes. Am J Hum Genetics 1974;26:400–411.
20. Horwitz R I. The experimental paradigm and observational studies of cause-effect relationships in clinical medicine. J Chron Dis 1987;40:91–99.
21. Hennekens C H, Buring J E. Epidemiology in Medicine. Boston: Little, Brown and Co., 1987.
22. Skrabanek P. The poverty of epidemiology. Persp Biol Med 1992;35:182–185.
23. Skrabanek P. The epidemiology of errors (commentary). Lancet 1993;342:1502.
24. Susser M. What is a cause and how do we know one? A grammar for pragmatic epidemiology. Am J Epidemiol. 1991;133:635–48.
25. Stallones R A. To advance epidemiology. Ann Rev Public Health, 1:69–82, 1980.
26. Waitzkin H. The social origins of illness. Int J Health Ser 11:77–103, 1981.
27. Eisenberg L. Rudolf Ludwig Karl Virchow, where are you now that we need you? Am J Med 1984;77:524–532.
28. Taylor R, Rieger A. Medicine as social science: Rudolf Virchow on the typhus epidemic in upper Silesia. Int J Health Ser 1985;15:547–559.
29. Coleman W. Death as a Social Disease: Public Health and Political Economy in Early Industrial France. Madison: University of Wisconsin Press, 1982.
30. Goldberger J. Goldberger on Pellagra. M Terris (ed.) Baton Rouge: Louisiana State University Press, 1964.
31. United Church of Christ Commission for Racial Justice. Toxic Waste and Race in the United States: A National Report on the Racial and Socio-Economic Characteristics of Communities with Hazardous Waste Sites. New York: United Church of Christ, 1987.
32. Mohai P, Bryant B. Race poverty and the environment: the disadvantaged face greater risks, EPA Journal 1992;18:6–8.
33. Brandt A. The cigarette, risk, and American culture. Daedalus 1990;119:155–176.
34. Pierce J P. International comparisons of trends in cigarette smoking prevalence. Am J Public Health 1989;79:152–157.
35. Schoenborn C A, Cohen B H. Trends in smoking, alcohol consumption, and other health practices among U.S. Adults, 1977 and 1983. Advance Data from Vital and Health Statistics 1986;118. DHHS Pub. No. 86–1250.
36. Masironi R, Rothwell K. Tendances et effets du tabagisme dans le monde. Wld Hlth Statist Quart 1988;41:228–241.
37. Council on Scientific Affairs. The worldwide smoking epidemic: tobacco trade, use, and control. JAMA 1990;263:3312–3318.
38. Chen T T L, Winder A E. The opium wars revisited as US forces tobacco exports on Asia. Am J Public Health 1990;80:659–652.
39. Hart N. Inequalities in health: the individual vs. the environment. J R Statist Soc 1986; 149A:228–246.
40. Barry M. The influence of the U.S. tobacco industry on the health, economy, and environment of developing countries. New England Journal of Medicine 324:917–20, 1991.

41. Wing S. Social inequalities in the decline of coronary mortality. Am J Publ Hlth 1988;78: 1415–16.
42. Amundson B. Scientific integrity and adequate health services: twin casualties of the nuclear arms race. PSR Quart 1992;2:210–215.
43. Davis S. Understanding the health impacts of nuclear weapons production in the southern Urals: an important beginning. PSR Quart 1992;2:216–220.
44. Caufield C. Multiple Exposures: Chronicles of the Radiation Age. Chicago: University of Chicago Press, 1988.
45. Kuhn T S. The Structure of Scientific Revolutions. Chicago: University of Chicago Press, 1970.
46. Harding S. Whose Science? Whose Knowledge? Ithaca: Cornell University Press, 1970.
47. Tesh S N. Hidden Arguments: Political Ideology and Disease Prevention Policy. New Brunswick, NJ: Rutgers Press, 1988.
48. Lewontin R C. Facts and the factitious in natural sciences. Critical Inquiry 1991;18:140–153.
49. Ratcliffe J W, Gonzalez-del-Valle. Rigor in health-related research: Toward an expanded conceptualization. Int J Health Ser 1988:18:361–392.
50. Holtzman E. Science, philosophy and society: Some recent books. Int J Health Ser 1981;11: 123–149.
51. Haila Y, Levins R. Humanity and Nature: Ecology, Science and Society. London: Pluto Press, 1992.
52. Keller E F. Between language and science: The question of directed mutation in molecular genetics. Persp Biol Med 1992;35:292–306.
53. Rorty R. Contingency, Irony and Solidarity. NY: Cambridge University Press, 1989.
54. Foster J B. Introduction to a Symposium on The Ethical Dimensions of Marxist Thought. Monthly Review 1993;45(2):8–15.
55. West C. The Ethical Dimensions of Marxist Thought. NY: Monthly Review Press, 1991.
56. Milio N. Promoting Health through Public Policy. Philadelphia: F A Davis, 1981.
57. Pearce N E, Matos E, Koivusalo M, Wing S. Industrialization and Health. In: Pearce N E, Matos E, Vainio H, Boffetta P, Kogevinas M (eds), Occupational Cancer in Developing Countries, Lyon: International Agency for Research on Cancer, 1994.
58. Population health looking upstream (editorial). The Lancet 1994;343:429–430.
59. Almeida-Fllho N. The epistemological crisis of contemporary epidemiology: Paradigms in perspective. Santé Culture Health 1991;8:145–166.
60. Brown P. Popular epidemiology challenges the system. Environment 1993;35(8):16–41.

QUESTIONS

1. Explain the relation of epidemiology to public health. In what ways do the goals and foci of public health differ from those of conventional medicine?

2. Explain what Wing means when he says exposures are often measured on the basis of convenience, rather than on the basis of biological models of disease process

3. Why does Wing argue that epidemiologists should be examining contextual factors, rather than hunting for universal dose-response relationships?

Chapter Three

Physicians' Knowledge, Attitudes, and Practice Regarding Environmental Health Hazards

Phil Brown and Judith Kirwan Kelley

This chapter examines the role of physicians in the detection and treatment of environmentally induced illness. It is important to investigate these issues because of increasing scientific knowledge of toxic health effects, mounting public interest in these effects, and the necessity for public policies to address environmental health hazards in the community and in the workplace.

Physicians are often called upon to advise on and participate in public health issues that may extend beyond their own traditional training and practice. Sometimes these issues seem as though they can be readily integrated with normal clinical practice, and they are even widely adopted, even if it takes time. Examples include prevention campaigns concerning smoking, diet, and seat-belt use. Other times, these concerns are only moderately well accepted and applied, as with examining children's lead exposure. In yet other cases, physicians have made little or no response to public health concerns, as with environmental health hazards. Although there is growing patient education about some environmental ills, such as ultraviolet radiation, we do not see much action on toxic substances in the environment.

Public Desire for Environmental Health Services

When people look to physicians for help with environmental disease, they are asking physicians to take on two roles that are not part of the routine life of medical practice. First, they are asking physicians to step outside the boundaries of medical practice and into the role of public health action. This may involve going beyond traditional limits of professional competency. Second, people are asking physicians to adopt a civic duty with political and economic overtones. Physicians thus face a difficult situation, given the complex scientific, legal, and ethical dimensions of environmental disease.

Public interest in the health effects of toxic exposure has increased dramatically in recent years (Castorina & Rosenstock, 1990; Cullen & Rosenstock, 1988; Frazier et al., 1991; Institute of Medicine [IOM], 1988, 1991; Rosenstock et al., 1991). Perceiving environmental hazards to be among their most pressing health concerns, the public expects that physicians will be able to provide diagnostic and treatment services for environmental illness (Castorina & Rosenstock, 1990). In 1989, Colorado 2000, a state interagency group funded by the Environmental Protection Agency (EPA), conducted a poll to assess the environmental health concerns of Colorado citizens. In that poll, people rated environmental concerns as the second most important facing the state, following the poor economy and high unemployment. Citizens were especially concerned about hazardous waste disposal, air pollution, solid waste disposal, and contamination of groundwater and surface water (Colorado Department of Public Health, 1990).

Physicians' Lack of Training and Knowledge

Many environmental health effects often go unrecognized, and thus untreated, in nonspecialist physicians' practices primarily because of a lack of knowledge of environmental health effects on the part of physicians. Despite the increasing public interest in environmental health issues, a number of studies have documented a critical shortage in the number of physicians who have been trained to recognize and treat environmental health effects (Castorina & Rosenstock, 1990; Frazier et al., 1991; IOM, 1991). To accurately assess the extent of the physician shortage in occupational and environmental medicine, the IOM commissioned a study that indicated a need for 3,100 to 5,500 additional physicians, including primary care physicians with special competence in occupational and environmental medicine. The deficit of occupational and environmental medicine (OEM)[1] specialists alone is estimated to range between 1,600 and 3,500 (IOM, 1991).

Although physicians are among the most trusted sources of medical information, they are among the least informed regarding environmental health effects. In addition, physicians often do not understand the importance of their roles in these matters. Physicians generally treat their patients' comments about environmental health threats as simple conversation, and they do not respond appropriately to the environmental concerns expressed (McCallum, 1992).

The general failure of physicians to deal with environmental health hazards is striking, given the extent of public concern. Case studies of toxic waste sites where community action created high public visibility have found a lack of physician involvement (Brown & Mikkelsen, 1990; Capek, 1987; Edelstein, 1988; Levine, 1982; Nelson, 1989). Local physicians near toxic waste sites (often rural areas, small towns, or small cities) are likely to be tied to sources of political, economic, and social power. They may therefore find it difficult to consider targeting environmental hazards that may pose challenges to local businesses and local government. Public health scientists have been chastised for taking the side of community groups in toxic waste cases (Freudenberg, 1984; Paigen, 1982), and physicians may fear they too will be seen as whistle-blowers.

In the absence of physician involvement, laypeople who have discovered toxic-waste-induced disease have a harder time legitimizing their claims. The newness of environmental epidemiology and the various controversies in that discipline already place citizens at a disadvantage in both disease detection and in making a causal argument about the relationship between toxics and disease (Lave & Upton, 1987; National Research Council, 1991). Physicians would be likely sources of aid in both detection and etiology, especially in clusters of low base rate phenomena such as childhood leukemia (Castorina & Rosenstock, 1990). Yet physicians usually examine, diagnose, treat, and refer patients as individuals, without understanding them as potentially part of an affected group (Goldsmith, 1986). This particularistic, rather than universalistic, approach has always made the clinical mentality quite unable to take a broader public health approach (Freidson, 1970, pp. 170–172).

For an example of this tendency, we may look at a Colorado Department of Health survey of physicians in the state to identify education needs regarding human exposure to environmental health hazards. Surveys were administered to 83 physicians through mailings and a newsletter. Additionally, three focus groups with a total of 75 physicians from three diverse groups were conducted to obtain a more complete assessment of the environmental education needs of physicians. Personal interviews were also conducted with 15 physicians, as well as with representatives of 21 health care organizations. The results of the surveys, focus groups, and interviews revealed some interesting points about lack of interest in general—physicians did not perceive environmental medicine concerns to play a significant role in their practice and were more interested in local rather than general environmental health issues (e.g., physicians in Greeley, Colorado, were interested in carbon monoxide air pollution in their area as opposed to the more global topic of health effects of air pollution). Another important finding was that physicians seemed noncommittal on most environmental medicine topics. This was attributed to a feeling of powerlessness over circumstances beyond the medical control of the physicians when patients presented with environmental health effects. One physician in the Colorado focus group asked, "What do I do? Tell them to move because their respiratory problem may be due to the factory located in the neighborhood (Colorado Department of Public Health, 1990, p. 42)?"

Physicians' failure to recognize environmental health effects may also be due in part to the generally negative opinion that most of traditional medicine holds toward "environmental illness" (EI; sometimes termed *multiple chemical sensitivity*). Medical societies and journals have argued that people complaining of EI are exhibiting symptoms of psychiatric disorder, not environmental contamination (Kroll-Smith & Ladd, 1993). It is easy to see how such antagonism might extend to cases of environmental contamination that did not resemble EI.

The Role of Organized Medicine and Government Health Agencies

Medical organizations do not play a significant role. No state or local medical societies have involved themselves in specific toxic waste sites. In only one case has a medical organization been involved—Physicians for Social Responsibility (PSR)—and this organization defines itself in terms of political action (usually concerning prevention of nuclear war). The Washington (State) PSR chapter has worked with the Hanford Educational Action League to examine the health effects of radiation releases from the Hanford Nuclear Reservation (Kaplan, 1992, p. 223). The national PSR organization organized a significant conference in October 1992, drawing people from all over the United States and abroad. Of the 750 attending, at least half were physicians. PSR believes that the end of the Cold War allows them to shift their emphasis, and they are doing so in the direction of environmental health. This large-scale involvement of a physicians' organization offers hope that more physicians will take up these concerns, yet it was also noticeable that the typical physician attending the PSR conference was an academic and/or clinician in a major teaching hospital. The average community primary care physician is still less likely to be aware of environmental health issues.

In terms of general medical practice and education, there are some recent initiatives. The American College of Physicians (1990) has published a position paper on the important role played by internists in occupational and environmental health. The American College of Pediatrics has an Advisory Committee on Environmental Health. The National Institute of Environmental Health Sciences (1994) has instituted an Academic Award grants program to support OEM physicians in developing and sustaining curriculum, and in providing leadership for faculty development. The Massachusetts Medical Society (MMS) has collaborated with the Massachusetts Department of Public Health (MDPH) on a survey of Massachusetts physicians' environmental knowledge. This is one of several state efforts funded by the Agency for Toxic Substances and Disease Registry (ATSDR) (MDPH, 1991) and will be discussed below in the Methods section. Yet even this type of action is more under the aegis of state public health agencies than the ongoing action of physicians or their organizations (Murdock, 1991).

In light of the documented deficits in physician education and interest in environmental health issues, this chapter presents a more detailed investigation of physician knowledge, attitudes, and practice regarding environmental health exposures. Data obtained from the 1990 Massachusetts Department of Public Health/Massachusetts Medical Society collaborative survey of physicians will be analyzed. We are particularly interested in whether physicians are more aware of environmental health issues if they practice in a community with a well-known toxic site or an actual or prospective toxic facility. We also present for comparison findings of physician environmental knowledge assessment surveys conducted by the state health departments of Rhode Island, Colorado, Connecticut, and New Hampshire.

Theoretical Considerations

The paucity of physician knowledge and action concerning environmental health effects, as noted above, certainly demands that we investigate this area further. But what broader perspectives can frame such an analysis? This section puts forth some key sociological theoretical points that can help that framing.

Freidson (1970), in arguing against Parsons, contended that physicians were more particularistic than universalistic, that they did not act for the common good as much as for their own needs and out of their own experiences. As Freidson notes, the physician's particularist thinking may be rational, but it is a rationality based largely on his or her direct, personal experiences. He continues:

> The difference between clinical rationality and scientific rationality is that clinical rationality is not a tool for the exploration or discovery of general principles, as is the scientific method, but only a tool for sorting the interconnections of perceived and hypothesized facts. (p. 171)

We may also apply the particularist-universalist distinction to the focus of clinician-patient interaction. We may view as particularistic the type of relation where the physician treats the individual patient. In focusing on a specific individual clinical challenge, the physician gains satisfaction from treating an individual's condition. We may view as universalistic the type of relation where the physician takes a more public-health, population-based approach to a broader community of actual and potential patients. Here, he or she would gain satisfaction from population-based preventive interventions and from influencing social policy and corporate practices.

Daniels's (1969) notion of the captive professional is useful here. Daniels found that military psychiatrists were captive, in that they were severely constrained in their practice: Rather than helping people identify and work on emotional problems, these physicians were charged with maintaining a disciplined military workforce. Like corporate physicians, these military psychiatrists served their employers rather than their patients. The general population of physicians are, of course, not captive in the same way as if they were employed by the military. However, they are captive to a larger surrounding community of local political/economic/social structures, and to a broader medical community that maintains relations with powerful social institutions. They are unlikely to risk local business or collegial censure, both possibilities if they become whistle-blowers. They are also captive to a worldview that lends little or no credence to the significance of environmental hazards. Their biomedical model, bereft of a view of social causation and social amplification of biological causation, keeps them from seeing the world in such a way that would appreciate the importance of environmental factors.

We can take a broader view, stemming from the above points, that sees the problems when physicians experience "sociological ambivalence" (Merton & Barber, 1976) in facing incompatible normative expectations. Following Merton and Barber, Coser (1979) argued that physicians often employ evasion to defend against status ambiguity. These physicians are acting in different roles: as servants of the local hospital and medical community, as members of a larger professional community, and as coresidents/neighbors of a community of citizens. As noted earlier, public desire for physicians to help with environmental disease asks physicians to step outside the boundaries of medical practice and into the role of (a) public health action and (b) civic activism with political and economic overtones. These different roles may seem quite incompatible, especially because the physicians typically lack sufficient training to be sure of their forays into environmental health. These multiple roles make it difficult for them to sort out the issues, especially if the result would challenge established knowledge bases and power centers. This is amplified by the high degree of uncertainty and conflict regarding environmental causation of disease.

Another consideration is the daily work culture of medicine. Becker, Geer, Hughes, and Strauss (1961) and Strauss, Schatzman, Ehrlich, Bucher, and Sabshin (1964) show that medical students and health professionals learn their work habits through a routine habituation to work, just as with any other form of work. Abstract ideals or normative roles are less important than simply getting the job done. For our purposes, physicians exposed to environmental health concerns in any significant amount will be launched on a trajectory that accepts the importance of such concerns. There will be a self-amplifying system that makes it second nature for these doctors to think in terms of environmental etiology, or at least to ask about environmental exposures.

Hypotheses

Based on the background material and theoretical framework, we are using the MDPH/MMS data to test four hypotheses:

Hypothesis 1:
Physicians whose practices are located within communities that have EPA National Priority List (NPL) sites (Superfund sites) within their borders will be no more knowledgeable about environmental health hazards than those physicians whose practices are not located in communities with NPL sites.

This hypothesis stems from the discussion of Freidson's (1970) distinction between the particularist and universalist concerns of medicine. It also follows from the prior discussion of sociological ambivalence, showing that physicians face conflicting normative expectations that make it difficult to act decisively and in ways that challenge established authority. It also flows from the "captive professionals" model, in that these physicians are captive to an established political-economic context and a conservative medical community. Empirically, this hypothesis is logical in light of the failure of local physicians to play a role in detecting health effects suffered by residents of many contaminated communities, and in light of the prior research on poor physician knowledge of environmental health. We believe the failure of medicine to recognize environmental hazards and to train physicians to deal with them is so serious that physicians practicing in areas with Superfund sites will be no different than those practicing elsewhere.

Hypothesis 2:
Physician knowledge and practice regarding environmentally induced disease will vary by specialty. In particular, those specializing in environmental/occupational health will

be the most knowledgeable. We also expect that those in obstetrics/gynecology will be the least knowledgeable.

We base this hypothesis as well on the particularist/universalist distinction and on the sociological ambivalence theme. In this hypothesis, the ambivalence is amplified by the fact that OB-GYNs face other significant conflicts that may make it harder to focus on environmental health concerns that may seem less provable. As well, we know that scarcely any medical schools offer even a brief curriculum in environmental medicine. Even when offered as an elective, only 1.4% of medical students study environmental medicine. Toxic waste activists and many environmental epidemiologists consider reproductive problems and birth defects to be especially related to toxic contamination. The generally poor education in environmental health effects leads to an individual clinical mentality that ignores the larger public health understanding, which is particularly dangerous in a medical area that ought to be more aware.

Hypothesis 3:
Self-rated knowledge of environmental health hazards will be related to knowledge about the Right-to-Know law[2] and also to the likelihood of asking questions about workplace and nonworkplace environmental exposures. Those who rate themselves as "very knowledgeable" will be more knowledgeable about medical interviewing, diagnostic criteria, and the state Right-to-Know law.

We base this hypothesis on the notion of daily work cultures of medicine, as discussed earlier. Like other forms of work, medical work relies on habituated forms of knowledge and experience. Once physicians have crossed the line where they have some knowledge about environmental health effects, they are likely to act more generally along those lines.

Hypothesis 4:
Physicians who perceive that nonworkplace environmental hazards are "very important" will be more knowledgeable about medical interviewing, diagnostic criteria for environmental disease, and the state Right-to-Know law than those who rate environmental health hazard knowledge as "irrelevant to my practice." Physicians who consider it important to be informed about nonworkplace environmental hazards will incorporate this perception into their practices.

We base this hypothesis as well on the notion of daily work cultures of medicine, as in Hypothesis 3.

Data and Methods

Data Source

On September 15, 1989, the ATSDR awarded cooperative agreements to 11 states to "support educational activities for physicians and other health professionals in the environment" (ATSDR, 1991, cover page). State health departments receiving the ATSDR awards were to develop educational materials on medical surveillance related to nonworkplace exposures to hazardous substances and negative outcomes. The states participating in the 1989 cooperative agreements were Arkansas, California, Colorado, Connecticut, Florida, Georgia, Illinois, Iowa, Kansas, Kentucky, Maine, Maryland, Massachusetts, Minnesota, Missouri, New Hampshire, New York, Oklahoma, Rhode Island, and Wisconsin (ATSDR, 1991). In 1990, the MDPH solicited the collaboration of the MMS in a study of physician knowledge of environmentally induced disease. A 21-item questionnaire mailed to all of the members of the MMS (approximately 14,500) sought information on knowledge, attitudes, and interest of physicians about environmental health hazards; their consideration of environmental health in daily practice; and their sources of information on environmental health.

A total of 1,609 physicians representing 23 different medical specialties completed the questionnaire, yielding a response rate of 11%. This completion rate, although low, was comparable to the rates of completion of similar surveys in Connecticut, Colorado, New Hampshire, and Rhode Island as shown in Table 3.1.

The survey completion rates in the five states ranged from 5% in New Hampshire to 42.5% in Rhode Island. Although the completion rate for Massachusetts falls at the lower end of the range, the overall number of surveys completed in Massachusetts is high enough ($N = 1,609$) that the results of the statistical analyses can be interpreted with confidence. Physicians often provide poor response rates on questionnaires. Mail survey responses of 20% to 30% are quite common, and even the AMA's Periodic Survey of Physicians recently garnered only 49% (Guadagnoli & Cunningham, 1989).

Given physicians' busy schedules, there is likely to be a greater response among those interested in the questionnaire subject, whether or not they are in OEM. If respondents of all specialties are at least somewhat interested, the differences between specialties will remain valid. It is possible that physicians have low response rates to surveys that do not deal with material that has been central to medical education. Nutrition is a comparable area: A recent national survey of nutrition knowledge obtained only an 11% response rate, and physicians were fairly unknowledgeable in this field, which is also typically ignored in medical education (Brody, 1993). Low response rates might also arise from touchy subjects. A recent survey of all physicians in Rhode Island sought answers to the question of what doctor the physicians would refer themselves or their loved ones to. Only 16% responded (Francis, 1994), probably because it seemed difficult to provide names. This is quite low when compared to the 42.5% Rhode Island response on environmental health effects.

The MDPH produced only a brief, descriptive report that presented the following findings: A majority of the physicians who responded to the survey considered environmentally related illnesses in differential diagnosis, but only about 12% had done so in the last 12 months. Most physicians did not ask patients questions about possible exposures to hazardous substances at work or at home, although about one half of the respondents rated themselves as "somewhat knowledgeable" about nonworkplace environmental health hazards. Whereas 86% of respondents felt that it was either "very important" or "somewhat important" to be informed about non-

TABLE 3.1
Completion Rates of Physician Environmental Health Surveys

State	Surveys distributed	Surveys completed	response rate (%)
Massachusetts	14,500	1,609	11.0
Connecticut	1,160	310	26.7
Colorado	595	83	7.2
New Hampshire	1,800	90	5.0
Rhode Island	306	130	42.5

workplace environmental health hazards, only 27% were aware of the protection provided by the Right-to-Know law. Specific survey questions asked about the health effects of radon exposure and methylene chloride exposure. Physicians were generally more knowledgeable about methylene chloride exposure than radon exposure. Only one quarter of the respondents reported that their patients had asked about pesticides in the past 12 months. A small percentage (12%) of physicians reported that patients raised questions about nonworkplace environmental health hazards. One questionnaire item asked, "To your knowledge, do any of your patients live in a community which has a hazardous waste site?" Seventy-two percent of the respondents either said "no" or did not respond to this item.

Analysis

In the MDPH/MMS, researchers grouped their 1,609 respondents into the seven categories presented in Table 3.2. It is interesting to note that "pediatrics" was not listed as a response in the survey instrument. Because elevated rates of childhood leukemia in Woburn, Massachusetts, in the 1970s called the attention of physicians and the public to environmentally caused disease clusters, it might have been possible to demonstrate pediatricians' heightened awareness of environmental health effects by including pediatricians as a category.

Table 3.2 indicates that there were 1,694 responses for the seven categories of medical specialty. Because there were only 1,609 respondents, it is likely that the total of 1,694, which exceeded the number of respondents by 85, was due to some physicians indicating more than one specialty (e.g., physicians who practice both obstetrics and gynecology may have checked off both "obstetrics" and "gynecology" on the survey). Because the analysis would have been affected by the low number of respondents in some categories, physician specialties were reclassified in this reanalysis of the data into four new categories, as shown in Table 3.3:

1. "primary care" = internal medicine and family practice
2. "occupational/preventive" = occupational medicine and preventive medicine
3. "ob/gyn" = obstetrics and gynecology
4. "other" = other medical specialties

It is important to note that when asked to indicate their medical specialty or type of practice, more than 50% of the respondents did not classify themselves by specialty. Further, the survey instrument contained a very limited list of specialties.

TABLE 3.2
Medical Specialty of Respondents

Specialty	n	Percentage of total
Internal medicine	415	25.8
Occupational medicine	20	1.2
Gynecology	112	7.0
Family practice	112	7.0
Obstetrics	72	4.5
Preventive medicine	10	0.6
Other	953	59.2
Total	1,694	100

TABLE 3.3
Medical Specialty of Respondents—Recoded

Specialty	n	Percentage of total
Primary medicine	510	31.7
Occupational/preventive medicine	26	1.6
Obstetrics/gynecology	120	7.5
Other medical specialty	15	0.9
Respondents who did not classify themselves by specialty	938	58.3
Total	1,609	100

Because the purpose of this project was to examine a number of questions with practice as a specific factor of interest, those unclassified by specialty were dropped from the data analysis.

The classification of physicians by practice site was done using the zip code of the practice location.[3] To determine if the presence of an NPL hazardous waste site in the vicinity of a physician's practice was related to his or her knowledge/interest/ attitudes about environmental health hazards, each physician received a designation of "NPLsite-yes" if their practice was located in a city or town that had any of the 22 NPL sites in Massachusetts within its borders; otherwise, the physician received a code of "NPLsite-no."

Chi-square and logistic regression were used, as appropriate.

Results

Comparison of Massachusetts with Other States

Out of the 11 states that were awarded cooperative agreements by the ATSDR to support environmental health education for physicians and other health professionals, 5 states (Massachusetts, Rhode Island, Connecticut, Colorado, and New Hampshire) developed survey instruments to assess physician knowledge of environmental health hazards. A comparison of questionnaire items and physician responses in those states is presented in Table 3.4.

As is evident in Table 3.4, the survey questions from Massachusetts most closely corresponded to those used in the Connecticut Health Department survey. Although the total number of physician respondents in Massachusetts ($N = 1,609$) was more

TABLE 3.4
Descriptive Statistics: Physicians' Responses and Percentage of Such Responses within Each State

Responses	MA (n = 1,600)	CT (n = 310)	CO (n = 83)	RI (n = 130)
Physicians are somewhat knowledgeable about environmental health hazards	49	52	—	—
Physicians routinely ask patients about possible hazardous exposures at work	45.4	49	23	—
Physicians routinely ask patients about possible hazardous exposures in the home/community	29.9	33	23	—
Patients have asked questions in the past 12 months regarding environmental health hazard exposures	48.4	61	61	—
Physicians consider environmentally related illness as part of differential diagnosis	85.3	60	89	—
Physicians have attended seminars on environmental health hazards in the past 12 months	8.0	—	—	—
Physicians have read a journal article on environmental health hazards in the past 12 months	60.2	—	—	—
Physicians know about the protection provided under the Right-to-Know law	28	—	—	—
Physicians believe that it is important to be informed about environmental health hazards	86	89	—	—
Physicians are aware of their patients who live in communities that have hazardous waste sites	18.8	12	—	—
Physicians would be willing to attend a seminar on environmental health hazard education	72	—	—	—
Physicians consider the awarding of Continuing Medical Education credits important in the decision to become more informed about environmental health hazards	72	44	45	73

than five times the number in Connecticut ($N = 310$), the proportion of affirmative responses was comparable in 5 out of 10 questions. Information obtained from the Colorado survey was similar to that reported in Massachusetts for some questions but closer to the responses of physicians in Connecticut for other questions. The survey data obtained from New Hampshire focused primarily on viable means of environmental health education for physicians and thus was not comparable to the substance of the Massachusetts, Connecticut, or Colorado instruments. Rhode Island survey data were available, but only one survey question was directly comparable to a survey question used in the other three states (Colorado Department of Public Health, 1990; Connecticut Department of Health Services, 1991; New Hampshire Division of Public Health Services, 1989, 1991; Rhode Island Department of Public Health, 1990; Szneke, Nielsen, & Tolentino, 1991).

Forty-nine percent of Massachusetts physicians and 52% of Connecticut physicians rated themselves as at least somewhat knowledgeable about environmental health hazards. Nearly half of the physicians in both states routinely asked patients questions about hazardous exposures at work, but only 23% of Colorado physicians routinely asked about possible work exposures. Thirty percent of the physicians in Massachusetts, 33% of Connecticut physicians, and 23% of Colorado physicians questioned their patients about hazardous exposures in the home or in the community. Almost half of Massachusetts physicians and three-fifths of the physicians in Connecticut and Colorado reported that in the last 12 months patients had asked them questions regarding environmental health hazard exposures. In all three states, a relatively high proportion of physicians (MA = 85.3%, CT = 60%, and CO = 89%) reported that they consider environmentally related illness as part of differential diagnosis. Eighty-six percent of physicians in Massachusetts and 89% of physicians in Connecticut believe that it is important to be informed about environmental health hazards, yet only 18.8% of Massachusetts physicians and 12% of Connecticut physicians reported that they are aware of which of their patients live in communities that have hazardous waste sites. Clearly, best intentions do not necessarily produce the requisite behavior and knowledge. Nearly three-quarters of the physicians in Massachusetts and Rhode Island and just less than half of those in Connecticut and Colorado consider the awarding of Continuing Medical Education credits as important in the decision to become more informed about environmental health hazards (Colorado Department of Public Health, 1990; Connecticut Department of Health Services, 1991; Rhode Island Department of Public Health, 1990; Szneke et al., 1991).

Results from Analysis of Hypotheses

From the MMPH/MMS survey instrument, 11 questions were selected for reanalysis in this study. These pertain to knowledge and information about nonworkplace environmental hazards, interviewing procedures, and diagnostic criteria; patient questions about environmental exposures; and perception of importance of environmental health hazards in daily medical practice. The four hypotheses previously specified were tested using these questions. The results are presented in Table 3.5 through 3.8.

Table 3.5 presents the results of the first hypothesis—that survey responses do not differ according to whether the physicians' practice sites were located within a community that had a hazardous waste site. This is the primary question in which we are interested. In 10 out of 11 survey questions, there were no significant differences by practice site. The only significant finding was that physicians whose practices were located within a community that had an NPL site reported that patients asked more questions about exposure to pesticide use in their homes than did the patients of physicians whose practices were not in communities with NPL sites.

The second hypothesis tested was whether physician knowledge and practice regarding environmentally induced illness varies by specialty. Those specializing in occupational/preventive health were assumed to be the most knowledgeable, whereas those practicing obstetrics/gynecology were assumed to be the least knowledgeable about environmental illness. The results are presented in Table 3.6. The findings indicate that there are very significant (.001 level) differences in survey responses by medical specialty in 10 out of 11 questions. The 11th question, which asked if physicians had knowledge of patients living in a community with an NPL site, also

TABLE 3.5
MDPH/MMS Survey Responses by Practice Location

	NPL site in practice area			
Responses	Yes %	No %	χ^2	df
Physician's self-rated knowledge of nonworkplace environmental health harzards				
Quite knowledgeable	10.7	5.7		
Somewhat knowledgeable	42.0	43.0		
Not knowledgeable	25.0	24.3		
Not relevant to my practice	22.3	27.0		
			5.15405	3
Physician routinely asks questions regarding possible exposures to hazardous substances at work				
Yes	42.7	45.6		
No	57.3	54.4		
			.34309	1
Physician routinely asks questions regarding possible exposures to hazardous substances in the home/community				
Yes	30.6	29.9		
No	69.4	70.1		
			.02083	1
Patients sought information about nonworkplace environmental hazards within the last 12 months				
Yes	55.9	47.8		
No	44.1	52.2		
			2.65321	1
Environmentally related illnesses are considered in differential diagnosis				
Yes	83.2	85.4		
No	16.8	14.6		
			.41445	1
Physician has attended a seminar on nonworkplace environmental health hazards within the past 12 months				
Yes	8.0	8.0		
No	92.0	92.0		
			.00082	1
Physician has read journal article on nonworkplace environmental health hazards within the past 12 months				
Yes	64.3	59.8		
No	35.7	40.2		
			.85941	1
Patients have asked about exposure to pesticide use in their homes				
Yes	33.3	24.7		
No	66.7	75.3		
			4.07593*	1
Physician knows about the protection provided by the Right-to-Know Law				
Yes	24.8	28.2		
No	75.2	71.8		
			.62360	1
Physician-rated importance of information about nonworkplace environmental hazards				
Very important	44.5	36.5		
Somewhat important	40.0	49.9		
Not important	6.4	4.6		
Irrelevant to my practice	9.1	9.0		
			4.45527	3
Physician is aware of patients who live in a community that has an NPL site				
Yes	23.7	18.5		
No	71.1	74.0		
Other responses	5.3	7.5		
			2.38011	2

*$p < .05$.

TABLE 3.6
MDPH/MMS Survey Responses by Specialty

Responses	Primary %	Occup/ prev %	Ob/ gyn %	Other %	χ^2	df
Physician's self-rated knowledge of nonworkplace environmental health hazards						
Quite knowledgeable	5.1	38.5	2.5	13.3		
Somewhat knowledgeable	42.5	30.8	35.8	66.7		
Not knowledgeable	24.7	30.8	25.8	20.0		
Not relevant to my practice	27.7	0.0	35.8	0.0		
					69.83228**	12
Physician routinely asks questions regarding possible exposures to hazardous substances at work						
Yes	60.6	88.0	30.3	93.3		
No	39.4	12.0	69.7	6.7		
					116.63337**	4
Physician routinely asks questions regarding possible exposures to hazardous substances in the home/ community						
Yes	28.7	65.4	13.6	100.0		
No	71.3	34.6	86.4	0.0		
					66.35005**	4
Patients sought information about nonworkplace environmental hazards within the last 12 months						
Yes	60.6	76.9	52.5	86.7		
No	39.4	23.1	47.5	13.3		
					75.32607**	4
Environmentally related illnesses are considered in differential diagnosis						
Yes	94.1	96.2	62.5	100.0		
No	5.9	3.8	37.5	0.0		
					90.08297**	4
Physician has attended a seminar on nonworkplace environmental health hazards within the past 12 months						
Yes	7.9	34.6	2.5	33.3		
No	92.1	65.4	97.5	66.7		
					42.94074**	4
Physician has read journal article on nonworkplace environmental health hazards within the past 12 months						
Yes	59.2	92.3	49.6	85.7		
No	40.8	7.7	50.4	14.3		
					20.71316**	4
Patients have asked about exposure to pesticide use in their homes						
Yes	27.6	61.5	35.3	53.3		
No	72.4	38.5	64.7	46.7		
					40.07110**	4
Physician knows about the protection provided by the Right-to-Know law						
Yes	31.0	73.1	20.3	20.0		
No	69.0	26.9	79.7	80.0		
					33.84033**	4
Physician-rated importance of information about nonworkplace environmental hazards						
Very important	44.2	69.2	33.5	85.7		
Somewhat important	52.0	30.8	55.6	14.3		
Not important	2.2	0.0	1.7	0.0		
Irrelavant to my practice	1.6	0.0	9.4	0.0		
					110.27422**	12
Physician is aware of patients who live in a community that has an NPL site						
Yes	19.6	30.8	16.7	40.0		
No	75.9	61.5	76.7	53.3		
Other responses	4.5	7.7	6.7	6.7		
					17.21309*	8

revealed significant differences (.05 level) by specialty. Overall, those in occupational/ preventive medicine rated themselves as more knowledgeable than all other specialties about nonworkplace environmental hazards; they were more likely to have attended a seminar or to have read a journal article about nonworkplace environmental hazards in the last 12 months than those in other medical specialties; and they routinely asked more questions about environmental exposures than did other physicians. Those who classified themselves as "other medical specialty" ranked second overall to those in occupational/preventive medicine in terms of knowledge, attitudes, and practice techniques regarding environmental health issues. As hypothesized, those who categorized their medical specialty as "ob/gyn" were the least knowledgeable about environmental health issues, asked the least questions of their patients about possible environmental exposures, and were the least informed about environmental health issues despite the fact that more than half of the "obstetrician/ gynecologists" reported that their patients sought information from them related to environmental exposures.

Table 3.7 presents the results of the MDPH/MMS questionnaire according to self-rated knowledge of environmental health hazards. The results clearly indicate that those who rate themselves as being "very knowledgeable" about environmental health issues are significantly more likely to incorporate this knowledge into their practice of medicine. The results presented show a perfect linear relationship between knowledge and practice for the 11 relevant questionnaire items—the higher a physician's self-rated knowledge, the more likely he or she is to be informed about environmental health issues, to consider environmental illness in differential diagnosis, to report that patients seek information from them about environmental exposures, and to rate environmental hazard knowledge as "very important" to their medical practice.

The fourth hypothesis tested is whether physicians who perceive that nonworkplace environmental hazards are "very important" are more knowledgeable about medical interviewing, diagnostic criteria, and the Massachusetts Right-to-Know law than those who rate environmental health hazard knowledge as "irrelevant to my practice." The results are presented in Table 3.8.

From Table 3.8 it can be seen that physicians' rating of the importance of environmental hazard knowledge is strongly related to knowledge, attitudes, and practice regarding environmentally induced illness. The responses to all of the relevant questions indicated that there were highly significant (.001 level) differences depending on the physicians' perceptions of the importance of this information. Relationships were perfectly linear on all but 2 of the 11 variables (self-rated knowledge and awareness of patient in NPL site), and in all cases, those responding that environmental hazard knowledge was "very important" had the highest response scores.

Logistic regression procedures were performed to examine the factors that influence knowledge and practice differences. The results of the logistic regression analyses are presented in Tables 3.9, 3.10, and 3.11.

The results presented in all the logistic regressions show that, controlling for all other relevant variables, practice site does not affect the likelihood of assessing hazardous exposures in the home or in the community. Table 3.9 indicates that

TABLE 3.7
MDPH/MMS Survey Responses by Self-Rated Knowledge of Environmental Hazards

Responses	High %	Moderate %	None %	Irrelevant %	χ^2	df
Physician's self-rated knowledge of nonworkplace environmental health hazards						
Quite knowledgeable						
Somewhat knowledgeable						
Not knowledgeable						
Not relevant to my practice						
Physician routinely asks questions regarding possible exposures to hazardous substances at work						
Yes	63.7	50.9	47.5	30.8		
No	36.3	49.1	52.5	69.2		
					56.31977**	3
Physician routinely asks questions regarding possible exposures to hazardous substances in the home/ community						
Yes	60.4	36.2	27.6	14.6		
No	39.6	63.8	72.4	85.4		
					100.16446**	3
Patients sought information about nonworkplace environmental hazards within the last 12 months						
Yes	68.8	55.5	48.7	33.5		
No	31.2	44.5	51.3	66.5		
					65.78395**	3
Environmentally related illnesses are considered in differential diagnosis						
Yes	95.8	89.6	87.1	74.0		
No	4.2	10.4	12.9	26.0		
					61.35437**	3
Physician has attended a seminar on nonworkplace environmental health hazards within the past 12 months						
Yes	31.3	9.1	6.9	2.4		
No	68.8	90.9	93.1	97.6		
					89.19705**	3
Physician has read journal article on nonworkplace environmental health hazards within the past 12 months						
Yes	83.9	69.9	60.4	38.2		
No	16.1	30.1	39.6	61.8		
					130.72756**	3
Patients have asked about exposure to pesticide use in their homes						
Yes	51.6	30.1	22.8	13.1		
No	48.4	69.9	77.2	86.9		
					77.06262**	3
Physician knows about the protection provided by the Right-to-Know law						
Yes	56.4	32.4	29.3	12.8		
No	43.6	67.6	70.7	87.2		
					92.34670**	3
Physician-rated importance of information about nonworkplace environmental hazards						
Very important	69.1	40.3	35.2	25.9		
Somewhat important	14.9	49.4	52.0	53.9		
Not important	5.3	3.3	6.0	6.1		
Irrelevant to my practice	10.6	7.0	6.8	14.1		
					90.16464**	9
Physician is aware of patients who live in a community that has an NPL site						
Yes	25.0	22.3	16.7	13.3		
No	64.6	70.0	76.8	79.6		
Other responses	10.4	7.7	6.5	7.1		
					20.64505*	6

$^*p < .01.$ $^{**}p < .001.$

TABLE 3.8
MDPH/MMS Survey Responses by Importance Rating of Environmental Hazard Knowledge

Responses	Very important %	Somewhat important %	Not important %	irrelevant %	χ²	df
Physician's self-rated knowledge of nonworkplace environmental health hazards						
Quite knowledgeable	11.4	1.8	6.7	7.1		
Somewhat knowledgeable	46.6	43.0	29.3	32.9		
Not knowledgeable	23.5	26.1	30.7	18.6		
Not relevant to my practice	18.6	29.1	33.3	41.4		
					90.16464*	9
Physician routinely asks questions regarding possible exposures to hazardous substances at work						
Yes	57.8	43.8	23.6	16.3		
No	42.2	56.2	76.4	83.7		
					95.94191*	3
Physician routinely asks questions regarding possible exposures to hazardous substances in the home/ community						
Yes	44.1	25.5	6.9	8.2		
No	55.9	74.5	93.1	91.8		
					109.810013	3
Patients sought information about nonworkplace environmental hazards within the last 12 months						
Yes	63.9	44.3	27.0	16.7		
No	36.1	55.7	73.0	83.3		
					129.55417*	3
Environmentally related illnesses are considered in differential diagnosis						
Yes	91.7	86.5	74.3	58.3		
No	8.3	13.5	25.7	41.7		
					107.99858*	3
Physician has attended a seminar on nonworkplace environmental health hazards within the past 12 months						
Yes	14.2	4.7	2.7	1.4		
No	85.8	95.3	97.3	98.6		
					53.66153*	3
Physician has read journal article on nonworkplace environmental health hazards within the past 12 months						
Yes	72.0	55.9	44.6	43.6		
No	28.0	44.1	55.4	56.4		
					62.60741*	3
Patients have asked about exposure to pesticide use in their homes						
Yes	39.2	20.3	6.8	5.0		
No	60.8	79.7	93.2	95.0		
					112.66900*	3
Physician knows about the protection provided by the Right-to-Know law						
Yes	32.8	27.0	24.3	16.3		
No	67.2	73.0	75.7	83.7		
					16.91134*	3
Physician-rated importance of information about nonworkplace environmental hazards						
Very important						
Somewhat important						
Not important						
Irrelevant to my practice						
Physician is aware of patients who live in a community that has an NPL site						
Yes	24.3	16.4	17.3	11.3		
No	68.5	77.7	78.7	78.2		
Other responses	7.2	5.9	4.0	10.6		
					25.27669*	6

TABLE 3.9

Regression Analysis: The Physician Routinely Asks Questions Regarding Possible Exposures to Hazardous Substances in the Home and in the Community

	Exp (b)	p
Practice located in a community that has an NPL site		
Yes	1.0	
No	.5495	.1836
Specialty		
Primary medicine	1.0	
Occupational/preventive	3.4577	.0074
Obstetrics/gynecology	.4018	.0047
Other	2920.116	.4246
Physician has knowledge of Right-to-Know law		
Yes	1.0	
No	1.6207	.0193
How important is knowledge concerning hazardous waste to medical practice?		
Very	1.0	
Somewhat	.5127	.0008
Not important	.4263	.2821
Irrelevant	.6305	.4918
Physician is aware of patients living in communities with NPL sites		
Yes	1.0	
No	1.6782	.0235

medical specialty was an important determinant of whether environmental exposures were assessed. Compared to primary care physicians, those specializing in occupational/preventive medicine were 3½ times as likely to ask patients questions regarding possible exposures to hazardous substances in the home and in the community. However, obstetricians/gynecologists were less than half as likely to assess hazardous substance exposures in the home and in the community. Physicians who considered knowledge concerning hazardous waste to be very important to their medical practice were twice as likely to ask questions regarding home and community exposures than were those who considered this information only somewhat important.

Two unexpected findings were that physicians who were not knowledgeable about the Right-to-Know law in Massachusetts were 62% more likely than those who knew about the Right-to-Know law to assess hazardous substance exposure in the home and community. Also, physicians who were not aware of which of their patients lived in communities that had NPL sites were more than 67% as likely to assess hazardous exposures in the home and in the community.

The second logistic regression (Table 3.10) attempts to predict the likelihood of physicians to routinely ask questions regarding possible exposures to hazardous substances at work. As the table indicates, the factors that predict the likelihood of a physician asking questions concerning environmental exposures in the workplace include medical specialty and knowledge of the protection provided by the Right-to-Know law. Compared to primary care physicians, those specializing in occupational/preventive medicine were more than four times as likely to routinely ask questions regarding possible exposures to hazardous substances in the home and in the community. Obstetricians/gynecologists, in contrast to primary care physicians, were 70%

less likely to assess exposures to hazardous substances at work. An unanticipated finding similar to that seen in Table 3.9 was that physicians who had no knowledge of the Right-to-Know law in Massachusetts were more than twice as likely as physicians who were aware of the Right-to-Know law to assess environmental exposures at work. Practice location, physicians' rated importance of environmental hazard information, and physicians' awareness of patients who lived in communities with hazardous waste sites had no effect on the likelihood of assessing possible exposures to hazardous substances at work.

Because 58.3% of respondents did not classify themselves by specialty, we must assume that our data on specialty differences may be weak. Occupational/preventive physicians may be likely to correctly fill in this item, given the fact that the topic of the survey is in their area. We have no reason to assume that any other specialty will be less likely to fill out the specialty item, so we can still accept the specialty differences we have found.

The final logistic regression (results presented in Table 3.11) attempts to predict the likelihood of a physician knowing about the protection provided by the Right-to-Know law. The independent variables in this procedure are practice site (whether the physician's medical practice is located in a community that has an NPL site), medical specialty, physicians' perception of the importance of environmental health hazards, and whether their patients live in communities that have NPL sites. As is indicated in the table, none of the variables except medical specialty predicts the likelihood of a physician knowing about the Right-to-Know law. Compared to primary care physicians, physicians practicing occupational/preventive medicine were 6½ times more likely to be informed about the Right-to-Know law in Massachusetts.

TABLE 3.10

Regression Analysis: Physician Routinely Asks Questions Regarding Possible Exposures to Hazardous Substances at Work

	Exp (b)	p
Practice located in a community that has an NPL site		
Yes	1.0	
No	.7386	.3854
Specialty		
Primary medicine	1.0	
Occupational/preventive	4.4008	.0503
Obstetrics/gynecology	.2962	.0000
Other	7.2937	.0591
Physician has knowledge of Right-to-Know law		
Yes	1.0	
No	2.0226	.0004
How important is knowledge concerning hazardous waste to medical practice?		
Very	1.0	
Somewhat	.7042	.0535
Not important	.5168	.2579
Irrelevant	.9633	.9448
Physician is aware of patients living in communities with NPL sites		
Yes	1.0	
No	1.3481	.1777

TABLE 3.11
Regression Analysis: Factors That Affect the Likelihood of Being Knowledgeable about the Right-to-Know Law

	Exp (b)	p
Practice located in a community that has an NPL site		
Yes	1.0	
No	1.0995	.7911
Specialty		
Primary medicine	1.0	
Occupational/preventive	6.5447	.0001
Obstetrics/gynecology	.6267	.0711
Other	.3840	.2205
How important is knowledge concerning hazardous waste to medical practice?		
Very	1.0	
Somewhat	.9747	.8906
Not important	1.4504	.5277
Irrelevant	.6021	.4424
Physician is aware of patients living in communities with hazardous waste sites		
Yes	1.0	
No	1.1293	.5793

Discussion

This reanalysis of the MDPH/MMS physician survey data supports the hypothesis that physicians whose practices are located within communities that have EPA NPL hazardous waste sites within their borders are no more knowledgeable about environmental health hazards than those whose practices are not located in communities with NPL sites. According to this data, there was no relationship between practice location and physician knowledge, attitudes, or interest in environmental health hazards. Although expected, this finding is very striking. We would assume that practicing in an NPL site area would heighten awareness of many educated people, especially physicians. In many locations, NPL sites were well known and publicized, and even the subject of intense political dispute. Physicians are not being sufficiently attuned to a public health perspective if they are not attentive to such a situation. Lack of attentiveness can stem from physicians' fear that they will suffer if they support claims of environmentally induced disease. Local businesses, civic leaders, and colleagues would likely look askance at such advocacy. Current and prospective patients might be wary as well. Even though the physicians are responding to a questionnaire, their lack of awareness may be deeply instilled by their fears and concerns.

A look at the size of the NPL towns and cities may help here. Of the 24 cities and towns with NPL sites, equal numbers were above and below the mean population for all cities and towns in the state. Looking more closely, only 2 were sizable cities—Lowell and New Bedford, with 1990 populations, respectively, of 103,439 and 99,922. All the rest were very small cities, or else towns of varying sizes. Fourteen of the remaining 22 had 1990 populations of less than 25,000. Hence the NPL locales were small enough that doctors should know about local affairs, yet also small enough that taking a stand would be potentially damaging.

There was also support for the hypothesis that physician knowledge and practice regarding environmentally induced disease varies by specialty. Physicians specializing in occupational/preventive medicine are the most knowledgeable about environmental health hazards whereas those specializing in obstetrics/gynecology are the least knowledgeable. We would expect occupational/preventive physicians to be more knowledgeable. As hypothesized, obstetrician/gynecologists (OB-GYNs) are rather unknowledgeable. This is unfortunate in light of the fact that many environmental health effects involve reproductive and birth disorders. These OB-GYNs are either unaware of this literature or know of it but believe it to be exaggerated.

OB-GYNs have perhaps less interest in environmental health because their specialty has been in the midst of significant conflicts. Consumer demand for less invasive obstetrical procedures and a more patient-centered birthing experience have been powerful forces of obstetrics/gynecology. Considerable malpractice litigation has also posed a problem, as in cases of birth defects alleged to be caused by medical error. The growth in infertility treatments, combined with new quasi-artificial technologies, has also produced much public concern. So this specialty has its hands full and may be less open to environmental health concerns.

As hypothesized, self-rated knowledge of environmental health hazards was found to be strongly related to knowledge about the Right-to-Know law and also to the likelihood of asking questions about workplace and nonworkplace environmental exposures. Those who rated themselves as very knowledgeable were noted to be informed about medical interviewing, diagnostic criteria, and the Right-to-Know law. The conclusion that physicians who perceive that nonworkplace environmental hazards are very important are more knowledgeable about medical interviewing, diagnostic criteria, and the Right-to-Know law than are those who rate environmental health hazard information as "irrelevant to my practice" can also be drawn from the data analysis. These last two findings lead us to affirm the importance of physician education as a way of improving appropriate practice styles. Once physicians are in the trajectory of environmental awareness, they are likely to continue to improve their knowledge, and hopefully to take active roles in remediation and prevention.

The regression analyses indicated that independent variables such as medical specialty, knowledge about the Right-to-Know law, and the physician's rating of the importance of environmental health hazard information could be used to predict the likelihood of certain practice outcomes. Specialty holds for all three logistic regressions. There are counterintuitive findings on the factors leading to routine questions on home/community and workplace exposures. In those two analyses, lack of knowledge of the Right-to-Know law was significant, and in the home/community exposure, lack of awareness of patients in an NPL site was significant.

How do we explain the odd findings about the Right-to-Know law? First of all, the MDPH/MMS questionnaire did not specify if it sought a response on the federal Right-to-Know law (under the 1986 Emergency Planning and Community Right to Know Act, EPCRA) or the state law. The federal law set up the Toxic Release Inventory, a national database available to all people through public libraries, which informs people of the *releases* from factories and other facilities. The state law, though rarely used, provides broader protection, allowing workers or citizens to find out what toxic substances are being *used* in a factory or other facility. Thus physicians may have been confused. Also, the fact that hardly any variables predict knowledge of the Right-to-Know law goes part of the way toward explaining the earlier mentioned unanticipated findings about it. If we refer back to descriptive statistics as shown in Table 3.4, we find that the two lowest percentage scores among all of the knowledge questions showed that only 28% of physicians know of the Right-to-Know Law and only 18.8% are aware of patients who live in an area with an NPL site. So many are so unknowledgeable that they make an effort to ascertain exposure data as a form of self-protection. Ultimately, we cannot have a satisfactory answer to these findings without directly interviewing physicians about such an apparent anomaly. It seems important for state environmental officials to join together with medical societies and public health agencies to provide specific outreach to health professionals on the Right-to-Know law. Left to their own devices, physicians will not likely be as aware of the law as they should be.

Physicians may be afraid of repercussions of recognizing and acting on environmental health effects. But even if this is not the case, they may be acting as technocrats, whereby they perform their medical tasks but lack perspective and critical understanding. They are not necessarily medically ignorant, because the profession of medicine has not come forward as a whole with a strong environmental health perspective. But the failure of organized medicine to do so makes medicine as a whole medically ignorant.

Conclusion: Implications for Medical Education and Social Policy

The results of this chapter are consistent with the findings of other studies that suggest that it is imperative to remedy the problem of inadequate physician education in the area of occupational and environmental medicine. However, even beyond the issues of inadequate training of physicians, there are other barriers to incorporating OEM into mainstream medical care and teaching. The IOM has identified the following three factors: (a) a shortage of trained OEM physicians to educate new physicians; (b) a lack of information on toxic substances present in the environment; and (c) the widespread perception among physicians that environmental disease is difficult to diagnose, and that complex legal and ethical issues arise when physicians identify environmental health hazards in the workplace, home, and community.

Physicians are looked upon as a source of valuable information on environmental issues and their effects on human health. However, the need of the lay public for a trusted and well-informed source of OEM information and care is not met due to the paucity of physicians trained in OEM. This training should not be restricted to brief rotations and to the rarely chosen specialty residency. Rather, medical schools need both to understand how environmental health issues are totally integral to all areas of medicine and to find clinical and academic material to teach in that manner. Medical schools may find it helpful to work together with environmental epidemiology programs in public health schools, which already have substantial research programs under way.

Medical education in environmental health alone is not sufficient. Given that physicians practicing in locations with Superfund sites are no different than others, we must assume a general lack of social concern and involvement. This will only be changed by a broader formulation of the social roles of the medical professional. Such a task requires both undergraduate and postgraduate medical education to pay more attention to public health issues and social concerns. National health reform proposals can help in this process because federal support for medical education can shape much of the curriculum.

Additional social, legal, and ethical factors further impede the willingness and ability of physicians to detect environmental illness. Hence medical education alone will not suffice. Federal health agencies need to fund more environmental epidemiology research, especially in contaminated communities, and to take strong defenses in favor of whistle-blowers. If public and professional support for a National Institute of the Environment is successful, such an institute could provide the kind of leadership needed to bring environmental health into more daily concern. Medical education reform must be combined with broad social reformulations of the importance of environmental health so that physicians can overcome the sociological ambivalence that they now face.

So we need to combine specific medical education reforms with more dedicated action by medical associations and more direction by state and federal policy. This would help reach the goal of educating "environmentally literate physicians," which Cortese and Armoudlian (1991) argue is one of the most important challenges for medicine:

> First, the physical environment is one of the most important determinants of human health. Second, protection of the environment and preservation of ecosystems are, in public health terms, the most fundamental forms of primary prevention of human illness. And, third, physicians should be the health practitioners most knowledgeable about the environmental factors that create health and cause disease, and should be prominent spokespersons in communicating with the public and policy makers about environmental hazards, and intervention strategies to prevent disease. (Pp. 77–78)

This range and breadth of environmental medicine is a valuable goal that can have great benefits for the whole society.

NOTES

We are grateful to Suzanne Condon, Judy Bygat, and Paul George of the Massachusetts Department of Public Health for sharing with us the data for this study. We thank Steve Kroll-Smith, Susan Masterson-Allen, and Craig Zwerling for offering helpful comments on the manuscript.

1. The following are definitions of occupational and environmental medicine (OEM) as set forth by the IOM: "Occupational Medicine is all aspects of the relation between workplace

factors (including physical, chemical, biological, social, and psychological) and health, with emphasis on the effects of work on health. Environmental medicine incorporates most but not all aspects of occupational medicine, and encompasses conditions caused by or aggravated by exposure to (1) toxic chemical substances, such as formaldehyde and asbestos, that are either man-made or become biologically available as a result of human activities; (2) physical agents, such as radiation or noise that occur naturally or as a result of human activities; and (3) biological substances, such as Legionella [the organism that causes Legionnaire's Disease] in heating and ventilation systems, that become problems as a result of human activities. The domain of environmental medicine also includes the psychological burden of anxiety and concern about environmental hazards—concerns that in some cases outweigh the direct biological threat. Environmental medicine excludes health effects of such behavior as active cigarette smoking, but includes exposure to a wide range of nonoccupational physical, chemical, and biological factors" (IOM, 1991, pp. 5–6).

2. The 1986 Emergency Planning and Community Right to Know Act (EPCRA) was part of the 1986 Superfund Amendments and Reauthorization Act, but it exists as a separate legal entity, Pub. L. No. 301–302, 42 U.S.C. §11001–11050. It established the Toxics Release Inventory under which businesses were required to make known how many toxic substances were released in the communities where they did business. Numerous states had already established Right-to-Know laws, some of which granted residents specific rights to query businesses about toxics on their premises.

3. It is believed that this method is more appropriate for comparative analysis by location than that used in Connecticut, in which a random sample of 1,148 physicians practicing in the 10 most populated Connecticut cities and towns was selected for inclusion in the survey. The towns included were Bridgeport, Bristol, Cheshire, Danbury, Hartford, Meriden, New Britain, New Haven, Norwalk, Stamford, and Waterbury. With the exception of Danbury, all of these towns lie within a 15-mile radius of an NFL site. Danbury is approximately 25 miles from a NPL site. Using this method, it is not feasible to accurately compare survey responses according to proximity to an NPL site.

REFERENCES

Agency for Toxic Substances and Disease Registry (ATSDR). (1991, September). *State cooperation agreements to develop educational programs for health care providers.* Atlanta, GA: Author.

American College of Physicians. (1990). Position paper: Occupational and environmental medicine: The internist's role. *Annals of Internal Medicine, 113,* 974–982.

Becker, H., Geer, B., Hughes, E. C., & Strauss, A. (1961). *Boys in white: Student culture in medical school.* Chicago: University of Chicago Press.

Brody, J. E. (1993, February 10). Doctors get poor marks in nutrition. *New York Times* p. C3.

Brown, P., & Mikkelsen, E. J. (1990). *No safe place: Toxic waste, leukemia, and community action.* Berkeley: University of California Press.

Capek, S. (1987, August 16). *Toxic hazards in Arkansas: Emerging coalitions.* Paper presented at the Annual Meeting of the Society for the Study of Social Problems. Chicago.

Castorina, J., & Rosenstock, L. (1990). Physician shortage in occupational and environmental medicine. *Annals of Internal Medicine, 113,* 983–986.

Colorado Department of Public Health. (1990). Colorado Department of Health "Physician Survey" report. Denver, CO: Author.

Connecticut Department of Health Services. (1991). *Environmentally related assessment surveys.* Hartford: Connecticut Department of Public Health.

Cortese, A. D., & Armoudlian, A. S. (1991). Fostering ecological and human healing. *Physicians for Social Responsibility Quarterly, 2,* 77–85.

Coser, R. L. (1979). *Training in ambiguity: Learning through doing in a mental hospital.* New York: Free Press.

Cullen, M. R., & Rosenstock, L. (1988). The challenge of teaching occupational and environmental medicine in internal medicine residencies. *Archives of Internal Medicine, 148,* 2401–2404.

Daniels, A. K. (1969). The captive professional: Bureaucratic limitations in the practice of military psychiatry. *Journal of Health and Social Behavior, 10,* 255–265.

Edelstein, M. (1988). *Contaminated communities: The social and psychological impacts of residential toxic exposure.* Boulder, CO: Westview.

Francis, S. (1994). Top doctors. *Rhode Island Monthly, 7*(1), 40.

Frazier, L., Cromer, J., Andolsek, K., Greenberg, G., Thomann, W., & Stopford, W. (1991). Teaching occupational and environmental medicine in primary care residency training programs: Experience using three approaches during 1984–1991. *American Journal of Medical Science, 302*(1), 42–45.

Freidson, E. (1970). *Profession of medicine: A study of the sociology of applied knowledge.* Chicago: University of Chicago Press.

Freudenberg, N. (1984). *Not in our backyards: Community action for health and the environment.* New York: Monthly Review Press.

Goldsmith, J. R. (Ed.). (1986). *Environmental epidemiology: Epidemiological investigation of community environmental health problems.* Boca Raton, FL: CRC Press.

Guadagnoli, E., & Cunningham, S. (1989). The effects of nonresponse and late response on a survey of physician attitudes. *Evaluation and the Health Professions, 12,* 318–328.

Institute of Medicine (IOM). (1988). *Role of the primary care physician in occupational and environmental medicine.* Washington, DC: National Academy Press.

Institute of Medicine (IOM). (1991). *Addressing the physician shortage in environmental medicine.* Washington, DC: National Academy Press.

Kaplan, L. (1992). *No more than an ordinary X-ray: A study of the Hanford Nuclear Reservation and the emergence of the health effects of radiation as a public problem.* Unpublished doctoral dissertation, Brandeis University.

Kroll-Smith, S., & Ladd, A. E. (1993). Environmental illness and biomedicine: Anomalies, exemplars, and the politics of the body. *Sociological Spectrum, 13,* 7–33.

Lave, L. B., & Upton, A. C. (Eds.). (1987). *Toxic chemicals, health, and the environment.* Baltimore: Johns Hopkins.

Levine, A. G. (1982). *Love Canal: Science, politics, and people.* Lexington, MA: D.C. Heath.

Massachusetts Department of Public Health (MDPH). (1991). *Summary of results of physician survey on environmental health.* Boston: Author.

McCallum, D. B. (1992). Physicians and environmental risk communication. *Health and Environment, 6,* 1.

Merton, R. K., & Barber, E. (1976). Sociological ambivalence. In R. K. Merton (Ed.), *Sociological ambivalence and other essays* (pp. 3–31). New York: Free Press.

Murdock, B. S. (1991). *Environmental issues in primary care.* Navarre, MN: Health and Environment Digest.

National Institute of Environmental Health Sciences. (1994). Training doctors in environmental and occupational medicine. *Environmental Health Perspectives, 102,* 150–151.

National Research Council. (1991). *Environmental epidemiology: Public health and hazardous wastes*. Washington, DC: National Academy Press.

Nelson, L. (1989). Women's lives against the industrial chemical landscape: Environmental health and the health of the environment. In K. S. Ratcliff (Ed.), *Healing technologies: Feminist perspectives* (pp. 347–369). Ann Arbor: University of Michigan Press.

New Hampshire Division of Public Health Services. (1989). *Environmental medicine questionnaire*. Concord: New Hampshire Department of Public Health.

New Hampshire Division of Public Health Services. (1991). *Data based intervention project physician survey*. Concord: New Hampshire Department of Public Health.

Paigen, B. (1982). Controversy at Love Canal. *Hastings Center Reports, 12*(3), 29–37.

Rhode Island Department of Public Health. (1990). *Survey of primary health care providers regarding environmental health*. Providence, RI: Author.

Rosenstock, L., Rest, K., Benson, J., Cannella, J., Cohen, J., Cullen, M., Davidoff, F., Landrigan, P., Reynolds, R., Hawes Clever, L., Ellis, G., & Goldstein, B. (1991). Occupational and environmental medicine: Meeting the growing need for clinical services. *New England Journal of Medicine, 325*, 924–927.

Strauss, A., Schatzman, L., Ehrlich, D., Bucher, R., & Sabshin, M. (1964). The hospital and its negotiated order. In E. Freidson (Ed.), *The hospital in modern society* (pp. 147–169). New York: Free Press.

Szneke, P., Nielsen, C., & Tolentino, N. (1991). Connecticut physicians' knowledge and needs assessment of environmentally-related health hazards—A survey. Hartford: Connecticut Department of Public Health.

QUESTIONS

1. Discuss how each of the following may inhibit physicians' consideration of environmental factors in diagnosing and treating illness: a) ties to business and industry; b) lack of adequate education; and c) physicians' feelings of powerlessness.

2. Relative to other medical specialities, how knowledgeable are obstetricians and gynecologists about environmental health threats? Why do the authors consider this a problem?

Chapter Four

Environmental Illness as a Practical Epistemology and a Source of Professional Confusion

Steve Kroll-Smith and H. Hugh Floyd

"Listen to the patient, he will tell you the source of his
disease. Listen more closely and he will likely tell you
how to cure him." I heard something like that once in
medical school.

—The first author's family physician

The confusing nature of MCS is reflected in the number of terms enlisted to describe
it: environmental illness, chemical sensitivity, cerebral allergy, chemically induced
immune dysregulation, total allergy syndrome, universal reactor syndrome, ecologic
illness, chemical hypersensitivity syndrome, universal allergy, and, more alarming,
chemical AIDS and twentieth-century disease. To simplify discussion we will use the
terms *multiple chemical sensitivity*, or MCS, and *environmental illness*, or EI, to refer
to the disease and the terms *chemically reactive and environmentally ill* to refer to the
people living with the disease.

While the terms describing this medical condition vary, they converge on a number of common premises that together describe a nascent theory of the body and its
relationships to the materials of modern life: office buildings, houses, shopping malls,
yards and gardens, common consumer products, and so on. Importantly, what
medical science knows about the etiology, pathophysiology, and treatment of EI is
derived from the stories the environmentally ill tell about their bodies. Stories are all
we have at the moment because there are no agreed-upon criteria for defining EI as
an official medical condition and, consequently, there is no consensus regarding
appropriate diagnostic protocols or treatment regimens (Ashford and Miller 1991;
Bascom 1989). On the second page of their recent collaborative report, the U.S.
Department of Health and the Agency for Toxic Substances and Disease Registry
(ATSDR) reported that the natural history of EI describes "diverse pathogenic mechanisms . . . but experimental models for testing them have not been established
(Mitchell 1995, 2).

Thus, medical researchers and physicians who accept the possibility that MCS may
be a legitimate physical disorder must listen closely to their patients' efforts to explain

what is wrong with their bodies. Attending to the stories of people in pain recalls the typical eighteenth-century dialogue between patient and doctor, which typically began with the question "*What* is wrong with you?" Today, however, as most of us know, a physician is more likely to ask "*Where* does it hurt?" reflecting her greater faith in sophisticated technology than in the commonsense reasoning of her patients (Foucault 1973, xviii).

But the symbols of medical technology are silent on the issue of EI. It is, rather, the phenomenology of MCS, the experiences and accounts of those living with the malady that are the primary source of knowledge about this nascent physical disorder.[1] A remarkable feature of the accounts collected are their similarities, in spite of the fact that with a few exceptions the people interviewed do not know one another. Interviews with plumbers, accountants, pharmacists, postal workers, homemakers, marine captains, insurance salespeople, sugarcane workers, college professors, and others from all fifty states, with little more in common than that they all happen to be alive at the same time, consistently reveal common patterns. Discrete people, without recruitment ideologies typical of social movements, are thinking about their troubles in an essentially similar manner.

One explanation for this uncoordinated convergence in the style and product of thinking about illness is the possibility that common changes in people's bodies are shaping common thought processes. Other, arguably less sympathetic, accounts of this unorganized collective pattern are found in several academic discussions of the MCS phenomenon, including arguments that it is a form of hysterical contagion (Brodsky 1984) or chemophobia (Brown and Lees-Haley 1992). Complementing these psychosocial constructions is the unsettling idea that MCS is a pandemic outbreak of one of a number of faulty thinking disorders, including conditioned responses, symptom amplification, or displacement/avoidance activities (Simon 1995, 45; Simon, Katon, and Sparks 1990; Terr 1987).

The environmentally ill talk about a polysymptomatic disorder that starts with an acute or chronic exposure to chemical agents. Many of these agents are found in ordinary household and work environments in amounts well below recognized thresholds for toxicity. Following the initial sensitization experience(s) to a single chemical irritant, the body begins to express intolerance to an increasing array of unrelated irritants. A person with EI, for example, can react to volatile organic compounds emitted from gas stoves, dry-cleaned clothing, ammonia found in paper products, boron in cosmetics, phenol in air fresheners, and ethyl chloride in plastics, at doses that are magnitudes below those known to be dangerous. Ann became ill when she was exposed to formaldehyde in the new carpet in her office. A few days after the onset of her initial symptoms, she noticed that her body reacted aversely to her husband's colognes, her housekeeper's cleaning solvents, the painted wooden baskets hanging in her den, her laundry soap, and so on.[2]

The body's increasing intolerance to ordinary, putatively benign places and mundane consumer products is a key feature of this illness and one that baffles most physicians. "We don't dismiss these people, they are truly ill," admits a prominent allergist and medical researcher who speaks for the majority of practicing physicians, "but batteries of chemical tests can't pinpoint any specific sensitivity. Some are definitely allergic and we all agree that they are suffering, but we simply don't understand the cause of the disease as determined by medical diagnosis" (Selner 1991, 2–3). Another sympathetic but discouraging assessment concludes that "there is no laboratory test that can diagnose MCS, no fixed constellation of signs and symptoms, and no single pathogen to isolate and transmit through a cell line. . . . Even worse, some chemicals are neurotoxic and may produce symptoms that resemble anxiety attacks or mood disorders" (Needleman 1991, 33). Still more pessimistic is a public health physician who concludes that at present what is known about MCS "is insufficient to recommend programs for preventive strategies" (Bascom 1989, 36).

Adding to an already complicated theory is a premise that bodies are vulnerable to extremely low levels of chemical exposures: "below exposure levels for various chemicals established by the government, and usually below exposure levels tolerated by most people" (Pullman and Szymanski 1993, 17). This a difficult premise to test, however. If exposure levels are orders of magnitude below those deemed medically permissible, measuring concentrations of chemicals in soil, air, or water is unlikely to yield any useful information. If the concentrations are lower than permissible levels, the question still remains, How are they adversely affecting these bodies? The question is currently unanswerable empirically, though MCS suggests a theoretical rationale: Is it not possible that some bodies are more sensitive than others? Is it reasonable to sort bodies into nonsensitive, sensitive, and "hypersensitive," where sensitive bodies are more reactive than nonsensitive bodies, and hypersensitive bodies "are more sensitive than sensitive"? (Bascom 1989, 10; Ashford and Miller 1991). At least one person with EI now sorts his world into new categories: "I used to think in terms of people who are good on the one hand and bad people. Now I'm more likely to wonder whether this person is supersensitive like me or able to tolerate everything."

Complicating an already complex theory, another premise of MCS is that each chemical irritant may trigger a different constellation of symptoms in each person and that every system in the body can be adversely affected. Thus, combinations of body systems and symptoms interact geometrically, creating, at least theoretically, a seemingly endless configuration of somatic miseries (Pullman and Szymanski 1993, 17; Ashford and Miller 1991; Cullen 1987). Consider, for example, an abbreviated list of EI symptoms distributed by the Chemical Injury Information Network, an MCS support group. Among the sixty-two symptoms listed are sneezing, loss of smell, nosebleeds, dysphagia (difficulty in swallowing), dry or burning throat, tinnitus (ringing in the ears), hearing loss, hyperacusis (sound sensitivity), coughing, shortness of breath, hyperventilation, high and low blood pressure, hives, constipation, thirst, spontaneous bruising, swelling of heart or lungs, night sweats, insomnia, poor concentration, and depression (Duehring and Wilson 1994).

Robert loses his balance and becomes disoriented when he is around fresh paint, while Diane is likely to become nauseated and tired. Both manifest different symptoms when exposed to different chemical agents, challenging the biomedical assumption that each disease is caused by a specific aversive agent affecting an identifiable body system (Freund and McGuire 1991). Symptoms simultaneously involving multiple body systems, but affecting each differently, violate a foundational assumption of biomedicine that diseases are classed as specific pathological configurations (Kroll-Smith and Ladd 1993). A physician-researcher who frequently testifies against plaintiffs who claim to be environmentally ill and sue their employers for negligence in the management of a chemical work environment writes, "The persistence of symptoms, worsening of symptoms, and appearance of additional new symptoms during therapy attest to a pattern of fear of the everyday environment engendered by an unfounded perception of an environmentally damaged immune system" (Terr 1987, 693). A theory of chemically damaged immune systems, however, is only one of several pathophysiology theories of MCS.

Finally, people with MCS are likely to ascribe to a treatment regimen that emphasizes avoidance and lifestyle changes rather than drugs, surgery, or other invasive therapies (Bascom 1989; Ashford and Miller 1991; Kroll-Smith and Ladd 1993). Healing the body is specifically not an invasive procedure. Rather, healing begins with removing the offending substances from the body and working to keep those substances at a safe distance. Avoidance and self-discipline are key elements of successful coping. Avoidance measures can be as subtle as moving away from a person wearing hair spray or cologne to moving into an environment built specifically to reduce chemical exposure. Wimberly, a small town in central Texas, has gradually become a chemically free refuge for people with extreme MCS. While only a small number of the chemically reactive move to such special environments, most are forced into some form of social and spatial exile to successfully manage their symptoms.

Avoidance can also be more proactive. Increasingly, people who theorize their bodies' relationship to environments using some variant of MCS try to persuade others to change their personal habits, approach employers with specific requests that would reduce their exposure to offending substances, and appeal to local, state, and national legislatures to create "safe zones" free of dangerous chemicals.[3]

A strategy of avoidance based on escape and one based on changing habits, ordinances, or the materials of production are effectively redrawing the boundaries between safe and dangerous places, though with varying social and political effects. Families who leave Los Angeles and move high into the Sierra Madres to escape a chemically saturated world are building alternative, "ecologically safe" communities; they are not, however, directly challenging society to change. A wife who refrains from wearing a "toxic scent," an employer who moves an offending copying machine from a nearby office, and a county board of supervisors that passes an ordinance establishing a "fragrance-free zone" in the local courthouse are examples of social and legal accommodations to the environmentally ill who petition others to change. When others change, the environmentally ill stand a chance of living within society rather than merely surviving by escaping from it.

Whether they manage their symptoms by escaping society or challenging it, or some combination of the two, the environmentally ill are forced to carve up the meaning of space in a manner unfamiliar to most people. Thus, while their behavior can appear strange and untoward, perhaps insulting, to others, for them it is a reasonable response to the management of their symptoms.

The exact number of people who claim to be environmentally ill is not known. The U.S. Department of Health and Human Services admits it cannot estimate their numbers (Samet and Davis 1995). Commonsense comparisons, speculation, and anecdotes are the fallback strategies for calculating the scope of the problem. The Labor Institute of New York notes: "While it is clear that a significant portion of the population is sensitive to irritants such as cigarette smoke, the percentage of individuals who are significantly affected by multiple chemical sensitivities appears to be much smaller" (Pullman and Szymanski 1993, 18).

Though it does not use the term multiple chemical sensitivity, environmental illness, or any of the other variants, the National Academy of Sciences (1987) suggests that between 15 and 20 percent of the U.S. population is allergic to chemicals commonly found in the environment, placing them at increased risk of contracting a debilitating illness. The National Research Council's Board on Environmental Studies and Toxicology (1991) reports that "patients have been identified with a condition of multiple and often diverse symptoms that have been attributed to chemical agents in the environment" (5), though it does not specify how many.

Complementing this anecdotal approach to determining the breadth of the problem are several additional facts and figures that suggest that EI is more than a minor medical annoyance. A nonrandom survey of people who identified themselves as having MCS found sixty-eight hundred respondents (quoted in Ashford and Miller 1991, 5). The Chemical Injury Information Network lists multiple support groups for people with EI in forty-four of the fifty states. Support groups also meet in Finland, Germany, Australia, Canada, Denmark, New Zealand, France, Mexico, Belgium, and the Bahamas. We identified twenty-nine newsletters circulating in the United States devoted to chemicals, bodies, and the environment.

The range of demographic groups reporting the symptoms of MCS suggest it is a pandemic problem:

> A review of the literature on exposure to low levels of chemicals reveals four groups or clusters of people with heightened reactivity: industrial workers, occupants of "tight buildings," . . . residents of communities with contaminated soil, water, and air, and individuals who have had . . . unique exposures to various chemicals. (Ashford and Miller 1991, 3)

This list implies that everyone is susceptible to the ravages of MCS. There is some evidence to support this unsettling idea.

Industry groups estimate that over a third of new and remodeled office and storage buildings harbor indoor air pollutants sufficiently toxic to increase employee absenteeism by as much as 20 percent (Molloy 1993, 3). In addition to the building materials themselves, the Occupational Safety and Health Administration counted a minimum of "575,000 chemical products . . . used in businesses throughout the U.S." (Duehring and Wilson 1994, 4; see also U.S. Department of Labor 1988). In 1989 the U.S. Environmental Protection Agency estimated that employers lose approximately sixty billion dollars a year to absenteeism caused by building-related illnesses (cited in Molloy 1993, 3). Not every victim of a "sick building" becomes environmentally ill, of course, but "bad air" at work is a common explanation for the origin of chemical reactivity among the environmentally ill.

But the workplace is not the only source of EI. Aerial pesticide spraying, incineration practices, and groundwater contamination are among the causes of MCS in neighborhoods and communities (Ashford and Miller 1991). In addition, the U.S. Environmental Protection Agency reported that one in four people in the United States live on top of, adjacent to, or near an uncontrolled hazardous waste site (1980; see also Szasz 1994).

Finally, consider a series of troubling statistics culled from several sources:

- In 1940 the annual production of synthetic organic chemicals in the United States was 2.2 billion pounds. By 1991 it had increased to over 214 billion pounds, an increase of 2000 percent in fifty years (National Research Council 1991, 21).
- "The EPA's Office of Toxic Substances is called upon to review approximately 2000 new chemical products a year" (Duehring and Wilson 1994, 4).
- The EPA can ensure the safety of only six out of six hundred active pesticide ingredients under its control (Duehring and Wilson 1994, 10).
- Less than 10 percent of the seventy thousand chemicals now in commercial use have been tested for their possible adverse effects on the nervous system and " 'only a handful have been evaluated thoroughly,' according to the National Research Council" (Duehring and Wilson 1994, 4).
- The EPA has identified over nine hundred volatile organic chemicals in ordinary indoor environments including offices and houses (reported in *Delicate Balance* 1992, 9).
- Finally, an EPA Executive Summary on chemicals in human tissue found measurable levels of styrene and ethyl phenol in 100 percent of adults living in the United States. The Summary also found 96 percent of adults with clinical levels of chlorobenzene, benzene, and ethyl benzene; 91 percent with toluene; and 83 percent with polychlorinated byphenols (Stanley 1986).

There is, in short, ample opportunity for individual exposure to a seemingly endless parade of chemicals whose effects on the body are simply not known.

While it is not possible to know with any certainty how many people claim to suffer from MCS, it is reasonable to assume the number is substantial and growing. At the very least, it is possible to imagine how a person might link an array of bizarre and debilitating symptoms to a disease theory based on a premise that the body is exposed to an extraordinary number of chemically saturated environments.

EI and the Profession of Medicine

People with MCS are theorizing what makes them sick, how specifically their bodies are changed (immune system, limbic system, and so on), and what can be done to decrease or manage their symptoms. When they speak of MCS, there is often a tone of certainty in their voices. While certain, they are not arrogant, however. The surety of knowing is typically accompanied by self-doubt, anger, fear of the future, and other troubling emotions. While a chemically reactive person is reasonably confident in his theory of what is wrong with his body, why, and how he can manage his symptoms, MCS is not recognized by the profession of medicine as a legitimate physical disorder.

Indeed, medical professionals are likely to admit that currently what they do *not know* about MCS is considerably more than what they *know*. A physician's report to the Maryland Department of the Environment on the problem of EI, for example, is primarily a list of things medicine does not know about this nascent disorder, herein called chemical hypersensitivity disorder, or CHS.

- There is no single universally accepted terminology for or definition of CHS.
- There is no known cause of CHS.
- There is no prognosis for individuals with CHS.
- There are no criteria or procedures for reporting sensitivity disorders as diseases.
- There are no prevalence studies of CHS.
- It is not known if the incidence or prevalence rate of CHS is increasing.
- A "risk profile" for CHS does not exist.
- Educational materials on the subject of CHS are limited, and it is not possible to determine the accuracy of the information that is available. (Bascom 1989, 2–19)

Not surprisingly, the author concludes her report by observing that not enough is known about CHS "to recommend programs for preventive strategies. . . . There is no consensus as to the cause of CHS, the appropriate medical treatment, or the appropriate policy approach" (36–37). The U.S. Department of Health and Human Services concurs: while an increasing number of people are defining themselves as environmentally ill, the definition of MCS "is elusive and its pathogenesis as a distinct entity is not confirmed" (Samet and Davis 1995, 1). An occupational medicine researcher expresses his frustration over this elusive problem: "If the question cannot be answered as to what MCS is, how can there be approval of research protocols or acceptance of investigative results? In order to appropriately address the controversies surrounding this phenomenon we must know where we're going!" (DeHart 1995, 38).

The first official recognition of MCS was probably a 1985 report by the Ad Hoc Committee on Environmental Hypersensitivity Disorders (1985) in Toronto, Canada. Two years later Dr. Mark Cullen, a medical researcher at Yale University, published a definition of MCS based on his observations of people exposed to chemical irritants at the workplace. While his definition is the most frequently cited in the biomedical literature, it clearly expresses biomedicine's uncertainty regarding this nascent disorder:

Multiple chemical sensitivities is an acquired disorder characterized by recurrent symptoms, referable to multiple organ systems, occurring in response to demonstrable exposure to many chemically unrelated compounds at doses far below those established in the general population to cause harmful effects. No single widely accepted test of physiologic function can be shown to correlate with symptoms. (Cullen 1987, 655)

The biomedical research community is divided over the meaning of MCS and the numbers of people who have it. For some researchers "evidence does exist to conclude that chemical sensitivity [is] a serious health and environmental problem and that public and private sector action is warranted at both the state and federal levels" (Ashford and Miller 1991, v). For others, however,

> a great deal more research is needed before there will even be a consensus on a definition of chemical hypersensitivity. It is premature to classify CHS [chemical hypersusceptibility] as a purely environmental problem. . . . Health related environmental standards are based on normally accepted exposure units. They do not take into account individuals who may be sensitive to chemicals at limits far below the norm, perhaps at undetectable limits given current technology. (Maryland Department of Environment, letter to Governor Donald Schaefer, in Bascom 1989)

In striking contrast to the difficulty of the biomedical research community in reaching agreement on the meaning of MCS, the clinical medical profession speaks with one voice in rejecting the legitimacy of this proposed disorder. From its perspective, MCS is a fugitive, hopefully transitory, concoction of beliefs with no rightful claim to legitimacy.

Local medical boards reportedly threaten to censure physicians who diagnose people with MCS (Hileman 1991, 27–28). National medical societies, including the American Academy of Allergy and Immunology (1989), the American College of Occupational Medicine (1990), and the American College of Physicians (1989) officially deny the reality of MCS as a physical disorder and caution physicians not to treat patients "as if" the disease existed. The executive committee of the American Academy of Allergy and Immunology could be said to speak for the other professional medical societies in its position statement on MCS:

> The environment is very important in the lives of every human being [*sic*]. Environmental factors, such as chemicals and pollutants, have been demonstrated to influence health. The idea that the environment is responsible for a multitude of human health problems is most appealing. However, to present such ideas as facts, conclusions, or even likely mechanisms without adequate support, is poor medical practice. The theoretical basis for ecologic illness in the present context has not been established as factual, nor is there satisfactory evidence to support the actual existence of . . . maladaptation. (Quoted in DeHart 1995, 36)

The California Medical Association reported that "scientific and clinical evidence to support the diagnosis of environmental illness is lacking" (1986, 239). The report went on to argue that evidence supporting the existence of a low-level chemical etiology to such health problems is based on hearsay and anecdote, not controlled clinical trials (243). A study published in the *New England Journal of Medicine* found the clinical testing for MCS to be seriously flawed and the typical environmentally ill patient to be unusually stressed and personally unhappy (Jewett, Fein, and Greenberg 1990). In a report prepared for the State of Maryland, a health policy analyst summarized the hostility of the medical profession toward a biomedical interpretation of EI, observing that the "controversy surrounding the chemical hypersensitivity syndrome begins with a debate as to its very legitimacy as a distinct entity" (Bascom 1989, 8).

Results from a survey of physician members of the Association of Occupational and Environmental Clinics—the one medical society most likely to be sensitive to people who claim they are suffering from MCS—are also worth considering. First, the survey found that only 9 percent of the physician population believe EI is predominantly physical in origin. Sixty-four percent, on the other hand, believe it to be a psychological disorder (Rest 1995, 61). With this bias toward a psychogenesis model of MCS, we should not be surprised to learn that occupational physicians were more likely to consult psychiatrists and psychologists when treating a patient who theorized his misfortune as MCS (63). Similarly, 64 percent of the occupational physicians reported referring people who claim to be chemically reactive to psychologists or psychiatrists. Fifteen percent did so "always," while 49 percent did so "at least half the time" (65).

A report in the *Annals of Internal Medicine* labeled people claiming to suffer from MCS a "cult" (Kahn and Letz 1989, 105).[4] Adding insult to injury, an allergist reports that he can reduce the symptoms of the disorder by "deprogramming" patients who internalize "environmental illness beliefs" (Selner 1991). A psychiatrist writes: "In the absence of objectively verified abnormalities detected in physical examination, the illness is subjective only. . . . Multiple Chemical Sensitivity constitutes a belief, not a disease" (Brodsky 1984, 742). A study of twenty-three people who identified themselves as environmentally ill found fifteen of them suffering from a mood, anxiety, or somatoform disorder (Black, Rathe, and Goldstein 1990). The authors of this study, published in the *Journal of the American Medical Association*, conclude that all people with EI "may have one or more commonly recognized psychiatric disorders that could explain some or all of their symptoms" (3166).

Finally, Gregory Simon, another psychiatrist and coauthor of a well-known article on MCS, "Allergic to Life: Psychological Factors in Environmental Illness" (Simon, Katon, and Sparks 1990), argues that MCS is simply a product of faulty reasoning. Recalling the classic anthropological question, "Can 'primitive' people distinguish fact from fancy or do they muck around in a hodgepodge of spirits, sprites, myths, and legends?" Simon and colleagues label the environmentally ill victims of, simply put, bad reasoning. Like Lévy-Bruhl's primitive, they cannot discern what is real from what is imaginary. Thus for some experts MCS is a result of behavioral sensitization. People associate a smell or taste with a physical symptom, in spite of the fact that there is no clinical relationship between the two. For others, MCS is a consequence of a tendency to react unreasonably to physical symptoms such as a sore throat or a rash. Investing too much attention in these symptoms, they search for causes and find them in the local environment. Finally, for still others MCS is a result of a faulty mode of reasoning perhaps best called "displacement confusion." Here a person avoids thinking about the "real" causes of physical distress, unhealthy lifestyles, excessive stress, and so on, and focuses instead on modern culture's overconcern with the environmental causes of disease (Simon, Katon, and Sparks 1990; see also Simon 1995, 45).

What are we to make of this confusing array of biological and psychological accounts of EI? Those in the medical research community are more sympathetic than their counterparts in clinical medicine to the idea that MCS is a legitimate medical disorder. But research on MCS is just beginning. Indeed, as we write this book, there is not even a commonly accepted case definition of the problem. Thus medical researchers are still debating the essential question: What *is* it? The clinical medical community appears to be ahead of its research colleagues, at least in knowing what MCS is *not*. It is not a legitimate physical disorder. While there is some confusion over what MCS might be—a belief, a cult, a psychiatric disorder, or a process of faulty reasoning—it is not recognized as a physical disease by the medical profession.

Thus, what happens when a person who has been closely monitoring his body, matching symptoms with environments, and organizing his local world to make some sense of his distress visits a physician trained to look beyond a patient's account and examine the body as the source of disease?

Doctors, Patients, and Paradigm Disputes

When physicians receive patients' complaints, it is their professional responsibility to translate them into a language that is created and controlled by the normal science model of medicine. Although they use the most sophisticated medical technology and are guided by the cultural authority of biomedicine to "define and evaluate their patients' condition" (Starr 1982, 16), most physicians who treat the environmentally ill fail to heal them.

Imagine the physician presented with a patient such as Howard, complaining of nasal obstruction, sinus discomfort, chest pain, flushing hives, itching eyes, loss of visual acuity, fatigue and insomnia, genital itch, and nausea. Imagine that no accepted tests of organ system function can explain the symptoms. Imagine also that the patient is nonreactive to any conventional treatment plan the physician prescribes. The complaints persist. Finally, imagine that the patient has a theory that explains the origins of the symptoms, but that such a theory does not correspond to any of the accepted etiologies within the biomedical model. It is not unreasonable to assume that patient and physician will tire of this cycle of frustration. The physician might suggest another doctor, or the patient might simply give up and go elsewhere. Whatever happens, the bioscience model of medicine has failed to provide the means for the patient to act like a patient and the doctor to act like a doctor; that is, the physician did not heal and the patient did not recover. If the enactment of biomedicine occurs at the moment its body of knowledge encounters a body, the body of the environmentally ill obscures that moment and effectively prevents the encounter.

Why is the profession of medicine unable to certify MCS as a legitimate physical disorder? Perhaps it isn't one. That is the simplest answer. It is more complicated and more interesting, however, to consider MCS as a theory of the body and the environment that contests both the medical profession's responsibility to define bodies and several of its paradigmatic assumptions about disease.

First, medicine works closely with the state to define and regulate bodies in the interest of cultural and capital production (Foucault 1973; Turner 1995). Capitalism in the waning years of the twentieth century is interested in bodies insofar as they are able to work and consume, and do so in a flexible manner (Martin 1990; Harvey 1989). The healthy body, in other words, is one that goes to work regularly, purchases and consumes the products of its or others' labors, and is capable of adapting quickly to changing modes of production and skill requirements. A putative somatic disorder that denotes change in the definition of the body in its relationship to common consumer products and domestic and workplace environments, therefore, is likely to be scrutinized closely before it is officially recognized as a disease. The environmentally ill body is, of course, anything but flexible. But something more basic than an abstract political economy is at work here.

Howard's unfortunate predicament suggests that a formidable problem for attending physicians is the result of the limitations of their diagnostic technologies in certifying something called MCS. Medical technology is built to measure and test the assumptions of the biomedical model. Among the many assumptions in this model are two that are particularly relevant to MCS. From classic toxicology comes the supposition that a relatively small number of individuals are sensitive to low, but nevertheless measurable, exposures to certain toxins. From allergy comes the classic IgE-mediated responses by atopic individuals with overactive antibodies that mistake ordinary environmental stimuli (ragweed, pollen, dust, and so on) for poison. What the biomedical model does not assume, however, is a third, entirely different, type of sensitivity.

A principal characteristic of MCS is that after the initial sensitization, there is no identifiable threshold or exposure level below which there is a negligible risk of becoming sick (Davis 1986). People who identify themselves as environmentally ill report that an acute or chronic exposure to chemicals sensitizes their bodies to respond adversely to extremely low, subclinical exposures to a seemingly endless array of unrelated chemical compounds. (The term *subclinical* is used here to denote the absence of a diagnostic technology capable of identifying the quantity of chemicals that purportedly change the bodies of the chemically reactive.)

Canada's Ministry of Health concludes in a report on MCS that "affected persons have varying degrees of morbidity and no single laboratory test including serum IgF is consistently altered" (Davis 1986, 34). Acknowledging this limitation, the National Research Council (1991) concludes quite simply that the "symptomatology related to multiple chemicals is a distinct feature of [EI] patients that is not classifiable by existing criteria used in conventional medical practice" (5). Multiple chemical sensitivity, in other words, is a medical anomaly; and like all scientific anomalies it is approached as an "untruth, a should-be-solvable-but-is-unsolvable problem, a germane but unwelcome result" (Mastermind 1970, 83).

But MCS is more than an awkward fact for the profession of medicine. Indeed, medical anomalies are common. At this time, for example, the etiologies of Sjögren's syndrome and idiopathic pulmonary fibrosis are simply unknown and treatments difficult to prescribe. A new strain of tuberculosis is resisting proven antidotes and spreading to dangerous levels in urban areas. And AIDS continues its deadly course, labeled but eluding cures. But most medical anomalies, including those just mentioned, are puzzles whose solutions will not change the cultural definition of the body. Multiple chemical sensitivity, on the other hand, is more a mystery than a puzzle. If a puzzle is a game to exercise the mind by encouraging a search for the solution, a mystery admits of no solution unless the rules of the game itself are changed. More than a puzzle or awkward fact, MCS would change the rules of the game by changing what is known about bodies and supposedly safe environments.

At the heart of this undecided battle are the environmentally ill, challenging the received wisdom about the body by linking their somatic disorders to rational explanations borrowed from the profession of medicine. It is not, in other words, the languages of the occult, New Age, or Eastern philosophy that are adopted by the chemically reactive to interpret their somatic misery. It is not crystal therapy, homeopathy, past-life regression, or obeisance to self-appointed gurus that serves as a resource for knowing. Rather, these individuals are apprehending their bodies using the rational, Enlightenment language of biomedicine. If Carl Sagan (1996) truly laments the modern revolt against science and the resurgence of a "demon-haunted world," he should be pleased to hear of ordinary people who are struggling to know something logical and reasonable about their bodies.

The environmentally ill are likely to apprehend their somatic misery using the technical language of biomedicine rather than some variation of New Age knowledge for at least one rather obvious reason: they experience their bodies changing in the presence of consumer items commonly regarded as safe and in ordinary environments commonly regarded as benign. Consider, for example, the following field note describing an incident that occurred during an interview with a person who claims to be environmentally ill:

> I sat roughly twenty feet from Jack. We were in his living room. Jack's house is set up for someone who is environmentally ill. Air-filtering machines are running in several rooms. Magazines, newspapers, and other printed materials are noticeably absent. A plastic housing covers the TV screen to block harmful low-level electromagnetic waves emitted from the picture tube.
>
> I am properly washed and attired. (That is, I showered without using soap and am wearing all cotton that has been washed dozens of times.)
>
> Shortly after starting the interview, Jack became visibly agitated, lifting himself from side to side and up and down in his chair. Red blotches appeared on his arms and face. He started to slur his words. He explained that he was reacting to something new in the house. Since I was the only new thing around, he started to ask me questions: Was I wearing a cologne? Was I wearing all cotton? Could I have washed my clothes using a fabric softener? And so on. With the exception of the cotton question, I answered "no" to each query.
>
> His symptoms were increasing in severity. He looked at my pen and asked if it contained a soy-based ink. I told him I bought it at a bookstore without checking the chemical composition of the ink. He smiled knowingly and asked me to put the ink pen outside. Within a few minutes his symptoms subsided.

The question is not whether Jack's body changed in front of me. It did. The question, rather, is how to interpret the change. Using a process of elimination, Jack concluded that the one foreign item in his house responsible for his somatic distress was an ordinary ballpoint pen. Remember, the distance between Jack and the pen was approximately twenty feet. I asked him to explain how he knew the cause of his symptoms was the pen and how an ink pen that was twenty feet away could affect him so seriously. He told me about the synthetic chemicals in ink and their particular effects on him. He explained how the air circulator in the living room was pointing at my back and facing him. Thus, it blew the offgassing ink from the point of my pen toward him.

Jack's carefully thought-out explanation of his somatic distress struck me as interesting, if debatable. Every move in his "first-this-and-then-that" style of reasoning is grounded in a testable assumption. And Jack was not surprised when his symptoms subsided after the pen was removed from the house. "What else could it have been?" he reasoned. Jack is in the habit of theorizing his illness by constructing what for him and, at least some, others are reasonable accounts of the causes of his misery. For Jack, theorizing his illness in a language of instrumental rationality allows him to explain his body to others and, importantly, allows him to live with some degree of self-respect in a very sick body.

For some people, however, Jack's story is questionable, indeed bizarre. He tells a fantastic tale about bodies and environments. Moreover, he requests that others modify and change what have always seemed benign, if not aesthetic or pleasurable, behaviors. If they do not do so, they are implicated in the exacerbation of his illness. His spouse, a friend, the teller at the corner bank, an office mate, a sociologist who requests an interview, and even a complete stranger become potential sources of acute, debilitating distress; once safe, innocuous places are now health risks. Jack approaches his new life as environmentally ill armed with an explanation of his body and its complicated relationship to common consumer items and local places.

For Jack, MCS is not only a chronic sickness; it is a vocabulary of motives, a type of "justificatory conversation" (Mills 1967). The "truth" of Jack's story can be measured in the degree of accommodation people make to his disabled body. The success of the environmentally ill in convincing others of the threat to health posed by mundane environments and ordinary consumer items, while also claiming the right to institutional recognition of their sickness, depends, as we will see, on the ability to borrow liberally from the vernacular of biomedicine to lobby for the transformation of their illness experiences into an official disease.

Environmental Illness as a Practical Epistemology

What is true for Jack is true for thousands of people living with bodies they believe are made sick by the environment. Multiple chemical sensitivity is a nascent theory of bodies and environments. It is a novel form of theorizing the relationships of people, bodies, and environments that unhinges an expert knowledge from an expert system and links it to historical and biographical experience to make a particularly persuasive claim on truth. It is a local knowledge, constructed in situ by people who believe they need to reorganize how they think about their bodies and the environments that surround them. Power may be a source of knowledge in a post-Enlightenment world, as Foucault announced, but rational knowledge nevertheless remains a powerful social resource. Indeed, if modernity has a commandment it is to *act in accord with reason*.[5] Rational knowledge is always an assertion of the correct, the logical, the appropriate. If something is accepted as true, then rational organizations and human beings are expected to organize their conduct to reflect this truth. Rational knowledge "is always a legitimating idea" (Wright 1992, 6). In fact, it is self-legitimating insofar as its claim to truth rests on the premise that "all that is real is rational, [while] all that is rational is real" (Lyotard 1992, 29). Thus, to accept someone's account as rational is to tacitly commit to the line of conduct and belief embedded in that account, or to risk the charge of behaving irrationally.

Society places a particular premium on the authority of rational knowledge to regulate nature and health (Wright 1992; Touraine 1995; Freund and McGuire 1991). Knowing nature, including the nature of the body, depends upon a detached observer trained to identify by means of calibrated instruments the intricacies of biological and physical systems. It is not surprising, therefore, that the privilege of theorizing the body and its relationship to the environment is limited to people educated and licensed by the state to speak the language of biomedicine.

It is the chemically reactive, however, and not the medical profession, who are classifying and explaining their anomalous medical condition. People who identify themselves as environmentally ill are shifting the social location of theorizing bodies and environments from medical professionals to nonprofessionals, from experts to nonexperts. When theorizing somatic distress in the language of biomedicine shifts from experts to laypersons, it enters a new social world, one governed by purposes other than institutional legitimation. Thus, when expert knowledge is separated from its institutional moorings and taken into another world, it is likely to be fashioned into a new cultural tool, or, as Geertz (1983) would have it, a "practical epistemology" (151). While Geertz leaves this term purposively vague, we will mean by it a technical, rational way of knowing that is responsive to the immediate personal and communal needs of nonexperts. A practical epistemology, in other words, joins the world of personal and biographical experiences to forms of instrumental rationality. Jack's story of a ballpoint pen is a good example of a practical epistemology at work. The state-sponsored owners of biomedical knowledge most likely would dismiss his account as nonsense, if not evidence of delusion. Jack, however, borrows liberally from biomedicine and common sense to conceptualize and organize a world of signs that allows him to explain and respond to a body his doctors cannot understand.[6]

It is not a desire to engage the medical profession in spirited debate, however, that is motivating the environmentally ill. A person who confiscates the privilege of physicians to explain bodies in relationship to environments is thinking about something more elemental than an epistemological dispute, to wit, simple survival. "We are always searching for ways of explaining to others what we have," acknowledges a woman with MCS, "and I guess . . . to explain to ourselves too." An engineer with a long history of the disorder recalls that "at first it was a search for a vocabulary that could express what I, or I guess my body, was going through. Crazy-sounding words like 'toxic toys' and 'VOC reactivity' became a standard way of talking for me; and still is." The efforts of the environmentally ill to find the words necessary to apprehend their misery constitute one part of this study; the specific ways they use these words to alter the social landscape and change their life circumstances constitute the other.

The environmentally sick use their theories of the body and environment to ask others to understand their misery, alter their behaviors, allocate time and money, and, generally, change the world to accommodate their illness. Specifically, rational theories of chemical reactivity become rhetorical idioms for assigning moral significance to previously amoral behaviors or habits and traditionally inconsequential environments and consumer products. When a chemically reactive husband requests that his wife of twenty years refrain from using her usual dry skin lotion, she will probably ask him why. If we listen to his reply, we are likely to hear a biomedical explanation of the effects of such chemicals as butylene glycol or phenoxyethanol on his immune system or his central nervous system. Whatever the particularities of his response, he is likely to make a causal link between chemicals in the lotion and his somatic troubles. In this fashion, what he knows about his illness becomes a lingual resource for both managing his somatic distress and critiquing behaviors, products, and environments that are routinely defined as appropriate, safe, and benign.

In theorizing the origins, pathophysiology, and effective management of their illness, the environmentally ill understand why their symptoms intensify and subside in accordance with the presence or absence of mundane consumer items and the personal habits and practices of people around them. Knowing what makes them sick and learning to avoid debilitating symptoms are cognitive resources for personal survival. With these resources these individuals can inhabit bodies that are routinely out of control with some degree of self-assurance.

Among its many manifestations, MCS is a dispute over the privilege to render a rational, in this case biomedical, account of a disabled body and the peculiar content of that account. It is a dispute over the ownership of expertise. It is a story about how institutions learn in a historical period wherein nonexperts wield languages of expertise to persuade influential others to modify their habits, regulations, and laws.

Narratives of the Environmentally Ill: A Word about Methods

It is said that human misery is bearable only if we can tell a story about it. Perhaps it is because each of us is a storyteller that our lives have a measure of coherence and clarity. Life without narrative would be discontinuous, formless, seemingly random. Narrators create story lines, linking occurrences and ideas into plots, and give time and space a linear order. Moreover, "Personal experience must be assigned a central role in accounting for the understandability," and, we would argue, origin, "of theoretical categories and concepts" (Calhoun 1995, 86).

Except for those whose symptoms are truly severe, who cannot write or talk without considerable discomfort, most people with MCS are willing to talk about their distress. To learn about the experiences of the environmentally ill, the first author attended an environmental illness support group for approximately ten months and conducted separate interviews with each of the four members who regularly attended the group. Each person was interviewed on several occasions, and a biography of his or her illness experience was constructed. Illness biographies were written in this fashion for twelve additional people with MCS who were not members of this support group.

To provide a rough check on the reliability of these illness biographies, we subscribed for two years to four nationally circulated newsletters distributed by organizations for the environmentally ill: *Our Toxic Times,* the *Wary Canary,* the *New Reactor,* and *Delicate Balance.* We searched these documents for personal accounts of the origins of the illness, its pathophysiology, and suggested treatment regimens. Comparing the newsletter accounts with our illness biographies, we found striking similarities in the interpretive strategies people use for understanding their bodies and environments. Next, we examined two biographies written by people with EI (Lawson 1993; Crumpler 1990) and again found considerable overlap in the types of explanations typically used to make sense of bodies unable to live in ordinary environments.

Reasonably confident that the patterns of theorizing MCS discovered in the initial interviews and confirmed in newsletter accounts and biographies were generalizable to the population of people who are chemically reactive, we obtained the membership directory of the Chemical Injury Information Network. While no list can be representative of the universe of the environmentally ill, this directory is the most exhaustive list we found, and perhaps the most exhaustive list in existence. It identifies people with MCS in every state of the Union and eleven foreign countries.

We constructed a simple, open-ended questionnaire designed to solicit information on how people experienced the illness and what specifically they thought about it. We mailed this questionnaire to seventy-five people listed in the membership directory. We also asked several newsletters to print a short notice announcing our study and directing people who were interested in participating to write or call. Between the seventy-five questionnaires mailed to directory addresses and the appeals in the newsletters, we obtained an additional 147 interviews. The quality of these interviews varied. Some people responded in short, curt sentences to each question, making it difficult to learn much from their answers. Responses to 42 interviews were too cursory to be of much help.

Other people wrote between ten and twenty pages—essays steeped in reflection and pain. Still others answered the questionnaire in five to ten pages. Narratives of this length were brimming with insights into how people organized their thoughts to apprehend their miseries. Through this technique we obtained 105 interviews. Combined with the 16 interviews we conducted during the first several months of work, we collected a total of 121 usable interviews.

In addition to the interviews, we searched Med File and other library databases for medical studies of MCS. We also purchased the Chemical Injury Information Network's bibliography on toxic chemicals and human health, which contains 1,106 entries. These secondary materials were also treated as stories of the illness.

Finally, we took our emerging conclusions back to several of the environmentally ill to ask for their comments. While a few people did not see the political importance of this type of work, expressing some disappointment that it was not a forthright call for public support, others found our story personally affirming, validating their hard-fought claim to know something important about modern bodies and environments. We are pleased to report that no one with EI who commented on our story disagreed with it.

While it is the stories of the environmentally ill that interest us, we are ever mindful of the importance of these stories to the identities of the narrators. And we are also mindful of the importance of these stories to the success of this project. The real strengths of our research are not found in our abstract musings (though we hope some readers find them useful) but in the compositions of the environmentally ill, their often insightful and always revealing accounts. We were privileged to hear and read these stories and report them.

NOTES

1. It is worth noting, however, that if EI is an anomaly for biomedicine, it is a tangible expression of the truth claims of a marginal and disputed body of medical knowledge commonly called *"clinical ecology"* or *"environmental medicine."* Clinical ecology is not recognized by the American Medical Association, in part because it assumes people can be made sick by ordinary environments, particularly petrochemical exposures. It is our impression, however, that comparatively few people who self-identify as environmentally ill have ever heard about clinical ecology, though they may learn something about it as they read, conduct research, and conceptualize their somatic troubles. (On the comparative insignificance of clinical ecology for EI, see Kroll-Smith and Ladd 1993.)

2. Illustrations like Ann's appear throughout this discussion. They are taken from newsletter accounts and interviews. Our research methods are explained at the end of this chapter.

3. To facilitate discussion, we will not always refer to both environments and products as sources of distress. When we use the word *environment*, we are implying both setting and products.

4. It is obvious to us, and we hope to the reader, that the demographic mix and areal distribution of people who claim to be environmentally ill suggest an organizational form considerably more complicated than a cult.

5. While no one disputes its commitment to rationality, it is doubtful the modern period will be remembered as a historical epoch guided by sensibility and wisdom.

6. To anticipate some semantic confusion over the words *epistemology* and *theory*, we are using the term *epistemology* to mean the nature of knowledge. Environmental illness is a way of knowing that combines abstract biomedical concepts with concrete, local, somatic experiences. The term *theory*, on the other hand, refers to the specific accounts of the environmentally ill who use biomedical knowledge to explain their somatic distress. Taken together, the theories of the environmentally ill constitute a practical epistemology, a way of knowing

their bodies and environments based on biomedical nomenclature. If epistemology means *how* one knows, theory means *what* one knows.

REFERENCES

Ad Hoc Committee on Environmental Hypersensitivity Disorders. 1985. "A Report to the Minster of Health." Toronto: Office of the Minister of Health.

American Academy of Allergy and Immunology, Executive Committee. 1989. "Clinical Ecology." *Journal of Allergy and Clinical Immunology* 78:69–71.

American College of Occupational Medicine. 1990. "ACOM Adopts ACP's Position on Clinical Ecology." *ACOM Report*, September 1.

American College of Physicians. 1989. "Clinical Ecology." *Internal Medicine* 3: 168–78.

Ashford, Nicholas, and Claudia S. Miller. 1991. *Chemical Exposures*. New York: Van Nostrand Reinhold.

Bascom, Rebecca. 1989. *Chemical Hypersensitivity: Syndrome Study*. Baltimore: Maryland Department of the Environment.

Black, Donald W., Ann Rathe, and Rise B. Goldstein. 1990. "Environmental Illness: A Controlled Study of 26 Subjects with '20th Century Disease.' " *Journal of the American Medical Association* 264:3166–70.

Brodsky, Carroll. 1984. "Allergic to Everything: A Medical Subculture." *Psychosomatics* 24:731–42.

Brown, Richard, and Paul R. Lees-Haley. 1992. "Fear of Future Illness, Chemical AIDs, and Cancerphobia: A Review." *Psychological Reports* 71:187–207.

Calhoun, Craig. 1995. *Critical Social Theory*. Oxford: Blackwell.

California Medical Association. 1986. "Clinical Ecology: A Critical Appraisal." *Western Journal of Medicine* 144:239–45.

Crumpler, Diana. 1990. *Chemical Crisis: One Woman's Story*. Sydney, Australia: Scribe.

Cullen, Mark. 1987. "The Worker with Multiple Chemical Sensitivities: An Overview." *Occupational Medicine: State of the Art Reviews* 2:655–61.

Davis, Earon S. 1986. "Ecological Illness." *Trial*, October, 33–34.

"Declaration of Rights for the Multiple Chemically Sensitive." 1981. *Delicate Balance*, March.

DeHart, Roy L. 1995. "Multiple Chemical Sensitivity: What Is It?" In *Multiple Chemical Sensitivity: A Scientific Overview*, edited by Frank L. Mitchell, 35–39. Washington, DC: Department of Health and Human Services and ATSDR.

Delicate Balance. 1992. August.

Duehring, Cindy, and Cynthia Wilson. 1994. *The Human Consequences of the Chemical Problem*. White Sulphur Springs, MT: Chemical Injury Information Network.

Foucault, Michael. 1973. *Birth of the Clinic*. London: Tavistock.

Freund, Peter E. S., and Meredith B. McGuire. 1991. *Health, Illness and the Social Body*. Englewood Cliffs, NJ: Prentice-Hall.

Geertz, Clifford. 1983. *Local Knowledge: Further Essays in Interpretive Anthropology*. New York: Basic Books.

Harvey, David. 1989. *The Condition of Postmodernity: An Enquiry into the Origins of Social Change*. Oxford: Blackwell.

Hileman, Betty. 1991. "Chemical Sensitivity: Experts Agree on Research Protocol." *Chemical and Engineering News*, April 1. Washington, DC: American Chemical Society.

Jewett, Don L., George Fein, and Martin Greenberg. 1990. "A Double-Blind Study of Symptom Provocation to Determine Food Sensitivity." *New England Journal of Medicine* 323:429–33.

Kahn, Ephraim, and Gideon Letz. 1989. "Clinical Ecology: Environmental Medicine or Unsubstantiated Theory." *Annals of Internal Medicine* 111:104–6.

Kroll-Smith, Steve, and Anthony Ladd. 1993. "Environmental Illness and Biomedicine: Anomalies, Exemplars, and the Politics of the Body." *Sociological Spectrum* 13:7–33.

Lawson, Lynn. 1993. *Staying Well In a Toxic World.* Chicago: Noble Press

Lyotard, Jean-Francois. 1992. *The Postmodern Explained.* Minneapolis: University of Minnesota Press.

Martin, Emily. 1990. "The End of the Body." *American Ethnologist* 19:121–40.

Mastermind, Margaret. 1970. "The Name of a Paradigm." In *Criticism and the Growth of Knowledge,* edited by I. Lakatos and A. Musgrave, 59–89.

Mills, C. Wright. 1967. "Situated Actions and Vocabularies." In *Symbolic Interaction: A Reader,* edited by Jerome G. Manis and Bernard N. Meltzer, 335–68. Boston: Allyn and Bacon.

Mitchell, Frank L., ed. 1995. *Multiple Chemical Sensitivity: A Scientific Overview.* Washington, DC: U.S. Department of Health and Human Services and ATSDR.

Molloy, Susan. 1993. "Measures Which Will Result in Greater Access for People with Environmental Illness/Multiple Chemical Sensitivity to California's Public-Funded Facilities: Steps toward Regulatory Remedies." Master's Thesis, University of California, Berkeley.

National Academy of Sciences. 1987. *Workshop on Health Risks from Exposure to Common Indoor Household Products in Allergic or Chemically Diseased Persons.* Washington, DC: National Academy Press.

National Research Council, Board on Environmental Studies and Toxicology. 1991. *Report.* Washington, DC: National Academy Press.

Needleman, Herbert L. 1991. "Multiple Chemical Sensitivity." *Chemical and Engineering News,* June 24, 32–33.

Pullman, Cyndey, and Sharon Szymanski. 1993. *Multiple Chemical Sensitivities at Work: A Training Workbook for Working People.* New York: The Labor Institute of New York.

Rest, Kathleen M. 1995. "A Survey of AOEC Physician Practices and Attitudes Regarding Multiple Chemical Sensitivity." In *Multiple Chemical Sensitivity: A Scientific Overview,* edited by Frank L. Mitchell, 51–66. Washington, DC: U.S. Department of Health and Human Services and ATSDR.

Sagan, Carl. 1996. *The Demon-Haunted World: Science as a Candle in the Dark.* New York: Random House.

Samet, Jonathan, and Devra Lee Davis. 1995. "Introduction." In *Multiple Chemical Sensitivity: A Scientific Overview,* edited by Frank L. Mitchell, 1–3. U.S. Department of Health and Human Services. Princeton, NH: Princeton Scientific Publishing.

Selner, John. 1991. "Chemical Sensitivity." In *Current Therapy in Allergy, Immunology and Rheumatology,* edited by Lawrence M. Lichtenstein and Anthony S. Fauci. Toronto: B. C. Decker.

Simon, Gregory E. 1995. "Epidemic Multiple Chemical Sensitivity after an Outbreak of Sick-Building Syndrome." In *Multiple Chemical Sensitivity: A Scientific Overview,* edited by Frank L. Mitchell, 41–46. Washington, DC: U.S. Department of Health and Human Services and ATDSR.

Simon, Gregory E., Wayne J. Katon, and Patricia J. Sparks. 1990. "Allergic to Life: Psychological Factors in Environmental Illness." *American Journal of Psychiatry,* 147:901–6.

Stanley, J. S. 1986. "Broad Scan Analysis of Human Adipose Tissue, Executive Summary, Vol. 1." EPA Contract—560/5–86/035. Springfield, VA: National Technical Information Service.

Starr, Paul. 1982. *The Social Transformation of American Medicine*. New York: Basic Books.

Szasz, Andrew. 1994. *Ecopopulism*. Minneapolis: University of Minnesota Press.

Terr, Abba J. 1987. "Multiple Chemical Sensitivities: Immunological Critique of Clinical Theories and Practice." *Occupational Medicine: State of the Art Reviews* 2:683–94.

Touraine, Alain. 1995. *Critique of Modernity*. Oxford: Blackwell.

Turner, Bryan S. 1995. *Medical Power and Social Knowledge*. London: Sage.

U.S. Department of Labor. 1988. OSHA 3111-Hazard Communication Guidelines for Compliance, 1–13. Washington, DC: OSHA.

U.S. Environmental Protection Agency. 1980. Hazardous Waste Facility Siting: A Critical Problem (SW-865). Washington, DC: U.S. Government Printing Office.

Wright, Will. 1992. *Wild Knowledge*. Minneapolis: University of Minnesota Press.

QUESTIONS

1. What is meant by the term "practical epistemology," and how is it used in this chapter to account for people's responses for their chemical sensitivities?

2. What are some of the social, legal, and political consequences of recognizing MCS as a legitimate medical disease?

Part 3

Measurement Disputes and Health Policy

In modern societies, governments are expected to take a leading role in protecting against actual or potential dangers that threaten large numbers of citizens. The armed forces stand ready to counter aggressive acts by terrorists or other countries. The National Weather Service provides warning of dangerous weather conditions, including blizzards, tornadoes, and hurricanes. Traffic regulations establish legal right-of-ways, and allow for revoking the driving privileges of individuals who pose a risk to themselves or the occupants of other vehicles.

The health hazards posed by postnatural environments constitute the kinds of broad-based threats to public welfare that have traditionally justified government action. Activists and oppositional professionals concerned about these hazards have been pushing for increased government protection for decades. These efforts have met considerable challenge, however. One formidable roadblock to government action is the uncertainty surrounding the exact degree of danger presented by many environmental hazards. In democratic societies, we are concerned not just about protecting the public welfare, but also about protecting individual freedoms. There are times when we allow government to infringe on these freedoms, to curtail certain actions that are harmful to others. Yet, as we saw in the previous section, at present there are many in the medical and epidemiological community who are not convinced environmental hazards pose a serious threat to human health and welfare.

Part 3 considers measurement disputes, an issue which is often the basis for the social and political conflicts that typically accompany modern environments and illness problems. Moreover, it examines the issue of measurement disputes in the larger context of health policy formation. In chapter 5, "Environmental Endocrine Hypothesis and Public Policy," Krimsky introduces us to what many activists and oppositional professionals consider one of the most pervasive environmental hazards of our time: synthetic chemicals that mimic human hormones. Once in our bodies, these substances can take the place of such naturally occurring hormones as estrogen, disrupting the delicate balance of the endocrine system that directs the carefully timed release of hormones, a process important to reproductive physiology. Our knowledge of the endocrine-disrupting capabilities of the tens of thousands of chemical compounds now in commercial use is woefully inadequate. It may take years before sufficient evidence is gathered for environmental endocrine disruption to be widely accepted as a valid scientific hypothesis. Protecting public health requires immediate regulatory action, however. Government officials do not always have the luxury of waiting for years or decades for scientists to make their final determinations about the degree of risk posed by environmental hazards.

Busch, Tanaka and Gunter's contribution, "Who Cares If the Rat Dies? Rodents,

Risks, and Humans in the Science of Food Safety," focuses on the role of scientists and governments in protecting the quality of a nation's food supply. In modern societies, where tens of thousands of individuals may consume products from a single processing plant, or where tens of millions may ingest such common food ingredients as preservatives and dyes, failure to ensure that food items are free of potentially hazardous substances may have devastating public health consequences. As these authors illustrate in chapter 6, however, the science of food safety is far from an exact one. In setting regulatory standards, government agencies must determine the level of a particular substance to which humans must be exposed before they start to experience health impairments. Frequently, these determinations are made on the basis of animal laboratory studies. Using the case of erucic acid, they illustrate how initially unrecognized metabolic differences between humans and laboratory rats resulted in a probable misdiagnosis of the toxicity of erucic acid to humans.

The chapter by Busch et al. provides an indication of the time and resources required for scientific investigation into the toxicity of a single substance. At present, government agencies lack the financial and personnel resources to exert this level of effort on each of the tens of thousands of chemical compounds on the market, not to mention the indeterminate number of naturally occurring substances that might pose health hazards. As a consequence, government regulators rely heavily on research conducted by others, including university and industry scientists, as a basis for setting exposure level standards.

Chapter 7 by Ziem and Castleman, "Threshold Limit Values: Historical Perspectives and Current Practice," addresses the problem of using research conducted by industry scientists as a basis for government regulatory standards. The focus in this chapter is on occupational exposures. Since the majority of adults in modern societies are employed outside the home, failure to identify and regulate environmental hazards in the workplace could place thousands, hundreds of thousands, or even millions of individuals at risk. In 1988, the U.S. Occupational Safety and Health Administration (OSHA) adopted threshold limit values (TLVs) for 376 airborne toxins that had been developed by the nongovernment entity, the American Conference of Government Industrial Hygienists (ACGIH). Industry representatives were active on ACGIH committees, and in many instances scientists employed by companies who manufactured a particular substance played a key role in setting the TLVs for that substance. Ziem and Castleman report on studies that document negative health consequences for exposure levels lower than that allowed for current TLVs for a number of chemical compounds.

Chapter 8, Allen's "Threshold Limit Values in the 1990s and Beyond, A Follow-Up," overviews developments on the use of TLVs since 1989 (the original date of publication of the Ziem and Castleman chapter). During the 1990s OSHA's TLVs were adopted as regulatory standards by other countries, and were also used by many state agencies in the United States to regulate ambient air quality. They have become important defense strategies in lawsuits, allowing companies to argue that workers were exposed to levels of particular substances that the federal government has deemed as "safe." Despite concerns about TLVs, their use over the last decade has become more pervasive, and more entrenched.

Chapter Five

Environmental Endocrine Hypothesis and Public Policy

Sheldon Krimsky

In the summer of 1991 a group of scientists representing over a dozen disciplines met at the Wingspread Conference Center in Racine, Wisconsin, to share research findings pertaining to the effects of foreign chemicals on the reproduction and sexual development of humans and wildlife. A consensus statement reached by participants (referred to as the Wingspread Statement) asserted, in part, the following.[1]

> We are certain [that]:
>
> A large number of man-made chemicals that have been released into the environment, as well as a few natural ones, have the potential to disrupt the endocrine system of animals including humans. Among these are the persistent, bioaccumulative, organohalogen compounds that include some pesticides (fungicides, herbicides, and insecticides) and industrial chemicals, other synthetic products, and some metals.
>
> We estimate with confidence that:
>
> Unless the environmental load of synthetic hormone disruptors is abated and controlled, large scale dysfunction at the population level is possible. The scope and potential hazard to wildlife and humans are great because of the probability of repeated and/ or constant exposure to numerous synthetic chemicals that are known to be endocrine disruptors.

The Wingspread participants underscored the urgency of elevating the importance of reproductive effects in evaluating the health risks of industrial chemicals. Subsequent to the Wingspread Conference there have been sporadic accounts of the potential dangers of environmental hormone mimics and antagonists in the print and electronic media, two Congressional hearings on reproductive hazards and estrogenic pesticides,[1] and follow-up scientific meetings.[2,3] In essence, a public and environmental health advisory has been raised by a group of scientists who study different aspects of environmental endocrine disruptors and reproductive physiology in humans and animals. I shall refer to this advisory as the environmental endocrine hypothesis (EEH). For those interested in the relationship between science and policy, some of the salient issues are: How will the popular culture respond to these warnings? How will the legislative and regulatory sectors respond? What will the proponents of the Wingspread Statement have to do, scientifically or politically, to raise the

issues to a priority concern on the public agenda? There are some historical prece-
dents and a body of social science research that may illuminate these questions.

This essay will address two aspects of the environmental endocrine hypothesis that
bears on public policy. (1) What are the factors that determine whether a risk
hypothesis gains a prominent place in the public agenda? (2) What are the current
prospects within the regulatory sector for addressing the risks to the human endocrine system from a class of synthetic chemicals?

Hypothesis Formation and Public Policy

It is rare for scientists to organize themselves to advance a public health or environmental risk hypothesis. There are some notable exceptions. In 1975 biologists con-
vened an international conference in California (the Asilomar Conference) to discuss
potential hazards of recombinant DNA molecule research.[4] The meeting resulted in
guidelines issued by the National Institutes of Health for genetic engineering research.

In another case, a group of atmospheric and environmental scientists, having first
posited a relationship between chlorofluorocarbons (CFCs) and the depletion of
atmospheric ozone, organized themselves as a force for changes in public policy.[5]
One of the outcomes of this largely science-directed initiative was the signing of the
Montreal Protocol in 1987, which included agreements on tighter quotas on the
amount of CFCs the signatory nations are allowed to produce in any given year as
well as heavy taxes on the continued use of the chemicals.[6] A third example of
science-initiated risk hypotheses arose in the 1950s when physicists, biologists, and
chemists (led by Linus Pauling) alerted the public that atmospheric testing of atomic
weapons posed significant health risks to populations hundreds, even thousands, of
miles from the test site.[7] Scientists went into the communities to alert citizens of the
health risks of radioactive fallout products like carbon-14 and strontium-90, the latter
of which was found in the milk of lactating mothers, as well as the bones and teeth
of young children.[8]

In each of these cases, the scientific findings were only one of several factors that
carried the hypothesis from scientific circles to the public policy arena, even while
there was uncertainty over the risks, and in some cases no concrete evidence that
there was a risk. The respective scientists who engaged in a public dialogue played
multiple roles as generators, interpreters, and purveyors of knowledge and as advocates for policy change. There are many other instances where scientists at America's
premier institutions trailed rather than led the advance of a risk hypothesis to a
national policy forum. A recent example is found in the Alar episode. The chemical
Alar (the trade name of daminozide) is a growth regulator used in preharvest fruits
to establish uniformity in ripening and color. Alar was removed from use by farmers
when the public stopped buying fruit treated with the chemical after a vigorous
media exposé of the cancer-causing risks.[9] The issue was brought to the media and
eventually to the public through the publication of a report assessing the cancer risks
proposed by Alar to children. This report was prepared by the Natural Resources

Defense Council, a national public interest group that has fought for greater controls over the use of pesticides.[10] Typically, scientists continue the debate over the risks of a banned chemical years after regulatory action or a consumer boycott has occurred. This suggests that there are different action thresholds within science and public policy for reaching closure on risk hypotheses.

In the progress of scientific discovery it is usual for the validity of a hypothesis to remain in debate for many years before it is either rejected or incorporated into the canons of established knowledge. After 50 years of skepticism, scientists eventually adopted the hypothesis that plants can obtain nitrogen from the air.[11] A discussion of nitrogen fixation by bacteria that reside in the root nodules of plants appears routinely in basic biology texts ending years of skepticism that plants could only obtain nitrogen from soils.

It took the scientific community about 16 years before consensus was reached that CFCs were responsible for the breakdown of the protective ozone shield, with a detectable "hole" identified over Antarctica.[12] One can chart a gestation period for hypothesis formulation. In many cases, there is a preestablishment phase. A hypothesis may be advanced in the gray literature or in popular science publications but has not yet been published in mainstream journals. Once a hypothesis reaches the established science literature, a small group of adherents makes the case, advances the thesis in symposia, enters into debates in the journals, and seeks additional forms of evidentiary support. During this transitional period, the astute minds representing antagonists and protagonists create the tension and self-criticism that provide the quality control in science. The emergence of adversarial camps is central to ensuring that a hypothesis undergoes careful scrutiny before its final disposition. Criticism is indispensable to the growth and certification of knowledge.

The final phase in the life cycle of a hypothesis can take several forms. The status of the hypothesis may emerge clearly and definitively to the vast majority of the scientific community. Usually this occurs after new evidence appears in the form of a crucial experiment or critical discovery that plays a deciding role in the acceptance or rejection of the hypothesis. At other times, the fate of a hypothesis is determined slowly and incrementally. Proponents grow in numbers until a functional consensus emerges among leaders in the field. Like the Kuhnian paradigm shift, certain hypotheses finally are accepted into the canons of science when older skeptics retire or are outnumbered.

There is an important difference between the ozone hole and nitrogen fixation hypotheses. For most of its gestation period, the latter hypothesis was of interest to scientists exclusively. In contrast, from the outset, the ozone hole hypothesis had important public health and environmental policy implications. A reduction in atmospheric ozone could result in global radiation imbalances that might endanger the survival of species or at the very least cause increases in skin cancer. When a scientific hypothesis has important implications for public policy, the time needed to acquire definitive evidence of validity takes on a special significance. Ordinarily, science probes aggressively but waits patiently for nature to reveal her secrets. Partial evidence may be the stimulus for accelerated investigation. The literature is filled with causal conjectures based on limited studies and inconclusive data. Some of these conjectures are eventually falsified, some remain dormant and fail to mobilize research interest from other investigators, and others serve as catalysts for focused research activity.

When a hypothesis describes a connection between a human activity and a public health effect of some consequence, it stands to reason it will draw concern from certain stakeholders and interest groups who seek a speedy resolution. Thus, a political context emerges that adds a trans-scientific value to the disposition of the hypothesis. Although nature is not apt to reveal her secrets on an earlier timetable because the knowledge obtained may reduce human suffering or protect the environment, public concerns may influence the social process of discovery. For example, additional resources may be allocated for accelerated data gathering. Certainly, the potential catastrophic consequences of ozone depletion brought widespread public pressure on resolving its conjectural status. Political pressure is inevitably reflected in the scientific community. Select constituencies begin to mobilize in support of mitigating action, even with limited evidence at hand. It may be rational to act "as if" the hypothesis were true even when the evidence is sparse. The assumption of taking action to diminish or eliminate a human activity that is weakly conjectured to have catastrophic effects is ostensibly an insurance policy.

However, from a public policy standpoint, it would be foolhardy to act on every hypothesis that describes an adverse outcome. First, the cost could be prohibitive. Some assurance is needed that there is a nontrivial probability of a significant effect. Second, competing hypotheses may confuse the plan of action and create social chaos. For example, there was the well-received hypothesis that chemical mutagens are likely human carcinogens. It has been reported in the scientific literature that ingredients in foods such as mustard, peanut butter, herb teas, and beer are mutagenic and therefore may be carcinogenic.[13] Policymakers acting on this hypothesis, who fail to consider alternative explanations, could bring about draconian regulations that erode confidence of the public in the integrity of science. Any scientific hypothesis is embedded in a wider system of beliefs. Similarly, an action principle in public policy must also consider multiple factors, such as the nature of the consequences if the hypothesis is true and no action is taken, the cost and effectiveness of strategies to mitigate the causal agent or agents, and the nature of the consequences if action is taken and the hypothesis is false.

The term "risk selection" refers to the social processes at work that elevate a risk hypothesis to the public agenda.[14] Rarely, if ever, is that accomplished exclusively by the aggregation of scientific knowledge. In 1948 dichlorodiphenyl trichloroethane (DDT) was cited in a widely publicized book as a potential hazard to the environment.[15] Fairfield Osborn was president of the New York Zoological Society, and the book had jacket reviews from Aldous Huxley, Robert Maynard Hutchins, and Eleanor Roosevelt. The author warned prophetically: "More recently a powerful chemical known as D.D.T. seems the cure all. Some of the initial experiments with this insect killer have been withering to bird life as a result of birds eating the insects that have been impregnated with the chemical. The careless use of D.D.T. can also result in destroying fish, frogs, and toads, all of which live on insects. This new chemical is deadly to many kinds of insects—no doubt of that. But what of the ultimate and net result to the life scheme of the earth?" This was 14 years before Rachel Carson[16] completed *Silent Spring*. It took a talented science writer and the serialization of her book in the *New Yorker* to bring the issue of DDT into the light of public inquiry and another decade of debate before it was banned in the United States.

The concern over radioactive fallout from nuclear weapons testing brought little government action until strontium-90 was detected in humans. Dramatic discoveries of this type can provide the selective pressure that captures public imagination and transforms popular opinion into a powerful political force. These and other examples help us begin to understand the public response to a risk hypothesis. Sometimes it is a significant media event usually around human catastrophe (Bhopal, Love Canal, thalidomide) that sets Congress and the regulators in motion. In other cases such as ethylene dibromide (EDB), lead, dioxins, asbestos, no single event or discovery explains why federal actions are finally taken. It may be the sheer preponderance of evidence, litigation, media perseverance, and the dedicated work of a small group of unyielding advocates.

Current Regulatory Policy and Endocrine Disruptors

The general hypothesis I refer to as the environmental endocrine hypothesis posits a link between an operationally defined set of industrial and agricultural chemicals (such as xenobiotic estrogens) and endocrine-disrupting activity in vertebrates, including mammalian species. Conjectures about the relationship between synthetic chemicals and the endocrine system in fetal development were raised as early as 1979 according to Stone.[17] The environmental endocrine hypothesis in its broadest form serves a unifying role for a disparate class of reproductive and hormone-related pathologies in a variety of species. The messages communicated to the public in the environmental advocacy literature and the press are provocative and profoundly disturbing. They include: intersex features of male/female organs found in marine snails, fish, alligators, fish-eating birds, marine mammals, and bears; decline in sperm count by 50%; increased risk of breast cancer; small phalluses in Florida alligators resulting from pollution; penises found on female mammals; undeveloped testes in Florida panthers, masculinized female wildlife attempt to mate with normal females.[18]

The public policy ramifications of even a small subset of these outcomes is quite significant. Each of the subhypotheses is associated with an evolving body of evidentiary support. Thus, the general hypothesis is supported by a lattice of interconnecting but independent hypotheses of association. To date, despite the growing evidence, the environmental endocrine hypothesis cannot claim a single dramatic episode or discovery capable of turning public opinion into a sudden force for political change.

Nevertheless, there are effects cited by scientists that, in their view, represent a clear and present environmental danger. Among the most well-documented effects of endocrine disruptors is the dramatic reduction in the alligator population of Lake Apopka, Florida, the state's fourth largest body of fresh water. The lake was contaminated from years of pesticide runoff, effluent from a sewage treatment plant, and a pesticide spill. Alligator eggs collected from the lake were found to have concentrations of endocrine-disrupting chemicals, such as DDE, 10,000 times greater than what is normally found in the blood of newborn alligators.[19]

Another documented effect of endocrine antagonists is the disappearance of trout in the Great Lakes. The source of the problem has been traced to dioxin-like pollutants from industrial effluent.[18] Both examples point to inadequacies in the system of environmental regulation. For some endocrine disrupters like DDT and polychlorinated biphenyls (PCBs) that were implicated in these cases, regulatory actions have been taken. However, many other chemicals with endocrine-disrupting effects remain widely in use and merit immediate attention.

United States regulatory agencies have begun to take notice of the environmental endocrine hypothesis. The Environmental Protection Agency (EPA) and the Fish and Wildlife Service have called on the National Academy of Sciences to undertake a study of endocrine-disrupting chemicals.[20] The agency is supporting some research that seeks to understand the linkages between ecological impacts and human health impacts of xenobiotic chemicals. It is also collaborating with other agencies like the National Institutes of Health in an effort to understand the causal effects of estrogen-mimicking chemicals. However, before any regulatory initiative begins, if that should happen, it might be useful to consider the statutory authority of the relevant agencies to regulate environmental hormone modulators and the structural limitations of the existing regulatory framework.

I have already discussed the problems associated with elevating a risk hypothesis into the policy arena. Let us assume that in the not-so-distant future the conditions are such that, despite the antiregulatory mood in our country, a decision is made to regulate endocrine disruptors. To what extent are we prepared? Are the existing laws sufficient to address the issues? Will we be able to add this responsibility to the already heavily burdened regulatory system?

Several things can be said at the outset about regulatory policy toward hazardous substances. It has, for the most part, approached the management of hazardous substances chemical by chemical. Of the tens of thousands that have been introduced into agriculture and industrial production, very few chemicals have been explicitly banned from use and it may have taken decades to remove them. A few groups of chemicals are banned or their use is severely restricted. For example, PCBs are a class of chemicals that were banned for all manufacturing processes in 1979 whereas CFCs are being phased out of many products. These are the exceptions, however. Most regulated chemicals may be used within permissible limits.

Chemicals are regulated differently according to when they were introduced into commerce (some have been grandfathered in), by the type of use (food additives versus insecticides), by their putative effects (cancer-causing versus neurotoxic effects), by who is exposed (children versus workers).

There are six general ways that governments respond to the control of hazardous or potentially hazardous materials: conduct research; establish economic incentives for substitution or reduced use; enact legislation; issue regulations; undertake public education; or use moral persuasion (voluntary guidelines) to change consumption or production patterns. The most publicized federal action on an estrogenic chemical occurred when the United States EPA banned the use of DDT in 1972 primarily because of its impact on the reproduction of birds. There was only conjectured evidence about the adverse effect of DDT on humans. The prohibition of DDT was rather unusual. Most pesticides that have been banned or severely restricted like chlordane or aldrin/dieldrin exhibited carcinogenic effects on mammals. Cancer and acute toxicity have been the dominant endpoints guiding the regulation of pesticides. In 1990, a report by the General Accounting Office (GAO) cited the insufficient attention given to reproductive toxicity by federal agencies. According to the GAO, at most 7% of the synthetic chemicals in use were tested to determine whether they harm the reproduction and development of animals.[21,22]

The EPA has two primary statutes that structure its regulatory course of action for chemical substances: they are the Federal Insecticide, Fungicide and Rodenticide Act (FIFRA) first passed in 1947 and amended in 1972 and the Toxic Substances Control Act (TSCA) enacted in 1976. The older and stronger of the two acts, FIFRA, is designed to regulate the use of pesticides. The 1972 amendments to FIFRA provided for premarket screening of pesticides. Manufacturers must demonstrate that a pesticide is not harmful to human health or to the environment prior to receiving a registration. However, there are several thousand formulations of pesticides already in use that have not been tested under the criteria established for new pesticides. Under a Congressional mandate, the EPA has been asked to reexamine the pesticides currently in use for their health and environmental impacts. Assessing chemicals for their effects on the endocrine system has not been an important part of the testing protocols for pesticides. However, it is within the authority of EPA under FIFRA to broaden the safety criteria according to which pesticides are evaluated. Prior to adding new criteria to its pesticide data requirements, the EPA would ordinarily prepare a scientific background paper justifying the action, convene a scientific advisory panel, and call for public comment to draft guidelines.

When TSCA was passed, there were about 30,000 chemicals in commerce. The EPA was given the authority to ban or restrict the manufacture, processing, distribution, commercial use, or disposal of any chemical substance or mixture that presents an unreasonable risk to human health or the environment. There are three main provisions of the act. The EPA may require testing of any chemical substance or mixture if it has reason to believe the chemical may present unreasonable risk. Prior notification is required for the manufacture of any new chemical substance (premanufacturing notice [PMN]) and for any manufacture or processing of an existing chemical substance for a significant new use. Finally, EPA may require record keeping and reports on chemical use and manufacture. The TSCA requirements are different for new and existing chemicals. Companies are required to issue a PMN before they can introduce a new chemical into production. The law gives the agency 90 days to review any new chemical that is being considered for industrial use. The agency uses limited criteria for the assessment of the chemical (for example, it frequently applies structure-activity analysis, which assumes chemicals with similar structures exhibit similar properties). The agency also maintains an inventory of about 70,000 chemicals currently in use. Many of these chemicals were grandfathered into use and were not subject to rigorous assessments. In the first half decade TSCA was in effect, less than 10% of the new chemicals were subject to rigorous review.[23] The EPA receives about 1500 PMNs of new chemicals being produced each year and requires some form of testing in 10% of the cases.[24] A GAO study of TSCA reported that an EPA committee recommended a total of 386 substances for testing over a 14-year period and that the agency had complete test data for only six chemicals.[25]

Although EPA has statutory powers under TSCA to remove or limit a chemical from use, it must meet a significant burden to exercise its authority. The agency does not currently require endocrine disruption data in its review of TSCA inventory chemicals. Given the limited powers of TSCA, it will be difficult without some amendments to the law to add effective endocrine function assessment criteria to the current screening protocols. Moreover, TSCA is probably more responsive to cost-benefit considerations and proprietary constraints than either FIFRA or the food additive laws. Portney[26] notes: "The legislation [TSCA] has not been very effective, largely because there are no clear cut testing procedures or standards to determine whether a chemical does indeed present an imminent hazard, and because many companies have claimed that their products are proprietary—that they and only they have a right to know what the chemicals are."

With the prospect that an undetermined percentage of tens of thousands of chemicals in current use might be endocrine disruptors, the government is faced with a serious policy challenge. Some prioritization must be developed for screening chemicals currently in use. Under current case-by-case approaches in much of chemical regulation, it would take decades and substantial sums of money to meet a new set of regulatory goals based on the hazards of endocrine disruptors. According to Hynes,[27] "On paper both laws [TSCA and FIFRA] purport to solve the problems of preventing dangerous chemicals from being used commercially. At best, they keep only the worst new chemicals off the market." If the testing of carcinogens is any example, the definitive studies are costly and time-consuming. The preferred mammalian studies may take from 3 to 5 years to complete and cost upwards of $1 million per chemical.[23]

In a few notable cases regulatory agencies have attempted to group chemicals for more efficient regulation, departing from the substance-by-substance approach for setting health standards. PCBs, which were regulated by Congress and banned in 1976 after 46 years in production, are defined by a class of chlorinated synthetic organic compounds, namely, biphenyls (Ashford N: Personal communication, 1995). In 1973, the Occupational Safety and Health Administration (OSHA) issued emergency temporary standards and ultimately permanent standards regulating occupational exposures to a group of 14 carcinogens, representing chemicals used in the photographic and dye industry. The rulemaking and litigation associated with the rulemaking took 40 months. Subsequently, OSHA proposed a set of regulatory actions on carcinogenicity based on four generic categories. For example, proven carcinogens were to be based on two positive animal tests or one positive animal test and positive evidence from short-term assays. OSHA eventually issued an emergency temporary standard for such chemicals that met the carcinogen test. The generic approaches to regulating chemicals spurred intense industry reaction and costly litigation. Eventually, the agency's approach to regulate groups of chemicals was brought into check by a new deregulatory mood.

More recently, bills have been introduced supported by environmental groups to phase out a large class of chlorinated organic compounds. Thus, the prospect of introducing a policy on regulating endocrine-disrupting compounds as a class is not unprecedented, but will be fought vigorously by manufacturers who would prefer substance-by-substance rulemaking. One early study noted that chemical regulation moves at a snail's pace "because the cumbersome statutory procedures provides a chemical with a full panoply of due process rights accorded to any individual in our constitutional system."[28] Our system of jurisprudence, which is grounded on individual rights, has found legitimacy for class action suits. Chemicals, as classes, may too gain status in regulatory decisions, but that has been the exception rather than the rule.

Even with a dependable, inexpensive, and short-term assay to identify xenobiotic hormone mimics or antagonists, along with evidence of causality, the decision on how they should be regulated could easily be mired in legal process. It is not sufficient to show that these chemicals can disrupt reproduction; it must be demonstrated that current exposures and doses pose an unreasonable risk to humans or wildlife. This may be accomplished either by demonstrating that this class of chemicals is the cause of past reproductive anomalies based on epidemiological data or through controlled laboratory experiments on human cells or animals. However, for most of the chemicals currently in commercial use the burden is on those who wish to restrict usage to demonstrate a causal effect.

Another problem facing regulators is how to address the by-products of chemicals that test negatively for endocrine-disrupting properties. Once emitted into the environment, these chemicals may recombine, in some instances producing metabolites that are hormone mimics or antagonists. The government's burden for regulating the commercial or industrial chemical sources of dangerous metabolites has always been high. For example, nitrosamine, a potent class of carcinogens, are a by-product of nitrites and nitrates which are used in processed meats. Sodium nitrate used as a preservative can react with amines in the stomach to produce nitrosamine. Regulators have argued that the benefits of the food additive in preventing botulism outweigh the risks of cancer. It is reasonable to assume that similar cost-benefit approaches will be applied to endocrine-disrupting chemicals. Unless regulators have a means to assess the potency of these chemicals on biological systems and their cumulative and combinatorial effects, the use of current laws may face insuperable obstacles.

Regulators will also be faced with the substitution dilemmas as they have had to address each time a product is banned or restricted. DDT and other pesticides developed in the 1940s and 1950s were replaced by chemicals that were less persistent in the environment but, in some cases, more toxic. Barring the elimination or phaseout of pesticides, regulators may see their role as being forced to choose between the lesser of two evils, a suspected carcinogen or an endocrine disruptor, a devil's gamble.

Restructuring the System of Chemical Regulation

The environmental endocrine hypothesis potentially covers a broad range and significant production quantities of structurally diverse chemicals. In some respects, the regulatory challenges for xenobiotics are like those for carcinogens, which also display wide variations in chemical structure. In contrast to carcinogens, for which there has been considerable attention within regulatory policy, far less attention has been given to managing endocrine disruptors. Although the knowledge of the causal mechanisms of disease would aid regulators, in the past chemicals have been restricted or banned when there was uncertainty about how a disease was caused. A case in point is asbestos regulation. Thus, for policymakers, the causal mechanism is less significant than some demonstration of causal pathology. Once there is consensus over cause and effect, policymakers must then assess the scope of the problem.

The process of chemical risk assessment has been slow and ponderous. It is estimated that toxicity information is unavailable for about 80% of the commercial chemicals.[23] The regulatory burden would be eased if the class of xenobiotics endocrine disruptors were small or if there were ample substitutes for the serious offenders. Currently, that class numbers about 45 synthetic chemicals, mostly pesticides. According to Wiles,[29] more than 220 million pounds of endocrine disruptors are applied to 68 different crops each year with atrazine (classified as a possible or probable human carcinogen by EPA) making up 29% of the total weight and 22% of the total acreage treated. Even without adding new demands on pesticide regulation, it is widely understood that the current system is inadequate to the task. Farmers, for example, are generally accorded a right to use an effective pesticide. Dorfman[30] notes: "the restrictions that are appropriate for any pesticide depend on the availability and effectiveness of substitutes."

It has become fashionable today to write about restructuring environmental regulation. Much of this trend is based on one or more of the conclusions that regulations are inefficient, irrational or illogical, unscientific, burdensome to industry, ineffective for protecting the public, and unresponsive to cost-benefit analysis. Many of the new critics support the conservative agenda that calls for downsizing federal regulations.

The emerging scientific evidence of endocrine-disrupting chemicals poses a challenge to the current regulatory system's capacity to manage a large-scale assessment of chemicals currently in commercial use that, ostensibly, have been grandfathered with regard to reproductive effects. It is also a challenge to an already overburdened system for testing and screening new compounds, and finally to international laws and treaties on transboundary disputes.

If we choose not to become mired in the slow pace of progress in the regulation of hazardous chemicals, some changes will have to be made in the fundamental way that our society evaluates new and existing synthetic compounds.

First, we will need to reach a consensus on effective and inexpensive assays for identifying endocrine disruptors and methods for assessing the human health risk of cumulative doses. In the past, agencies used different action criteria for deciding to regulate or ban a chemical. Once assays are developed and validated, they should be applied uniformly to the 70,000 plus chemicals used in industrial production and the 600 plus active ingredients in pesticides.

Second, the standards of regulation should not depend on the route of exposure but rather on the amount of exposure. Endocrine disruptors that enter the food chain as pesticides should be regulated like food additives if their effects, such as reproductive toxicity, are comparable. Currently, there are two sets of de facto regulations, one applied to food additives and another to pesticide residues that place less controls on the latter. The Natural Resources Defense Council joined the State of California in filing suit against the EPA that argues the agency has failed to apply the 1958 Delaney amendment to dozens of pesticides that are known to cause cancer in animal studies. A recent settlement to the suit may remove 36 pesticides, but leaves open both the future of the Delaney amendment and its general application to pesticide residues.[31]

Third, we need an effective way to deal with endocrine disruptors that have become part of the waste stream through industrial effluent or pesticide runoff and that enter the food chain by absorption and concentration in marine organisms. This will require a reexamination of criteria pollutants under the Clean Water Acts where much of the emphasis has been on heavy metals, carcinogens, nitrates, and phosphates, and more recently PCBs and dioxins. Efforts to reduce the total environmental load of endocrine disruptors will also require amendments to FIFRA if it is shown that pesticides are a significant source of human exposure. Current information reveals a regulatory problem of significant scale, since it is estimated that more than 220 million pounds of pesticides known to be endocrine disruptors are applied to 68 different crops each year covering 225 million acres.[29]

Fourth, we must begin to address the problem of transnational food shipments where endocrine-disrupting chemicals may be introduced into United States markets at levels above those permitted in this country. Just a few years ago, an EPA administrator acknowledged that a minuscule number of imported bananas treated with benomyl (classified as an endocrine disruptor and a possible carcinogen) was inspected at the Mexican border. It is widely recognized that border inspectors, who are supposed to monitor pesticide residues on foreign produce, cannot meet the growing demand of imports as funds for inspectional services decline.

Fifth, TSCA must be strengthened if PMN requirements are to include mandatory screening for endocrine disruptors. The reporting and inventory requirements under TSCA could be amended to provide information on whether the chemicals currently in use are potentially endocrine disruptors in vertebrates.

Sixth, environmental regulations must address the issue of cumulative xenobiotic hormones from multiple chemical sources. The accumulated or lifetime exposures of individuals to particular chemicals have been factored into risk assessment models.[32] However, a system of regulation and risk assessment may have to take account of total xenobiotic chemical load originating from *different* agents if the additivity effect is confirmed in scientific experiments. This is a new challenge for regulators and will require an integrated look at chemicals. The closest analogy we have is with radiation standards. Annual and lifetime exposure limits have been established for workers in the nuclear industry. A common metric for all sources of ionizing radiation along with the assumption that radiation risk is cumulative makes this regulation possible.

ACKNOWLEDGMENTS

Special thanks to Dr. Nicholas Ashford for his suggestions. This research was supported in part by a grant from the National Science Foundation, Ethics and Value Studies Program, Grant #SBR-94/2973. Any opinions, findings, and conclusions or recommendations expressed in this material are those of the author and do not necessarily reflect the views of the National Science Foundation.

REFERENCES

1. Colborn, T., and C. Clement, eds. 1992. *Chemically Induced Alterations in Sexual and Functional Development: The Wildlife Human Connection.* Advances in Modern Environmental Toxicology. Vol. XXI. Princeton: Princeton Scientific Publishing Co.
2. U.S. Congress: Senate Committee on Governmental Affairs. *Governmental Regulations of Reproductive Hazards*, October 2, 1991. Washington, DC: U.S. Government Printing Office, 1992.
3. U.S. Congress: House Committee on Energy and Commerce. Subcommittee on Health and the Environment, *Health Effects of Estrogenic Pesticides.* Washington, DC: U.S. Government Printing Office, 1994.
4. S. Krimsky. *Genetic Alchemy: The Social History of the Recombinant DNA Controversy.* (MIT Press, Cambridge, MA, 1982).
5. S. Roan. *Ozone Crisis.* (John Wiley & Sons, New York, 1989).
6. A. Gore. *Earth in the Balance.* (Houghton Mifflin, Boston, 1992) pp. 318–19.
7. M. Bookchin. *Our Synthetic Environment.* (Harper, New York, 1974), pp. 175–87.
8. B. Commoner, *Science and Survival.* (Viking, New York, 1971).
9. J. D. Rosen. *Issues Sci. Tech.* 7:8 (1990).
10. National Resources Defense Council. *Intolerable Risk: Pesticides in our Children's Food.* (Natural Resources Defense Council, New York, 1989).
11. P. W. Wilson. *The Biochemistry of Symbiotic Nitrogen Fixation.* (University of Wisconsin Press, Madison, 1940).
12. J. P. Glas. Protecting the ozone layer: A Perspective from Industry, in J. H. Ausubel and H. E. Sladovich, eds. *Technology and Environment.* (Washington, DC: National Academy Press, 1989), p. 137.
13. B. N. Ames, R. Magaw, and L. Gold. *Science* 236:271 (1987).
14. S. Krimsky. The role of theory in risk studies, in S. Krimsky and D. Golding eds., *Social Theories of Risk.* (Praeger, New York, 1992), pp. 19–20.
15. F. Osborn. *Our Plundered Planet.* (Little, Brown, Boston, 1948), p. 61.
16. R. Carson. *Silent Spring.* (Houghton Mifflin, New York, 1962).
17. R. Stone. *Science* 265:308 (1994).
18. National Wildlife Federation. *Hormone Copy Cats.* (National Wildlife Federation, Ann Arbor, MI, 1994).
19. I. J. Guillette. Testimony. U.S. Congress Subcommittee on Health and the Environment.

Committee on Energy and Commerce. *Health Effects of Estrogenic Pesticides*. October 21, 1993.

20. W. K. Stevens. *New York Times* C1, C6, August 23, 1994.

21. A. Gibbons. *Science* 254:25 (1991).

22. General Accounting Office. *Toxic Substances: EPA's Chemical Testing Program Has Made Little Progress*. (U.S. General Printing Office, Washington, DC, 1990).

23. L. B. Lave and A. C. Upton, eds. *Toxic Chemicals, Health, and the Environment*. (The Johns Hopkins University Press, Baltimore, 1987), pp. 282, 283.

24. S. Lewis, Federal-statutes, in G. Cohen and J. O'Conner, eds. *Fighting Toxics*. (Island Press, Washington, DC, 1990), p. 199.

25. S. Breyer. *Breaking the Vicious Circle*. (Harvard University Press, Cambridge, MA. 1993), p. 19.

26. K. E: Portney. *Controversial Issues in Environmental Policy*. (Sage, Newbury Park, CA, 1992), p. 137.

27. H. P. Hynes. *The Recurring Silent Spring*. (Pergamon Press, Elmsford, NY, 1989), p. 102.

28. Environmental Defense Fund and R. H. Boyle. 1980. *Malignant Neglect*. (Vintage, New York, 1980), p. 131.

29. R. Wiles. Testimony. U.S. Congress. Subcommittee on Health and the Environment. Committee on Energy and Commerce, *Health Effects of Estrogenic Pesticides*, October 21, 1993.

30. R. Dorfman. The lessons of pesticide regulation, in W. A. Magat, ed. *Reform of Environmental Regulation*. (Ballinger, Cambridge, MA, 1982), p. 17.

31. J. H. Cushman. E. P. A. Settles Suit and Agrees to Move Against 36 Pesticides. *New York Times*, October 13, 1994, p. A24.

32. J. J. Cohrssen and V. T. Covello. *Risk Analysis: A Guide to Principles and Methods for Analyzing Health and Environmental Risks*. (Council on Environmental Quality, Washington DC, 1989), p. 85.

QUESTIONS

1. What is the environmental endocrine hypothesis (EEH)?

2. Describe the system currently used in the United States to regulate chemicals, paying particular attention to the ways in which the system regulates, or fails to regulate, endocrine disruptors.

Chapter Six

Who Cares If the Rat Dies?

Rodents, Risks, and Humans in the Science of Food Safety

Lawrence Busch, Keiko Tanaka, and Valerie J. Gunter

Food as a possible source of human contamination is an increasingly common and troublesome issue. Alar on apples, *E. coli* in meat, mercury in fish, and herbicides and insecticides on fruits and vegetables are some of the more well-known cases of food that becomes a public health risk. One of the methods used by scientists to determine the potential health risks of these substances is to administer them to laboratory animals. The use of animals as models for the study of human problems of nutrition, disease, and toxicity, however, has recently been the subject of considerable questioning.

Some of these concerns have focused on the conditions of generalizability of such studies to humans. For example, Gold, Manley, and Ames (1992) have argued that a positive test in one species will be repeated in another species in 50 percent of the cases. They have noted three forms of "extrapolation" required in such cases: between individuals within any one species, between species, and from high doses to low doses. Others have argued that no extrapolation appears adequate. As a recent article in *Science* puts it,

> The difficulties in inferring causation from epidemiological studies are well known to scientists (if not to laymen), and the relevance of high dose animal studies to evaluating risks of low exposures to humans is unclear at best. (Foster, Bernstein, and Huber 1993, 1509)

Yet, the central premise for the use of laboratory animals is that they *are* valid models for humans. Scientists have been willing to assume that in some (usually unspecified) way, laboratory rats (or other species) are the same as human beings. Indeed, without that premise the legitimacy of such studies would be undermined. Moreover, the ethical underpinnings for the very use of animals would be diminished if not destroyed.

Furthermore, the careers of thousands of scientists, millions of dollars spent for laboratories and equipment, and the accountability of regulatory agencies have been founded on the presumed validity of animal models. In short, the metaphor "rats are miniature people" has been institutionalized. In this chapter, we examine carefully the virtually unnoticed use of this metaphor in the transformation of rapeseed

(*Brassica napus* and *B. rapa*) into canola in Canada. Furthermore, we attempt to illustrate how that metaphor may have created a problem that was not there at all.

The removal of erucic acid (C_{221n-9}) from rapeseed through plant breeding efforts took place based on studies with laboratory animals that suggested its possible toxicity for human consumption. As a result, millions of dollars were spent, the behavior of farmers, processors, and consumers was altered, patterns of international trade were modified, and the crop itself was radically transformed. Yet the question of the toxicity of erucic acid reached closure despite the fact that not a single study ever linked human consumption of it with any harm. Furthermore, once the "toxic" compound had been removed through plant breeding, new findings that suggested its benign or even positive character became irrelevant to scientists in developed countries.

In the following section we describe the context for research on rapeseed in Canada, and the evidence marshalled for the toxicity of erucic acid. Next, we examine the challenges raised against the conventional wisdom that rats were valid stand-ins for people in toxicological laboratory experiments. We then discuss the breeding program that removed most of the erucic acid from rapeseed oil. We conclude by offering some observations on how similar situations might be avoided in the future.

Rapeseed in Canada and Allegations of Toxicity

Rapeseed is an Old World crop, first cultivated in Asia and North Africa several thousand years ago, and in Europe about 600 years ago (Dupont et al. 1989). Rapeseed oil remains the preferred cooking oil in several Asian countries. Its initial use in North America, however, was industrial. It was first grown in Canadian prairie provinces during World War II as a replacement for marine lubricants, which were becoming increasingly hard to find due to wartime disruptions in international trade (Busch et al. 1994). With the resumption of traditional supplies following the cessation of hostilities, the rapeseed market collapsed. By the early 1950s, however, explorations were under way for creating a new market niche for rapeseed. Spearheaded by the newly organized Associate Committee on Fats and Oils of the Canadian National Research Council, the possibility that rapeseed might be used as a source of edible oil was raised in 1952 (Associate Committee 1952). By the 1956–57 season, rapeseed oil was being used for edible purposes in Canada in such foods as salad oil, shortening, and margarine (Beare-Rogers 1957).

However, a potential problem with consumption of rapeseed oil loomed on the horizon, that of the possible toxicity of a naturally occurring substance in the plant, erucic acid ($C_{22:1n-9}$). Erucic acid is a fatty acid, a hydrocarbon chain with an organic acid at one end.[1] It was known by the 1950s that rapeseed contained erucic acid. However, few scientists at the time really knew much about long-chain fatty acids, including erucic acid. Most fatty acids in humans and in the human diet contain 16 to 18 carbons in the carbon chain. Because of its much longer length, erucic acid, with 22 carbons, was defined by scientists as "abnormal."

The first reports of adverse nutritional effects of rapeseed oil came in the 1940s.

The initial studies were not concerned with human food and nutrition but with substances apparently found in the meal of *Brassicas* that caused abnormal function and growth of the thyroid gland (Woodward 1954). Meal, the vegetative matter left after the oil has been pressed, is used as a foodstuff for farm animals such as cows and hogs. Changes were noted in the adrenal cortex, delay in ovary development in females, and histological changes in the pituitary gland in animals fed the meal.[2] Additional studies noted that young rats fed with High Erucic Acid Rapeseed (HEAR) had lower growth rates than those fed with other edible oils (Kennedy and Purvis 1941; Carroll 1951, 1953, 1957a; Carroll and Noble 1952, 1956; Thomasson and Bolding 1955).

The Canadian Food and Drug Directorate (FDD) also supported experimental research, largely confined to rats. In 1957, for example, two researchers from the FDD presented a report to the Associate Committee on Fats and Oils on rat studies (Beare and Campell 1957). Their study suggested that reduced weight gain and litter size were attributable to rapeseed consumption and specifically to erucic acid. Also during the 1950s, Kenneth Carroll of the University of Western Ontario traced the rise in adrenal and fecal cholesterol in rats to erucic acid in rapeseed (Carroll 1992). He also presented a brief paper on his research findings to the Associate Committee on Fats and Oils in 1957 (Carroll 1957b).

By 1960 the evidence against erucic acid was growing rapidly. A paper published that year by Roine et al. showed that "a 70% calorie HEAR oil diet caused myocarditis [an inflammation of heart muscle] and myocardial necrosis [the outright death of heart muscle] in rats" (cited in Sauer and Kramer 1983, 285). Through the 1960s further rat studies confirmed these findings (e.g., Beare-Rogers 1964).

But it was not until 1970, at the first International Conference on Rapeseed and Rapeseed Products in Quebec, that apparently conclusive results were reported. Two researchers at Unilever in The Netherlands reported that three-week-old male Wistar rats fed a diet high in HEAR oil not only had abnormal adrenal glands, but also developed myocarditis, myocardial necrosis, and fatty deposits around the heart (Abdellatif and Vles 1970), suggesting abnormal fat metabolism. Equally important, control groups fed sunflower oil and Canbra (later known as Low Erucic Acid Rapeseed, or LEAR) oil did not experience the same damage. At the same time, a French team, also using male Wistar rats, reported myocardial lesions when significant amounts of both HEAR and Canbra oil were used in their diets (Rocquelin et al. 1970). However, they noted that "[c]onsidering all the data we possess so far on this subject, it seems that lesions are more frequent and severe among animals fed rapeseed oil high in erucic acid than among animals receiving Canbra oil" (1970, 414–15).

These findings by two groups of scientists were considered conclusive evidence on the erucic acid issue. In 1973, Health and Welfare of Canada restricted the maximum content of erucic acid in processed edible fats and oil to 5 percent of the total fatty acid present (Food and Drugs Regulations 1974). Other countries followed suit. To this point studies of health impairments resulting from consumption of erucic acid were limited to animal studies, primarily on rats, and especially on one sex of one strain of rat specially bred for laboratory studies, the male Wistar rat. In the following section we discuss the reason for this limitation, and consider its implications for determining potential human health impacts from the consumption of erucic acid.

Toxicological Animal Studies

Scientific efforts to determine the possible toxicity of substances are conducted within two disciplinary specialties, toxicology and epidemiology (Whorton 1974). Toxicology uses experimental procedures where test subjects are purposefully exposed to a substance under highly controlled conditions, while epidemiology conducts health inventories on already exposed populations. In Canada, however, rapeseed was not used as an edible food product until the 1956–57 season. Therefore, in that country already exposed populations on whom epidemiological investigations might be conducted did not exist, at least initially.

Populations that had consumed rapeseed oil did exist in other countries, and indeed some epidemiological studies were under way elsewhere (described below). However, despite the fact that the Associate Committee on Fats and Oils discussed the possibility of supporting epidemiological studies on European populations at their 1956 meeting, this suggestion was never pursued by Canadian scientists. In addition to the logistics problem of studying distant populations, the toxicologists and other scientists involved with the erucic acid issue were more comfortable with highly controlled laboratory experiments than the field investigations undertaken by epidemiologists.

It was of course conceivable that scientists could have created an exposed human population by having individuals consume products containing rapeseed oil. Two problems confronted the use of humans as experimental subjects in this case, however. First, it would have been difficult, if not impossible, to create a highly controlled environment with humans as the subjects since humans could not be kept in laboratory settings for weeks or months on end to ensure that they stuck with the rigorous high-fat dietary regimen the experiments required. Second, the use of human subjects in a situation in which they might develop serious illness raised a number of ethical questions. Accordingly, scientists turned to the next-best option, the use of animal subjects.

When required to rely on animal subjects, toxicologists would ideally make use of species most similar to humans. However, in order to assess toxicity of a substance to humans using an animal model, the experiment requires large numbers of animals, often at least two generations, who are exposed to the substance by the same route of administration (dermal absorption, inhalation, ingestion) as experienced by humans. To make such tests feasible in a world where researchers face time and resource constraints, test species that have a short life-cycle and require little initial and maintenance cost per animal are used. In short, the rationale for the use of small animals is based less on theoretical, and more on time and economic, concerns.

But for this practice to be defensible, especially as a basis for formulating policy, it must be assumed that rats can indeed constitute a valid stand-in for people. Toxicological animal studies, in other words, require the metaphor "rats are people" to become more than a rhetorical trope. Indeed, toxicology must assume that, at least with respect to the question of toxins in the food chain, rats and people are interchangeable. To question the validity of this assumption is to risk the entire edifice built upon this metaphor, including academic departments, research laboratories equipped with expensive equipment, and government regulatory agencies that depend on findings from toxicological studies to legitimate and justify official exposure-level standards.

The institutional framework thus served to mask the metaphor, encouraging everyone to proceed as if rats really were miniature, oddly shaped, unusually furry humans. This occurred despite the fact there were some early indications of interspecies variation in response to consumption of rapeseed oil, and despite the fact that there are some good reasons to question the very validity of the "rats (or other laboratory animals) are people" metaphor.

With respect to the latter concern, the biological functions of human organisms are more complex than those of smaller species. Moreover, humans are exposed to many naturally or artificially occurring toxins on a highly irregular basis. In contrast, laboratory animal populations are highly inbred; all individuals within a species are nearly identical. Their response is fairly uniform. An additional problem with extrapolating from animal species to human populations is that in laboratory studies animals are typically exposed to very high doses of a substance (Harte et al. 1991); for substances that are consumed in food products, these levels are often far beyond what would be found in the diet of average persons (e.g., hundreds of cans of soft drinks per day). Humans in the real world also eat a much more varied diet, live in a more stimulating environment, and engage in more extensive and diverse types of physical activity than animals in laboratory cages.

With respect to indications of interspecies variation, Deuel et al. noted in the late 1940s (1948, 1949) that "species difference does exist between man and the rat with respect to the utilization of rapeseed oil" (1949, 377). By 1956, Carroll, whose biochemical analyses with rats had been a key factor in raising concerns about erucic acid, was also aware that there appeared to be species differences in response to erucic acid in the diet. At the time he was conducting work that would show that rapeseed oil increased the adrenal cholesterol of rats and mice but not that of rabbits, guinea pigs, chickens, or dogs (Carroll 1957a).

Explaining interspecies variation requires an understanding of the physiological mechanisms and processes that are activated when a particular substance is introduced into an organism. At the time of the early studies on erucic acid, however, scientists were operating without a clear model of exactly how erucic acid damaged the hearts of experimental subjects. Indeed, as previously stated, scientists at this time knew little about any of the long-chain fatty acids. Obviously, knowledge of the exact means by which a substance inflicts damage on a living organism is the preferable situation. Toxicological studies, however, can proceed in the absence of such knowledge. It has only been in recent years that scientists have offered an explanation for the physiological and biochemical processes activated by the consumption of erucic acid.

Numerous studies, including those of Joyce Beare-Rogers (Beare-Rogers et al. 1971; Beare-Rogers and Nera 1972) confirmed that rats fed diets high in rapeseed suffered from myocardial lipidosis. These studies identified erucic acid as the central causative agent. However, at that time few chemists and nutritionists were familiar with the function of peroxisomes in decomposing fatty acids. Peroxisomes are subcellular structures containing enzymes which metabolize certain types of molecules, including fatty acids. They get their name from hydrogen peroxide, which serves as a free-radical donor for some of the enzymatic reactions. Before peroxisomes were described and understood, it was thought that all fatty acid breakdown occurred through oxidation in mitochondria, subcellular structures which play a primary role in cellular metabolism.

Furthermore, it was later found that several long-chain monounsaturated fatty acids were implicated in myocardial lipidosis. While short-chain fatty acids go directly to mitochondria where they are oxidized, long-chain fatty acids must first have their hydrocarbon chain shortened in the peroxisomes before they can be oxidized in mitochondria. The differences between species of laboratory animals in the extent of lipidosis is apparently due to the difference in the activity of the peroxisomes. Rats, which were the animal of choice in studies of erucic acid, have inactive peroxisomes, and thus a low level of oxidation of erucic acid. On the other hand, dogs and monkeys with active peroxisomes have a faster rate of oxidation of long-chain fatty acids, and thus higher levels of digestibility (Sauer and Kramer 1983).

Even before these theoretical advances occurred, epidemiological studies from other countries also suggested problems with the "rats are people" metaphor transformed into fact in the toxicological investigations. For millennia, the Indians and Chinese, and later the Japanese, had consumed rapeseed oil on a regular basis. Moreover, during World War II all cooking oil in Germany, Poland, and Switzerland was obtained from rapeseed. Indeed, in Sweden the law required (presumably for economic reasons) that margarine contain *at least* 35 percent homegrown rapeseed oil. Hence, at the time rapeseed was becoming an edible oil in Canada there existed a European population with similar food consumption patterns that had been exposed to the oil for an extended period of time.

At the 1957 meeting of the Associate Committee on Fats and Oils, a report was made on epidemiological studies of rapeseed oil consumption that had been conducted in some of these countries (Hulse 1957). These studies found no evidence of adverse effects on humans from the consumption of HEAR oil and some indications of positive effects (reduction of excessive calorie intake and increased excretion of cholesterol). Still further revelations were presented at the 1958 meeting of the Associate Committee on Fats and Oils. The results of a Finnish study were conveyed as follows:

> When rapeseed oil is being used as human food it has been proved to be absorbed from the alimentary tract at the same rate as olive oil and butter fat. It also leaves the blood stream at the same time as both fats mentioned above. Rapeseed oil seems to cause a similar drop in the cholesterol content of blood as do all the other vegetable fats containing unsaturated fatty acids (Associate Committee 1958, G1).

The report went on to note (1) that for as yet unexplained reasons rats were particularly susceptible to deleterious effects whereas swine and humans were not, and (2) harmful effects on people were only likely if consumption were to exceed 25 percent of total calories, a virtually impossible figure.

The Breeding Program

The institutionalization of the "rats are people" metaphor was not without political and economic ramifications. In 1973, Department of National Health and Welfare Canada restricted the maximum content of erucic acid in processed edible fats and oils to 5 percent of the total fatty acids present. In 1985, the United States ruled that, for granting of Generally Recognized as Safe status (GRAS), the maximum level of erucic acid in canola oil (the modern version of rapeseed oil) was to be 2 percent. In part as a result of U.S. pressure, Canada reduced the maximum erucic acid content in canola oil to 2 percent in 1986. Other countries have followed suit.

Long before these actions occurred, however, it was apparent to agricultural scientists that erucic acid was a suspected compound, and the future success of rapeseed as an edible oil likely depended on its removal. Indeed, rapeseed oil had already been briefly banned in Canada over a decade and a half prior to the 1973 action. On July 23, 1956, the Canadian Food and Drug Directorate of the Department of National Health and Welfare disapproved the use of rapeseed oil for human consumption in Canada and ordered immediate discontinuance of sales of the oil for edible purposes (Reynolds 1975). This ruling was based on the various studies that linked it to abnormalities arising in animal studies.

Leaders of the rapeseed industry were outraged. They noted that no studies had linked rapeseed oil consumption with *human* health problems. As a result, within a few days the FDD withdrew its objection to the use of rapeseed as an edible oil. While successful at this juncture, the rapeseed industry could see the writing on the wall. The unfavorable results from the experiments with laboratory animals and the widespread suspicion these generated were considered sufficient reasons for attempting to remove erucic acid from rapeseed through plant breeding.

Both Agriculture Canada and the Canadian National Research Council began in the late 1950s to support the breeding program, a full two decades before the erucic acid "problem" had become conclusive. Soon, breeders were hard at work developing a low erucic acid rapeseed (LEAR) oil, which eventually was renamed "canola." As early as 1960 individual plants had been identified with 0 to 1 percent erucic acid, although they were not agronomically viable. By 1963 the low acid plants were being backcrossed into the high yielding varieties. Moreover, as the new LEAR plants became available within the scientific community, they were shared with the FDD. As such, in 1964 Joyce Beare-Rogers (1964) was able to report that rats fed with LEAR oil registered weight-gains similar to those fed olive oil. The clear implication was that erucic acid was the sole compound responsible for the poor performance of rapeseed oil in earlier rat studies.

By 1965 plant breeder Keith Downey was able to report to the Committee that as a trial approximately 430,000 pounds of LEAR seed was to be produced (Downey 1965). This would be sufficient to allow processors to determine whether the new oil would be superior to the HEAR versions. It would also permit the industry to determine if it would pay a premium price for the new oil.

In 1968 the first LEAR cultivar, Oro (*B. napus*) was released by the Agriculture Canada Research Station at Saskatoon. It was licensed and 10,000 acres of certified seed were grown. By 1970 the Canadian government had made the decision to convert all rapeseed production to the new low erucic cultivars. By planting 1000 acres of seed in California during the winter of 1970, it was possible to produce enough seed to convert the entire Canadian crop to LEAR varieties by 1972. Moreover, by prohibiting the registration of new HEAR varieties, the new LEAR varieties were soon the only ones available for sale to farmers in Canada. In 1973, when the new law limiting erucic acid in rapeseed to 5 percent was enforced, nearly all of the Canadian crop was already transformed.

As a result, further study of the effects of erucic acid soon became a purely academic matter in Western nations. In addition, the Canadians were able to capture a very significant share of world trade in LEAR seed and later canola. Canola oil has become a healthy choice for edible oils as it is characterized by a relatively high concentration of oleic acid, a moderate concentration of polyunsaturated fatty acids, and a low concentration of saturated fatty acids.

However, it should be understood that the switch from HEAR to LEAR oils in Canada did not please everyone. The initial cultivars were not as high yielding and they were more prone to lodging, that is, they would fall over in the field. Two early LEAR cultivars had lower oil content than earlier ones. Furthermore, the margarine from LEAR oil tended to be grainy. Finally, for reasons that are unclear, LEAR oil requires preheating in processing (Unger 1990).

Conclusions

In the case of determining the toxicity of erucic acid in rapeseed, we can observe how the metaphor "rats are people" was used by scientists at three stages of the research process: (1) defining the problem, (2) testing the problem, and (3) challenging the findings of early investigative studies.

First, the problem was defined in such a way as to make animal studies essential to its resolution. The metaphor "rats are people" was *institutionalized*. Because of their training and commitments, scientists working on the erucic acid question were uncomfortable with the vagueness and statistical (as opposed to experimental) procedures associated with epidemiology. This was compounded by the fact that human populations engaged in regular consumption of rapeseed oil lived far away, in places like India and China. Thus they opted for laboratory studies. Since laboratory studies of humans were ruled out for ethical reasons, they naively embraced studies of rats— studies that were precise for the rats but that required metaphorical extrapolations of a sort far more tenuous than those required in epidemiological studies. It was not merely that scientists preferred to work with rats, but that they built an entire infrastructure around rats. Laboratories, cages, measuring devices, and other para-phernalia were designed for use by rats. Even occasional use of larger animals like pigs or monkeys was expensive and cumbersome.

Second, unfavorable results from studies with laboratory animals (largely rats) led the FDD to conclude that HEAR oil would have caused similar results if the subjects were humans. Scientists rationalized the elimination of erucic acid from rapeseed as follows:

> These studies disclosed no major differences in the nutritive value of the common vegetable oils. . . . High erucic acid rapeseed oil constitutes an exception to the concept. . . . When fed in *large amounts*, this oil was shown to retard growth and to induce changes in various organs. . . . Although high erucic acid rapeseed oil *has never been shown* to be a hazard to *human health*, it was prudent to effect a changeover to new, low erucic acid varieties. (Vles and Gottenbos 1989, 71, emphasis added)

Despite this, it is obvious that humans do eat more than rapeseed oil, and do not consume rapeseed oil in large quantities at any time. Moreover, humans engage in far more diverse activities than rats contained in laboratory cages. If humans had been fed any vegetable oil in the same proportion and in the same physical conditions as were the rats, they would have undoubtedly become ill. Furthermore, these "min-iature humans" helped real humans justify the expenditure of millions of dollars and untold hours of labor time to remove erucic acid from rapeseed. As a result of toxicological laboratory studies on rats, industries were reorganized, fortunes were made and lost, and consumption patterns of millions of people were altered.

Finally, recent findings about erucic acid by lipid chemists and epidemiologists have been used to challenge the metaphor "rats are miniature people" that guided early laboratory studies. However, this effort to demystify erucic acid has been relatively unsuccessful, partly due to the lack of communication among the scientists from different disciplines and research groups.

Recent findings with fish oils suggest that long-chain fatty acids like erucic acid may actually serve a positive role in human nutrition (Fritsche, Huang, and Cassity 1993; Turley and Strain 1993; Bailey, De Lucca, and Moreau 1989). Erucic acid may be an effective treatment for certain forms of Adrenoleukodystrophy (ALD), a rare genetic disorder which causes dysfunction of the brain due to abnormal development of nerve cell membranes. Scientists were originally skeptical about using it, given the results of the rat studies cited above. An insistent parent of a child afflicted with ALD, however, was able to obtain some in purified form and demonstrate its effect-iveness (this was the subject of the movie *Lorenzo's Oil*; see also Moser 1993).

In sum, there are at least four possibilities that must be considered when using animal models. These include compounds which are:

1. toxic to both animals and humans;
2. toxic to animals but not humans;
3. not toxic to animals but toxic to humans; and
4. not toxic to animals or humans.

In order to determine which of these four cases holds using only animal studies, one has to assume precisely that which one seeks to find out! Only if prior knowledge about similarities between humans and animals for a particular class of compounds is available can a proper determination be made. Moreover, even then, there may be exceptions. Hence, we are forced to conclude that epidemiological studies and human experimentation (where ethically appropriate) are likely to be superior to animal studies in most cases.

NOTES

The research reported in this paper was supported by a grant from the (U.S.) National Science Foundation (#SBE-9212928) and by the Michigan Agricultural Experiment Station. We would like to thank Will Aiken, Katherine Clancy, Richard Levins, Richard Lewontin, and Steve Kroll-Smith for their comments on earlier drafts. We would also like to thank Joel Gunter, MD, and Teresa Gunter, RN, for their assistance with the medical terminology. The views reported herein are those of the authors.

1. Three fatty acid molecules combined with a glycerol molecule form a triglyceride, or fat molecule. Fatty acids come in various lengths and may be saturated (containing only single carbon-carbon bonds), monosaturated (containing a single carbon-carbon double bond), or polyunsaturated (containing multiple carbon-carbon double bonds). The length of the hydrocarbon chain and the degree of saturation determine the physical and biological properties of fats made with the fatty acid, including the fluidity of membranes made with the fat (which may affect cellular function) and the pathway(s) used by an organism to metabolize the fat.

2. Histological changes are changes in the appearance of cells or tissues, usually microscopic.

REFERENCES

Abdellatif, A. M. M., and R. O. Vles. 1970. "Physiopathological Effects of Rapeseed Oil and Canbra Oil in Rats." Pp. 423–33, in Proceedings of the International Conference on the Science, Technology and Marketing of Rapeseed and Rapeseed Products (St. Adele, Quebec, Canada, September 20–23).

Associate Committee on Fats and Oils. 1952. Proceedings, First Annual General Meeting. Canadian Committee on Fats and Oils Ottawa: National Research Council of Canada.

———. 1958. Proceedings, Eighth Meeting. Canadian Committee on Fats and Oils. Ottawa: National Research Council of Canada.

Bailey, A. V., A. J. De Lucca, II, and J. P. Moreau. 1989. "Antimicrobial Properties of Some Erucic Acid-Glycolic Acid Derivatives." *Journal of the American Oil Chemists' Society.* 66: 932–34.

Beare, J. L., and J. A. Campbell. 1957. "Summary of Rapeseed Oil Studies Carried Out in Food and Drug Laboratories." Pp. T1-T2, in Proceedings, Seventh Meeting, Canadian Committee on Fats and Oils. Ottawa: National Research Council of Canada, Associate Committee on Fats and Oils.

Beare-Rogers, J. L. 1957. "Rapeseed Oil as a Food." *Food Manufacture* 32: 378–84.

———. 1964. "Report from Nutrition Section, Food and Drug Directorate." Pp. J1-J2, in Proceedings, Fourteenth Meeting, Canadian Committee on Fats and Oils. Ottawa: National Research Council of Canada, Associate Committee on Fats and Oils.

Beare-Rogers, J. L., and E. A. Nera. 1972. "Cardiac Fatty Acids and Histopathology of Rats, Pigs, Monkeys and Gerbils Fed Rapeseed Oil." *Comparative Biochemistry and Physiology* 41: 793–800.

Beare-Rogers, J. L., E. A. Nera, and H. A. Heggtveit. 1971. "Cardiac Lipid Changes in Rats Fed Oils Containing Long-Chain Fatty Acids." *Institute of Canadian Food Technology Journal* 4: 120–24.

Busch, L., V. Gunter, T. Mentele; M. Tachikawa, and K. Tanaka. 1994. "Socializing Nature: Technosocience and the Transformation of Rapeseed into Canola." *Crop Science* 34: 607–14.

Carroll, K. K. 1951. "The Growth-Retarding Effect of Erucic Acid and Its Relation to Requirements for Fat-Soluble Vitamins." *Canadian Journal of Biochemistry and Physiology* 37: 731–40.

———. 1953. "Erucic Acid as the Factor in Rapeseed Oil Affecting Adrenal Cholesterol in the Rat." *Journal of Biological Chemistry* 200: 287–92.

———. 1957a. "Rape Oil and Cholesterol Metabolism in Different Species with Reference to Experimental Atherosceloris." *Proceedings of the Society for Experimental Biology and Medicine* 94: 202–5.

———. 1957b. "Feeding Trials with Erucic Acid." Pp. U1-U2, in Proceedings, Seventh Meeting, Canadian Committee on Fats and Oils. Ottawa: National Research Council of Canada, Associate Committee on Fats and Oils.

———.1992. Personal interview. University of Western Ontario.

Carroll, K. K., and R. L. Noble. 1952. "Effects of Feeding Rape Oil on Some Endocrine Functions of the Rat." *Endocrinology* 51: 476–86.

———. 1956. "Erucic Acid and Cholesterol Excretion in the Rat." *Canadian Journal of Biochemistry and Physiology* 34: 981–91.

Deuel, H. J., S. M. Greenberg, E. E. Straub et al. 1948. "The Comparative Nutritive Values of Fats. X: The Reputed Growth-Promoting Activity of Vaccenic Acid." *Journal of Nutrition* 35: 301–14.

Deuel, H. J., R. M. Johnson, C. E. Calbert, J. Garner, and B. Thomas. 1949. "Studies on the Comparative Nutritive Value of Fats. XII. The Digestibility of Rapeseed and Cottonseed Oil in Human Subjects." *Journal of Nutrition* 38: 369–79.

Downey, R. K. 1965. "Rapeseed Research at the Canada Agriculture Research Station, Saskatoon." Pp. C1-C3, in Proceedings, Tenth Meeting, Canadian Committee on Fats and Oils. Ottawa: National Research Council of Canada, Associate Committee on Fats and Oils.

Dupont, Jacqueline, Pamela J. White, Kathleen M. Johnston et al. 1989. "Food Safety and Health Effect of Canola Oil." *Journal of the American College of Nutrition* 8: 360–75.

Food and Drugs Regulations, Consolidated Regulations of Canada. 1974. C. 870, B.07.035 and B.09.022.

Foster, Kenneth R., David E. Bernstein, and Peter W. Huber. 1993. "Science and the Toxic Tort." *Science* 261: 1509, 1614.

Fritsche, Kevin L., Shu-Cai Huang, and Nancy A. Cassity. 1993. "Enrichment of Omega-3 Fatty Acids in Suckling Pigs by Maternal Dietary Fish Oil Supplementation." *Journal of Animal Science* 71: 1841–47.

Gold, L. S., N. B. Manley, and B. N. Ames. 1992. "Extrapolation of Carcinogenicity between Species: Qualitative and Quantitative Factors." *Risk Analysis* 12(4): 579–88.

Harte, John, Cheryl Holdren, Richard Schneider, and Christine Shirley. 1991. *Toxics A to Z: A Guide to Everday Pollution Hazards.* Berkeley: University of California Press.

Hulse, J. D. 1957. "Rapeseed Oil." Pp. V1-V3, in Proceedings, Seventh Meeting, Canadian

Committee on Fats and Oils. Ottawa: National Research Council of Canada, Associate Committee on Fats and Oils.

Kennedy, T. H., and H. D. Purvis. 1941. "Studies on Experimental Goitre. I: The Effects of Brassica Seed Diets on Rats." *British Journal of Experimental Pathology* 22: 241–54.

Moser, Hugo. 1993, February 27. "Lorenzo's Oil." *The Lancet* 341:344.

Reynolds, John Robert. 1975. *The Commercial Development of a Low Erucic Acid Rapeseed Variety*. Saskatoon: unpublished M. S. Thesis, University of Saskatchewan, 3.

Rocquelin, G., B. Martin, and R. Cluzan. 1970. "Comparative Physiological Effects of Rapeseed and Canbra Oils in the Rat: Influence of the Ratio of Saturated to Monounsaturated Fatty Acids." Pp. 414–15, in Proceedings of the International Conference on the Science, Technology and Marketing of Rapeseed and Rapeseed Products (St. Adele, Quebec, Canada: September 20–23).

Sauer, F. D., and J. K. G. Kramer. 1983. "The Problems Associated with the Feeding of High Erucic Acid Rapeseed Oils and Some Fish Oils to Experimental Animals." Pp. 253–92, in *High and Low Erucic Acid Rapeseed Oils*, ed. J. K. G. Kramer, F. D. Sauer, and W. J. Pigden. Toronto: Academic Press Canada.

Thomasson, H. J., and J. Bolding. 1955. "The biological value of oils and fats. II: The growth-retarding substance in rapeseed oil." *Journal of Nutrition* 56: 469–75.

Turley, E., and J. J. Strain. 1993. "Fish Oils, Eicosanoid Biosynthesis and Cardiovascular Disease: An Overview." *International Journal of Food, Science, and Nutrition* 44: 145–52.

Unger, Ernie H. 1990. "Commercial Processing of Canola and Rapeseed: Crushing and Oil Extraction." Pp. 235–49, in *Canola and Rapeseed: Production, Chemistry, Nutrition and Processing Technology*, ed. Fereidoon Shahidi. New York: Van Nostrand Reinhold.

Vles, R. O., and J. J. Gottenbos. 1989. "Nutritional Characteristics and Food Uses of Vegetable Oils." Pp. 63–86, in *Oil Crops of the World: Their Breeding and Utilization*, ed. Paulden Knowles, Gerhard Robblen, Keith Downey, and Ashram Ashri. New York: McGraw Hill.

Whorton, James. 1974. *Before Silent Spring: Pesticides and Public Health in Pre-DDT America*. Princeton, NJ: Princeton University Press.

Woodward, J. C. 1954. "Progress Report on Studies Relating to the Nutritive Effects of Rapeseed Oil Meal." P.F.-1, in Proceedings, Fourth Meeting, Canadian Committee on Fats and Oils. Ottawa: National Research Council of Canada, Associate Committee on Fats and Oils.

QUESTIONS

1. What eventual revelations suggested that rats were not valid "stand-ins" for people, at least as far as determining the toxicity of erucic acid is concerned?

2. In the previous chapter, Krimsky argued that evidence of endocrine disruption in animal populations provides sufficient justification for government regulation of possible xenobiotic hormone mimics or antagonists. The authors of the present chapter argued that, in the case of erucic acid, reliance on animal studies resulted in probably needless, and certainly costly, government regulation. In your estimation, how should government regulators decide at what point they have sufficient indications of potential harms to justify controlling or eliminating human exposure to a particular substance?

Chapter Seven

Threshold Limit Values
Historical Perspectives and Current Practice

Grace E. Ziem and Barry I. Castleman

Before reviewing the subject of occupational exposure limits, certain basic issues bear mention. First, the medical profession has a fundamental importance in the investigation and evaluation of the harmful effects of industrial materials. Second, the necessary medical resources need to be provided by industry to make possible the medical surveillance and care of workers exposed. Third, there is a need to develop and train the professional resources to meet the needs of the millions of places of employment. The adoption of standards with specified numerical exposure limits accomplishes nothing unless the necessary professional resources are provided to gather and evaluate information on exposures and health effects.

The Occupational Safety and Health Administration (OSHA) Standard for Air Contaminants

In the closing days of the Reagan Administration, OSHA adopted new permissible exposure limits for 376 substances.[1] Virtually all of these limits came from the 1987 list of Threshold Limit Values (TLVs) published by the American Conference of Governmental Industrial Hygienists. Industry was required to be in compliance by September, 1989.

In developing this standard, OSHA disregarded recommendations by the National Institute for Occupational Safety and Health (NIOSH) for stricter limits for 68 specific substances. The idea of adopting the TLVs had been suggested in 1983 by the Synthetic Organic Chemical Manufacturers' Association, and the chemical industry's response to this OSHA rulemaking was unusually favorable (C. B. Mackerron; *Chemical Week*, January 25, 1989; and comment from the Dow Chemical Company on OSHA's proposed air contaminants rule [July 1988]). The AFL-CIO and at least 16 industrial parties have gone to court over the standard.

OSHA has recently announced that it is considering additionally requiring medical monitoring and air monitoring in industries where regulated substances are used.

Physicians in industry have good cause, therefore, to wonder how protective the new limits are. In fact, the scientific quality of the process for developing the TLVs has been critically examined, and evidence of "corporate influence" in developing the TLVs has figured in the debate over the new OSHA rule. A paper on these issues, published in May 1988, has engendered a lively discussion, including more than 20 commentaries by the end of 1988 in the *American Journal of Industrial Medicine*.[2]

The purpose of this paper is to summarize the critique of the TLVs, referring to data already published as well as presenting information gathered since OSHA proposed its new rule in mid-1988. The material here will be addressed primarily to physicians in industry.

The story begins by recounting how the medical responsibility of relating working conditions and health was in large part assumed by a group of industrial hygienists, the American Conference of Government Industrial Hygienists (ACGIH).

Origins of the TLVs

ACGIH traces its history as a professional organization back to 1938. An ACGIH committee compiled a listing of state government exposure limits for various chemicals in 1942.[3] In 1946, ACGIH published its first annual list of recommended "Maximum Allowable Concentrations" (MACs) for 144 substances.[4] The primary sources for this were ACGIH's 1942 compilation and a 1945 paper by industrial hygienist Warren Cook.

It is interesting to recall what was said of the safety of these limits at the time, in view of later developments. ACGIH said in 1942, "The table is not to be construed as recommended safe concentrations." As if to underline that point, the text went on to say, "The material is presented without comment."[3] Cook,[5] whose paper supplied 118 of the exposure limits adopted, emphasized that his intent was "to provide a handy yardstick to be used as guidance for the routine control of these health hazards—not that compliance with the figures listed would guarantee protection against ill health." Cook went on to advise that maintenance of the limits he suggested should not be considered a substitute for medical monitoring.[5]

By 1946, most state governments had industrial health units, and so did some cities and counties. The MAC values reported by 27 of these agencies were quite variable for some chemicals. For *n*-butanol, the limits varied from 25 to 300 ppm in air, depending on where the workplace was located. For turpentine, the range was 100 to 700 ppm; for methanol, from 100 to 300 ppm; for nitrobenzene, 1 to 5 ppm. On the other hand, there was substantial agreement among the health agencies for many other chemicals.[6]

Up to this time, MACs in use had "on the whole [caused] no serious handicap to industry," according to J. J. Bloomfield, one of the leading industrial hygienists of the US Public Health Service.[6] However, there was a desire among the government people to harmonize their MACs and thus avoid the health and economic impacts of having divergent conditions for industry.

ACGIH acknowledged in 1946 that no uniform definition of MACs existed, citing three concepts then in use: no safety margin against known health effects, some safety margin against health effects, and protection from objectionable but not harmful concentrations. ACGIH initially declined to define what its MACs were or to state whether they were limits not to be exceeded for 30 minutes, 1 hour, 8 hours, or longer.[4] For example, the MAC for chlorine was 5 ppm, which compared unfavorably with 4 ppm recommended by Henderson and Haggard[7] for 30-minute to 1-hour exposure periods; the 100 ppm MAC for carbon monoxide had been recommended by Henderson and Haggard for "several hours."

The 1947 list included 155 MACs. The chairman of the Committee on Threshold Limits, chemist L. T. Fairhall, expressed great confidence that the industrial hygienist was well placed to set health standards: "He is in contact with the individuals exposed and therefore soon learns whether the concentrations measured are causing any injury or complaint."[8] Up to this point, the five-man TLV committee still did not include one physician member.

In 1948, the MACs were renamed Threshold Limit Values. Despite the very different emphasis of this new nomenclature, the term TLV was not defined at the time of its introduction. The TLV committee noted that, "People vary greatly in their response to drugs and toxic substances." To this irremovable obstacle was added the acknowledged difficulty of trying to protect the worker while not imposing an "impossible burden on the manufacturer."[9]

In 1953, a preface was added, wherein the TLVs were described as "maximum average concentrations of contaminants to which workers may be exposed for an 8-hour working day (day after day) without injury to health."[10] Both the term used and its definition now promoted the TLVs as health-hazard thresholds for exposure to chemical and mineral substances, many of which were known to have serious, irreversible effects. The TLV committee now sought to offer a guarantee where Cook had explicitly said no guarantee was warranted. Most of the exposure limits on the list were the same values recommended in 1945 by Cook.[5] Despite the accompanying preface assertion that the TLVs were based on the best available information, there is no evidence that any review was done or new rationale offered to justify this sweeping disregard for the uncertainties underlying the TLVs.

The TLV committee chairman, industrial hygienist Alan Coleman, used more qualified language at the ACGIH meeting of 1954. He described the TLV as the concentration of a substance that "should cause no significant injury to the health of the large majority of persons" exposed daily.[11] The committee itself tempered its description in 1958: "[TLVs] represent conditions under which it is believed that nearly all workers may be repeatedly exposed, day after day, without adverse effect."[12]

Precisely because the TLV committee had taken a very difficult technical, political, and economic problem off the shoulders of state and local agencies, the TLVs were uncritically welcomed as uniform limits across the country. State and local agencies reduced their proportional employment of medical personnel and all but stopped issuing MACs on their own in the early 1950s.[13] As it became clear that the state and local officials on the TLV committee were not issuing terribly burdensome limits for their local plants to meet, industry adjusted to this state of affairs without protest.

Meanwhile, the commercialization of new chemicals by industry far outstripped the capabilities of a volunteer committee to keep up. As new chemicals were widely introduced by the hundreds, the TLV committee struggled to add to its list less than 10 per year. Revisions of the TLVs, once listed, were fewer still, and it is evident that, after the first few years, the primary focus of the TLV committee was on expanding the list. Until 1962, this responsibility was handled by a committee of only four to eight people.

Many pitfalls of reliance on the TLVs had been anticipated from the time they were launched. W. P. Yant, first President of the American Industrial Hygiene Association, told members of the Industrial Hygiene Foundation that monitoring average concentrations of air, contaminants would not take account of several factors: peak exposures that could be very harmful, synergistic effects of multiple exposures, and the great increases in respiration rates arising from high levels of physical activity and work in hot environments.[14]

Yant also observed that lists of limits gain "prestige and authority through mere copying and repetition." He warned that mandatory requirements, which the TLVs were clearly destined to become, were usually minimum requirements, "representative of the worst permissible conditions." Such requirements, he said, could "stifle progress and freeze endeavor at the established minimum."[14] Yant's apprehensions proved well founded, as the TLV committee fell further behind in its efforts to keep pace with innovation. As the list of TLVs grew longer, more of the limits would tend to be based on reviews and information not updated for years.

British authorities criticized American practice for its heavy emphasis on measurement and reliance on reference limits. Noting that TLVs were almost always amended in the downward direction, "reducing the concentration formerly accepted as safe," United Kingdom factory inspector Bryan Harvey[15] preferred to call them "theoretically allowable maximum concentrations." Medical Inspector of Factories A. I. G. McLaughlin[16] derided the very idea of threshold limits as reflecting an assumption that "man is a standardised machine." The indoctrination of American industry and professionals with a preoccupation with taking samples and designing controls to meet reference exposure limits was in turn seen as the basis of another serious shortcoming: a peculiarly American tendency to consider substitution of dangerous materials as the last line of approach to health hazard control instead of the first.[15]

Industrial physicians in the United States were also dismayed at the growing acceptance of the TLVs, issued by a committee dominated by industrial hygiene engineers, chemists, and toxicologists. Initially, not a single physician was on the TLV committee; at most, physicians comprised only a small minority of the committee members. Never had the chairman of the TLV committee been a physician (this would not happen until 1985). At a 1952 meeting of leaders of the Industrial Medical Association, clinician Frank Princi said: "[M]ost of the [TLVs] are picked out of a hat, 95 percent are on the basis of animal experiments only, incorporated into state codes, and we are faced with ridiculous standards. Is there a doctor among the group that puts out these standards?"[17]

After all, what industrial hygienist sees the workers' health status the way the plant doctor does? What toxicologist is intimate enough with his rats to learn whether they feel pain or are suffering from reduced mental acuity? What did these government engineers, chemists, and toxicologists read or know of the medical literature, even just what is imparted in *JAMA* or the *Lancet*? It would have been malpractice if a council of doctors had prescribed such a list of exposures as approved for consumption by all the workers in the country.

ACGIH nonetheless went on to make the essentially medical evaluations on which new TLVs were based. The industrial physicians' group did not undertake the task of either publicly criticizing the TLVs or proposing its own workplace exposure limits. Only occasionally did individual industrial physicians pass on information to the TLV committee through the 1950s and 1960s.

Led by toxicologist Herbert Stokinger of the Public Health Service, the TLV committee expanded its membership and output in the early 1960s. ACGIH also published for the first time a volume entitled *Documentation of Threshold Limit Values*,[18] where the basis for about 250 TLVs was stated, with references, in the space of 112 pages.

Stokinger approached the Manufacturing Chemists' Association (now Chemical Manufacturers Association) for increased input from member firms starting in 1964. This met with limited response. The companies had no statutory duty to disclose new knowledge about chemicals used in general industry before the passage of the Toxic Substances Control Act in 1976. In the years before the Occupational Safety and Health Act (1970), regulation of workplace health hazards by the states was minimal, and manufacturers were about the only parties capable of knowing what the exposures were in their plants and whether there were adverse medical consequences. ACGIH's annually republished claim that the TLVs were based on the "best available" information thus sidestepped the reality that the TLV committee was left begging for data. A committee of the Industrial Medical Association acknowledged that unpublished data was in the possession of companies that could contribute to the establishment of "realistic TLVs."[19]

Corporate Influence on the TLVs

The recommendations of corporate officials and consultants were given great weight by TLV Chairman Stokinger. Massachusetts health official and longtime TLV committee member Hervey Elkins complained in a letter to fellow committee member William Frederick in 1966: "It annoys me no end, that any action that could possibly adversely affect a certain chemical company is immediately objected to by a consultant to said company, and the objection is always accepted by the Chairman." The companies themselves, individually and under the auspices of such groups as the Industrial Hygiene Foundation, also had periodic meetings with members of the TLV committee and communicated their concerns both orally and in writing.

Just as some corporate communications to the TLV committee delayed or prevented action, other unpublished information was accepted as the basis for setting TLVs. The growing reliance on unpublished corporate communications, some of which were phone calls and most of which do not survive in written form today, is reflected in the *Documentation*.[2,20]

By 1986, unpublished corporate communications were important in supporting

TLVs for 104 substances out of less than 600 listed in the *Documentation* (5th ed). For twenty-five of these key communications from corporate employees (out of 104), the corporate affiliations of the correspondents were not stated in the *Documentation*. Of the 104, the 37 cases of unpublished "industrial experience" reflect a pattern of uncritical acceptance of assertions from financially interested parties, based on scant data of poor quality. These assertions, absent explanations of materials and methods used, would never be accepted for publication in medical or other scientific literature. Moreover, they include many evaluations of a medical nature that were reported by industrial hygienists and other nonphysicians.[2,20]

Some had expected that the story of the TLVs would have ended with the adoption by OSHA of most of the 1968 TLVs as its first set of exposure limits. Congress established and funded NIOSH for purposes of conducting research and making recommendations to OSHA for health standards. But the TLV committee not only remained as active as ever after 1970, it even permitted full-time employees of chemical companies to become centrally involved in the development of the TLVs.

The participation of industry representatives on the TLV committee began with the addition of Dow toxicologists V. K. Rowe and Theodore Torkelson as "liason members" in 1970. These men were assigned responsibility for developing "documentations" on which new or revised TLVs would be based. The chemicals assigned to them initially were *all* Dow products (2,4,5-T, vinyl chloride, ethylene glycol, methyl bromide, and propylene glycol methyl ether). No objection was evidently made over the fact that many of the chemicals assigned to these employees of Dow and industrial hygienist James Morgan of DuPont (starting in 1972) were products marketed by the firms that employed them. The chemical assignments of these individuals were regularly recorded in the minutes of the TLV committee, although publicly it was stated that these "consultants" did not actually vote on adoption of TLVs.[21]

Perhaps because the corporate representatives were paid for their work on the TLV committee and provided with the ample resources and support of their employers, they were among the most active contributors to the TLVs. TLV committee minutes and other records show that Torkelson, individually and as chairman of key subcommittees, was assigned at least 30 of Dow's halogenated solvents, pesticides, and other industrial chemical products between 1970 and 1988 (since 1977, including perchloroethylene, trichloroethylene, 1,3-dichloropropene, divinylbenzenes, carbon tetrachloride, chlorine, propylene dichloride, ethylene diamine, methylene chloride, and acrylamide). Similarly, Morgan and his successor Gerald Kennedy of DuPont obtained the task of documenting many DuPont pesticides, chlorofluorocarbons, and other products (since 1977, including hydrogen cyanide, acrylonitrile, hexafluoroacetone, *p*-nitrochlorobenzene, trichlorotrifluoroethane, and dimethyformamide).[2]

Dr. Georg Kimmerle has been listed in TLV booklets since 1981 simply as "German MAK Commission liaison." He is a physician employed by the German chemical producer, Bayer, whose US subsidiary Mobay makes pesticides, isocyanates, and other chemicals. Shortly after joining the TLV committee, Kimmerle was primarily responsible for the decision to double the TLV for one Mobay pesticide (fenthion). He then drafted documentations for new TLVs for three other pesticides made in the US solely by Mobay (fenamiphos, metribuzin, and sulprofos). The other five chemicals recorded as assigned to Kimmerle (amitrole, thiram, xylidine, perchloromethyl mer-captan, and phenylene diamine) are produced by Bayer in West Germany. Kimmerle's handling of TLV committee work, unlike his duties with the German MAK [Maxi-mum Workplace Concentration] Commission, were not required to be handled as a confidential, separate matter from his job at Bayer.

In all, corporate representatives were given primary responsibility for developing TLVs on more than 100 substances between 1970 and 1988, including at least 36 classified as carcinogens by official bodies. (Complete information on the chemical assignments is not even available from ACGIH.) There is no question that the economic impact of the TLVs on the chemical industry generally and on Dow, DuPont, and Bayer in particular, has been enormous. There seems no reason to doubt that chemical industry employees working on the TLV committee were imple-menting corporate policies of their firms. This view is consistent with Dow's recom-mendations to OSHA to adopt the TLVs instead of stricter NIOSH recommendations for seven Dow products, at least six of which had been handled by Torkelson on the TLV committee.[5]

Industry employees aside, practically nothing has been disclosed about the docu-mentations assigned to committee members who had part-time consulting relation-ships with chemical producers. ACGIH has never required members of the TLV committee who do corporate consulting to either disclose these business connections or excuse themselves from development of TLVs on chemicals of importance to their clients. The TLV booklets have listed such persons only by affiliations they had with universities.

TLV committee minutes fleetingly mention meetings of the committee and its subcommittees with dozens of representatives of companies and trade associations. Nothing in writing relating to most, if not all, of these meetings is in the chemical files at ACGIH. Robert Spirtas, a member of the committee since 1981, wrote of his impression of these encounters in a 1987 letter to TLV Chairman E. Mastromatteo: "[P]resentations by outside groups have, in my experience, always been allied with the management point of view. The majority of the presentations have been personal interpretations of the published literature. In my opinion, these presentations have been attempts to lobby the committee, with very little new data." This state of affairs led ACGIH Secretary-Treasurer Philip Bierbaum to urge, in a 1988 memorandum to the ACGIH Board, that no presentations by outside groups to committee members be allowed unless they are publicly announced and open to the public.

As of late 1988, the TLV committee would not even permit interested scientists to attend its meetings as observers. Only after OSHA had proposed to adopt hundreds of the TLVs in 1988 did ACGIH allow researchers to examine relevant ACGIH files (e.g., recent TLV committee minutes).

The TLV committee has stoutly refused to publicly disclose members' paid sources of corporate consulting work. This might be extensive, as the present 22-member committee includes only six full-time government employees. The TLV committee has also resisted recommendations by its only labor representative and members of the ACGIH Board that industry employee members of the committee be precluded from drafting TLV documentations.

Although some reforms may finally be instituted in 1989, the many TLVs just adopted as OSHA standards are a legacy of an earlier era. Former NIOSH director John Finklea has remarked that the TLVs were "the result of a process that would currently be viewed as seriously flawed."[22]

Medical Inadequacy of the TLVs

The information base upon which the TLVs were developed was severely limited. For standards intended for a potential lifetime of exposure, chronic data are critical. However, for at least 90% of the TLV chemicals, sufficient data on long-term effects are unavailable, either from animal studies or studies of industrial workers with long-term exposure to known concentrations of the substances.

The very concept of a *daily average* exposure limit has been attacked as being inconsistent with what is known about toxicity, and evidently originating more from economic than scientific considerations.[23,24] Atherley's analysis concluded that the time-weighted average index "cannot be viewed as a scientific idea underpinned by either empirical evidence or plausible scientific hypothesis."[24] In view of this, the TLV committee's deletion in 1984 to 1986 of most of the short-term exposure limits was particularly harmful. ("STELs" for nearly 200 substances were dropped prior to OSHA's adoption of the TLVs.)

The TLV committee's heavy reliance on animal data (mostly acute and subacute toxicity studies) raises a number of unavoidable problems. One cannot elicit a medical history from an animal, and symptom data can be missed that could be severe enough in a human to interfere with productive function at work. In addition, the animal data gathered were very limited in scope as well as duration. Typically, no study was done of neurologic and neurobehavioral function beyond meager observations of animal behavior such as lethargy, fighting, etc. Thus, animal studies are unable to evaluate cognitive changes such as we now know can occur from exposure to many solvents and other chemicals.

Animal studies also have not included an evaluation for pulmonary function, despite the fact that many chemicals are irritants and/or chemical allergens, and repeated exposure to such agents could well reduce pulmonary function, nor was immunologic function evaluated for the vast majority of chemicals. Endocrine function was at best evaluated by an occasional blood glucose test, typically ignoring the potential for endocrine alterations in other organs. Animal studies often did include information on the appearance of many but not all organs at death by gross and light microscopic analysis. However useful structural information is, though, it is not a substitute for evaluating the function of organs.

Another shortcoming of the TLVs was the frequent failure to use information that was available. For example, no systematic literature search was done in preparing documentations on hundreds of chemicals. References used are often very dated, and more recent information is often missing. Information in the international medical literature does not appear to have been reviewed for the vast majority of chemicals. In fact, little reference is made to the basic US medical literature in the TLV

TABLE 7.1
Health Effects at and below the TLVs**

Substance	Effects reported/exposure	TLV	Ref
Acetone	Neurobehavioral performance effects after 4 h at 250 ppm	750 ppm	25
Benzene	Bone marrow changes, leukemia, at or below 1 pm	10 ppm	26, 27
Beryllium	Respiratory sensitization below 2 $\mu g/m^3$	2 $\mu g/m^3$	28
Cadmium	Changes in renal function at 0.007–0.039 mg/m^3	0.05 mg/m^3	29
	Changes in renal function at 0.003 mg/m^3		30
	Changes in renal function at cumulative exposures equal to 20–22 yr at 0.05 mg/m^3		31
2-Ethoxyethanol and 2-Methoxyethanol	Lowered sperm counts and reduced red and white (granulocytes) blood cell counts at combined exposure to 9.9 mg/m^3 2-EE and 2.6 mg/m^3 2-ME	19 mg/m^3 (2-EE) 16 mg/m^3 (2-ME)	32, 33
Ethylene diamine	Respiratory sensitization at 1–10 ppm	10 ppm	34
Ethylene oxide	Increase in sister chromatid exchanges in lymphocytes at 0.35 ppm	1 ppm	35
Formaldehyde	Respiratory cancer, allowing for 10 yr latency, 0.1–1 ppm	1 ppm (1.5 mg/m^3)	36
	Pathologic changes in nasal mucosa at 0.1–1.1 mg/m^3		37
Glutaraldehyde	Increased respiratory symptoms and headache below 0.04 mg/m^3	0.7 mg/m^3	38
Lead (naphthenate)	Blood lead concentrations over 60 $\mu g/dL$; mean ZPP† of 265 $\mu g/dL$, at 0.96 $\mu g/m^3$	150 $\mu g/m^3$	39
Manganese dust (inorganic)	Fatigue, trembling, tinnitus, irritability at 1 mg/m^3	5 mg/m^3	40
Mercury vapor	EEG changes at 25 $\mu g/m^3$	50 $\mu g/m^3$	41
Methyl chloroform (1,1,1-trichloroethane)	Behavioral perfromance deficits with 3 h at 175 ppm and 350 ppm	350 ppm	42
	Decreased DNA concentrations in the brains of gerbils after continuous exposure to 70 ppm for mos		43
Oil mist, mineral	Cross-shift decline in 1-sec FEV at 0.2 mg/m^3	5 mg/m^3	44
Silica	Lung scarring in bricklayers exposed to dust (2.1% free silica) 0.5–2.0 mg/m^3	2 mg/m^3	45
Styrene	Occupational asthma at 62.7 mg/m^3	215 mg/m^3	46
Sulfur dioxide	Bronchoconstriction among asthmatics after 3–5 min at 0.5 ppm	5 ppm‡	47
Sulfric acid mist	Laryngeal cancer at 00.2 mg/m^3	1 mg/m^3	48
Toluene	Neurobehavioral effects in rats exposed 20 min to 125 ppm	100 ppm; 150 ppm‡	49
	Impairment in human performance after 6–6.5 h at 100 ppm		50
	Increased fatigue, short-term memory changes, reduced concentration at 11.5 ppm and 41.8 ppm		51
	Neurobehavioral changes (visual vigilance) after 4 h at 100 ppm		52
Toluene diisocyanate	Asthma developed within 1 yr in 9 workers, at 0.002 ppm	0.005 ppm	53
Triethylamine	Blue haze (foggy vision), corneal edema, eye irritation at 18 mg/m^3	40 mg/m^3	54
Zinc oxide fume	Lung function changes in guinea pigs exposed 3 h on 6 consecutive days to 5 mg/m^3	5 mg/m^3	55

* This table was compiled from review of the contents of 33 months (January 1987 to September 1989) of 4 journals: *Journal of Occupational Medicine, American Journal of Industrial Medicine, Scandinavian Journal of Work Environment and Health,* and *British Journal of Industrial Medicine.*

† ZPP, Zinc protoporphyrin.

‡ Short-term exposure limit.

documentation. Thus, contrary to the TLV booklet's persistent claim, the TLVs are not "based on the best available information."

The medical inadequacy of the TLVs is evident from a review of four occupational medicine journals since the start of 1987 (Table 7.1). From this limited sampling, it would appear that further evidence of harm at and below the TLVs appears in the literature almost monthly. This review did not include industrial hygiene, toxicology, and general medical journals. Other scientists are encouraged to review these for sub-TLV effects.

The development of TLVs and evaluation of relevant scientific literature have mainly been done by industrial hygienists and other nonphysicians. Although industrial hygienists are vital to developing control strategies for chemicals, most lack training in the biomedical sciences to interpret health effects data reliably and independently. Yet this is exactly what they had to do as volunteers on the TLV committee. Copies of reviewed articles were generally not provided to the entire committee: the sole responsibility for accurate interpretation of the articles typically fell to the committee member assigned each chemical. The result was a list of exposure limits produced almost entirely by hygienists, chemists, and toxicologists, most of whom lacked the necessary training, let alone clinical experience with humans.

Toxic Torts and the TLVs

It is ironic that doctors may now be asked to confer legitimacy-in-retrospect on the TLVs.

An increasing number of persons are appearing before the courts with conditions medically attributed to chemical exposures. The courts are interested in knowing the state of medical knowledge when these people's exposures to chemical products and wastes occurred. A rationale often used to parry charges of negligence and assessment of liability is known as the "TLV defense." This amounts to: We thought that the exposures here would be below the TLVs, and we also thought the TLVs were *safe*, so what happened is not our fault.

But although those in other professions may say, "we thought the TLVs were safe and sound," it is the opinion of the medical profession (i.e., *not* the medical opinion of the industrial hygiene profession) that is most often sought to test such claims today. Consequently, doctors may be asked to appraise the TLV for one or more chemicals, the TLVs in general, and possibly even the TLV committee, too.

It is hard enough to look back on the withholding of medical expertise and medical knowledge that left so much to the TLV committee for so long. But it is professionally humiliating when doctors are asked to dignify the medical stature and safety of guidelines that were never really a product of industrial medicine. Lawyers defending chemical liability cases may find that the TLV defense is easier for them to raise in an opening argument than it is to support with credible medical testimony.

Alternatives to the TLVs

There is an urgent need to compile the information that is available but has been ignored in the TLV development process. For example, the New Jersey Department of Health recently utilized chronic health effects data from the Environmental Protection Agency known as the Integrated Risk Information System (IRIS) data base. Workday air concentrations were calculated for noncarcinogens and carcinogens, reportedly corresponding to no risk of chronic health effects or (for carcinogens) a one-in-a-million lifetime risk of cancer. The resulting exposure limits, even for noncarcinogens, were markedly lower than the TLVs, not infrequently by 3 or more orders of magnitude (R. T. Zagraniski, 1988 testimony of the NJ Department of Health at informal hearings on OSHA's proposal to update permissible exposure limits for toxic substances). IRIS data exist on many more chemicals and could be used to supplement our understanding.

Further use needs to be made of the international literature, particularly for the industrialized countries. Much information on chemical effects is available in English from the Scandanavian countries. In addition, the Soviet Union has exposure limits on more than 1400 chemicals. These limits were reportedly developed to prevent physiologic alteration, not just clinical disease. Critics of the Soviet exposure limits have sometimes raised a separate presumption that the Soviets in practice follow less stringent limits. But the medical issue is not the state of Soviet engineering practice. Physicians need to have as many data as possible on long-term effects of chemicals to understand what levels could cause harm. Philosophic and political differences have not prevented scientific cooperation in other health-related areas, and international relations now offer hope for expanded USA-USSR contact on matters of importance in industrial medicine.

Exposed workers themselves are a potentially vast source of data. The US experience in occupational health is that medical and environmental monitoring that is not legally required is often not conducted. We are thus missing an enormous amount of potentially useful health-effects data on early functional changes in workers. OSHA's expected medical and air monitoring standards may soon help stimulate industry to generate this dose-response data. However, doctors should not wait for legally mandated monitoring to begin to conduct medical evaluations for potential health effects below the TLVs.

Industry needs to provide adequate resources to allow physicians to visit workplace areas regularly, to monitor all exposed workers medically, and to update their knowledge about toxicologic effects regularly. Doctors need adequate computer and other literature access for all substances used, released, and produced in the workplace.

Doctors also need to be provided with sufficient time to do a thorough "review of systems." Experience has shown that a great deal of knowledge on chemical health hazards is not in the books, and clusters of adverse effects can be clinically identified sometimes before one of our busy medical colleagues has gotten the problem into print. Clinical cases are frequently the first evidence of occupational disease phenomena, and the patients themselves are an indispensable source of information. With the deficiency of published literature on the chronic effects of most chemicals in use, the need for doctors to take the time to listen to patients is underscored. Occupational medicine is a demanding field, and a full evaluation of a single person with illness potentially related to chemical exposure can take several hours.

Much better use can be made of industrial nurses, especially in small plants, where it is nurses who are the first to see problems and hear workers' health complaints. Industrial nurses and physician assistants, working with physicians, are capable of playing a more sophisticated role in occupational disease assessment than they have been offered in the past. To be most effective, however, these professionals will need to develop additional skills in occupational disease recognition. For example, they will require further training regarding the toxic effects of chemicals, taking medical and occupational exposure histories, and conducting physical examinations. Their preliminary assessments can then be useful in a team approach, working with the physician.

Industrial hygienists can expand their reconnaissance capability far beyond the generation of numbers on exposure levels. Industrial hygienists, as well as nurses and worker health and safety representatives, need specific training on the toxic effects of chemicals on the body and on how to interview workers for health effects in the intervals between medical monitoring. These health effects interviews, although not a substitute for medical evaluation, can nonetheless assist in the early detection of effects from irritants, sensitizers, and nervous system toxins.

The New Jersey Department of Health is developing a Guide to Workplace Inspection that could facilitate this process. That agency's Hazardous Substance Fact Sheets, now available for about 1000 chemicals, include target-organ toxicity information that can help to focus workplace health-effects interviews.

Because the TLVs lack scientific validity, the role of air monitoring should be a different one. A specific air concentration should never be relied upon as indicating safety. Rather, air monitoring should be used to assess the effectiveness of controls. In addition, because at this time there are no known safe exposure levels, physicians as part of the management team should insist on controls that reduce all exposures to the *maximum extent technically feasible.* To advocate a lesser degree of protection would violate the dictum to "do no harm."

Similarly, workplace inspectors may be better off not using the TLV booklet as now written, because of the misleading assertions stated in its preface. These inaccurate claims ("based on the best available information"; "nearly all workers may be repeatedly exposed day after day without adverse effect") provide a false sense of security to nonmedical personnel. Unless such claims are deleted from TLV booklets, industrial physicians would be prudent to instead obtain or encourage development of other sources of information.

Despite laws and regulations giving workers the "right to know," most hazard communication training programs are general and prepackaged, and do not address the specific toxic effects of the substances used in the workplace. If workers are not properly informed about such dangers, they will not be prepared to alert the medical department when early symptoms develop. Industrial physicians should ensure that all hazard communication training fulfills the legal requirement to include hazards of the specific chemicals used. The New Jersey Hazardous Substance Fact Sheets are particularly useful in this regard, as they discuss early symptoms and effects in lay English that the worker and supervisor can understand.

Of course, it is primarily industry's responsibility to provide and encourage adequate, coordinated occupational health programs. Just as management provides engineers with flow charts to monitor the industrial process, doctors need to be informed about the materials used and created in every department. It is management's responsibility to encourage cooperation between health professionals, to provide the resources and a framework for monitoring exposures and health of workers, and to grant industrial physicians the authority to fulfill their professional obligations to the people at work. Industrial physicians will not be able to do their job well unless and until industry respects the importance of industrial medicine and makes the commitment to prevent, not ignore, occupational diseases.

The judgment of how much exposure can cause disease in humans is ultimately a medical decision. Industrial physicians, as a profession and through the American College of Occupational Medicine, have the obligation to step forward and assert their responsibilities in assessing health hazards in industry. In fact, to not do so could be viewed as malpractice by some. Workers have been ill served by having critical decisions about their health delegated to engineers carrying TLV booklets.

REFERENCES

1. Air contaminants; final rule. *Federal Register* January 19, 1989;54:2332–2983.
2. Castleman BI, Ziem GE. Corporate influence on threshold limit values. *Am J Ind Med.* 1988;13:531–559.
3. Report of the Subcommittee on Threshold Limits. *Transactions of the 5th Annual Meeting of the National Conference of Governmental Industrial Hygienists*; 1942;163–170.
4. Report of the Subcommittee on Threshold Limits. *Proceedings of the 8th Annual Meeting of the American Conference of Governmental Industrial Hygienists*; 1946;54–55.
5. Cook WA. Maximum allowable concentrations of industrial atmospheric contaminants. *Ind Med.* 1945;11:936–946.
6. Bloomfield JJ. Codes for the prevention and control of occupational diseases. *Industrial Hygiene Foundation Transactions Bulletin No 8, Eleventh Annual Meeting*; 1946;71–79.
7. Henderson Y, Haggard H W. *Noxious Gases*. New York, NY: Reinhold, 1943;132, 168.
8. Report of the Committee on Threshold Limits. *Proceedings of the 9th Annual Meeting of the American Conference of Government Industrial Hygienists*; 1947;43–45.
9. Report of the Committee on Threshold Limits. *Transactions of the 10th Annual Meeting of the American Conference of Government Industrial Hygienists*; 1948;29–31.
10. Report of the Committee on Threshold limits. *Transactions of the 15th Annual Meeting of the American Conference of Government Industrial Hygienists*; 1953;45–47.
11. Coleman AL. Threshold limits of organic vapors. *Transactions of the 16th Annual Meeting of the American Conference of Government Industrial Hygienists*; 1954;50–53.
12. Threshold limits for 1958. *Arch Ind Health.* 1958;18:178–182.
13. Trasko V. Industrial hygiene milestones in government agencies. *Am J Public Health.* 1955; 45:39–46.

14. Yant WB. Industrial hygiene codes and regulations. *Industrial Hygiene Foundation Transactions of the 13th Annual Meeting*; 1948;48–61.

15. Harvey B. Some personal observations on industrial health in the United States of America. *Br J Ind Med*. 1954; 11:222–226.

16. McLaughlin AIG. The prevention of the dust diseases. *Lancet* 1953∞:49–53.

17. Meeting of Committee Chairmen, Industrial Medical Association, April 24, 1952. Archives of the American Occupational Medical Association.

18. *Documentation of Threshold Limit Values*. Cincinnati, Ohio: *American Conference of Government Industrial Hygienists*; 1962.

19. Golz HH, et al. Report of an investigation of threshold limit values and their usage. *J Occup Med*. 1966;8:280–283.

20. *Documentation of Threshold Limit Values and Biological Exposure Indices* (5th ed), Cincinnati, Ohio: *American Conference of Government Industrial Hygienists*; 1986.

21. Operational guidelines and procedures. Threshold limit values (TLV) committee for chemical substances in the work environment. *Appl Ind Hyg*. 1988;3:R5–R8.

22. Finklea J A. Threshold limit values: a timely look. *Am J Ind Med*. 1988;14:211–212.

23. Halton DM. A comparison of the concepts used to develop and apply occupational exposure limits for ionizing radiation and hazardous chemical substances. *Regul Toxicol Pharmacol*. 1988;8:343–355.

24. Atherley G. A. critical review of time-weighted average as an index of exposure and dose, and its key elements. *Am Ind Hyg Assoc J*. 1985;46:481–487.

25. Dick R, Setzer J, Taylor B, et al. Neurobehavioral effects of short duration exposures to acetone and methyl ethyl ketone. *Br J Ind Med*. 1989;46:111–121.

26. Infante P. Benzene toxicity: studying a subject to death. *Am J Ind Med*. 1987;11:599–604.

27. Landrigan P. Benzene and leukemia. *Am J Ind Med*. 1987;11:605–606.

28. Cullen M, et al. Chronic beryllium disease in a precious metal refinery. Selected Reviews from the Literature. *J Occup Med*. 1988;30:6–8.

29. Chia K, Ong C, Ong H, et al. Renal tubular function of workers exposed to low levels of cadmium. *Br J Ind Med*. 1989;46:165–170.

30. Kawada T, Koyama H, Suzuki S. Cadmium, NAG activity, and βa-microglobulin in the urine of cadmium pigment workers. *Br J Ind Med*. 1989;46:52–55.

31. Mason H, Davison A, Wright A, et al. Reactions between liver cadmium, cumulative exposure, and renal function in cadmium alloy workers. *Br J Ind Med*. 1988;45:793–802.

32. Welch L, Schrader S, Turner T, et al. Effects of exposure to ethylene glycol ethers on shipyard painters: II Male reproduction. *Am J Ind Med*. 1988;14:509–526.

33. Welch L, Cullen M. Effects of exposure to ethylene glycol ethers on shipyard painters: III Hematologic effects. *Am J Ind Med*. 1988;14:527–536.

34. Aldrich F, Stange A, Geesaman R. Smoking and ethylene diamine sensitization in an industrial population. *J. Occup Med*. 1987;29:311–314.

35. Sarto F, Clonfero E, Bartoluci G. et al. Sister chromatid exchanges and DNA repair capability in sanitary workers exposed to ethylene oxide: evaluation of the dose-response relationship. *Am J Ind Med*. 1987;12:625–637.

36. Kauppinen T, Partanen T. Use of plant- and period-specific job-exposure matrices in studies on occupational cancer. *Scand J Work Environ Health*. 1988;14:161–167.

37. Edling C, Hellquist H. Ödkvist L. Occupational exposure to formaldehyde and histopathological changes in the nasal mucosa. *Br J Ind Med*. 1988;45:761–765.

38. Norbäck D. Skin and respiratory symptoms from exposure to alkaline glutaraldehyde in medical services. *Scand J Work Environ Health*. 1988;14:366–371.

39. Goldberg R, Garabrant D, Peters J, et al. Excessive lead absorption resulting from exposure to lead naphthenate. *J Occup Med.* 1987;29:750–751.

40. Roels H, Lauwerys R, Buchet J, et al. Epidemiological survey among workers exposed to manganese: effects on lung, central nervous system, and some biological indices. *Am J Ind Med.* 1987;11:307–327.

41. Piikivi L, Tolonon U. EEG findings in chlor-alkali workers subjected to low long term exposure to mercury vapor. *Br J Ind Med.* 1989;46:370–375.

42. Mackay C, Campbell L, Samuel A, et al. Behavioral changes during exposure to 1,1,1-trichloroethane: time-course and relationship to blood solvent levels. *Am J Ind Med.* 1987; 11:223–239.

43. Karlsson J, Rosengren L, Kjellstrand P, et al. Effects of low-dose inhalation of three chlorinated aliphatic organic solvents on deoxyribonucleic acid in gerbil brain. *Scand J Work Environ Health.* 1987;13:453–458.

44. Kennedy S, Greaves I, Kreibel D, et al. Acute pulmonary responses among automobile workers exposed to aerosols of machining fluids. *Am J Ind Med.* 1989;15:627–641.

45. Myers J. Respiratory health of brickworkers in Cape Town, South Africa. *Scand J Work Environ Health.* 1989;15:198–202.

46. Moscato G, Biscaldi G, Cottica D, et al. Occupational asthma due to styrene: two case reports. *J Occup Med.* 1987;29:957–960.

47. Balmes J, Fine J, Sheppard D. Symptomatic bronchoconstriction after short-term inhalation of sulfur dioxide. *J Occup Med.* 1989;31:303, 307. Selected Reviews from the Literature.

48. Steenland K, Schnorr T, Beaumont J, et al. Incidence of laryngeal cancer and exposure to acid mists. *Br J Ind Med.* 1988;45:766–776.

49. Kishi R, Harabuchi I, Ikeda T, et al. Neurobehavioral effects and pharmacokinetics of toluene in rats and their relevance to man. *Br J Ind Med.* 1988;45:396–408.

50. Baelum J, Andersen L, Lundqvist G, et al. cited in ref 49.

51. Dick R, Setzer J, Wait R, et al. cited in ref 49.

52. Ørbaek P, Nise G. Neurasthenic complaints and psychometric function of toluene-exposed rotogravure printers. *Am J Ind Med.* 1989; 166:67–77.

53. Venables K. Epidemiology and the prevention of occupational asthma. *Br J Ind Med.* 1987; 44:73–75.

54. Akesson B, Bengtsson M, Floren I. Visual disturbances after industrial triethylamine exposure. *J Occup Med.* 1988;30:201. Selected Reviews from the Literature.

55. Farrell F. Angioedema and urticaria as acute and late phase reactions to zinc fume exposure, with associated metal fume feverlike symptoms. *Am J Ind Med.* 1987;12:331–337.

QUESTIONS

1. What kinds of criticisms have been raised about TLVs?

2. What recommendations made by Ziem and Castleman suggest sources other than ACGIH OSHA might consult for information about possible health consequences from exposures to particular substances?

Chapter Eight

Threshold Limit Values in the 1990s and Beyond
A Follow-Up

David Allen

The workplace is safer in the 1990s, the decade following Ziem and Castleman's revelations regarding the politics of threshold limit values. Indeed, in 1997, the rate of illness and injury fell to its lowest point since the Bureau of Labor Statistics began recording this information in the 1970s (Jeffress 1999). The decline in work-related illness and injury, however, appears to be unrelated to the persistent influence of threshold limit values (TLVs) in setting a variety of chemical exposure standards on the job. This follow-up chapter to Ziem and Castleman will examine trends in the application of TLVs in the 1990s. Literature on this topic suggests that a faulty industrial health standard is being applied to an increasingly diverse array of situations as if it were an accurate, reliable assessment of risk. It appears scientific, and therefore is approached as a legitimate, defensible standard. In fact, however, it remains a corporate sponsored level of health risk that bears little reality to the role of chemicals in workplace illnesses. A short discussion of the TLV controversy will preface a longer discussion of the trends in TLV application.

Revisiting and Adding to the Debate

Castleman and Ziem (1988) touched off a controversy by asserting a causal relationship between reliance on corporate data and the imposition of standards too low to protect the health of workers. While they documented the influence of industry representatives who served on committees charged with setting the standards, they did not attempt to establish systematically that the standards were unsafe to workers. In the early 1990s, however, studies documented the adverse effects of chemical exposures on the job and pointed to the inadequate protection of workers caused by the use of TLVs. These studies suggest three observations critical to understanding the current problems with TLVs.

First, TLVs were originally established on the basis of inadequate and substandard scientific data, systematically disregarding existing evidence of toxicity (Roach and Rappaport 1990). It is generally conceded by the Occupational Safety and Health

Administration (OSHA) that the scientific standards employed in setting and main-taining exposure limits were poor at best (Jeffress 1999; Castleman and Ziem 1988). Moreover, while the 1990s have seen an adjustment downward of permissible limits for many substances, the OSHA admits that "in many instances these adopted limits are not sufficiently protective of worker health" (Jeffress 1999).

Second, setting official TLVs, revising workplace air standards, and setting levels for substances not previously covered was (and remains) a political, rather than a clinical or epidemiological, process (Vladeck and Wolfe 1991; Paustenbach 1997). TLVs were set to avoid inconveniencing a particular industry. Health of the work-force was a secondary consideration. The goal was not genuine safety. Indeed, most of these standards matched the levels industries were already using (Roach and Rappaport 1990).

Third, subsequent studies typically show that the levels were set too high. In a systematic study of adverse effects of regulated substances, Roach and Rappaport (1990) found that a substantial number of cases of illness had been diagnosed at levels well below the TLVs. Reviewing a large number of studies, they concluded that the majority of them showed adverse effects at sub-TLV levels of exposure. Virtually no correlation was found between compliance with TLVs and the incidence of adverse symptoms. Moreover, OSHA adopted, and continues to use, many of the same limits set in 1946, preferring these more industry-sponsored standards to the more restrictive standards set by the National Institute for Occupational Safety and Health (see also Robinson, Paxman, and Rappaport 1991; *RACHEL* 1994a, 1994b). Incorporating TLVs into government regulatory policy meant in effect that exposures falling below the official standard were assumed to be risk free. This unfortunate assumption was and most likely remains a serious source of workplace contamina-tion.

These three interrelated critiques of TLVs strongly suggest that the numeric in-dexes are calculated with very little regard to the health of workers. Ironically, while TLVs are said to represent something real about the human body and toxins, they are more like public fictions that work to persuade us that the world, or a part of it, can be trusted. Started as a preliminary means to determine the risks of exposures to industrial chemicals, TLVs have become powerful tools of corporate control over the workplace and the bodies of the workers themselves.

The Effects of Misplaced Confidence

The idea of standards expressed as numerical coefficients is a powerful rhetoric in a culture organized around rational inquiry and a realist assumption of truth. Sociol-ogists who study risk often point out the fallacy of separating risk assessment from risk perception and acceptance. Corporations and governments often argue that risk assessment is the outcome of the scientific method applied to the problem of the distribution of danger while risk perceptions and acceptance are subjective and emotional responses to danger (Clarke 1992). On the contrary, however, formal risk assessments themselves are determined by value-related choices, profit incentives, and other nonscientific criteria (Freudenburg 1988; Clarke 1992). The history of TLVs in the 1990s is a good illustration of what can happen when we act on the faulty assumption that if it looks like science, reads like science, and is presented with the aura of scientific respectability, it must be science.

The reality is that setting TLVs, revising them, and extending their reach to previously unregulated substances has been a political exercise and not an objective, value-free inquiry into the relationship of chemicals to workplace diseases. The OSHA, as previously noted, acknowledges the inadequacy of its standards. A reasonable question in light of this admission is, "Why on earth were they adopted in the first place?" While Ziem and Castleman address this question in the preceding chapter, an additional, more modest, question also begs discussion: "What role do these faulty standards of workplace safety continue to play in industrial culture in the 1990s?"

Securing a Role for TLVs: A Questionable Alliance

A strong pro-business climate has encouraged the OSHA and related regulatory agencies to join the setting of air standards with the estimated costs of those standards for businesses. The OSHA encourages what it calls a "partnership" between management and government (and other interested parties) in setting safe exposure levels (Jeffress 1999; see also Dear 1995). The OSHA's review procedures include the use of an estimate of "average annual cost per establishment to achieve compliance, and total costs by industry sector" (Jeffress 1999). The agency thus bases proposed revisions not only on scientific evidence of clean air standards, but also on the assessment of economic impact.

Systematic review of standards is slow in coming, in part because of "OSHA's outdated record keeping system" (OSHA 1996). The fact is that industries are far more sophisticated in their capacity to track and study chemicals in the workplace than federal or state government. There are some 500,000 specific substances emitted into the air. Conducting systematic studies, reviewing the existing research data, developing a convincing case for updating, and affording due process to affected businesses takes more resources than a beleaguered federal or state agency can muster.

Adhering to its theme of partnership, the OSHA generally seeks to maintain and revise standards by employing a process of "negotiated rule making" using "national consensus" (Jeffress 1999). If corporations with vital financial interests in maintaining existing TLVs exercise all of their rights, the process can be tied up for years. One can envision that every time a review takes place it will be challenged by a company which perceives itself as involved in a life-or-death struggle and thus invests heavily in several strategies, including legal help, more research, petitions, and appeals. This "gumming the works" strategy may serve the interests of stockholders, but it is not likely to decrease adverse chemical exposures on the job.

Absence of Organized Labor in Setting TLVs

While there was virtually no participation by labor unions in the setting of original TLV levels (Ziem and Castleman 1989; see chapter 7 in this volume), today the OSHA acknowledges its readiness to work with labor (Jeffress 1999). Indeed, the idea of partnership, at least in theory, includes labor participation in setting standards. This, however, has not yet happened.

As of 1994, only 12 states required any employee participation in the process of setting and monitoring workplace environmental standards (Reich 1994). Unions have been very much on the defensive in the last few decades and have many competing interests, such as maintaining jobs and benefits. Threats of tougher clean air standards often bring counterthreats of plant relocations and closures. Since many nations (and many states) with more permissive standards—or no standards at all—may be happy to welcome a profitable company, keeping business "at home" may deflect unions and other interested parties from contesting too vigorously a company's standard.

The increasing percentage of the workforce employed in the service sector of the economy may in fact account for much of the decline in job-related illness. As "dirtier" jobs are moved to poorer countries, the problem of workplace pollution is simply exported. The air in service-sector workplaces, however, is with few exceptions unregulated. Indeed, office environments are increasingly recognized as sources of risk (Reich 1994). The problem of "sick building syndrome" is quickly becoming a major public health issue (Roston 1998).

An additional, unanticipated problem for labor is the dangers their work poses for the family. Airborne particles may stick to workers' clothes or skin and be transported home. Home contamination, created when workers carry hazardous materials from their workplaces to their houses, is not monitored or regulated (*Applied Environmental Training* 1996). The problem has caught the attention of the National Institute of Occupational Safety and Health (NIOSH), however. The NIOSH acknowledges that the use of air threshold limits in the workplace as the sole measure of worker safety fails to take account of or measure in any way these additional hazards. A summary of a NIOSH report states, "There are no information systems to enable tracking of illnesses and health conditions resulting from [work-related home contamination]." Thus, testing workplace air, but not the people who unwittingly carry contaminants off the worksite (Goldsmith 1991), may result in unseen risks to family members living far outside the plant gates.

Bias of the Courts

Existing TLVs have also tended to take on a stubborn legal reality because of their long history. More often than not, the courts side with those who challenge attempts to establish more conservative standards. Numerous court decisions reveal a preference for existing standards, placing the burden of proof squarely on anyone wishing to establish lower limits. According to Reich (1994), "the courts have placed such a difficult burden of proof on OSHA that it has become nearly impossible to regulate more than a few chemicals at a time." In 1992, the eleventh Circuit Court of Appeals struck down an OSHA-sponsored proposal to modify and upgrade workplace air standards (Reich 1994). Corporations are also enjoying success with what has been called the "TLV defense"—avoiding liability for an employee's illness because its emissions fell below the standard level (Ziem and Castleman 1989, 915). Once again we see the rhetorical authority of statistics, even dubious ones: although TLVs were designed only as temporary starting points, the assumption of their validity and accuracy extends beyond their intended use and can even influence the decisions of another branch of government.

Problems of State Regulation

The 1990s has seen an economic climate in which corporate downsizing and plant closures have left workers, municipalities, and states in fear of losing their livelihoods and economic bases. This exerts a competitive pressure on states and cities to maintain standards of industrial pollution acceptable to industry. In the 1980s, the Environmental Protection Agency began to delegate to state agencies the task of controlling toxic air emissions (Tarr 1999). TLVs have been used by states in setting many clean air standards (Jeffress 1999; Reich 1994). In 1989, almost half of all states, including many of the most populous (California, Florida, Michigan, New York, Ohio, Pennsylvania, and Texas), used TLV levels to regulate air quality (Tarr 1999).

Moreover, while TLVs are intended to represent, however inaccurately, the level that will protect most workers exposed on a job site 40 hours per week, states have frequently used the very same standards to evaluate the safety of the ambient air concentration outside the workplace. In other words, the TLV levels were elevated from the status of a standard for the workplace, where individuals are exposed only eight or so hours a day, to one used in the evaluation of air standards in *neighborhoods* near the industries. In extending the reach of these questionable standards beyond the gates of industry, we witness, once again, improperly calculated statistics applied to a realm far removed from the shop floor.

Globalizing the Reach of TLVs

The public fiction of TLVs is now reaching beyond the U.S. Other nations who seek to monitor chemicals in their workplaces perceive a system from afar which may appear to be working acceptably well (Hamm 1990). Developing their own standards would require endless research and lead to the sorts of battles now typical in the United States; TLVs, therefore, are simply adopted wholesale by both industrial and industrializing countries with little regard for their limitations and biases (Castleman and Ziem 1988, 531). The widespread acceptance of TLVs by other countries does not indicate their general adequacy, but only the utility of having *some* standard which has a record of "successful" use. Moreover, as more countries have adopted these standards, they have become increasingly irresistible to nations who discover the need to develop a process for regulating workplace air.

A Concluding Word: The Future Is in Doubt

The OSHA is charged with the task of regulating the workplace levels of thousands of potentially hazardous substances. Given the constantly increasing complexity of industrial production, the number of such substances present in the workplace will continue to grow. Saddled with an inadequate set of standards and a cumbersome process for changing them, the OSHA's task is daunting. With limited resources and limited power, it has been difficult to develop and update standards in a way that takes account of both the increase in knowledge about the dangers of existing substances and the potential dangers of new substances. TLVs have been an obstacle to, rather than a tool for, a rational and just process for regulating workplace air standards.

The OSHA's partnership strategy has done little to increase the role of labor in setting workplace emission standards. Industry representatives, however, continue to be quite visible in this legislative process. The absence of labor and the presence of corporate interests has meant a continuation of the *politics* of regulation and hampered the implementation of scientifically based levels of air contaminants.

While they are sometimes modified to take account of new knowledge, TLV levels have become a touchstone for polluters who argue against any new standards which might hamper their regular methods of production or of monitoring worker safety. These levels have also hampered the OSHA's ability to enforce regulation. Courts have not been sympathetic to the OSHA's attempts to change its procedures for updating TLVs without Congressional mandate (Reich 1994).

The influence of TLVs has far surpassed the intentions of those who originally codified them. From a set of guidelines, they have become, first, standards taken to be an objective measure of workplace safety and, second, a standard for air safety generally. The history of TLVs is the particular history of a public fiction that increases its authority both within the workplace and beyond, primarily because it appears factual to a culture enamored with the "obvious reality" of numbers. It is fair to assume that this history will extend well into the twenty-first century.

REFERENCES

Applied Environmental Training. 1996. "Worker Home Contamination a Worldwide Problem." Http://www.otrain.com/1996/X0015_Worker_home_contamin.html.

Castleman, B. I., and G. E. Ziem. 1988. "Corporate Influence on Threshold Limit Values." *American Journal of Industrial Medicine* 13:531–59.

Clarke, L. 1992. *Acceptable Risk.* Berkeley, CA: University of California Press.

Dear, J. A. 1995. "The Department Experience with the Regulatory Flexibility Act of 1980

(RFA): OSHA Congressional Testimonies." Http://www.osha-slc.gov/OshDoc/Testimony_data/T19950210.html.

Freudenburg, W. R. 1988. "Perceived Risk, Real Risk: Social Science and the Art of Probabilistic Risk Assessment." *Science* 242 (Oct.7): 44–49.

Goldsmith, J. R. 1991. "Perspectives on What We Formerly Called Threshold Limit Values." *American Journal of Industrial Medicine* 19: 805–12.

Hamm, R. Douglas. 1990. "Troubled Times for TLVs?" *A.O.M.* 8(1): 1–5.

Jeffress, Charles N. 1999. Testimony presented to OSHA's House Committee on Education and the Workforce, Subcommittee on Workforce Protection, March 23.

Occupational Safety and Health Administration (OSHA). 1996. *Federal Register: Unified Agenda*. November 29. Washington, D.C.: U.S. Government Printing Office.

Paustenbach, D. J. 1997. "Updating OSHA's Permissible Exposure Limits: Putting Politics Aside." *American Industrial Hygiene Association Journal* 58(12): 845–49.

RACHEL. 1994a. "The Scientific Basis of Chemical Safety—Part 1: Limits on Workplace Chemical Exposures." *RACHEL's Environment and Health Weekly*. Electronic Edition No. 415. Http://www.enviroweb.org/publications/rachel/rehw416.htm.

RACHEL. 1994b. "The Scientific Basis of Chemical Safety—Part 2: Standards that Kill." *RACHEL's Environment and Health Weekly*. Electronic Edition No. 416. Http://www.enviroweb.org/publications/rachel/rehw416.htm.

Rappaport, S. M. 1993. "Threshold Limit Values, Permissible Exposure Limits, and Feasibility: The Bases for Exposure Limits in the United States." *American Journal of Industrial Medicine* 23 (5): 683–94.

Reich, Robert B. 1994. "Testimony before Committee on Labor and Human Resources U.S. Senate." OSHA Congressional Testimonies, February 9. Washington, D.C.: U.S. Government Printing Office.

Roach, S. A., and S. M. Rappaport. 1990. "But They Are Not Thresholds: A Critical Analysis of the Documentation of Threshold Limit Values." *American Journal of Industrial Medicine* 17: 728–753.

Robinson, J. C., D. G. Paxman, and S. M. Rappaport. 1991 "Implications of OSHA's Reliance on TLVs in Developing the Air Contaminants Standards." *American Journal of Industrial Medicine* 19: 3–13.

Roston, Jack. 1998. *Sick Building Syndrome*. London: Routledge.

Tarr, J. 1999. "Ethics, Threshold Limit Values, and Community Air Pollution Exposures." *Stone Lions Environmental Corporation Home Page*. Http://www.stonelions.com/article.htm.

Topping, M. D., C. R. Williams, and J. M. Devine. 1998. "Industry's Perception and Use of Occupational Exposure Limits." *Annual of Occupational Hygiene* 42(6): 357–66.

Vladeck, D. C. and S. M. Wolfe. 1991. "The Politics of OSHA's Standard-Setting." *American Journal of Industrial Medicine* 19(6): 801–4.

Ziem, G. E., and B. I. Castleman. 1989. "Threshold Limit Values: Historical Perspectives and Current Practice." *Journal of Occupational Medicine* 31: 910–18. Reprinted in chapter 7 in this volume.

QUESTION

1. Provide examples of how TLVs are being applied to environmental conditions they were not initially intended to measure.

Part 4

Toxins in the Workplace

It is in the workplace, in all of its many manifestations, that humans produce the physical necessities of life, plus luxury items that make our lives more comfortable and enjoyable. "Getting a living" has always required extensive interaction with the environment, whether one is mining iron ore, cutting down trees for a log cabin, or planting yams. Furthermore, humans have been exposed to work-related environmental hazards since the time the earliest hunter-gatherers had to contend with large predators. But with the advent of industrialization, workers confronted occupational dangers of a qualitatively different nature. Factories concentrate hazards in confined geographic spaces: noisy machinery that damages hearing, airborne particulate matter that damages the respiratory system, and deadly chemical vapors that sicken or kill. Modern technology has allowed human intrusions into the environment on an unprecedented scale: we dynamite mountains to rubble, drench acres of crops with pesticides, and mine for coal and precious metals miles underneath the earth. Even the world of the white-collar office worker is far from hazard-free: carpets may off-gas formaldehyde, molds may grow in heating and air-conditioning ducts, and hours spent working at a computer keyboard may result in carpal tunnel syndrome.

In the previous section, the chapter by Ziem and Castleman examined government efforts to regulate occupational exposures to airborne toxins. The selections in the present section take readers inside the workplace, to provide insights into the kinds of hazards many workers face on a daily basis, and to convey some sense of the kinds of worker-management conflicts that surround efforts to reduce occupational exposures to environmental hazards.

Chapter 9 by Siskind, "An Axe to Grind: Class Relations and Silicosis in a 19th-Century Factory," provides a historical introduction to the kinds of work conditions that existed under early industrial capitalism. With increased specialization, some workers spent their entire work days doing a single, highly dangerous job. In the nineteenth-century axe factory the most hazardous part of the production process was grinding down the tools, an activity that exposed workers to stone dust. After a number of years of continuous inhalation of this dust, many grinders became ill with the deadly respiratory disease silicosis. The owner of the factory expressed little concern about the risks to health posed by grinding. Indeed, he switched over to the less-dust-producing technology of shaving only when a shortage of skilled grinders made shaving the most cost-effective option.

In chapter 10, "From Dust to Dust: The Birth and Re-Birth of National Concern about Silicosis," Rosner and Markowitz bring the discussion of worker exposure to silica particles into the twentieth century. While factories were still fairly rare in the early 19th century period examined by Siskind, by the early 20th century the United States had a thriving industrial economy. As a result, an increasing number of workers found themselves in occupations where they were exposed to silica particles. Silicosis became a focus of national concern during the 1930s, but by the 1940s and 1950s improvements in working conditions were seen to make silicosis a disease of the past. By the late 1960s, however, an increased incidence of silicosis was being noted among workers in shipyards in Louisiana. By the 1970s, silicosis was on the rise among oil field workers. In response, the National Institute of Occupational Safety and Health recommended lowering the Permissible Exposure Limit for crystalline silica. This recommendation was eventually defeated by a well-organized, industry-sponsored opposition campaign.

The contribution by Quandt, Arcury, Austin, and Saavedra, chapter 11 on "Farmworker and Farmer Perceptions of Farmworker Agricultural Chemical Exposure in North Carolina," introduces readers to hazardous work conditions in the world of industrialized agriculture. Chemicals are used extensively on modern farms, but a continued focus by government regulators, owners, and workers on acute rather than long-term, low-level exposures presents obstacles to adequate health protection. Migrant farmworkers employ a number of "common-sense" notions to determine when it is safe to return to fields after spraying. Unfortunately, many of these notions, such as the belief that residues can always be detected through sensory perception (sight, touch, smell), are inaccurate. Farm owners, on the other hand, believe the major risk posed by pesticides occurs during mixing and application, activities the owners frequently do themselves. They see little health threat from the minimal quantities of pesticides that reside on the fruits and vegetables picked by migrant workers.

In chapter 12, "Competing Conceptions of Safety: High-Risk Workers or High-Risk Work?", Draper reports on an industry campaign to "individualize" the blame for work-related illness. The view promoted is that some individuals have inherited genetic traits that make them more susceptible to environmental hazards than the average person. If successful, this campaign would result in the designation of current work practices as safe for the vast majority of workers. The gap between management and workers' definition of a safe working environment remains substantial. Management wants to increase safety by removing at-risk individuals. Labor, on the other hand, wants to increase safety by reducing the number of hazards routinely encountered by workers.

Chapter Nine

An Axe to Grind

Class Relations and Silicosis in a 19th-Century Factory

Janet Siskind

Occupational disease is a class issue (Berman 1978; Gersuny 1981; Navarro 1986 [1980]). The level of safety or hazard in the workplace depends upon the relative strength of labor versus management. Some of the environments in which occupational diseases are produced predate both industrialization and capitalism (cf. Kaprow 1985:344), but the Industrial Revolution increased unhealthy environments greatly, and capitalist relations of production dictated their structure. In 19th-century New England, textile mills and new manufacturing industries produced new goods, new wealth, and new or more widespread diseases.[1] At the Collins Company of Connecticut, founded in 1826, part of the process of manufacturing axes resulted in deaths from silicosis. The documents that remain from the early years of this company, while specific to this factory at a particular time (1826–46), describe the interwoven strands of class relations and technology that created the "inorganic environment" (Mumford 1963 [1934]:69) in which many of the men who ground axes became sick and disabled, and died.

The development of class relations in early industrializing New England has been the focus of a number of recent case studies and discussions (e.g., Dawley 1976; Dawley and Faler 1976; Grimsted 1985; Hirsch 1978; Kulik 1978; Licht 1986; Prude 1983). These studies, which challenge "the proposition that antebellum America's passage into the industrial age was remarkable for its smoothness" (Prude 1983:xiii), carefully describe the conflict inherent in transforming artisans or farmers' sons and daughters into "hands," as well as the continuing struggle between managers and workers over the conditions of work. This struggle over working conditions took place within each factory, but it was powerfully affected by external forces, such as the labor supply and competition or collusion among manufacturers. The Collins documents add additional data to this ongoing investigation of class relations and how they structured management's and workers' perceptions and reactions to the hazards of occupational disease.

The Collins documents provide a picture of early capitalist industrialization from the viewpoint of one of its founders and the manager of the manufacturing side of the enterprise, Samuel Collins. Like archaeological evidence, the letters and memoirs in the collection are less than complete and must be placed with the documents of other early manufacturing companies to fill out our knowledge of this major transition from mercantile to capitalist production.

For the period 1826 to 1846, approximately 50 letters remain of the daily correspondence from Samuel Collins at the plant in Collinsville, Connecticut, to his brother David in the business office in Hartford.[2] Samuel is the major informant, and it is through his words and perceptions that we learn about his beliefs, his rationalizations, and his struggles with his workforce. Through his words we can envision the working conditions at the factory, the availability of work, and some of his workers' responses and actions, even though there is no similar collection of workers' letters from the Collins Company.[3]

In addition to the letters, Samuel Collins wrote memoirs about the company in 1866, called "Memorandums," which must be appraised as the work of an older man looking back on his earlier days: a respected manufacturer speaking to a current and future public. Within them Collins wrote down, apparently verbatim, copies of speeches that he had made to the workers and notices that had been posted in the factory. Two other documents in the collection provide important information: "Terms with Workmen," a book of contracts signed by the workers for the period December 1834 to November 1839, and a small notebook entitled "Accounts Payable 1832–1833," which contains scattered notes.

Background

When the Collins brothers started their factory in 1826, mercantile capitalism was old but was not yet the main way of life in a still predominantly rural nation (Clark 1979; Henretta 1978; Merrill 1977). The first industrial capitalist enterprises in New England, the textile mills of Rhode Island and Massachusetts, had been operating for 20 to 35 years; the shopkeepers of Lynn were already manufacturing shoes; and in nearby Waterbury along the Naugatuck River, workers had already begun rolling brass and making buttons.

In 1820, "some 12 percent of the nation's labor force was engaged in manufacturing and construction" (Montgomery 1968:4), and the potential for a "free" labor force was expanding. Land for farms was becoming scarce because of the combination of population increase and an economic climate that did not promote preservation of the fertility of soil. By the end of the 18th century in New England, land was already in short supply, and "there was a gradual increase in the numbers of landless and land-poor, who attempted to make their livings either by selling their labor or by engaging in trade" (Clark 1979:176). The new industrial enterprises provided a promising alternative for such landless men and women who otherwise faced dependency, tenancy, or possibly a move west.

In most communities the economic domination of the new industrialists had not yet been translated into political power (Prude 1983:163), but the law was changing as both new legislation and judicial interpretation of common law consciously chose private over public development (Horwitz 1977). By 1815 the value system and the institutional infrastructure to promote capitalist industrialization were in place: a banking network for short-term credit and for amassing domestic capital (Henretta 1973:184); tariffs and water rights that supported private entrepreneurs in the enterprises they described as development. As capitalist relations of production began to dominate the society, both using mills and water power to produce for a distant market, rather than a community or local region, and charging interest on borrowed money became acceptable practices.

History of the Company[4]

The entrepreneurs Samuel and David Collins came from a mercantile background. They had behind them an early stage of manufacturing in which their wealthy uncles, the Watkinsons, supplied iron to a skilled blacksmith, Charles Morgan, paid him for axes made on his own forge, and sold them in their large hardware store in Hartford (Wittmer 1977:25–26). In 1826 David and Samuel Collins formed a partnership which included their cousin, William Wells. Each partner invested $5,000 in the enterprise. "They purchased a saw-mill and gristmill and water privileges," as well as some land on the banks of the Farmington River in the town of Canton, a small farming community two miles south of what became Collinsville (Memorandums 2). The river was dammed, canaled, and bridged to deliver a steady and strong source of power (Memorandums 15).

Beginning with eight men (including Charles Morgan, who had been the axe-maker for the Watkinsons' hardware store, and two of his relatives), the factory rapidly increased the workforce by hiring workers from the neighboring towns of Canton and Simsbury, as well as from more distant parts of Connecticut and nearby states. It offered workers "wages of twelve to sixteen dollars per month, depending upon their skill and experience, for a twelve-hour day, six-day week" (Wittmer 1977: 28). By 1832 the Collins workforce had expanded to 300 men employed for 12 hours per day (U.S. Treasury Department 1969[1833]). In answer to a nationwide inquiry, the Collins Company reported having a capital investment of $250,000 and estimated the value of their manufactures at $294,000 (U.S. Treasury Department 1969[1833]: 996). It was the first large-scale manufacturer of axes, nearly ten times the size of its nearest competitor.

At the time the Collins brothers started their factory, all axes in the northern states were made by blacksmiths who also shod horses. In the South, on the other hand, cheap unground axes were imported from England and readied for use after half a day's grinding. From the inception of the Collins Company, the manufacturing process was divided roughly into four activities: forging, grinding, tempering, and polishing. At first, men tempered the axes they had forged, after which someone else ground and polished them (Memorandums 3). Within a few years, however, the process was structured so that each man performed only one operation in the manufacture of the finished product. In the period 1835–39, for which men's signatures on contracts are available, it is possible to estimate the proportion of workers in each branch of the industry. Roughly, this was 29% forgers, 29% strikers, 6% temperers, 20% grinders, 7% polishers, and about 12% employed at a variety of other tasks. Additional personnel (whose signatures are not in "Terms with Workmen") would have included overseers, other supervisory personnel, or anyone on a long-term contract, as well as a large staff of maintenance men.[5]

Forging involved hammering and shaping hot metal to obtain the correct shape of the axe head or other tool. It was carried out by a foreman and an assistant, called a "striker," who received half the amount of pay (whether by piece or day) of the foreman or "forger." Grinding involved shaping the tools by holding them against a large rotating stone wheel. Water was dripped onto the wheel to limit the spread of stone dust. Axe heads were then heat-tempered to retain an edge, and a final polishing finished off the axe (Wittmer 1977:28).

Grinding was the hazardous operation, and despite the attempt to diminish the amount of stone dust by wetting the wheels, the inhalation of free silica particles began a progressive disease that frequently debilitated and eventually killed the worker.

Silicosis

Silicosis is caused by particles of silica that are retained in respiratory organs, first poisoning and then destroying the lymphatic system, thus rendering it incapable of defending the lungs from additional silica particles and other infections. Nodules form in the lung, causing a continual cough, then lung function deteriorates rapidly, and any disease, even a cold, can be deadly. Prior to 1940, tuberculosis was the most common complication (Foster 1985:274). Zaidi, an expert on the disease, indicates that tubercle bacilli are usually involved, though not always easily identifiable (Zaidi 1969:226).

Silicosis was described by Pliny in the first century A.D. and "this characteristic lung disease of miners [was] linked to dust from the beginning" (Foster 1985:268). Medical science in the 1860s began to clarify the difference "between phthisis (the 19th-century term for tuberculosis), pneumoconiosis (the generic name for all lung diseases resulting from dust inhalation), and silicosis (the most common and most virulent form of pneumoconiosis)" (Foster 1985:269). In 1910, Haldane and R. A. Thomas isolated "silica dust as the cause of most mining disease" (Foster 1985:272).

In 1830 in England, at the same time that the Collins Company was being developed, Dr. Arnold Knight, physician to the Sheffield General Infirmary, published his observations about the disabling and killing disease that affected the cutlery grinders. He clearly stated that the disease had recently appeared in Sheffield, following upon the concentration of grinding in a small area and the division of labor in manufacturing.

> Until the beginning of the last century, grinding was not a distinct branch of business, but was performed by men who were also employed in forging, and hafting; hence they were exposed but seldom, and then only for a short time. [Knight 1830:86]

Dr. Knight's publication is widely quoted as one of the earliest references to silicosis in an industrial setting. He called it "consumption, which prevails amongst the workmen, who are employed in grinding the different kinds of cutlery goods manufactured in Sheffield and the neighbourhood" (1830:85). Knight distinguished two kinds of grinders, depending upon whether their work utilized dry or wet grindstones.

> Some of these are ground on dry grind-stones, others on wet grind-stones, hence the grinders are divided into two classes, the dry, and the wet grinders—and there is a third class, who grind both wet, and dry—altogether they amount to about two thousand five hundred. [1830:86]

He states that dry grinding rapidly destroyed the workers; wet grinding was slower but equally deadly:

> about one hundred and fifty, viz. eighty men and seventy boys are fork grinders—these grind dry and die from twenty-eight to thirty-two years of age. The razor grinders grind both wet, and dry, and they die from forty to forty-five years of age. The table-knife grinders work on wet stones, and they live to betwixt forty and fifty years of age. [Knight 1830:86]

In his study Knight cites several medical reports that implicate the inhalation of dust as causing deadly respiratory diseases. One that connects grinding and mining is that of a Dr. Forbes, who stated that

> it seems hardly to admit of question, that the habitual inhalation of dust of various kinds, is a fruitful source of bronchial inflammation, among various kinds of artisans, and more especially, in this country, needle-grinders, leather dressers, and, I can add from my own experience, *miners*. An immense proportion of the miners in Cornwall are destroyed by chronic bronchitis; one of the principal, though by no means the sole cause of which, I consider to be the inhalation of dust. [Quoted by Knight 1830:169, italics in original]

Knight goes on to describe the course of the disease in great detail.[6] The first symptoms—shortness of breath at any exertion and coughing—appear within six years of work. Respiratory difficulties, weakness, and discomfort continue to increase until the worker is incapacitated (generally within 14 years for dry grinders, 24 for wet grinders), develops tuberculosis, and dies.

Hunter, in his classic text *The Diseases of Occupations* (1976[1955]), cites Knight's observations and also those of Charles Turner Thackrah (1832), whom he credits with beginning industrial medicine. Thackrah commented on the common association of dust inhalation and tuberculosis in both mining and grinding. "Miners rarely work for more than six hours a day, yet they seldom attain the age of forty. A parallel case . . . occurs in the grinders of Sheffield" (quoted by Hunter 1976:125).

Even though the toxic effect of silica on the lymphatic system was not known until the 1920s and its etiology not worked out until the early 1960s, the understanding of how to prevent silicosis by avoiding or limiting the breathing of certain kinds of dust dates back to before the Industrial Revolution (Foster 1985:268). *What was done or not done to protect workers in these unhealthy environments depended upon class relations, not on the availability of knowledge.*

Samuel Collins had a remarkable grasp of every element of the manufacturing process. He was responsible for skillfully planning the water system to supply power (Gordon 1985); he located and bought the essential materials, eventually even producing the steel that was used to give a hard edge to the axes; and he managed the workforce. Given his generally high level of industrial knowledge, it is almost certain that from the very beginning he knew something of grinder's disease and the hazards of stone dust. The fact that the grinding shop employed wet grinding indicates such an understanding. The documents give a clear indication that Collins knew about the consequences of grinding in 1830 if not earlier, when he wrote to his brother, "Encourage stout [strong] men to come on, I want to see our gang improved in muscle—I want to hire 3 stout fellows to grind nights[7] as long as they live" (SWC: November 8, 1830). Whether or not he shared this knowledge with his workers is unknown.

Class Relations and Grinder's Disease

While capitalism was not a new mode of production in 1826 when the Collins Company began, it was new both to the Collins brothers and to most of the men who came to work for them. The Collinses came from a family of merchants and lawyers, and their workers for the most part from Yankee farm families. The transformation to industrial capitalism made farmers' sons producers of surplus labor and the Collinses into appropriators of that surplus.

Class Relations

The signatures on work contracts indicate that until the 1840s almost all of the men were Yankees (Terms with Workmen). Some were blacksmiths; a few had previous factory experience (Accounts Payable 1832–1833). Most were farmers' sons, and even the blacksmiths were probably not full-time craftsmen but farmers whose specialty was shoeing horses and providing other smithing services. These would undoubtedly have included making crude axes to sell to merchants in a way similar to their fathers, who had bartered axes for a neighbor's chickens, barrels, beef, or soap (Clark 1979; Henretta 1978; Merrill 1977).

Collins complained that many of his workers left the factory for farm work for part of the year: "numbers of our men have gone home to maying [sic] and harvesting" (Notice of 1832). As late as 1846 he complained, "it is summer now and you forget that you were very glad of a place here during the winter" (Memorandums 37). Because of the pull to farm work the company paid more in summer, "when men [were] tempted out to haying and other jobs," and less in winter (Memorandums 39).

In the early years Collins had to train most of his men. Many would have had some experience using and fixing farm machinery, which helped them to acquire the new skills. In contrast, factory discipline was an abrupt change and, in the Collins's operation as elsewhere, was difficult for management to establish (cf., for example, Dawley 1976; Prude 1983; Smith 1977; Wallace 1972). At the Collins Company a reproachful note put up in the factory in 1832 attests to the problem.

Our rules have not been enforced. We were willing to see how men would act if left to themselves. The experiment has been tried and the question settled. From small infringements our workmen have gone on to large, until our patience is exhausted, and we cannot suffer the evil to exist any longer.

He continues with a list of their misdeeds:

Some men commence work fifteen or twenty minutes, and even half an hour after bell ringing; and some leave their work during bell hours for that length of time, and even leave the shops and go up to the houses and stores. [Notice of 1832]

Relations within the workplace were strongly affected by the outside environment of job opportunities and alternatives. The pull back to the farm was obviously still strong, and other employers were also eager to hire trained men. Collins complained frequently that as soon as he had succeeded in training his workers, other axe manufacturers tried to steal them and make use of their knowledge to compete against him.

Marble's time is out and I am quite at a loss whether to keep him or not . . . I have steadier better men, but Marble poor as he is will be an acquisition to Simmons, Hunt, or any other axe concern, because he has seen good work and his ideas of work are raised—Wm Brainerd's time is out in October and I ought not to keep him without it is from such motive as keeping him away from other folks—he is a mischief maker and not a profitable man he does not earn his wages. . . . If I did not fear competition I should discharge Marble, Brainerd and Spencer. [S W C: August 12, 1830]

Marvin Gages has applied for a discharge, says Hunt of Douglass have [sic] offered him $600 per year and the privilege of forging axes @ $1 each. . . . I assured him that I would prosecute him for damages as long as he lived if he broke his contract. [S W C: August 27, 1830]

Collins tried various strategies to keep his workers:

I am very glad you are going to bring C. S. Hubbard out here. Depend upon it we ought to bring every man out here and pay him all the attention in our power (lay him under some obligation, if possible) that we wish to have engage in our article in *earnest*. [S W C: August 7, 1830, italics in original]

The contracts that men signed stipulated that they would receive only part of their pay during the first part of their contract and the remainder only after completing it. A typical contract read as follows:

It is agreed that when three months work is performed, all the money earned shall be paid except twenty dollars, and that a final settlement shall be made, when the whole six months work is performed and the money paid in full. [Terms with Workmen 38]

The company housing at Collins was clearly an inducement both to come to Collinsville and to remain, particularly for men with families. Collins wrote,

I find married men and most docile and manageable, it is considerable of a job for them to pull up stakes and move elsewhere and having a family to support they are not willing to be discharged and take their chance of finding another place. [S W C: December 16, 1830]

Despite these contracts and Collins's efforts, men left because of other opportunities and dissatisfaction with the job. In the two-year period before the Panic of 1837, the rate of nonrenewal of contracts, as judged by men's signatures on three- or six-month contracts, averaged 27% each quarter. It is clear from Collins's complaints that numerous workers also left before fulfilling their contracts. There is some indication that at Collinsville, as elsewhere, the least skilled workers left at a higher rate than the others (cf. Prude 1983: 146). Their higher turnover rate suggests that factory life itself was not the problem, nor were workers simply resistant to a change from an older style of working to a newer one. Their resistance was to particular conditions of work. As Prude has pointed out, as long as jobs were readily available, one of the workers' strongest weapons against the demands and discipline of management was their ability to leave and find another job (Prude 1983: 144).

Collins, of course, deplored this state of affairs. "I want very much to see more applicants for employment in every department that I may make some changes for I think it would have a salutary influence on some (even on those who wish to leave)" (SWC: September 10, 1830). Prude's analysis of this kind of "complex, back-and-forth tugging between these working people and the men and rules they faced" is that it signifies class conflict: "what occurred in local mills during the generation after 1810— and what the workers' opposition both reflected and helped create—was class" (Prude 1983:157).

Class relations did not remain fixed. There were many factors, both internal and external to the Collins Company, that influenced the shifting balance of power. The Panic of 1837 and the depression that followed put more power in the hands of management as jobs became scarce. Collins was, of course, not slow to take advantage of the situation. A notice posted on January 17, 1843, read as follows:

Having frequent applications for work, and some offers to work for *less* than we now pay, we have concluded to receive proposals for forging bitts of Axes, and also for hammering off the heads for this year ensuing. As we are not willing to contract for a larger quantity than we are now making, those who are now at work will do well to put in proposals unless they are willing to lose their places. [Terms with Workmen 96, italics in original]

In the pursuit of profit, management continually attempted to decrease its need for skilled, trained labor. Employers' vulnerability to their workers' decisions to leave made them strive to increase the skills of their machines and decrease their dependency upon workers' skills. Increased productivity was, of course, desirable, but much more so when it was due to the use of a machine owned by the employer, rather than a skill carried in the head and hands of a worker.[8] Wallace describes the efforts of Delaware mill owners to replace their skilled and independent mule-spinners with a docile machine (Wallace 1972:193). Some degree of scientific knowledge and technical ability are, of course, components of the invention process, but manufacturers' decisions about where to place their money and hopes are also a critical component of the inventor's imagination and success.

The Collins Company was founded on the inventive skills of Charles Morgan, a craftsman. Within a few years the division of axe-making into several procedures made it possible to teach these skills to reasonably adept workers who had prior general skills in handling tools and machinery. In 1833 Collins complained that "this want of good workmen and the time it took to learn [teach] them was a bar to such success as we had counted upon" (Memorandums 16). By 1846, however, the transfer of skills from workers to machinery had progressed to a point where Collins responded to the men's petition for a raise by telling them, "The fact is that axe-making has *become so simplified* that any Yankee farmer boy of average ingenuity and strength can as soon as he is of age and starts for himself soon get hold of any branch of the business with very little practice" (Memorandums 37, emphasis added). Even though one would suspect Collins in part of downplaying the men's skills to weaken their demand for a raise, this statement reflects the effects of mechanization in decreasing the importance of workers' skills and training and strengthening the hand of management.

Grinders' and Management's Responses to the Disease

Conflict between labor and management was embedded in every nut and bolt of the industrial environment. It indelibly colored decisions and perceptions that affected the health and life of the workers. The relative strength of workers as a class in the early years would have made a job such as dry grinding unthinkable, since its lethal effects were rapid and apparent. Had Collins attempted to utilize this method, he would have lost his men at a period when there was no other labor supply available capable of replacing them. Wet grinding may or may not have been the cheapest or quickest method for grinding edged tools, but it delayed the disabling stage of silicosis and allowed Collins to retain his workers.

In Collins's letters of 1830 there are indications that some of the grinders were beginning to fall sick. From Knight's description of the disease, it seems early for the onset of visible symptoms, since he estimated six years before they usually appeared, and in 1830 the longest a man could have worked at the Collins Company was three to four years. However, conditions were not identical, and it is suggestive that the references to sickness in the factory apply only to the grinders. Referring to them, Collins wrote in a letter to his brother, "my present set of hands will not work nights any longer—pretend to be sick, work sometimes half the night and sometimes not at all" (SWC: November 11, 1830). The following day he again mentioned, "the grinders feel cross and take the opportunity to be sick many of them" (SWC: November 12, 1830), and a few days later:

> This falling off in the grinding is attributable to several causes—firstly they have for sometimes worked very hard and steadily by the piece and were about ready to think of a respite—when our refusing to accept the axes, in connexion with the bad weather

which made it very dark, and occasioned some colds, turned the scale, and they are sick and missing and what are left might almost as well be. [SWC: November 16, 1830]

By 1845 there is conclusive evidence that the men not only knew of the danger of their occupation but refused to continue with the work. Collins wrote in his memoirs that during this year,

There had been so many deaths among the grinders that no Yankee would grind, and the Irish were so awkward and stupid that we did not get the quantity needed even by having extra men working at night. [Memorandums 30]

Neither compassion nor any sense of responsibility for the men's deaths appears to have occupied Collins's mind. His statement was written in 1866, as part of a public self-assessment of his long career. Collins apparently assumed, undoubtedly correctly, that no one of his class would find fault with this attitude.

Collins, like many of his fellow New England manufacturers, was concerned publicly about avoiding the well-known horrors of the English factory system and the obvious degradation of the English workers (Sanford 1958). The focus of their enlightened concern was the moral well-being of their workers, which was primarily to be achieved through temperance. Every contract signed at the Collins Company included a statement forbidding alcohol, not only in the workplace but in the village as well: "no man will be retained . . . who is guilty of disorderly conduct, or carries spiritous liquors to the shops, or drinks them daily and habitually elsewhere" (Terms with Workers 2). Collins bought up and closed down the local tavern (Memorandums 18) and fired several men who did not meet the standards of the contract.

The medical profession in the United States colluded with manufacturers in blaming workers' habits for the destruction of their lives, rather than the manufacturers' industrial environment. "When such physicians did not ignore endemic disease entirely, they blamed any symptoms they found on the workers' 'improvident' way of life" (Rosen 1944, cited by Kaprow 1985: 346). Although this approach to workers' health is reflected in the statement of a British physician, who studied "phthisis" among French millstone manufacturers in 1861, his honest scientific observations contradict his preconception. He stated that the millstone makers, who worked with siliceous stones, "rarely live beyond the age of 40," and though he ascribed "much to the inhaling of gritty particles," he thought that "the unhealthiness of the occupation is greatly aggravated by other unfavorable hygienic influences, especially the habits of intemperance in the men" (Peacock 1861:40). In the same paragraph, however, he notes that men employed by the same manufacturer but in a different process, wire-weaving, reach "the full duration of human life" (Peacock 1861:40).

From Collins's statement, quoted above, it is clear that by 1845 there was no doubt in anyone's mind as to the cause of the grinders' deaths. The Yankees' refusal to grind thus seems perfectly logical. Yet it was dependent on the fact that they were still (or again after the depression that started to lift after 1843) in a favorable position to find other jobs. A comparison of their responses to their working conditions with those of laborers in Sheffield, as described by Knight, reveals the importance of the labor market in influencing workers' behavior.

Grinders' asthma is the name given by the Grinders themselves, to that form of consumption which prevails amongst the workmen, who are employed in grinding the
different kinds of cutlery goods manufactured in Sheffield, and the neighbourhood. The
name conveys but a very imperfect idea of the disease, since the grinders' asthma bears
scarcely any resemblance to asthma, properly so called, except in those symptoms which
are common to almost all affections [sic] of the chest. It is probable however that it has
been so designated not only from this vague resemblance, but also from *a natural desire
on the part of the grinders to conceal the fatal character of their complaint* by assigning to
it the name of a disease, which does not necessarily interfere with the duration of life.
[Knight 1830:85, emphasis added]

Unlike Sheffield grinders in the 1820s, Connecticut Yankees in the 1840s were in a
position to refuse this hazardous occupation. The Irish, who were brought in to
replace them, were at this time a less skilled and more vulnerable workforce. Earlier
Collins had boasted to his employees of "the policy I have always adopted and
advocated, viz. to employ no foreigners, none but Americans, believing them to be
not only more ingenious and industrious than foreigners, but more enlightened and
consequently more rational and reasonable" (Speech of 1833, quoted in Memorandums 11). In the 1840s, however, Collins along with his fellow manufacturers eagerly
hired the Irish and put them to work at, among other things, the hazardous job of
grinding that the Yankees refused to do.

Because the "awkwardness" of the Irish at grinding the axes (which may have been
resistance to the work) limited production, "Mr. Root invented a process for shaving
the axes as a substitute for grinding but it was not put in operation in time to aide
[sic] us much this year," wrote Collins, referring to 1845 (Memorandums 30). The
startling inference to be drawn from this statement is that an alternate process was
possible or, at the very least, conceivable. Yet Collins did not ask or encourage Root
to invent a new machine until production was threatened, despite his knowledge over
a period of years that grinding was deadly. When the Yankees refused to grind, there
was still no reason to look for an alternate process, since the Irish could be hired to
take on the work. While Collins's writings do not suggest any concern or sense of
responsibility for the health of his workers, they do indicate concern over lowered
production and the well-being of his machines. "The effect of running our grinding
with green Irishmen became visible by the increased 'wear and tear' of machinery"
(Memorandum 32). Only when production fell did Mr. Root invent a new process.

Mr. Root's Invention

Mr. Root had been employed at Collins Company since 1832, starting as a journey-
man mechanic. He had invented two of five machines that were installed in the
factory between 1830 and 1845. One was a power-driven device that formed hot iron
between cast iron dies (Gordon 1983:20), and the other was a specialized machine for
axe production (Uselding 1974:550). Root was, in fact, one of the foremost inventors
and mechanics of his time. He became superintendent of the Collins Company and
in 1849 superintendent of the Colt Company (see Uselding [1974] for a discussion of
Root's career and inventions).

The shaving machines and entire manufacturing process were described by an anonymous writer for *Scientific American,* in an article entitled "Our Visit to the Collinsville Ax Works." The writer also noted that

> these machines are . . . the product of Mr. Root's busy brain, and are peculiar to this establishment. Prior to their application the axes were all ground by which tedious, unhealthy, and disagreeable operation, the surplus metal was all washed away with the grit, but the shavings are now all saved and sold for scrap-iron. [Scientific American 1859:36]

Grindstones were still used for shaping flat tools, such as picks, hoes, and machetes (1859:36), which were probably less difficult to fashion than the axes. As the article suggests, the machine also introduced an economy in that axe shavings were no longer simply wasted. As Collins made clear, however, this machinery was not the result of an autonomous technological progress or evolution, or solely the product of a brilliant inventor, but of a combined set of social factors and relations that included the independence and solidarity of the Yankee grinders and the vulnerability and lack of skill of the Irish newcomers. Concern for the men for whom grinding was a deadly job seems not to have played a role in this invention. Had the Irish been better at grinding, Mr. Root would probably not have invented his shaving machine.

Conclusion

Arnold Knight, the physician at Sheffield, ended the first part of his path-breaking article with a suggestion:

> Fork grinding [dry grinding] is the most destructive kind of grinding carried on in this neighbourhood, and, as it requires the least skill, it is the most easily learnt. Might not criminals, who have grossly violated the laws of their country, be condemned for life to fork-grinding, as for minor offences they are sentenced to the treadmill? [1830:91]

This suggestion may seem strange, but according to Mumford, the similarly hazardous job of mining was carried out in this way until the 16th century (1963[1934]:67). Dr. Knight apparently did not accept the wages paid for this kind of work as just recompense for the destruction of human life. Employers, however, have long argued that in accepting a job, workers accept the risks of the workplace. The conditions of work, particularly the machinery and industrial processes, are represented as being independent of the will and desires of the employer.

In his own words Collins tells us that he directed the construction of the buildings, machinery, and power for manufacturing axes, machetes, and other edged tools, thus creating the environment in which silicosis occurred. His first statement about grinder's disease appeared in a letter to his brothers in 1830. The second, appearing in the Memorandums and written in 1866, was a public statement that he made to posterity with no sense of incongruity or shame, a statement that he could assume

to be acceptable to his audience. Machines were outside the responsibility or inten-
tionality of the manufacturer, even though he ordered them to be built and maintained, and he profited by their use. Only when Collins faced a decline in production
did he make the effort to modify this technology, though he had known for at least
15 years that its utilization shortened his workers' lives.

In other industries, of course, the same type of decisions were made.[9] Collins
shaped and was shaped by the developing class relations of early industrial capitalism.
By 1866, when Collins described the Yankee workers' objections to grinding, he had
fully developed both a capitalist mentality—the ability to separate workers as one of
the factors of production from workers as fellow humans—and a capitalist ethic that
defined profit as management's concern, health as the workers'.

NOTES

I would like to thank Annstress Paine and Robert Gordon for their help in directing me to
documentary sources. I am also indebted to the librarians of the Canton Historical Society
and of the Connecticut Historical Society. I am grateful for the opportunity of interviewing
three men who had worked for much of their lives at Collins Manufacturing Company: Mike
Gotaski, packer; Miles Henry, machinist; and Guy Whitney, superintendent. In addition,
Constance Berggren, whose father had worked at Collins for more than 30 years, and Thomas
Perry, who is now in charge of the property, were extremely informative. I found Paul
Wittmer's help and his work on the company to be invaluable.

1. No statistics exist as to the overall incidence and mortality of occupational diseases in
the 19th century, as indeed none exist on a national level to this day (Goldsmith and Kerr
1982:28). This absence of record keeping graphically illustrates government's collusion with
industry. In her excellent critique of Douglas and Wildavsky's *Risk and Culture*, Kaprow points
out how government has supported industry in its resistance to accepting responsibility for
both occupational disease and environmental destruction: "Industry and government have
cooperated to the detriment of the general public, and manufacturing danger has been as
much a political activity in the past as now" (Kaprow 1985:346).

2. In citing the letters, the abbreviation SWC (Samuel W. Collins) will be used, followed
by the date it was written. All letters are from Samuel to David. The letters are listed under
"Primary Sources" along with the archives in which they may be found.

3. There are collections of workers' letters for this period from the textile mills (Dublin
1979) however.

4. In the late 1960s, when I was carrying out an ethnographic study in the tropical forest
region of eastern Peru, my friends and informants frequently asked me for gifts. The gift that
men valued most was a machete and, they specified, *not* a Brazilian one but a *Collins* machete.
I found, as have many other anthropologists, that Collins machetes were manufactured in
Connecticut. I next encountered the name in a book called *Machines that Built America*
(Burlingame 1953), which stated that the Collins axe had "cleared the West." It was the same
company. It was founded in 1826 and had remained profitably in operation until it was
liquidated in 1966. The machetes I had bought in Peru were probably made in one of the
factories holding Collins's patents in Mexico, Colombia, or Guatemala. Collinsville, Connec-

ticut, is no longer a bustling industrial town, but many of the factory buildings, waterways, and houses that were built for workers are still there.

The company's name was originally Collins and Company. In 1833 when it was incorporated, its name officially became Collins Manufacturing Company. In Collinsville, however, where people still talk about it, the plant is simply called the Collins Company, which is the name that will be used here.

5. Guy Whitney, who had been the superintendent of the Collins Company for more than 30 years, described it as consisting of two companies: one producing tools and the other maintaining the machinery and plant (Whitney interview, 1984).

6. In the description of the onset of this disease that follows, Knight assumed that boys started to work at age 14.

> Grinders, who have good constitutions, seldom experience much inconvenience from their trade until they arrive at about twenty years of age: about that time the symptoms of their peculiar complaint begin to steal upon them, their breathing becomes more than usually embarrassed on slight exertions, particularly on going up stairs, or ascending a hill; their shoulders are elevated in order to relieve the constant and encreasing [sic] dyspnoea; they stoop forward, and appear to breath the most comfortably in that posture in which they are accustomed to sit at their work, viz.: with their elbows resting on their knees. Their complexions assume a dirty, muddy appearance; their countenance indicates anxiety; they complain of a sense of tightness across the chest; their voice is rough, and hoarse; their cough loud, and as if the air were driven through wooden tubes: they occasionally expectorate considerable quantities of dust, sometimes mixed up with mucus, at other times in globular or cylindrical masses enveloped in a thin film of mucus. Haemoptysis frequently occurs; the blood is seldom florid or in large quantities; there is frequently a perceptible thickening about the larynx or trachea, with tenderness and cough on pressure. . . . About thirty years of age the dry grinders become incapable of performing their usual labour, the wet grinders are similarly affected about forty. The sense of *fastness* under the sternum now grows very distressing; the lungs feel as if they were choked up with dust; the cough becomes incessant; the expectoration copious, and purulent; and the pulse quick; dropsy, in some form or other, not unfrequently [sic] supervenes. Haemoptysis, inability to lie down, night sweats, colliquative diarrhea, extreme emaciation, together with all the usual symptoms of Pulmonary Consumption at length carry them off. . . . The habits of the grinder, and his daily exposure to wet and cold, subject him to attacks of acute inflammation, particularly in the respiratory organs; hence acute bronchitis, pneumonia, and pleurisy, are complaints of frequent occurrence amongst this class of operatives. [Knight 1830:170–171, emphasis in original]

7. In 1829 Collins reduced the 12-hour day to 10 hours, finding that "they accomplished just as much during the shorter hours and burned less coal" (Wittmer 1977:33). However, at times the irregularity of water-based power meant that work piled up when the river was low, and power was wasted when the river flow was strong. At times, therefore, Collins attempted to get men to work at night, paying them by the piece.

8. At the textile mills of Rockdale, Delaware, for example:

> The tendency in the 1830's and 1840's, then, was for the cotton mill machinery to become progressively more specialized and intricate, while the cotton mill operative became progressively more standardized and indifferently skilled. [Wallace 1972:383]

9. Cotton mill owners, for example, nailed the windows of their spinning rooms shut to keep the cotton from drying out. The air was filled with dust and fibers which caused a variety of serious respiratory ailments, including pneumonia (Arlidge 1892:360).

REFERENCES CITED

Primary Sources

Collins, Samuel, 1866–1867, Memorandums. Connecticut Historical Society, Hartford. Hand-written; pages cited in the text are those of the library.

Terms with Workmen, Collins Manufacturing Company, December 1834 to November 1839. The Connecticut Historical Society.

Accounts Payable 1832–1833, Collins and Company. The Connecticut Historical Society.

Notice of 1832, signed by Samuel W. Collins, Collins and Company, July 28, 1832. Canton Historical Society, Collinsville, Connecticut.

Letters from Samuel Collins, Collinsville, to David Collins, Hartford (cited as "SWC" in the text)

August 7, 1830, Connecticut Historical Society, Hartford
August 12, 1830, Connecticut Historical Society
August 27, 1830, Connecticut Historical Society
September 10, 1830, Connecticut Historical Society
November 8, 1830, Canton Historical Society
November 11, 1830, Canton Historical Society
November 12, 1830, Canton Historical Society
November 16, 1830, Canton Historical Society
December 16, 1830, Canton Historical Society.

Interview with Guy Whitney, June 9, 1984, Collinsville, Connecticut.

Secondary Sources

Arlidge, John. 1892. *The Hygiene Diseases and Mortality of Occupations*. London: Percival.

Berman, Daniel. 1978. *Death on the Job*. New York: Monthly Review Press.

Burlingame, Roger. 1953. *Machines that Built America*. New York: Harcourt Brace.

Clark, Christopher. 1979. The Household Economy, Market Exchange, and the Rise of Capitalism in the Connecticut Valley, 1800–1860. *Journal of Social History* 13:169–189.

Dawley, Alan. 1976. *Class and Community: The Industrial Revolution in Lynn*. Cambridge, MA: Harvard University Press.

Dawley, Alan, and Paul Faler. 1976. Working-class Culture and Politics. *Journal of Social History* 9:466–479.

Dublin, Thomas. 1979. *Women at Work*. New York: Columbia University Press.

Foster, James. 1985. The Western Dilemma: Miners, Silicosis, and Compensation. *Labor History* 26:268–287.

Gersuny, Carl. 1981. *Work Hazards and Industrial Conflict*. Hanover, NH: University Press of New England for the University of Rhode Island.

Goldsmith, Frank, and Lorin Kerr. 1982. *Occupational Safety and Health*. New York: Human Sciences Press.

Gordon, Robert. 1983. Material Evidence of the Development of Metalworking Technology at the Collins Axe Factory. *Journal of the Society for Industrial Archaeology* 9:19–28.

———. 1985. Hydrological Science and the Development of Waterpower in Manufacturing. *Technology and Culture* 26:204–235.

Grimsted, David. 1985. Ante-Bellum Labor: Violence, Strikes and Communal Arbitration. *Journal of Social History* 19:5–28.

Henretta, James. 1973. *The Evolution of American Society, 1700–1815*. Lexington, MA: D.C. Health and Company.

———. 1978. Families and Farms: *Mentalité* in Pre-industrial America. *The William and Mary Quarterly* (3rd ser.) 35:3–32.

Hirsch, Susan. 1978. *Roots of the American Working Class*. Philadelphia: University of Pennsylvania Press.

Horwitz, Morton. 1977. *The Transformation of American Law, 1780–1860*. Cambridge, MA: Harvard University Press.

Hunter, Donald. 1976[1955]. *The Diseases of Occupations*. 6th edition. London: Hodder and Stoughton.

Kaprow, Miriam Lee. 1985. Manufacturing Danger: Fear and Pollution in Industrial Society. *American Anthropologist* 87:342–356.

Knight, Arnold. 1830. On the Grinders' Asthma. *North of England Medical and Surgical Journal* 1:85–91, 167–179.

Kulik, Gary. 1978. Pawtucket Village and the Strike of 1824: The Origins of Class Conflict in Rhode Island. *Radical History Review* 17 (Spring):5–37.

Licht, Walter. 1986. Norman Ware's "The Industrial Worker" Revisited: Reflections on Recent Writings on Early American Industrialization. *Radical History Review* 17 (Fall):137–148.

Merrill, Michael. 1977. Cash Is Good to Eat: Self-sufficiency and Exchange in the Rural Economy of the United States. *Radical History Review* 3 (Winter):42–71.

Montgomery, David. 1968. The Working Classes of the Preindustrial American City, 1780–1830. *Labor History* 9:1–22.

Mumford, Lewis. 1963[1934]. *Technics and Civilization*. New York: Harcourt Brace Jovanovich.

Navarro, Vicente. 1986[1980]. Work, Ideology, and Science: The Case of Medicine. In *Crisis, Health and Medicine: A Social Critique*. Pp. 144–182. New York and London: Tavistock.

Peacock, T. B. 1861. Millstone Makers Phthisis. *Transactions of the Pathological Society of London* 12:36–40.

Prude, Jonathan. 1983. *The Coming of Industrial Order*. Cambridge, MA: Cambridge University Press.

Rosen, George. 1944. The Medical Aspects of the Controversy over Factory Conditions in New England, 1840–1850. *Bulletin of the History of Medicine* 15:483–497.

Sanford, Charles. 1958. The Intellectual Origins and New Worldliness of American Industry. *Journal of Economic History* 18:1–16.

Scientific American. 1859. Our Visit to the Collinsville Ax Works. *Scientific American* (n.s.) 1:36–37.

Smith, Merritt Roe. 1977. *Harpers Ferry Armory and the New Technology*. Ithaca: Cornell University Press.

Thackrah, C. T. 1832. *The Effects of Arts, Trades and Professions*. London: Longmans.

United States Treasury Department. 1969[1833]. *Documents Relative to the Manufacturers in the United States, Collected and Transmitted to the House of Representatives*, Vol. 1. New York: A. M. Kelley.

Uselding, Paul. 1974. Elisha K. Root, Forging and the "American System." *Technology and Culture* 15:543–568.

Wallace, Anthony. 1972. *Rockdale*. New York: W. W. Norton.

Wittmer, Paul. 1977. Early History of the Collins Company. In *Bicentennial Lectures, 1976*. Pp. 21–50. Canton, CT: Canton Bicentennial Commission.

Zaidi, S. H. 1969. *Experimental Pneumoconiosis*. Baltimore: John Hopkins Press.

QUESTIONS

1. Why does Siskind argue that the minimal efforts to protect factory workers from exposure to stone dust in the early industrial capitalist era was the result of class relations, rather than lack of knowledge about the causes of silicosis?

2. What evidence does Siskind provide of the factory owner's lack of concern for the health of his workers? Do you find this evidence convincing?

Chapter Ten

From Dust to Dust

The Birth and Re-Birth of National Concern about Silicosis

David Rosner and Gerald E. Markowitz

John Farmer, an African American sandblaster, was born during the Depression along the Gulf Coast of Texas. He spent two years in college, joined the army, and then worked primarily in a shipyard as a laborer and sandblaster, until 1982, when he retired. Because of his short, thin stature (5 foot 6 inches tall, weighing less than 125 pounds), Farmer was often sent into the poorly ventilated holds and double-bottoms of ships, where he sandblasted asbestos and other residue. He usually used a "desert hood" against ricocheting particles, a cartridge respirator to partially filter the silica-laden air he breathed, and sometimes an air-fed hood: a cumbersome space suit that supplied him with relatively pure air for the short time he was able to wear it in the hot and humid environment of this southern shipyard. In 1988, when he was 53 years old, a doctor diagnosed him with "massive progressive fibrosis." Three years later, he had deteriorated to the point where he "was no longer able to walk and [could] only stand briefly while using supplemental oxygen." After considering him for a lung transplant, the physician noted that this 56-year old man's future looked "bleak both for his longevity as well as quality of life."

Lawrence Brown was born in Louisiana in 1946 and following discharge from the army in 1977, he began working as a sandblaster and painter for a local paint company that contracted non-union workers out to virtually every major refinery in the Port Arthur, Texas, area. He usually wore a desert hood and sometimes a paper dust mask as he blasted the inside of storage tanks and other vessels, preparing them for painting. In 1988, at the age of 41, he developed night sweats, violent coughing, and shortness of breath, and went to the veterans' hospital where he was diagnosed with TB. Two years later, this 44-year-old white unemployed veteran had lost 20-25 pounds, had persistent coughing with phlegm production, and suffered intermittent episodes of vomiting and shortness of breath. He had difficulty exerting himself to take a shower. In June of 1990 his X-rays were re-evaluated and he was diagnosed with silicosis. Two years later, he was dead, at the age of 46. He had only sandblasted for ten years (Dwight Nd).

These men suffered from silicosis, once considered the plight of older workers who became disabled after 30 to 40 years of working in the "dusty trades" of potteries, foundries, construction, and mining and which, once again, has emerged as a major public health concern. From West Texas to West Virginia, from California to New York, in industries from oil refining and coal mining to foundries and shipyards, America is experiencing an epidemic. In the last three years much attention has been paid to this problem. There have been two international conferences to evaluate the scientific evidence of the link between silica and cancer, and the International Agency for Research on Cancer (IARC) named silica as a probable human carcinogen. There have been numerous panels at scientific and public health conferences, and a National Conference to Eliminate Silicosis sponsored by the National Institute of Occupational Safety and Health (NIOSH), the Mine Safety and Health Administration (MSHA), and the Occupational Safety and Health Administration (OSHA), and the American Lung Association. Former Secretary of Labor Robert Reich announced a campaign on silicosis that would seek to eliminate this disease. The American National Standards Institute (ANSI) has recommended a ban on the use of sand in indoor abrasive blasting as a means of reducing the risk of disability and death for sand blasters, and, most recently, Secretary of Labor Alexis Herman reaffirmed OSHA's and MSHA's National Campaign to Eliminate Silicosis.

How can we understand the re-awakening of national concern about silicosis? How can we begin to disentangle the social, economic, and epidemiological factors that have led scores of scientists, policy makers, industrial hygienists, labor unions, and industry representatives to reassess the danger that silica sand poses to the health of an estimated two million workers? In part the answer lies in the changing American workplace, and the increased dust exposures that workers face. In part the answer lies in the efforts of a new generation of industrial hygienists, government officials, and labor activists. In part it has to do with diseased workers themselves bringing lawsuits against employers, sand suppliers, and equipment manufacturers. Together, these factors have created a financial burden for industry, and a political and social problem for government and labor.

This chapter examines how we, as a culture, proceed to define disease and how disease itself has been understood in the postwar era. It is our contention that contemporary popular, professional and governmental awareness of disease is not necessarily due to medical advances or epidemiological changes. Rather, it is shaped by social, political, and economic forces as well as technical and scientific innovation. Further, we maintain that it is impossible to understand the history of disease without understanding the social context and the social constraints that allow for its emergence or disappearance as a problem. Since the establishment of OSHA and NIOSH in 1970, a contentious debate has emerged about the role of government in regulating private industry's workplace practices. The case of silicosis allows us to test both the strengths and weaknesses of an approach that has emphasized cooperation and voluntary compliance with health and safety standards that can sometimes dramatically affect the way that industries conduct business.

Background to the Modern Crisis

Silica is a ubiquitous mineral in our environment, the chief component of sand. The earth's crust is approximately 90 percent silica. Every time we go to the beach or climb through the mountains we tread on silica in combination with other minerals. The material is so common that it is difficult for us to see it as a threat to our health and well-being, but in crystalline form, as quartz crystals, it most assuredly is. Although silica dust has been perceived as a health risk for workers since ancient times, it was only in the twentieth century that this material was defined as a widespread hazard. The emergence of silicosis as a major threat to the American workforce can be traced to the technological innovations that accompanied the country's rise to world power at the turn of the century. The enormous movement of people from the open air of the nation's farms and into the confined spaces of factories, mines, and mills was the precondition for popular sensitivity to the relationship between work and disease. The introduction of all the power equipment that drove those mines, mills, and factories created vast quantities of dust that greatly affected the health of the millions of industrial workers.

Silicosis assumed importance in the first third of the twentieth century because professionals and the public alike recognized that it represented a new kind of disease, unlike those such as lead or phosphorous poisoning, that had posed major threats in earlier times. Here was a disease that was chronic and irreversible in nature, that developed years or even decades after first exposure, whose symptoms were often ambiguous, that was rooted in work and factory production, and that had enormous social and economic implications for industries at the heart of American industrial might.

The story of silicosis is more than the story of a particular crisis in American medicine and public health. Rather, it is the story of the discovery of chronic industrial disease and its relationship to industrial society. At different moments, the crises around workers and their health have transcended narrow professional bounds and become part of a broader discussion of the relationship of illness to dependence and social welfare.

The story of the current attention to silicosis really begins in the 1930s and the Great Depression. In 1935 it was discovered that about 700 workers had died after drilling tunnels for Union Carbide at Gauley Bridge, West Virginia, a tragedy often referred to as the "Hawk's Nest disaster" (Cherniack 1986). Investigations by the U.S. Congress revealed that the very company that was killing them had buried hundreds at the side of the road outside the tunnel in unmarked graves. Less well known, but equally important, were the thousands of silicosis lawsuits that threatened the economic stability of foundries, metal mines, potteries, construction companies, among many other industries. As a result of these lawsuits from Depression-era workers, silicosis was considered the "king of occupational diseases," threatening millions of American workers and their families. During this time silicosis would lead workers to use the issue of industrial illness as a means of achieving social welfare objectives. As workers lost their jobs, they turned to the courts for redress, maintaining that this terrible disease, caused by the industries in which they worked, had led to their disability and deaths. The social conditions of the Depression forced a blurring of the line between health and welfare, disease and dependence.

During the Depression, with millions of workers unemployed, this medical problem was transformed into a national crisis as out-of-court settlements and jury decisions threatened the economic stability of thousands of companies and hundreds of communities throughout the nation. Throughout the 1930s, books and films, popular articles and magazines, scores of reports in medical journals, and news articles in such weeklies as *Business Week* and *Newsweek* proclaimed that this ever-present substance threatened the health of millions of Americans. In 1936, the Federal Government called together Public Health Service and Labor Department officials, industry leaders, and organized labor to the first National Silicosis Conference in Washington, D.C., to assess the means of preventing silicosis among the nation's workforce (Rosner and Markowitz 1994).

The industry responded to the crisis by convincing state governments to incorporate silicosis into state workers' compensation schedules after 1935. This proved to be an important factor in removing silicosis from public attention in the 1940s and 1950s. The workers' compensation statutes were generally quite limited in that they provided very little assistance to individual workers and also took away workers' ability to take their case to court. Instead of juries of their peers making decisions about culpability for disease, workers' lives were in the hands of panels of experts, many of who viewed their cases skeptically. In New York, for example, the bill provided for no compensation for partial disability from silicosis and a maximum of $3,000 for total disability (Andrews 1936). Between 1936, the year compensation for silicosis became law, and the end of the decade, only 79 workers had been compensated for silicosis, receiving a total of $99,594 ("Silicosis" 1940). Along with other actions taken by industry to limit the visibility of silicosis, by the 1940s the disease began to fade from public view.

By the 1940s, the professional and business communities, despite continuing documentation of cases, declared silicosis a "disease of the past," whose current victims were a legacy of the unhygienic and primitive conditions of work of a bygone era (Rosner and Markowitz 1994). Despite the controversy and conflict over the nature of the silicosis hazard, by the 1950s silicosis was virtually forgotten in the mass media. The same symptoms that had forced physicians in the 1930s to look for an occupational history of work in the dusty trades now led to diagnoses of asthma, emphysema, or other non-occupational lung conditions. The popular and professional press no longer published the large number of articles that had made silicosis a national issue. Silicosis was no longer integrated into plots of books and movies and workers no longer saw it as a major threat to their health. Private practitioners, when forced to confront an undeniable case of silicosis, now saw the patient as a mere curiosity, someone that was a reminder of an earlier era.

Ironically, at the very moment silicosis was declared dead by business and the industrial hygiene community, the booming the postwar economy was leading to dangerous dust exposures for hundreds of thousands of other workers. In the Gulf Coast region of Louisiana, east Texas, and Mississippi thousands of workers found jobs in the booming shipyards and offshore oil rigs and oil refineries of Gulf Coast communities. As workers were sent to clean the hulls of ships, oil storage tanks, and refineries, and given the dirty job of sandblasting a variety of objects, including piping and the inside of tanks covered with residues, they would become a new generation of victims.

The two men portrayed at the beginning of this chapter illustrate the movement of the silicosis epidemic from the shipyards of Louisiana to the oil refineries of Texas's Gulf Coast cities. Despite the years of assurance that silicosis was a disease of the past and that current workers could be adequately protected through the use of proper ventilation, substitution, and protective equipment, the reality was that workers continued to be exposed to excess amounts of silica and that silicosis never really vanished. It is important to note that death certificates often never even listed silicosis as a cause or contributing factor. Given the general complacency of industry, industrial hygienists, and the medical community regarding silicosis, and given the inability of local physicians and county health departments to recognize the disease, it was virtually impossible to develop reliable statistics concerning its prevalence in the decades following World War II.

NIOSH, OSHA, and the Debate over Banning Sand

By the late 1960s, as sandblasters, painters, and others who worked in the shipyards in Louisiana began to come forward, complaining of constricted breathing and terrible pain, Morton Ziskind, Hans Weill, and Bezhad Samimi began a series of epidemiological studies at Tulane University that documented widespread silicosis. This documentation of the silicosis epidemic in Louisiana corresponded with the passage of the Occupational Safety and Health Act of 1970, creating the National Institute of Occupational Safety and Health (NIOSH) in the Department of Health, Education, and Welfare and the Occupational Safety and Health Administration (OSHA) in the Department of Labor. Among the first activities of the NIOSH was to write reports that would provide a scientific justification for Occupational Safety and Health Administration's regulatory activity.

Among the first substances that NIOSH examined was silica. Since silicosis was one of the oldest and presumably best-documented occupational diseases, it was believed that developing the scientific base for an enforceable standard would be politically less controversial than establishing standards for newer toxic substances. In the early 1970s, NIOSH contracted with Austin Blair, an industrial hygienist from the Boeing Aerospace Company in Seattle, to survey working conditions in sandblast operations. His report proved to be an indictment of silica exposures in general and the protection—or more accurately, the lack of protection—that respirators and protective equipment afforded workers in particular. For the previous two decades, it was assumed that silicosis could be prevented if workers used respirators that lowered inhaled dust to levels below the Threshold Limit Value (TLV). But, the report raised questions about whether this was a valid assumption. Overall, he found, "the protection afforded these workmen is, on the average, marginal to poor" (Blair 1974).

The indictment of protective equipment provided by Blair's report was serious enough. But, shortly thereafter, NIOSH received evidence that silicosis, far from being a historical entity, was a serious problem in the New Orleans steel fabrication yards where sandblasting was practiced. In mid-1974, two other NIOSH-supported projects reported cases of silicosis among shipyard workers and steel fabricators in New Orleans and elsewhere (Ziskind 1971–74).

The response of officials in NIOSH was swift. In 1974, NIOSH issued a Criteria Document Recommendation for a Crystalline Silica Standard that recommended that the Permissible Exposure Limit (PEL) be cut in half to "50 micrograms per cubic meter of air." The Document further recommended that silica be banned as an abrasive in blasting (NIOSH 1974).

The NIOSH recommendation provoked a storm of protest and a massive organizing effort by industry groups especially in Texas and California to make sure that the Occupational Safety and Health Administration did not adopt the recommendation. In February 1975, just after the NIOSH document was published in the Federal Register, more than fifty people representing these various industries gathered together in Houston to form the "Silica Safety Association (SSA)." Their purpose, they proclaimed, was to "investigate and report on possible health hazards involved in [the] use of silica products and to recommend adequate protective measures considered economically feasible" (SSA 1975). Shortly thereafter, SSA wrote to various industries requesting financial support for the organization, and encouraged them to write letters to the OSHA Docket Officer requesting delay in the public hearings on NIOSH's proposed standard. In these appeals, it was clear that the health of workers was secondary to other considerations. The primary purpose of the organization was "to represent interested parties in the attempt to assure the continued use of sand in abrasive blasting operations (Sline 1975).

In the course of the next few months, the SSA developed its argument justifying the continued use of sand. The organization held that if workers used "proper protective devices" there was little danger of excessive exposure. Therefore, if equipment was capable of lowering exposure to "safe" levels, there was no need to change the existing Threshold Limit Value for silica. The SSA also argued that there was little evidence that silicosis was a serious problem and that there was no data to support federal regulation of the industries that employed sandblasting. L. L. Sline, the president of SSA, argued that sandblasting was safe (*Materials Performance* 1975). The SSA mobilized industry and employed a consulting firm to organize its lobbying efforts in Washington. The Association, distrustful of NIOSH and its staff's strong advocacy of a stringent standard, centered its efforts on convincing the Occupational Safety and Health Administration (OSHA) to delay adoption of the standard (Rosner and Markowitz 1994)

In public, the Silica Safety Association maintained a "firm position that the best available technology could now protect workers" and the reason that workers in the past had come down with silicosis was that they "had no air-fed hoods" (NACE Regional Meeting 1977). What the SSA officials did not reveal, however, was a private study completed nine months earlier that showed that "under conditions considered good work practice" in a plant of one of the officers of the Association, nearly half of all air samples were above the TLV, indicating danger for the worker (*Industrial Hygiene Survey* 1977).

SSA was successful in delaying OSHA's adoption of the NIOSH recommendation.

Despite the appointment of Eula Bingham to head OSHA under President Jimmy Carter, during whose tenure more occupational safety and health standards were promulgated and adopted than in any similar period before or since, no new silica regulations were adopted. In 1981, the SSA claimed credit for forestalling and delaying the adoption of the proposals (Wright 1981)

The final blow to the NIOSH proposal to ban sand was dealt by the change in the Federal government when Ronald Reagan was elected President. As the Executive Director of the SSA commented, "with the change in administration, the ever increasing avalanche of government regulations have been reversed. Economic impact studies are *now* a required part of every regulatory process. As a result, OSHA's proposed abrasive blasting standard has been moved from a top priority 'target' regulation to the back burner" (Wright 1981). Also, the Reagan Administration's dismemberment of OSHA included drastically reducing the number of inspectors and inspections nationwide.

By 1982, the anti-regulatory and pro-business environment in Washington had all but killed the efforts to lower the silica standard and made lobbying efforts unnecessary. The Silica Safety Association found its contributions drying up and faced its own economic crisis. In February 1982, the Silica Safety Association Newsletter noted that "it's been a while since our last newsletter [because] Federal regulations have also been few and far between; so, as we say about sleeping dogs—" (*Silica Safety Association Newsletter* 1982). A few months later, a special meeting of the SSA's Board of Directors concluded, "the association should be put on hold" (Wright 1983). The records of the organization were placed in storage and the offices closed.

Paradoxically, at the very time that the SSA was winding down its operations in east Texas, the epidemic spread from the Gulf Coast region to the western oil fields in the Permian Basin around Odessa and Midland. As a result of the OPEC oil crises in the mid-1970s, the west Texas oil fields, which had been in a long period of decline, began to boom as domestic oil prices rose, making marginal wells profitable again. In the process of restarting the fields, miles and miles of piping, scores of small and large oil storage tanks and an enormous amount of equipment were reconditioned and cleaned. Men hired by small, non-union contracting companies sandblasted tar and oil residues off of oil pipelines and tanks that were used to store the raw oil product.

It was not until November 1988, however, that the epidemic brewing for the past decade in west Texas became public. At that time, the Ector County, Texas, Health Department was informed by a physician in Odessa, the nearby working-class oil town, that he had examined and diagnosed three Hispanic American men who had been working in a nearby plant as suffering from acute silicosis. Within a year seven other workers had been identified as victims of silicosis, for a total of ten. All were Mexican Americans, seven of them under the age of thirty, and all had worked as sandblasters, cleaning tanks and pipes used in the local oil fields. By the early 1990s, scores of silicotic workers, mostly Hispanic and mostly young, had come down with the disease.

Most of those employed to do the dirty and dangerous job of sandblasting were migrants who had recently arrived in the boom towns of west Texas and were quickly put to work in these low-paying, extremely hazardous jobs. Many of the workers were never provided adequate protection from the silica sand they blasted and, in the early and mid-1980s, young men began to appear in doctors' offices, complaining of shortness of breath, coughing, and sweating, all symptoms of silicosis. Despite a plethora of lawsuits against sand providers and manufacturers of equipment, the anti-Mexican feeling among the conservative Anglos in west Texas served to mask a developing epidemic. Contractors made little effort to protect the workforce because they assumed that many of the Hispanic workers would either move to other jobs in other areas of the country or return to Mexico when the sandblasting operations ceased or their health deteriorated.

However, the prolonged oil boom in the region transformed what might have been a little-noticed outbreak of disease into a mini-epidemic that would eventually attract the attention of occupational health physicians and federal and state occupational safety and health agencies. In the late 1980s and early 1990s, workers exposed to silica began to sue sand providers, and equipment manufacturers, rather than their employers who were protected by workers' compensation statutes developed in the 1930s to limit liability suits. Sand providers and equipment manufacturers, unprotected by workers' compensation laws, were vulnerable to liability suits because, under Texas common law, those selling dangerous products had an obligation to adequately warn users of potential hazards. The anti-Mexican feeling among the conservative Anglos in west Texas made it difficult for the plaintiffs to win in jury trials where largely Anglo jurors had little sympathy for Mexican migrants. But, occasional victories put industries on notice. Plaintiffs' lawyers, using legal strategies developed in the asbestos litigation of the mid-1970s and 1980s, began to reach substantial settlements. As a result, today, most companies in west Texas have substituted non-silica abrasives for the deadly sand.

This growing medico-legal controversy corresponded with the election of a Democratic President and the attempt to revitalize the Department of Labor in general and NIOSH and OSHA in particular. In OSHA, individuals who had long been concerned about silicosis, such as Mike Connors and Richard Fairfax, were given positions of real authority. In NIOSH, Gregory Wagner was head of respiratory diseases and in the Mine Safety and Health Administration, Davitt McAteer (the son of a miner himself) and Andrea Hricko were named Assistant Secretary and Deputy Assistant Secretary, respectively. All three cooperated to call the National Conference to Eliminate Silicosis in March 1997. This conference, aimed at focusing attention on the disease, brought together industry, labor, academics, and government. Following this conference, OSHA commenced a Special Emphasis Program devoted to eliminating silicosis and MSHA brought attention to a new arena where silicosis threatened a new group of workers.

Silicosis, long considered a danger to hard rock miners in the west, but not to coal miners in the east, was identified among surface workers in East Coast coal mines. Fundamental changes in the mining of coal and the political economy of coal mining in West Virginia, Pennsylvania, and Kentucky had led to severe silica exposures and resulting disease. Once the richest and most easily accessible source of coal in the nation, in recent decades, the veins of coal have become thinner and deeper in the ground. This has meant that miners have had to drill through silica laden rock to reach the coal. As a result, they were now endangered not only by coal workers pneumoconiosis, which had plagued anthracite miners for decades, but silicosis as well. In the words of MSHA's Davitt McAteer, the miners were now using "better drills that can drill through rock that they couldn't get through before, but that means that they are drilling through silica" (McAteer 1997). McAteer's fears were confirmed when a study prompted by an insurance company's letter to MSHA found that of 150 miners screened, eight had silicosis. The insurance company said it was paying claims on young surface miners in Pennsylvania who had died from silicosis— and had questioned what MSHA could do to prevent the disease. In response, an MSHA manager, Jack Kuzar, spearheaded an MSHA-NIOSH X-ray screening and out-reach program in Johnstown, Pennsylvania, to investigate the magnitude of the problem and to educate coal miners and mine operators on ways to reduce exposure to dust containing silica (Hricko 1997).

This recent activity gives us reason to believe that the silica standard will once again become an issue. Although the arguments will be cast primarily as epidemiological and technical debates, the historical record indicates that much more is at work in framing the question of when and at what amount silica is safe or dangerous. In the 1930s, with a severe liability crisis forcing industry to act, standards were established that reflected the economic and technical capabilities of equipment manufacturers, industries that used sand, and insurance companies that paid liability claims. During the postwar years, with a generally conservative political environment, business efforts to downplay the seriousness of the silica hazard, along with the incorporation of silicosis into state workers' compensation laws, led to the cessation of efforts to revise the silica standard. Despite continuing studies documenting silicosis as a continuing health hazard, few voices called for legislative action. In the 1970s, following the passage of the Occupational Safety and Health Act and the establishment of NIOSH and OSHA, the issue of the standard once again became important. It appears that only the concerted efforts of affected industries and the conservative triumph of the 1980s forestalled the banning of sand as an abrasive in blasting.

In the aftermath of the silica education program, the National Silicosis Conference, and the flood of lawsuits from diseased workers that are beginning to take their toll in Texas, Louisiana, Florida, and other states, industry is faced with a quandary about which way to proceed. On the one hand, ANSI has called for banning the use of sand in indoor abrasive blasting. On the other hand, some companies are already planning for a replay of the 1930s and 1970s when industry combined to stifle government activity in the area of regulation and legislation. One such example of the direction that many in industry are heading is the formation of the Silica Coalition, "a diverse coalition of trade associations and companies involved in the mining, processing, production, and use of silica and silica-containing materials," established in 1997 in anticipation of "OSHA rulemaking to control worker exposure to crystalline silica dust in the not-too-distant future" (*Inside OSHA* 1997). While ostensibly aimed at providing "sound science" and legal resources to foundries and other industries potentially affected by any change in government regulation of silica, it is also clear that the increased awareness of the dangers of silica and the resulting threat of litigation hangs over the heads of industry executives. One special demonstration of this was the meeting held in Washington just before the National Silicosis Conference, where it was made clear that the scope of the potential liability issue was terrifying. In an abstract entitled, "Toxic Tort Litigation Overview," Jean McHarg of the Washington law firm Patton Boggs noted that "approximately 2,000,000 workers are exposed to respirable silica annually" and that this posed an enormous litigation problem. "If only 10% of occupationally exposed workers (or their heirs) believed their lung cancer is due to their occupational silica exposure," then there was a potential for enormous claims (McHarg 1998).

At present it appears that the MSHA, NIOSH, and OSHA, along with private industry, are approaching a moment of decision. On the one hand, the cooperation between the government and industry that led to the National Silicosis Conference in March 1997, has, in marked contrast to the past fifty years, created enormous attention and activity around the issue of silicosis. On the other hand, this attention has the potential to create enormous conflicts. Attempts to reduce the PEL, for example, could pit the regulators in governments against the regulated in industry. Similarly, attempts to ban sand as an abrasive, one of the longest-standing goals of NIOSH, will antagonize end users and providers of sand as well. As more publicity about the dangers of silica develops, it is likely that more lawsuits from diseased workers will ensue.

In recent years, a new dynamic is affecting the debates over silica exposure. IARC has deemed silica a probable human carcinogen, leading to new concerns in government and industry. In the past year, OSHA has begun a process aimed at reducing by half the Permissible Exposure Limit for workers. Industry has countered by proposing a "negotiated rule making for a comprehensive standard" rather than a new PEL. In the words of Robert Glenn, chairman of the Silica Coalition and president of the National Industrial Sand Association, industry is seeking a regulation that would protect workers' health while, at the same time, making sure "that this important material [sand] can continue to be used safely" (*Silica Safety Association Newsletter* 1998). Just as in the past, government has sought to impose stricter regulation through the reduction of the PEL and the banning of sand as an abrasive in sandblasting. Once again, industry has responded by seeking to shift regulators' focus from the inherent dangers of the material to means of reducing exposure through the enforcement of protective codes among the workforce.

Conclusion

We readily accept that the social and political activities of each generation build on the traditions of earlier generations: the New Frontier and Great Society programs of the 1960s were part of a long tradition that incorporated elements of both Progressive Era and New Deal programs; Medicare and Medicaid are linked directly to the struggles for social and health insurance during World War I and the Depression. But it is more difficult to understand the ways that disease reflects this process of historical reformulation and redefinition. We tend to think of disease as an objective reality whose existence transcends the boundaries of time and subjectivity.

The case of silicosis raises questions regarding the exceptional nature of disease as social and political. The case of silicosis shows how the social and political environment shapes the variety of questions traditionally seen as the province of science and the laboratory. In the course of the postwar era, scientists, engineers, NIOSH and OSHA officials, and business executives have contested the very definitions of disease, responsibility for risk, and even what constitutes pathology. The boundaries of environmental and occupational disease are questions that the silicosis debate has helped redefine, re-evaluate, and answer anew. These issues emerge and take shape in the context of changing social, political, and economic circumstances and allow for this process of rediscovery.

The history of silicosis reflects a broad discourse about the nature of and responsibility for risks in an advanced industrial society. Throughout the twentieth century, the programmatic questions revolved around the relative degree of responsibility for the dangers at work of individual or organized workers, managers, and government officials. As people grow increasingly aware of the importance of chronic disease, and are sensitized to the impact of environmental change and industrial pollution as factors in producing cancers, heart disease, and stroke, the case of silicosis emerges as more than just a narrow and isolated condition affecting a relatively small group of workers. Rather, it emerges as emblematic of a new class of conditions previously ignored or identified as diseases of particular occupational groups.

The crisis of the 1930s forced a reformulation of the issue. The conflict over disease was conflated with a more general crisis in labor-management relations. Just as the New Deal sought to defuse the crisis between labor and capital by establishing federal boards and mechanisms to aid in the resolution of strikes, so too did the government seek to find the means by which to promote compromise in the arena of industrial disease. On both the state and federal levels, government and industry acted to finesse the question of responsibility by incorporating silicosis and other chronic conditions into the workers' compensation system and by moving the discourse over disease into the arenas of expert panels and professional organizations. Implicit in this solution lay the assumption that chronic disease was not the responsibility of management, labor, or government per se, but was simply an inherent risk in the new industrial society. In the case of the pneumoconiosis, the very boundaries between diseased and healthy lungs became blurred as many argued that to live in an industrial society changed the very standards of normalcy that had previously been used. The presence of chemicals or dusts in the tissue of human lungs was no longer enough to define pathology. In the 1950s some medical experts even asked whether it was appropriate to consider pneumoconiosis as a disease, since so many people throughout the society showed some evidence of lung damage. Meanwhile, unions and some labor officials argued that the presence of any foreign substance in the human body was reason for concern. Implicit in the various historical debates about silicosis was an evolving argument about the boundaries delineating the personal, occupational, and environmental sources of disease.

In the years following World War II, the silicosis issue fed an increased awareness of the importance of chronic disease in American society. Concern about chronic diseases developed at the same time that antibiotics, vaccines, and other therapies seemed to augur the decline of the threat from infectious disease. In some sense, the solutions devised to cope with the immediate crisis of silicosis would prove to be the seeds of our generation's problem. Silicosis was hidden from public view in professional journals. But the very conception of disease was changing for professionals as well as for the larger public. By the 1970s, when silicosis would once again emerge as a public issue, labor's long-time contention that chronic diseases were rooted in the workplace merged with the wider population's suspicion that there was a link between disease and the chemicals, pollutants, and waste that came from the factory. By the 1970s, the postwar consensus that industrialization carried with it implicit trade-offs with regard to human health was no longer acceptable.

Today, the relative importance of the workplace and the wider environment are hotly contested as policy makers, research scientists, and others work to establish criteria for distinguishing environmental and occupational sources of cancer. The major contention has not been over medical researchers' competing claims, but rather over who should bear the costs of treating those who become ill. In the absence of a comprehensive social and health insurance system, the distinction between occupational and environmental origins of illness becomes more than an abstraction: once again, the legal, governmental, and labor communities seek redress for disease through the courts.

Today, we are still grappling with questions of long-standing historical significance: what proportion of the new chronic diseases arising in the modern industrial community can be properly ascribed to industry itself? What part of the new epidemics of cancer, heart disease, and stroke should be understood as part of a broader social cost? What burden should we put on industry to make the work environment safe for workers? What responsibility should we place on workers to protect themselves from industrial poisons? Some in the labor community argue that, in the absence of a comprehensive social and health insurance system, it is essential that management assume the financial responsibility for workers who become ill. They maintain that in the absence of a clear distinction between environmental and industrial sources of chronic disease, the burden of proof should be on management to show that disease is not job-related: if the labor force is made to work with toxic materials and years later workers become ill, the assumption should be that their suffering is rooted in their work experience. In contrast, as the silicosis story indicates, industry has continually maintained that poisons are an intrinsic element of the job, and that any material can be handled safely, in spite of an impressive array of contrasting evidence.

REFERENCES

Andrews, Elmer F. 1936. *Memorandum for Dr. Greenburg*. April 13, National Archives, Record Group 174. Office of the Secretary of Labor. Folder: State Labor Department, New York.
Blair, Austin. 1974, April. *Abrasive Blasting Respiratory Protective Practices*. U.S. Department of

Health, Education and Welfare, National Institute for Occupational Safety and Health, Division of Laboratories and Criterion Development. Cincinnati, Ohio.

Cherniack, Martin. 1986. *The Hawk's Nest Incident: America's Worst Industrial Disaster*. New Haven: Yale University Press.

Dwight, Diane. Nd. Case Histories. Counsel to Provost and Umphrey. Beaumont, Texas.

Hricko, Andrea. 1997. Personal communication to the author, July 16, 1997. Arlington, Virginia.

Industrial Hygiene Survey. 1997, February 18. Deer Park, TX: Courney and Company (and attachments).

Inside OSHA. 1997, June 30. "New Coalition Will Seek 'Sound Science' on Silica Health Hazards," p. 1.

Materials Performance. 1975, May. "More on Proposed OSHA Standard for Crystalline Silica Use," pp. 41–42.

McAteer, Davitt. 1997. Personal communication to the author, July 16, 1977. Arlington, Virginia.

McHarg, Jean. 1998. "Toxic Tort Litigation Overview." In Patton Boggs, "Silica in the Next Century: The Need for Sound Public Policy, Research and Liability Prevention Efforts." March 24, 1997, ANA Hotel, Washington, DC. Mimeo. October 29.

NACE Regional Meeting. 1977, October 5. B. C. Wright, "Future of Abrasive Blasting."

"NIOSH Criteria Document Recommendation for a Crystalline Silica Standard." 1974. *Occupational Safety and Health Reporter*, pp. 716, 718. See also, Federal Register, Dec. 27, 1974, v. 39, #250 (29CFZ Part 1910).

Rosner, David and Gerald Markowitz. 1994. *Deadly Dust: Silicosis and the Politics of Occupational Disease in Twentieth Century America*. Princeton: Princeton University Press.

"Silicosis." 1940, February. *Survey Mid Monthly*, 76, n.p.

Silica Safety Association (SSA). 1975, February 5. "Report of Preliminary Meeting." *Silica Safety Association Exhibit*, v. 8, pp. 222–23.

Silica Safety Association Newsletter. 1982, February 22. "SSA Papers, Cook and Butler."

———. 1998, April 20. "Presidents Address."

Sline, L. L. 1975, March 21. "To Dear Sir." *Silica Safety Association Exhibit*, v. 1, n.p.

Wright, B. C. 1981, May 26. "Silica Safety Update—History—Current Endeavors—Future Plans. *Silica Safety Association Exhibit*, v. 9, p. 184.

———. 1983, June 22. "SSA Papers, Cook and Butler." *Silica Safety Association Newsletter*.

Ziskind, Morton. 1971, June 1–1974, August 31. "Accelerated Silicosis in Sandblasters," 5 RO1 CH 00387. *Terminal Progress Report*. New Orleans: Tulane University School of Medicine.

QUESTIONS

1. Why do the authors argue that there are no reliable statistics on the prevalence of silicosis in the United States in the decades following World War II?

2. In what regions of the country, and in what kinds of jobs, did we start to see increased incidence of work-related silicosis beginning in the 1960s?

Chapter Eleven

Farmworker and Farmer Perceptions of Farmworker Agricultural Chemical Exposure in North Carolina

Sara A. Quandt, Thomas A. Arcury, Colin K. Austin, and Rosa M. Saavedra

Modern American agriculture depends on the use of a wide range of chemicals to maintain its current levels of productivity. These chemicals include pesticides such as insecticides, herbicides, and fungicides; fuels; fertilizers; and ripening agents. Pesticides and other agricultural chemicals come in different forms, including gas, liquid, dust, and granular. Approximately one billion pounds of chemicals per year (Aspelin 1997; Gianessi and Anderson 1995) are applied through spray, in irrigation water, and from the air. Although new techniques (e.g., no-till agriculture, integrated pest management) and more effective chemicals are being developed to reduce the amounts of chemicals used, as well as their potential environmental and human health effects, chemicals remain important and widely used.

All persons in the farm environment, including farmers, farm families, and farmworkers, can come in contact with chemicals. Despite the benefits of chemicals in terms of farm production, they can pose a health hazard to people. Epidemiological evidence indicates that there are both short-term and long-term negative health consequences of such exposure (Arcury and Quandt 1998a; Savitz, Arbuckle, Kaczor, and Curtis 1997; Zahm, Ward, and Blair 1997), so it is considered prudent to minimize exposure.

Some chemical exposure is obvious. Spills of concentrated chemicals can occur while chemicals are being mixed or applied. Contact with the skin, inhalation of vapors, or unintentional ingestion of such chemicals results rapidly in poisoning, with severe symptoms such as tachycardia, profuse sweating, pin-point pupils, vomiting, and loss of consciousness. If untreated, death can occur. Other exposure to chemicals is less apparent, as it may be relatively asymptomatic. Low-level exposure such as that produced by the airborne drift of chemicals during application may not be apparent to those exposed, if they experience no immediate ill effects. Likewise, contact with the residues left by previously applied chemicals on foliage, tools, farm equipment, and soils can result in exposure. However, because the long-term cumulative low-level effects of such exposure can be subtle (e.g., neurological deficits), non-specific (e.g., dermatitis), or delayed (e.g., cancer or sterility), exposure to residues receives relatively less attention in safety regulations and safety training than does acute poisoning (Arcury and Quandt 1998a).

Exposure to chemical residues is of particular concern for those working in the fields. Over 85% of the fruits and vegetables produced in the United States today are harvested or cultivated by hand (Oliveira, Effland, Runyon, and Hamm 1993), an activity that brings the worker into contact with chemicals. In large-scale agricultural enterprises, this work has traditionally been done by a seasonal labor force composed of local and migrant farmworkers. Over time, this group of workers has included persons from a wide range of ethnic backgrounds; but the workers have in common a life characterized by its harshness, deprivation, and disease. As President Truman pointed out in 1951, "[the United States] depend[s] on misfortune to build up our force of migratory workers and when the supply is low because there is not enough misfortune at home, we rely on misfortune abroad to replenish the supply" (Migratory Labor in American Agriculture 1951).

Current estimates place the number of seasonal and migrant farmworkers and their dependents at 4.2 million. 1.6 million of these are classified as migrants (HRSA 1990), meaning that they are persons whose principal employment is in agriculture on a seasonal basis and who, for the purposes of this employment, establish temporary homes. They are employed in 42 out of the 50 states. As late as the 1980s farmworkers were ethnically diverse, including African American, Native American, Mexican, Haitian, and white workers (Mines, Gabbard, and Boccalandro 1991). In the 1990s the farmworker population has become largely Hispanic and foreign-born (Mines, Gabbard, and Steirman 1997), although such a generality conceals considerable variation in nation and state of origin, as well as language and cultural diversity (see, for example, Grieshop 1997).

The current understanding of how to minimize exposure to agricultural chemicals is based on a behavioral model coupled with data on routes by which chemicals enter the body. Persons can come in contact with chemicals by handling crops or tools, by working in soil, by experiencing airborne drift of chemicals or their residues into housing, or by having others carry residues into housing on their clothes and skin. To be harmful, chemicals must be absorbed through body openings or the skin. Only a portion of chemicals on the skin will be absorbed, and this varies by site. For example, only 10% of chemicals on the palm of the hand will be absorbed, while 99% of those coming in contact with the scrotum will be absorbed (Feldman and Maibach 1967). Skin compromised by rashes or cuts will absorb more (Wester and Maibach 1983). Heat and humidity increase transdermal absorption, as well (Blank, Scheuplein, and MacFarlane 1967; Meuling, Franssen, Brouwer, and van Hemmen 1997). Several behavioral steps can reduce the amount of chemicals that a farmworker comes in contact with and the amount absorbed. These include wearing protective clothing (e.g., long sleeve shirts and wide brimmed hats), maintaining personal hygiene (e.g., frequent hand washing and showering after work), and laundering work clothes separately from other household laundry. In recognition of the importance of such behavioral steps, the U.S. Environmental Protection Agency designed the Worker Protection Standard (WPS), a set of regulations enacted in the early

1990s (U.S. Environmental Protection Agency 1992). The WPS requires that all farmworkers be taught the dangers of chemical exposure and how to protect themselves from agricultural chemicals. It also requires farmers to follow safe practices and provide facilities for personal hygiene and emergency medical care. The Worker Protection Standard provides a unique opportunity to educate and empower farmworkers to protect themselves from chemical exposure, but it is not consistently enforced. Although WPS training undoubtedly increases farmworker awareness of risks and of preventive behaviors, there has been little testing of the WPS provisions to know whether they will actually prevent exposure and reduce its harmful consequences (Langner 1997).

Farmworkers generally lack control over their exposure. They bear a disproportionate share of the occupational and environmental health risks from chemical use in agriculture. For this reason, attention has been focused on farmworkers as part of the effort to increase environmental equity (Moses, Johnson, Anger, Burse, Horstman, Jackson, Lewis, Maddy, McConnell, Meggs, and Zahm 1993). As part of the effort, the National Institute of Environmental Health Sciences has funded the PACE Study (Preventing Agricultural Chemical Exposure in North Carolina Farmworkers) to develop, test, and disseminate culturally appropriate interventions for farmworkers. In this chapter, we describe the theoretical basis for the PACE project and focus on the formative research that describes and compares beliefs held by farmers and farmworkers about agricultural chemicals. Existing research on farmer and farmworker beliefs about chemical exposure is quite limited, but emphasizes the importance of power relations and cultural health beliefs for farmworkers' approach to chemical safety. In a comparison of farmers and farmworkers in California, Grieshop, Stiles, and Villanueva (1996) found that farmworkers tended to place control over workplace safety outside of themselves (e.g., in God, luck, or supervisors). While they thought about ways to stay safe in the workplace, they also had a cognitive strategy of accepting danger. In contrast, farmers emphasized their own personal control over safety and therefore made plans to stay safe, rather than simply accept danger. Vaughan (1993a, 1993b) also found that many farmworkers perceived little control over exposure to chemicals and their negative health effects, and this was associated with non-use of protective behaviors to prevent or reduce exposure. Those farmworkers in better economic circumstances were more likely to perceive themselves as having control over exposure (Vaughan 1995). Baer and Penzell (1993) found that pesticide exposure-related symptoms were interpreted within a cultural framework. Following an incident in which a large number of Mexican migrant farmworkers in Florida were treated for pesticide poisoning, many attributed residual symptoms to the Mexican folk illness, *susto*. While this may have represented somatization of the psychological trauma of the poisoning, those individuals who self-diagnosed *susto* did report more residual symptoms. This indicates that they may indeed have had more negative effects from pesticide exposure and expressed illness in a culturally specific way. While the body of literature on farmworker beliefs concerning chemical exposure is small, it suggests a number of directions for the present research.

Research Design

The data reported here come from PACE, a community participation health project designed to reduce exposure to pesticides and other agricultural chemicals by developing, testing, and disseminating culturally appropriate interventions. This is a four-year project in which the major goal of the first year was to conduct formative research to be used in the development of the intervention. This paper will focus on this formative research. We will first describe the theoretical framework of PACE, then present the methods used in the formative research, and finally describe our findings.

As a study to change health behavior, the PACE study draws on existing theory in the field of behavior change. Consistent with current health behavior and health education research (Glanz, Lewis, and Rimer 1997), PACE is based on a multi-theory approach, using an overall planning framework for the study as a whole that incorporates individual, inter-personal, and community-level theoretical approaches. The overall planning framework for the study is the PRECEDE-PROCEED model of Green and Kreuter (1991). This model is intended to provide a structure for applying theories to identify and implement the most appropriate and effective interventions to change health behaviors (Gielen and McDonald 1997). Key components of the model that comprise the PRECEDE aspect are predisposing, reinforcing, and enabling factors. All need to be considered in planning a strategy to change health behavior. Predisposing factors relevant to the PACE project include personal characteristics such as age, gender, experience in farm work; beliefs and perceptions about susceptibility to effects of chemicals; and knowledge of chemical exposure and its prevention and treatment. Reinforcing factors include characteristics of the social environment (e.g., employer or family concerns about health), experience and those observed in others, and community support for behavior change. Enabling factors are the availability of appropriate resources for behavior change (e.g., hand-washing water and laundry facilities) and the availability of skills training. This chapter will concentrate on predisposing and reinforcing factors, as these are most closely linked to health beliefs and knowledge.

Field Sites

The PACE project is based in an eight-county region of central North Carolina. Agricultural production includes tobacco, cucumbers, sweet potatoes, cotton, and a number of other fruit and vegetable crops. PACE focuses on farmers and farmworkers involved in tobacco or cucumber production, as those involve large amounts of chemicals and hand labor throughout the agricultural season. North Carolina ranks fifth in the nation in number of farmworkers. Recent estimates place the number of migrant workers and dependents at 140,000, with twice as many seasonal farmworkers (North Carolina Employment Security Commission 1995). The ethnic composition of the North Carolina farmworker labor force has changed in the last decade from mostly African American workers to Mexican and Central American workers. The eight-county study area includes the counties with the state's highest concentrations of farmworkers.

TABLE 11.1
Number of Individuals Participating in In-Depth Interviews and Focus Groups, by Gender and Ethnic Group

Ethnic Group	Female		Male	
	Interviews	Focus groups	Interviews	Focus groups
African American	2	6	4	0
Hispanic White	3	6	16	32
Non-Hispanic White	1	0	0	0

Data Collection

In order to understand predisposing, reinforcing, and enabling factors related to preventing agricultural chemical exposure, we undertook formative research that included both traditional ethnographic fieldwork of observing and conducting casual interviews in the community, and a set of more structured interviews. These latter included individual in-depth interviews (27 with farmworkers and 7 with farmers) and focus group interviews (7 with a total of 44 farmworkers). Farmworker in-depth and focus group interviews included both Hispanic and African American farmworkers, and both males and females (Table 11.1). Most of the Hispanic farmworker interviews were conducted in Spanish. Informants were recruited with the assistance of our community-based organization partner, the North Carolina Farmworkers' Project. Efforts were made to include a diverse group of farmworkers to help understand the range of beliefs held. Farmworkers were natives of Mexico, Puerto Rico, Guatemala, and the southeastern United States. Some of those of Hispanic origin had lived for a number of years in the United States. Their time spent as a seasonal or migrant farmworker ranged from less than one year to more than twenty years. The farmers interviewed were also a diverse group, including farmers with different size operations and from different North Carolina counties. In-depth and focus group interviews were conducted by trained interviewers using a standard interview guide. This included topics such as personal experiences with chemicals and beliefs about health effects of exposure and its prevention.

Data Analysis

All interviews were tape recorded. Most interviews were transcribed verbatim. Spanish interviews were then translated by a professional translation service and edited by bilingual project staff. For a few of the farmworker interviews with limited relevant content, notes were taken from the tape recording in lieu of a verbatim transcription. A systematic text analysis plan was developed and implemented (Arcury and Quandt 1998b). The analysis was designed to derive common "themes," or generalizations, from the interview sets. Codes were developed to label beliefs, knowledge, and practices related to chemical exposure. Each transcript was coded by more than one coder to reduce bias. After segments were identified and coded, segments were retrieved and reviewed by the authors to identify common themes representing predisposing and reinforcing factors. The

Ethnograph (Version 4.0) computer software (Seidel, Friese, and Leonard 1995) was used for search and retrieval of text segments.

Results

We orient the presentation of results for both farmworkers' and farmers' beliefs around the central themes found in the interviews. Six themes emerged from the analysis of farmworker transcripts. The diversity and amount of comments vary considerably across these themes. In some cases there are diverse ideas held. For some themes, farmworkers have many comments, indicating highly salient themes to which they have given a significant amount of thought. For other themes, farmworker comments are far more cursory, indicating that the theme holds limited salience for them. Only two themes emerged from the farmer interviews. These reflect greater consensus of opinion about the issue of agricultural chemicals and health.

Farmworkers Theme 1: Sensory Detection of Chemicals

The senses are important to farmworkers in detecting exposure. They expect to *feel* the chemical on the plant or to *taste* it or to *smell* it.

> I have been exposed to it, 'cause like, you know, you can work in tobacco or whatever and then you can go eat lunch, and you can wash your hands. But then when you eat, you can still taste it. 'Cause I mean, like the stuff they use to grow tobacco now, you know, it just don't wash off with the rain, you know, it just. . . . it's there . . . You can still taste it. And you know it. You can tell tobacco, like tobacco juice. You can tell how bitter, you know. You can tell the differences. Distinguish the difference. Yeah. And you know it's chemicals. Sometimes, you know, you can see chemicals on the tobacco. On the tobacco leaves. You can still see the chemicals on the leaves. Most of the time you keep working and hope you don't get it. (FW006)

> Right in the morning time, in the morning time. They'll spray in the morning time. And then we'll go right behind and start cropping tobacco. Or we might be cropping on one set of rows . . . And he might be spraying right next door to it. And by the wind blowing, if it's blowing in our direction, you can feel that coolness, like mist hitting you. 'Cause of the wind blowing it on you, you know. It's according to which way the wind blowing sometimes . . . So you can actually feel mist from the spray when the wind blow. (FW005)

> Inhaling a chemical, you can't get rid of that smell. What they mix it with, the particles of it. Well, you can smell it. (FW019)

For all but one worker, visual detection of chemicals required seeing wet chemicals on the plants. This is important because there is considerable evidence that chemicals leave a dislodgeable residue on plants after drying that may constitute a significant exposure source for field workers (Fenske 1997). Only one farmworker reported seeing chemical residue that had dried.

I said, oh, they had sprayed. It was some blue stuff that was—blue residue stuff. When you're picking fruit and the leaves—when they spray some pesticide, you see some of it still on the leaves and stuff. I know when I seen that color on the leaves and the trees, I know that's what it was then. (FW002.FG)

While sensory detection of chemicals seems to be important for farmworkers to know the chemicals are present, some farmworkers seem to believe that smelling a chemical does not constitute exposure.

I remember, we was in a field and they were spraying once while we was in there. But it was—you could smell it—but it was so far, you know, so many rows over, it wouldn't be that close to you to harm you . . . I have been that close you could smell it, but as far as being close to it [for exposure], [the crew leader] won't let us come out there like that. (FW004)

Thus, in general, farmworkers believe that they will be able to detect chemicals with their senses. The important converse implied is that if chemicals are *not* seen, felt, tasted, or (in some cases) smelled, the chemicals are not there.

Farmworkers Theme 2: Body Openings as Exposure Routes

These farmworkers present a variety of conflicting opinions about how chemicals enter the body. There is a general understanding that chemicals can get into the body in several ways, but the emphasis is on lungs and mouth, with little mention of absorption through the skin. The lungs are seen as a primary route by which chemicals enter the body.

In the fumigated fields you breathe the smell from the fumigation and it is through the nose that it goes to the brain. You feel just like that, as if you were drunk or dizzy, and I think that if I were to turn around, I am going to make myself sick, vomit, 'cause that's what causes it. (FW001.FG)

And when you think about it, if you're breathing those fumes, what's going on inside your lungs. You don't know what's taking place and what effect it gonna have. (FW002.FG)

Inhaling a chemical [fertilizer], it's in your lungs. It's all through your body. You don't know if it's gonna hurt your blood, you don't know if it's gonna do damage to your heart. You don't know if it's gonna do damage to your lungs. (FW019)

For exposure through the mouth, the farmworkers' descriptions of exposure include instances where their sweat ran into their mouths while working in hot fields, and the chemicals tasted bitter.

The [chemical] they put in the fields does cause an effect because it is too hot. You eat the sweat coming from your body, and many times, especially when we are in tobacco, there are many pesticides. (FW003.FG)

We are sweating a lot, and it is the sweat that goes into our mouths and it is bitter, because of the tobacco leaves. All of that bitterness. And sometimes we swallow our saliva and it is so bitter. That's what hurts us. (FW018)

Most farmworkers believe that pesticides on the skin must get into a natural body opening to be absorbed. For a few, the pores in the skin are such an opening.

> The chemicals when you are wet is when they penetrate your body better. When you get wet with the water on the plant is when it penetrates you, not when the plant is dry. (FW004.FG)

However, most see the skin as a barrier and believe that chemicals on the skin must get into the mouth to be absorbed.

> If you get it on your hands and then you can't wash your hands, then you go to get a drink of water and cup your hands, that's how the poison gets in you. (FW003)

> Chemicals can affect you if you do not wash your hands. At lunch there is no water; you have to eat without washing your hands. The chemicals can get inside your system. (FW005.FG)

> When you are drinking water, like this, your hands are dirty. Some kind of infection from the chemical can go into the glass, and sometimes you don't even notice. (FW021)

Farmworker Theme 3: Coming in Contact with Chemicals

Consistent with the emphasis on water, the most commonly reported type of exposure occurs early in the morning. Some workers think this is because the plants are wet from early morning dew, but most think that the farmers spray early, making the plants wet with chemicals.

> So when we go early in the morning and they have just sprayed them, that's when it penetrates your hands and everything. (FW004)

> When you start working the plant is all wet and all that moisture rubs on you. It dries on you, on your skin and that is what affects you. Even if it dries out, the illness is already in you. And that is why we are concerned because sometimes we are bending down and we feel the pesticide in our eyes. You start thinking, what can I do, because the boss is always in a hurry, so we get wet, and then dry out, and then wet again. (FW004.FG)

In this farmworker population, there are only a few reports of instances of direct spraying of farmworkers. Some believe that workers are exposed to chemicals by being made to work in fields in which the no-entry period has not expired or by drift from spraying in nearby fields.

> I have worked many times after they have sprayed—here as well as in [another place]. They would spray and the spraying machine would go, and they'd bring in a crew leader and they would tell us that we were going to pick. And we would start to pick . . . Sometimes the boss tells us that he wants us to finish filling up the barns, and sometimes they spray and we are behind picking. (FW021)

> I have never been sprayed directly, except when the wind blow. I've seen the guy come. He'll come right behind you and spray you, like toward the end of the day, cause he's trying to get that field done. He'll wait for us to get a set of rows out, then he'll come

round and spray, you know, like toward the end of the day cause he catches up with us while he's spraying. It doesn't happen every day, you know, but it might happen pretty regular. (FW005)

However, others workers report that growers do not allow workers into fields until the no-entry period expires.

There are no times that I have worked in fields that have been recently sprayed. (FW026)

We never work in fields that have been recently sprayed. The patron does not want us to do that. We have worked with him a long time and he has never treated us badly. I have never been accidentally sprayed. (FW025)

I have not worked where they just sprayed. I have always worked several days after they have put on the liquid. Then we go and work there. (FW018)

No, I have never worked in fields where there were still chemicals in the plants. The times they were going to spray, they let us know and the inspectors go and take all the people out. (FW007.FG)

In all of these instances—entering a field too soon, being sprayed, or being kept out of a field by a farmer—the common variable is the farmer. He or she (not the worker) has the power to determine whether or not the worker is exposed.

Farmworker Theme 4: Susceptibility Is Individualized

There is a widespread belief that farmworkers vary in their susceptibility to the effects of chemicals; some are sensitive and experience ill effects, and others are inherently more resistant.

There are some workers that have a resistant body. There are some that can stand two days [wearing the same contaminated clothes], some others three. But it is rare if a person can stand three days with the same clothes. But sometimes we have a very strong body, like with me, I can stand them for two days. (FW007.FG)

And I done heard two or three people say, "Hey, man, I can't do that kind of work, it breaks me out." I'm not saying it's the spray, the pest control they are putting on it. It might just be the tobacco . . . It ain't never got me, like I done seen it with my own eyes how, you know, the rash—it's a rash, you know—people scratch it, this big one day, next day it's all way up the whole arm or whatever. I can understand about that. But luckily me, myself, I never had it. I am lucky not to be—not to have the kind of skin that breaks out like that. (FW004)

It hurts some people and it doesn't hurt others, and I don't know what else to tell you. (FW001.FG)

[My adult daughter was sprayed.] She is very thin, right? so it affected her. I had her at home for a week, vomiting and with fever. (FW006.FG)

[When they are spraying, the smell gives you] headache, dizzy, your head hurts fast, you feel pain, you feel dizzy in your head. Some don't feel as much, but others are more delicate. (FW015)

The medicine [chemical] goes directly to the plant, and for the person, if he is a weak one, it will affect him. But for the person who has strong health, it doesn't do anything. (FW009)

For those who smoke, the pesticides won't hurt them. (FW003.FG)

Chemicals affect people differently, depending on their strength and constitution. It would also be easier for a woman, remember that a woman has her time of the month, those are the days that can have the most effect and it is easy to become contaminated by pesticides. (FW013)

The liquid [chemical] is very strong for women, for some. For others they are very strong and also their blood is strong because they don't get sick. (FW018)

Farmworkers use oppositions such as strength and weakness, old and young, male and female to try to categorize susceptibility. However, while some variation can be attributed to these, they have no general explanation for susceptibility. People are just different.

Farmworker Theme 5: Acute Exposure is the Problem

Most farmworkers are concerned with the immediate or acute effects of exposure. They connect the cause (agricultural chemicals) with a variety of effects, symptoms that can range from those that are annoying to those that are temporarily debilitating. In fact, many infer as evidence of exposure these symptoms, most of which subside when fieldwork stops. They include nausea and vomiting, headaches and dizziness, and a variety of skin rashes, welts, and pimples. Farmworkers are aware that working in tobacco and cucumbers produces some of these symptoms. There is no clear distinction for some, when talking about these crops, whether the ill effects are from agricultural chemicals or the plants themselves.

Many have gotten sick there in the work. Some get sick with a headache, vomit, all of that you suffer, but it is all because of the chemicals. (FW018)

That chemical on it. Oh, baby, when that sun comes out or you get wet and you start sweating, that stuff'll set you on fire. And it just burns the heck out of your skin. I mean you just swell up . . . That stuff make you swell. Your eyes'll swell, too. Your face'll swell and you get real tight. It's like you got high blood pressure, you know. I did that and I got sick off of it. (FW019)

Those scars are from the chemicals that we used. When they get in your clothes and they burn your skin, little blisters come up. Naturally, you're wondering what that little blister come from. It gets all over your body. I mean, there's no other explanation for it. You play in the dirt all your life and then all of a sudden you begin to get these things popping out on you and your face swell up . . . I don't get pimples, and it's just—I put two and two together. (FW019)

I only worked four months, and then I started feeling ill, I started getting a lot of little welts and little pimples that were itching. I went to the doctor and he told me it was because of the tobacco, because replanting you had to bend over, then you would get a lot of the odor and that's how I got dizzy . . . What happened was that when you are replanting the plant, you are getting from the soil the chemicals. I think they put fertilizer . . . (FW015)

Very few workers are aware of potential long-term effects of exposure to chemicals, and none of them link these to chronic exposure or to residues. Most described these effects only when pressed, and they could not elaborate. They did not seem to know of anyone who had suffered from any of these health outcomes from chemical exposure.

From the chemicals I know you can get real bad chemical burns, and sometimes it makes bone cancer. Sometimes it can cause spots to come on your lungs from inhaling it.(FW019)

Well, [the chemical] reaches your brain and you become retarded. You don't think the same. (FW016)

Pesticides and fertilizers and all this, herbicides and all that stuff. It's dangerous and causes cancer. . . . It causes birth defects, you know, that stuff is dangerous. (FW006)

Maybe it would even cause skin cancer. I don't know what it is. It just stick to your skin, you know, make you itch. (FW005)

Farmworker Theme 6: Chemical Exposure Is Not a Problem

Farmworkers are divided on whether or not chemicals are potentially dangerous. There are some farmworkers who believe that the chemicals are not dangerous, and that they can only hurt insects or kill weeds. Some state that farmers would not use chemicals if they were dangerous to farmworkers.

I don't think the pest control, what it is, the pesticide, I don't think it that dangerous. I mean, they just spraying to keep the insect. It ain't nothing harmful to the human being, the worker. It's for to keep the insect away, see. To keep the insects from eating the crop. The farmer they wouldn't put it out there if it, if it that harmful, if it that poison you couldn't breathe it, you know. I don't think they'll put it out there, no. I don't think it harmful. (FW004)

Other farmworkers, even when pressed, had no answers to questions that would have allowed them to link chemicals with health outcomes. The belief that agricultural chemicals can be harmful is by no means universal.

Farmer Theme 1: There Is No Problem

Farmers are unanimous in asserting that farmworkers are *not* exposed to chemicals because they do not mix or apply chemicals. Because the farmers themselves or their certified applicators are the ones mixing and applying, it is they themselves at risk, not the farmworkers. Farmworkers are *not* exposed to chemicals, they believe, because they do not go into the fields until after the re-entry period. None of the farmers consider chemical residues to be present or to be a source of exposure. As far as they are concerned, the only significant sources of exposure on farms are spills during mixing and application.

> Our responsibility is to keep [farmworkers] out of the fields until the re-entry period is safe. And so that's how we handle that. We feel as long as they don't go back in before they should, then, really, there's not a problem. (GR001)

> Those [farmworkers] aren't applying chemicals. The guys that we show [videos about chemical safety] to are primarily our equipment operators and applicators, where they're actually handling the chemicals on a daily basis. We feel like there's a different exposure to those guys in that our responsibility is to keep them out of the field during the re-entry period. Then their exposure should be minimal. Once the re-entry period is over, they're pretty safe anyhow. (GR001)

> There's no use showing [training films] to hand labor that's not gonna be exposed to [chemicals]. 'Cause they don't go in the fields where it's at. They're harvesting and nothing's been put out there for that . . . No, there's no need for us 'cause the fields we plant in are treated and we wait for the evaporation time period for to be out. So they don't have any exposure to it just riding on the [planting] machine, except maybe the dust from the ground. They don't never touch the ground except where they drag their feet on the ground on the planter. (GR003)

> Other than the people applying the stuff, we don't have the problem [of contamination] with our tobacco. (GR002)

Farmer Theme 2: Risks from Chemicals Have Decreased

Farmers believe chemicals are safer now than they used to be. They argue that smaller amounts of chemicals are used today than in the past. Also, they believe that herbicides are not as harmful as insecticides; more herbicides than insecticides are used today.

> From the standpoint of twenty years ago to now, I think that we are using less chemicals. You talk about fertilizer, we're using higher concentration of fertilizer so you don't have to apply as much. Twenty years ago you may have put a ton of fertilizer to the acre. Now you use less than half that. They're higher concentrations of chemicals and the methods of activation are a lot more efficient and a lot better. (GR001)

> If proper precautionary methods are taken, I don't think there's any health risk to the family. There's no problem. (GR001)

> This herbicide stuff will drift and kill the roses, stuff like that. And that's still not harmful to anyone other than the plant. That's a foliage thing . . . makes it really white, the foliage and stuff. It doesn't hurt the human. (GR002)

> Basically, all the chemicals we're using are what you would say are relatively safe, although they'll burn you up . . . We're using some herbicides. Your herbicides are not the ones that create the problem, healthwise, most of the time. (GR003)

> When migrant laborers are working, I do not use any chemicals at that point that should create any problem. The only chemicals that I use are something that's safe enough that, I'm told, you can eat it if you wanted to and it wouldn't hurt you. (GR004)

If there *is* risk from chemicals, it is not in the crop a particular farmer grows. Said one tobacco grower:

> I think farm working people got most of that covered, cause we know you've got to wait [to re-enter] in tobacco. It's more in produce, stuff like that, that's really bad on [farmworkers] to get it on. I think we pretty well got our tobacco thing settled down. (GR002)

However, produce farmers have the opposite opinion.

> We're harvesting cucumbers so we're not allowed to use any chemicals that are gonna be detrimental to anybody's health. So most of [the chemicals] are relatively safe that we use. (GR001)

Discussion

It is notable, though not surprising, that farmers are relatively uniform in their beliefs, and farmworkers are not. Farmers are culturally similar, having learned about chemicals in government mandated and administered courses, as well as through experiences as North Carolina farmers with a common set of crops and chemicals. Farmworkers, in contrast, have varying experiences with agriculture, particularly tobacco production. Although all were employed in North Carolina, they vary in ages, farmwork experience, and national origins. They have contradictory opinions on whether or not they are exposed to chemicals and on the dangers associated with exposure.

Beyond these within-group differences, there is a striking difference between farmers and farmworkers on the question of whether or not farmworkers are exposed to agricultural chemicals. Farmworkers believe they have been exposed because they experience symptoms. The temporal association of field work and symptoms reinforces the cause and effect relationship they perceive. Because many of these symptoms appear after work hours and because workers are unlikely to inform a farmer about them, farmers may well not be aware of symptoms experienced by farmworkers unless they become life-threatening, as would be the case for acute poisonings. Farmers base their belief that farmworkers are not exposed to chemicals on the knowledge acquired in chemical applicator classes that, if re-entry rules are followed, workers will be safe. However, these re-entry rules are designed to prevent acute poisonings, not low-level chronic exposure (Woodruff, Kyle, and Bois 1994). In addition, as farmworkers point out, re-entry rules are broken when there is time or economic pressure to get the fieldwork finished. This is consistent with Perry and Bloom (1998) who found that, although farmers know safety regulations, they engage in risky behaviors (and allow farmworkers to do so, as well) when under pressure.

Farmworkers believe susceptibility to chemicals is inherent, beyond one's control. This is an important finding, as across health behaviors, ideas about control predict health behavior (e.g., Janz and Becker 1984; Wallston and Wallston 1982. That is, persons who feel they have control over a health outcome are more likely to adopt the appropriate preventive behaviors when faced with a health risk. Vaughan (1993b) found that most of the farmworkers she interviewed in California felt that they had little control over experiencing the negative consequences of pesticides, and this feeling was associated with failure to use protective behaviors. For the North Carolina farmworkers, the ideas they expressed of inherent susceptibility indicate a key barrier to be overcome in an intervention.

Tables 11.2 and 11.3 summarize farmworker and farmer themes and their content, as well as scientific evidence on the themes. Many of the ideas held by both farmworkers and farmers, particularly those concerning residues, are inconsistent with scientific evidence on chemical exposure. For example, farmworkers believe that chemicals can be sensed when exposure occurs. Farmers, too, seem to expect that chemicals can be detected with the senses. Both groups believe that danger is present when chemicals are wet; residues are not a problem. One farmer, for example, refers to the re-entry period as the "evaporation period," as though the chemicals themselves simply evaporate with no residue remaining. While other farmers did not use this term, other statements they made suggest that it fits their cognitive models of chemical exposure.

Because it is unlikely that farmers will stop using agricultural chemicals, farmworker protection from chemical exposure must be based on behaviors that lower

TABLE 11.2

A Summary of Farmworker Themes Concerning Chemical Exposure of Farmworkers, with Scientific Position on the Themes and Classification for PRECEDE-PROCEED Model

Theme	Main ideas	Scientific evidence	PRECEDE
Sensory detection of chemicals	If a chemical cannot be sensed, it is not there.[1]	Dislodgeable residues constitute an important but invisible exposure source for farmworkers.	Predisposing
Body openings as exposure routes	Chemicals enter the body primarily through lungs and mouth; skin serves as a barrier against absorption.[2]	Lungs, mouth, nose, and eyes are all exposure routes. Absorption through the skin is an important exposure route.	Predisposing
Coming in contact with chemicals	Working in wet plants causes exposure.	Moisture on skin and plants facilitates exposure.	Predisposing
	Airborne drift and re-entering fields early because farmers are in a hurry cause exposure.[2]	These are esposure sources, but residues are also important.	Reinforcing
Susceptibility is individualized	Some people or classes of people are inherently more susceptible to health effects of chemicals.[2]	There is inter-individual variation in absorption of chemicals, but all persons will absorb some.	Predisposing
Acute exposure is the problem	Most effects are immediate, causing discomfort, but they are not life threatening.[1]	While immediate effects do occur, long-term exposure to low levels of chemicals can lead to chronic neurological injury, birth defects, and cancer.	Predisposing
Chemical exposure is not a problem	Chemicals do not cause health problems.[1]	Chemicals can have both acute and chronic health effects.	Predisposing

[1] Inconsistent with scientific evidence
[2] Partially inconsistent with scientific evidence

TABLE 11.3
A Summary of Farmer Themes Concerning Chemical Exposure of Farmworkers, with Scientific Position on the Themes and Classification for PRECEDE-PROCEED Model

Theme	Main Ideas	Scientific Evidence	PRECEDE
There is no problem of chemical exposure	Once the re-entry period has expired the danger of chemical exposure is over.[1]	Residues remain that can have cumulative health effects.	Reinforcing
	Farmworkers do not go into fields until re-entry period has expired	Residues remain that can have cumulative health effects.	Reinforcing
Risks from chemicals have decreased	Smaller amounts of chemicals are used and more herbicides, which are not dangerous.[1]	Dangerous chemicals are still used. The cumulative health effects of many are not yet known.	Reinforcing

[1] Partially inconsistent with scientific evidence

the amount of contact farmworkers have with chemicals (e.g., protective clothing) and facilitate removing the chemicals from the skin and clothing before they can be absorbed. The model guiding the PACE project, the PRECEDE-PROCEED planning model, suggests that these behaviors will be more likely if predisposing factors such as knowledge and beliefs are consistent with the behavior and if reinforcing factors such as attitudes encountered in interpersonal interactions support the protective behavior. Tables 11.1 and 11.2 classify themes according the PRECEDE-PROCEED model. There are several areas of farmworkers' health beliefs that must be addressed as predisposing factors. These could be included in pesticide safety training programs that farmers are required by the Worker Protection Standard to provide for farmworkers (U.S. Environment Protection Agency 1992). At present, the educational materials developed for the WPS vary considerably in quality, in the depth of information presented, and in how well they cover topics such as long-term health effects and exposure via chemical residues (Quandt, Austin, Arcury, Summers, and Martinez 1998). The factors labeled as "reinforcing" indicate that protection of farmworkers from chemical exposure hinges on power relations between farmers and farmworkers, and will depend on changing knowledge and behaviors of farmers. Because farmworkers are dependent on farmers for employment, they cannot object to work conditions that expose them to chemicals. The existing WPS training puts much of the responsibility for preventing exposure on farmworkers and assures them that it is their legal right to refuse to work in unsafe conditions (e.g., entering a field before the re-entry period has expired). Yet, farmworkers, many of whom are undocumented workers, have little confidence in such assurances of rights. Farmers are trained that no exposure occurs after the re-entry period has expired. This reinforces their belief that farmworkers are not exposed. Breaking the re-entry period rules is rationalized as necessary when time is short.

These findings suggest that reducing exposure of farmworkers to chemicals will require a multi-layered intervention. While better education of farmworkers concerning risks of exposure and appropriate preventive behaviors should be part of this, farmers must be included as well because they do not believe a problem exists. Our experience to date in North Carolina indicates that attempting to change knowledge and beliefs of farmers will not be easy. This group already feels burdened by regulations and are concerned with the costs (in time and money) of additional interventions with farmworkers. In addition, the kind of information needed by farmworkers challenges the accuracy of knowledge gained in classes farmers are already required to take. All these factors constitute formidable obstacles to designing and carrying out an effective intervention.

REFERENCES CITED

Arcury, Thomas A., and Sara A. Quandt. 1988a. Chronic Agricultural Chemical Exposure among Migrant and Seasonal Farmworkers. Society and Natural Resources 11: 829–43.

————. 1998b. Qualitative Methods in Arthritis Research. Sampling and Data Analysis. *Arthritis Care and Research* 11:66–74.

Aspelin, Arnold L. 1997. *Pesticides Industry Sales and Usage: 1994 and 1995 Market Estimates.* Office of Prevention, Pesticides and Toxic Substances, Washington, DC: U.S. Environmental Protection Agency.

Baer, Roberta D., and Dennis Penzell. 1993. Susto and Pesticide Poisoning among Florida Farmworkers. Culture. *Medicine and Psychiatry* 17:321–27.

Blank, Irvin H., Robert J. Scheuplein, and Dorothy J. MacFarlane. 1967. Mechanisms of Precutaneous Absorption. III. The Effect of Temperature on the Transport of Non-electrolytes across the Skin. *Journal of Investigative Dermatology* 49:582–89.

Feldman, Robert J., and Howard I. Maibach. 1967. Regional Variation in Percutaneous Penetration of 14C Cortisol in Man. *Journal of Investigative Dermatology* 48:181–83.

Fenske, Richard A. 1997. Pesticide Exposure Assessment of Workers and Their Families. *Occupational Medicine: State of the Art Reviews* 12:221–37.

Gianessi, L. P., and J. E. Anderson. 1995. *Pesticide Use in U.S. Crop Production: National Summary Report.* Washington, DC: National Center for Food and Agricultural Policy.

Gielen, Andrea Carlson, and Eileen M. McDonald. 1997. The PRECEDE-PROCEED Planning Model. In *Health Behavior and Health Education: Theory, Research and Practice* 2nd Edition. Karen Glanz, Frances Marcus Lewis, and Barbara K. Rimer, eds. Pp. 359–83. San Francisco: Jossey-Bass.

Glanz, Karen, Frances Marcus Lewis, and Barbara K. Rimer, eds. 1997. *Health Behavior and Health Education: Theory, Research and Practice,* 2nd Edition. San Francisco: Jossey-Bass.

Green, Lawrence W., and Marshall V. Kreuter. 1991. *Health Promotion Planning: An Educational and Environmental Approach.* Mountain View, CA: Mayfield.

Grieshop, James I. 1997. Transnational and Transformational: Mixtec Immigration and Health Beliefs. *Human Organization* 56:400–407.

Grieshop, James I., Martha C. Stiles, and Ninfa Villanueva. 1996. Prevention and Resiliency: A Cross Cultural View of Farmworkers' and Farmers' Beliefs about Work Safety. *Human Organization* 55:25–32.

HRSA. 1990. *An Atlas of State Profiles Which Estimates Number of Migrant and Seasonal Farm Workers and Members of Their Families.* Washington, DC: Health Resources and Services Administration.

Janz, N. K., and M. H. Becker. 1984. The Health Belief Model: A Decade Later. *Health Education Quarterly* 11:1–47.

Langner, G. 1997. *A National Dialogue on the Worker Protection Standard.* Office of Pesticides

Programs, Office of Prevention, Pesticides and Toxic Substances, U.S. Environmental Protection Agency, EPA 735-R97-001.

Migratory Labor in American Agriculture. 1951. Report of the President's Commission on Migratory Labor.

Mines, R., S., Gabbard, and B., Boccalandro. 1991. *Findings from the National Agricultural Workers Survey (NAWS) 1990: A Demographic and Employment Profile of Perishable Crop Farm Workers.* Research Report No. 1, Office of the Assistant Secretary for Policy, Office of Program Economics. Washington, DC: U.S. Department of Labor.

Mines, R., S. Gabbard, and A. Steirman. 1997. *A Profile of U.S. Farm Workers: Demographic, Household Composition, Income and Use of Services. Based on Data from the National Agricultural Workers Survey (NAWS).* Office of the Assistant Secretary for Policy, prepared for the Commission on Immigration Reform. Washington, DC: U.S. Department of Labor.

Meuling, W. J., A. C. Franssen, D. H. Brouwer, and J. J. van Hemmen. 1997. The Influence of Skin Moisture on the Dermal Absorption of Propoxur in Human Volunteers: A Consideration for Biological Monitoring Practices. *Science of the Total Environment* 199:165–72.

Moses, Marion, Eric S. Johnson, W. Kent Anger, Virlyn W. Burse, Sanford W. Horstman, Richard J. Jackson, Robert G. Lewis, Keith T. Maddy, Rob McConnell, William J. Meggs, and Shelia Hoar Zahm. 1993. Environmental Equity and Pesticide Exposure. *Environmental and Industrial Health* 9:913–59.

North Carolina Employment Security Commission. 1995. *Estimates of Migrant and Seasonal Farmworkers During Peak Harvest by County.* Raleigh, NC.

Oliveira, Victor J., Anne B. W. Effland, Jack L. Runyon, and Shannon Hamm. 1993. *Hired Farm Labor Use on Fruit, Vegetable, and Horticultural Specialty Farms.* Washington DC: USDA.

Perry, Melissa J., and Frederick R. Bloom. 1998. Perceptions of Pesticide Associated Cancer Risks among Farmers: A Qualitative Assessment. *Human Organization* 57(3):350–57.

Quandt, Sara A., Colin K. Austin, Thomas A. Arcury, Mandi E. Summers, and H. Nolo Martinez. 1998. Pesticide Safety Training Materials for Farmworkers. An Annotated Bibliography. Working Paper 98-01, Center for Urban and Regional Studies, University of North Carolina at Chapel Hill, Chapel Hill, NC.

Savitz, David A., Tye Arbuckle, Diane Kaczot, and Kathryn M. Curtis. 1997. Male Pesticide Exposure and Pregnancy Outcome. *American Journal of Epidemiology* 146:1025–36.

Seidel, J. V., S. Friese, and D. C. Leonard. 1995. *The Ethnograph: A User's Guide (Version 4.0).* Amherst, MA: Qualis Research Associates.

United States Environmental Protection Agency. 1992. *Pesticide Worker Protection Standards Training 40cfr Part170.130.*

Vaughan, Elaine. 1993a. Chronic Exposure to an Environmental Hazard: Risk Perceptions and Self-protective Behavior. *Health Psychology* 12:75–85.

———. 1993b. Individual and Cultural Differences in Adaptation to Environmental Risk. *American Psychologist* 48:673–80.

———. 1995. The Socioeconomic Context of Exposure and Response to Environmental Risk. *Environment and Behavior* 27:454–89.

Wallston, K. A. and B. S. Wallston. 1982. Who Is Responsible for Your Health? The Construct of Health Locus of Control. In *Social Psychology of Health and Illness.* G. S. Sanders and J. Suls, eds. Pp. 65–95. Hillsdale, NJ: Erlbaum.

Wester, Ronald C., and Howard I. Maibach. 1983. Cutaneous Pharmacokinetics: 10 Steps to Percutaneous Absorption. *Drug Metabolism Reviews* 14:169–205.

Woodruff, Tracey J., Amy D. Kyle, and Frédéric Y. Bois. 1994. Evaluating Health Risks from Occupational Exposure to Pesticides and the Regulatory Response. *Environmental Health Perspectives* 102:1088–96.

Zahm, Shelia Hoar, Mary H. Ward, and Aaron Blair. 1997. Pesticides and Cancer. *Occupational Medicine: State of the Art Reviews* 12:269–89.

QUESTIONS

1. What is the Worker Protection Standard (WPS)? Describe the behavioral model employed in the WPS to reduce worker exposure to agricultural chemicals.

2. Identify the two themes about agricultural chemical exposures that emerged from interviews with farmers. What kinds of challenges do these beliefs present to changing farm practices in ways that will reduce farmer exposure to low levels of agricultural chemicals?

Chapter Twelve

Competing Conceptions of Safety
High-Risk Workers or High-Risk Work?

Elaine Draper

The dance of death is not random. . . . The adage, "what is one man's meat is another man's poison," is a truism wrapped in genetic foil.

> —Alexander G. Bearn
> (Senior Vice President for Medical and Scientific
> Affairs, Merck Sharp and Dohme International)

Now when worker health problems develop, some of us are being told that it's not the dirty workplace that causes the problem, but it's who our ancestors are. The company explains that they're going to protect us by imposing a genetic scarlet letter on us, which will bar us because supposedly we're more susceptible to these poisons that the employer himself introduces into the workplace. So we have this susceptibility factor, which is part of the whole thrust of saying it's the victim whose fault it is by virtue of their [sic] genes.

> —Anthony Mazzocchi
> (as Director of Health and Safety, the Oil, Chemical,
> and Atomic Workers International Union)

In *Illness as Metaphor*, Susan Sontag (1977) gives an eloquent account of the social use of myths about illness—tuberculosis in the nineteenth century, and cancer in the twentieth. As we approach the end of this century, the worker susceptible to chemical hazards may be seen as a powerful symbol of the quest for an individual, genetic explanation for diseases suffered by many. But that quest is also an ideological extension of a social and political belief in the safety and benefit of industrial chemicals. The social ills of chemical hazards are projected onto susceptible individuals, in patterns that reflect dynamics of the workplace and the broader social structure.

The genetic paradigm in occupational health has been more appealing to industrial managers and occupational physicians than to industrial workers and union officials. An important factor contributing to the rise of the susceptibility thesis in industry, and to its selective appeal, is the conflict in ideological orientations toward the extent and locus of risk on the job. Contrasting assumptions about workplace safety have been central to the controversy over susceptibility.

Genetic testing and exclusionary practices reflect management's interest in finding individual rather than industrial causes of problems. Identifying individuals in the work force who are at risk can be another way of saying that conditions are in fact safe for most workers. This safety component of susceptibility policies is the basis for some of the most sweeping claims by advocates and some of the strongest objections by critics.

The Workplace Is Safe: Employers Minimize Occupational Risk

The controversy over genetic technologies parallels the debate over the nature of occupational health hazards themselves. Corporate managers and other proponents of susceptibility policies tend to believe that health hazards have been overestimated and overregulated.[1] They generally believe that risk at work accounts for only a small proportion of chronic illness in the United States. Thus we see a major effort by industry to show that cancer, for example, is due not to jobs or the environment but to life-style, genetic factors, and naturally occurring carcinogens.[2]

Tracing cancer to its causes—whether in occupation, smoking, or diet—is charged with controversy. Methods for establishing occupational causality vary widely. For cancer incidence, estimates of the proportion due to workplace conditions range from 1 percent to up to 40 percent (see Table 12.1).[3] Epidemiological and other evidence linking disease to work exposure is complex and equivocal. Doll and Peto, for instance, refer even to their own estimates of work-related cancer as "informed guesses," arguing that deficiencies in data on current exposure and health effects necessarily lead to "rather arbitrary" estimates for environmental carcinogens.[4] Like many others, however, they nevertheless adhere to specific percentages in their arguments for disease etiology.

Analysts who maintain that work exposure is a relatively insignificant factor (1–5 percent) in producing cancer generally use models and other research practices that contribute to the low figures they give for workplace hazards. Three of these practices involve highlighting smoking effects and excluding nonwhites and old people from cancer studies.

First, researchers who cite low percentages of cancer due to occupation typically trace a strong causal link between cancer and tobacco smoking while ignoring ways in which occupation confounds the impact of smoking. There is evidence that smokers are more heavily concentrated in occupations with exposure to carcinogenic hazards. Those who believe work contributes little to cancer causation seldom build such confounding effects into their models. Therefore, they are likely to hold smoking responsible for cancers of the lung and bladder, which are vulnerable to both industrial and nonindustrial exposures.[5]

TABLE 12.1
Estimates of the Contribution of Occupational Hazards to Cancer
Incidence

Percent of cancer incidence attributed to work hazards	Source
1 or 2 percent	Handler (1979)
5 percent	Peto (1980)
1 to 10 percent	Reinhardt (1978)
Greater than 20 percent	Swartz and Epstein (1981)
From 10 to 33 percent or perhaps higher	Stallones and Downs (1979)
Up to 20 to 40 percent	Bridbord et al. (1978)

Second, scientists who see a low incidence of occupational cancer often exclude data on nonwhites from their analysis. Peto (1980) argues that limiting cancer studies to whites is a good idea because improved medical care and cancer identification have made cancer mortality rates appear high for blacks and other "previously poor people" whose cancers have gone undetected due to inferior health care. However, findings by Percy and others (1981) indicate that race does not affect the accuracy of cancer mortality statistics. Improvements in medical care have been uneven, with the quality of health care facilities and services actually declining for many nonwhites. Further, nonwhites perform a higher proportion of hazardous jobs, so that restricting studies to whites falsely depresses occupational cancer rates.[6]

The third research practice is excluding individuals over 65 from calculations of cancer causation on the grounds that mortality figures for individuals under 65 are likely to be more accurate.[7] Yet approximately half of all U.S. cancer occurs in individuals over age 65. Excluding people over 65 removes a large group in which work-related cancer is particularly likely to occur, because of longer exposure histories and latency periods for tumor induction.[8] Since the 1960s, the chemical production of carcinogens and other substances capable of causing chronic health effects has grown substantially. Age-adjusted cancer incidence and mortality rates also have risen over this period. Smoking accounts for only part of these increases.[9] Given the time span between carcinogenic exposure and disease manifestation, the effects of recent production increases of environmental carcinogens will probably not be clear until the late 1990s and beyond, and those effects will be most pronounced in older people. By omitting people over 65 from analysis, excluding nonwhites, and ignoring occupational factors that confound disease rates among smokers, researchers can produce statistics that suggest low work-induced disease rates. Reversing these omissions reveals higher levels of occupational cancer.

With a broad range of estimates available to them, then, corporate officials generally accept the lowest estimate of cancer traceable to workplace hazards.[10] They do the same with regard to birth defects. For example, two scientists at Dow Chemical (Rao and Schwetz, 1981) contend that there has been no significant increase since 1970 in the proportion of birth defects related to chemicals and industrial exposures, which they claim has always been small. Powerful economic, legal, and regulatory pressures impel company physicians and others in industry to minimize the extent of occupational illness, preferably by identifying a worker's ailment as stemming from personal habits or genetic constitution. In general, company managers believe that chemical risks are less dangerous than does the public. According to a 1980 Harris survey, corporate executives are three times more likely than either the public or federal regulators to say that there is less risk from chemicals today than twenty years ago.[11] This pattern holds true for scientists in industry as well. Lynn's survey (1986) found that scientists in industry are three times more likely than university and government scientists to believe that Americans are overly sensitive to risk and want to be protected from nearly all dangers. This belief that risks from chemical exposures at work are minimal and declining bolsters the susceptibility thesis that the industrial workplace is safe for all but a few.

Safety and Susceptibility

In 1938, the geneticist J. B. S. Haldane wrote: "There are two sides to most of these questions involving unfavourable environments. Not only could the environment be improved, but susceptible individuals could be excluded" (1938, p. 180). While highlighting genetic susceptibility, he nevertheless argued that industrial hygiene standards were "shamefully low." A few decades later, however, most advocates of genetic screening would conclude that the environmental dimension of workplace risk was fairly unimportant. They typically would find that the environment had been improved sufficiently, leaving only abnormally susceptible individuals in danger.

The idea that normal working conditions are safe for everyone except the susceptible can be found in scientific journal articles and books that argue the merits of genetic screening, in conferences and hearings at which medical testing is debated, and in statements by the numerous types of social actors who recommend susceptibility policies for workplace use. This conception of safety also can be seen in the classical medical definition of special susceptibility to illness, sometimes called "hypersusceptibility": "A condition of inordinate or abnormally increased susceptibility to chemicals, infective agents, or other agents which in the normal individual are entirely innocuous."[12] Note that here, chemicals to which only "susceptible" workers are considered vulnerable are assumed to be innocuous. Indeed, this assumption pervades much of the discussion about genetic testing and exclusionary practices in industry.

Proponents of Susceptibility Screening

Scientists and company officials who advocate genetic screening say that identifying susceptible workers will help protect the whole working population, since only these few individuals are at risk from exposure levels below those that protect everyone else. A former corporate medical director of Celanese, at congressional hearings on genetic testing, articulates this view of the high-risk worker's unique vulnerability to otherwise tolerable industrial exposures: "Concern for an individual who is known to be hypersusceptible to adverse effects from any agent requires protection from exposure beyond that which normal individuals may tolerate readily."[13] A physiology professor at Harvard Medical School concurs with this emphasis on the unusual worker at risk:

> [W]hat we are trying to do, I think, is to explore ways in which genetic information can be used to recognize the occasional person who is outside the normal range of susceptibility and therefore should have any of another kind of job, but not exposure to that particular chemical to which he reacts peculiarly.[14]

The key safety ideas are: the *occasional* person who is outside the *normal* range of susceptibility who reacts *peculiarly*.[15]

An important implication of genetic screening is pointed out by two of its earliest advocates, Stokinger and Mountain: genetic tests enable occupational physicians to rule out work exposures as the cause of disease among employees. They wrote in 1967:

> With specific tests available, the industrial physician is in an advantageous position to decide whether a claimed case of emphysema is job-connected or not. . . . Hypersusceptibility tests are useful in predicting the direct, or indirect but additive effects of some of these chemicals, or in distinguishing between effects due to chemical exposure vs. congenital defect.[16]

They went even further, suggesting that genetic susceptibility will largely explain employee illnesses that arise: "As our understanding of the relationship of disease manifestations to genetic abnormalities grows, a large part of our past bewilderment and surprise at finding workers claiming injury to health under conditions known to be protective will disappear."[17] Thus, injured workers' claims linking their diseases to occupational exposures need not challenge the assumption that current conditions protect normal employees. Greater attention to genetic susceptibility can allow us to remain confident that working conditions are safe.

Chemical Safeguards and Workers at Risk

The focus on the susceptible worker is part of a larger effort to demonstrate that the workplace is safe without stricter regulation. Screening advocates believe that existing government standards such as "threshold limit values" adequately protect all but the unusual worker at risk. Threshold limit values (TLVs) are industry standards for chemicals set by the private ACGIH health organization and in many cases adopted by the Occupational Safety and Health Administration (OSHA).[18] The assumption behind TLVs is that they establish concentration levels of workplace substances to which "healthy" workers may be exposed for a lifetime without becoming ill. Employers, in particular, maintain that most workers are safe once these standards are maintained, and that restricting industrial exposures any further is therefore unnecessary. It was early scientific proponents of screening who proposed this link between TLVs and susceptibility policies. In 1967 Stokinger and Mountain wrote: "Medical advances in the last few years have revealed ways of identifying those presently not fully covered by the threshold limits."[19] Thus it is assumed that if workers *do* develop adverse health effects, these must have been caused by the less protective conditions of many years ago, by nonindustrial causes, or by special susceptibility.

Threshold limit values have been the subject of considerable debate, however. Supporters have argued that carcinogens and other disputed substances are safe for normal workers at or below the already low TLV exposure levels. Opponents have argued that no safe exposure levels can exist for a carcinogen; that zero exposure should be the goal; and that all persons, not only the susceptible, are at risk from these chemicals.[20] Representatives of organized labor have generally been among the opponents. As a national health and safety official with the AFL-CIO has argued: "Because there are no thresholds for carcinogens, employers are not eliminating the risk by excluding workers considered susceptible."[21]

In recent years, support for the TLV concept has declined, due in part to mounting evidence of cancer risks at very low exposure levels.[22] Lynn's (1986) survey of scientists in industry, universities, and government reveals that whereas 80 percent of scientists in industry agree that thresholds exist for carcinogens, only 37 percent of government scientists and 61 percent of university scientists agree. Some industry officials, in semantic retreat, now maintain that while effective threshold levels may not exist, at least workers will suffer no "significant risk" unless they are susceptible.

Others who support the susceptibility thesis shift their argument to the economic realm. They argue that in the case of toxic chemicals, it is economically and technologically impossible to eliminate exposure. They discount the possibility of substituting safer chemicals or adopting new control technology. Industry, they claim, cannot afford to protect the vulnerable and survive in a competitive corporate environment. There is no doubt that in certain cases, employers might indeed be strained financially if required to reduce exposure levels.[23] But the industry claim is often made as a sweeping general argument: Government regulation and industry standards cannot become more restrictive without putting a great many companies out of business. Employers assert that if they pay increased attention to the few susceptible workers, the coverage of current safeguards could become complete.

Cases of Safety Claims

If management considers the workplace safe for "normal" workers, then if employees become sick, it must be because they were individually susceptible to that sickness. The focus is on individuals who get cancer, not on industrial poisons that may cause the disease. For example, when industrial workers in a petrochemical plant in Texas discovered eighteen brain tumors within their workforce, one scientist remarked: "Why those eighteen? You need to identify groups of workers at risk."[24] This scientist did not ask why there were eighteen instead of none; he asked why *those* eighteen. In fact, nobody has yet determined why those eighteen individuals, and not others, developed brain tumors. Nevertheless, proponents of susceptibility testing assume that those people must have had something in common beyond the workplace exposure. And if you believe that only certain types of workers will have a problem, you can believe that the others are safe.

Another example of the safety emphasis implicit in the susceptibility perspective may be taken from a large textile firm. Cotton dust is widely believed to be hazardous, and the federal OSHA has set exposure limits. But a corporate official in

Burlington Industries, in discussing the company's medical program, stated that only 2 percent of the work force are vulnerable to cotton dust, and that they are protected by not being allowed to work in textile industry jobs with exposure to cotton dust.[25] Clearly, this is also intended to mean that 98 percent of the workers are safe when exposed to cotton dust.

A third example is the case of susceptibility policies affecting fertile women. Companies can identify such women as high risk, just as they screen for sickle cell trait and G-6-PD deficiency. Besides expressing concern for protecting the fetus and avoiding lawsuits for fetal damage, saying that women of childbearing capacity are at risk is another way of saying that the remaining work force is safe once fertile women are eliminated. Yet there are male reproductive hazards—including sperm damage, low fertility, and birth defects—which are in no way removed by excluding women.

Problems with the Focus on High-Risk Workers

Susceptibility tests and exclusionary policies go beyond the scientific evidence on which proponents claim these practices are based. Problems of evidence, accuracy, and industrial application reduce the medical effectiveness of genetic screening. They also limit the effectiveness of susceptibility policies for fetal hazards. However, the weak association between worker characteristics and occupational hazards has not prevented companies from testing and excluding workers. Proponents of genetic screening and fetal exclusion in industry tend to overlook or attempt to explain away several important characteristics of these practices.

Tautological Arguments

Screening advocates claim that existing working conditions assure protection of all workers except the susceptible. This perspective sometimes takes the form of tautological arguments: Those who are susceptible develop disease, and those who do develop disease must have an abnormal predisposing factor and therefore must be susceptible.

Polarized Categories

The second limitation of susceptibility policies is the difficulty of differentiating low-risk from high-risk groups. Even when the spectrum of response suggests no clear method for doing so, supporters of genetic screening and fetal exclusion tend to adopt polarized categories of risk and safety. Yet these categories are neither polarized nor distinct. Susceptibility policies that focus on heterozygous carriers frequently rely on vague associations between biological traits, disease, and chemical exposures. Differentiating workers who are at risk from those not at risk, or those who are sick from those who are well, or even those who have a trait from those who do not, is typically a murky process.

Only a few genetic characteristics allow for nearly binary judgments as to whether or not a person has a specific condition. Sickle cell anemia, or hemoglobin S, is such a case. Hemoglobin electrophoresis testing allows one to determine with considerable confidence that a person has sickle cell anemia or does not. But even in these relatively clear, usually homozygous cases, ambiguities abound.[26] For example, the extent to which a person will be affected, and whether testing or interpretation errors affect the outcome, are often unclear.[27]

Most genetic traits present even greater difficulties in making distinctions. Testing for single gene disorders generally relies on quantitative, not qualitative, differences in genetic material, with uncertainties and overlap between those considered heterozygotes and those considered normal. The distinctions rely on statistical probabilities. The best the tests can do is show that someone may be at somewhat greater risk for one type of ailment when exposed to one chemical agent or group of chemicals. What if 49 percent of workers are at higher risk or sicker than 51 percent because of biological characteristics, work exposures, medical history, health background, and other factors? How can one differentiate clearly between low risk and high risk? Where is the dividing line in the continuum of vulnerability? The susceptibility perspective frequently avoids the complexity of these questions by arbitrarily identifying high-risk workers and assuming that they alone are at risk.

Limited Predictiveness

Current screening tests suffer from both narrow applicability and low levels of predictiveness for disease.[28] Many of the traits screened for are rare. Those who are identified as high-risk workers infrequently develop ailments. The relationship between specific traits and the exposure problems that workers confront is often unclear. Indeed, most current tests are insensitive to industrial health hazards. A university scientist who is a specialist in the health effects of mineral dusts such as asbestos and silica highlights the difficulty of establishing relationships between biological factors and susceptibility to occupational hazards. In the case of mineral dusts, he explains:

> While certain immunologic markers are noted with increased prevalence in workers who have mineral dust diseases, there is no evidence that these serve to predict the condition rather than appearing after the disease is well established. . . . [T]his is a problem in a number of instances. There may be a marker. The question is, does that marker pre-exist—that is, predate the onset of exposures, or is it simply an attendant consequence of the effects of exposure?[29]

Another factor limiting the usefulness of screening tests is that many of the tests produce high rates of false positives. This leads individuals who in fact do not have the predisposing factor to be classified inaccurately as high risk and then excluded from a job on that basis.

At Risk for Specific Diseases and Exposures

Workers screened out of jobs who may be somewhat more at risk for one ailment when exposed to a substance may be less at risk for another substance that is present in the same workplace. They might be at lower risk of developing emphysema, but they may develop bladder cancer from the same substance or from other chemicals to which they are exposed. Further, those who are identified as susceptible to adverse effects from working conditions may be less vulnerable to infectious agents, or drugs, or contaminants in air, water, and food. Certainly, some workers are more vulnerable to particular substances than others are. But the susceptibility perspective tends to be very inclusive. It assumes that this variation is significant and relevant to a wide range of illnesses and working conditions. Thus, employers may screen individuals out as generally unfit for employment, even though the workers' suspected susceptibility may in fact be confined to very specific exposure conditions.

Spreading Disease through Job Rotation

Susceptibility policies can promote harmful job rotation practices. Rotating workers out of jobs and replacing them with workers who do not show damage can spread disease or exacerbate hazards by exposing more people to toxic chemicals.[30] Genetic screening, by facilitating the replacement of workers with signs of damage, can multiply cases of birth defects, mutation, and cancer.

Noncompliance with Government Standards

Company officials claim that current industrial and government safeguards offer workers adequate health protection. Yet some companies that have screened workers thought to be susceptible have been found to be out of compliance with occupational health regulations.[31] These employers are not in a position to assert convincingly that government regulations and susceptibility policies help ensure that normal workers are safe if the regulations are neither followed nor enforced effectively. In addition, employers' noncompliance with government rules calls into question the meaning of their stated commitment to employee health as motivation for their screening programs.

Stigmatization: The Genetic Scarlet Letter

Testing for particular genetic traits focuses attention on the inborn features of individuals, which can stigmatize workers.[32] It can also improperly shift occupational health concerns away from more general exposure problems. By locating the cause of occupational risk in the worker's unusual biological vulnerability, susceptibility policies shift the blame for disease to the workers. Identifying workers as genetically at risk can diminish their self-esteem, damage their family life and other social relationships, and threaten their ability to find work. For example, railroad companies ask job applicants whether they previously have been rejected from jobs for medical reasons. The burden of occupational hazards thus falls more heavily on employees.[33]

Ambiguous Testing Data and the Benefit of the Doubt

Employers' selective caution in assessing evidence of risk and safety is notable. The benefit of the doubt is given to chemicals, to existing company practices, and to exclusion, not to the possibility of health risks to the work force from industrial chemicals. This is a matter of ideology and a question of power.

Management advocates great caution and strict criteria for restrictions on chemicals. For example, in 1981 congressional hearings, the Chemical Manufacturers Association argued that under the proposed OSHA labeling standard, chemicals would be falsely accused of being carcinogens. Corporate representatives referred to evidence of negative health effects as unproven "allegations." They criticized labeling chemicals with health risk information because not all the evidence was in.[34] In this and other cases, employers have used terminology that suggests chemicals are on trial and should be presumed innocent until proven guilty beyond a reasonable doubt. For management, lack of data becomes an argument for safety or for screening workers.

Employers show considerable tolerance for "beta errors," false negatives that inaccurately indicate that chemical exposures are safe; but they display no similar tolerance for false positives. Managers frequently demand human evidence rather than data obtained from animal studies to prove damage at low exposure doses, but they have used animal evidence to clear chemicals for industrial use.[35]

Employers who have screening programs tend to assume that any abnormal biological characteristics revealed by testing will interfere with work or lead to medical problems. In the cases of employers screening out black military cadets and fertile women, women and blacks have been excluded despite slim evidence. For example, American Cyanamid's Lederle Laboratories barred fertile women from specific production jobs, on the basis of limited and ambiguous data, because they claimed that although the workplace chemicals were safe for adults, they were unable to prove that "more vulnerable fetuses" could tolerate exposure to them.[36] The company rejected evidence, that their working conditions posed a threat to workers' health. Like many employers, they chose to exclude workers they considered susceptible rather than take the chance of leaving themselves vulnerable to high medical expenses or aggressive legal action on behalf of damaged fetuses.

In the case of *genetic monitoring*—as opposed to *screening*—the same bias in favor of presumed workplace safety applies. Employers who conduct monitoring tend to interpret test results showing abnormalities as uncertain findings of little significance. Other employers reject monitoring outright, for that reason; or, in the process of targeting possible exposure problems, they express concern about studies that might suggest that a chemical they use is hazardous.

While employers respond to ambiguous data by trusting in chemical safety, they deny people jobs based on preemployment and periodic screening tests that have undetermined or low levels of predictiveness for disease. The same standards of evidence and proof do not apply to data in both cases. It is chemicals that enjoy the benefit of the doubt, not workers with questionable test results.

Monitoring and Environmental Hazards

Both those who support and those who oppose genetic monitoring in industry, in order to identify high-risk workers and jobs, agree on two points. First, chromosomal abnormalities that the tests detect are not clearly associated with subsequent clinical health effects. In other words, individuals who have high levels of abnormalities may be no more likely to develop disease than those who do not. Second, genetic monitoring is a useful research tool. But the consensus stops there. Beyond that point, opponents and supporters of monitoring have conflicting concepts of risk and safety in the workplace. The two perspectives may be characterized as follows.

Opposition to Monitoring. Opponents of genetic monitoring in the workplace argue that all chromosome studies of worker populations have had major and perhaps insurmountable methodological problems that make useful interpretation impossible. Monitoring, therefore, should not be incorporated into industrial surveillance programs, should not be used for risk assessment, and should not affect regulatory action. Further, advocates of chromosomal testing overstate exposure hazards and too willingly undertake tests of limited value. Accordingly, chromosome studies should be limited to research purposes.

Corporate officials, particularly, subscribe to this view. They emphasize problems of interpreting uncertain results and explaining preliminary findings of high rates of chromosomal abnormalities to workers or the public. An industry scientist states,

> *All* of these short-term tests must be considered *research tools.* They are *not* established clinical tests. . . . Genetic monitoring should not be used to set standards for safe exposures, or to establish conditions for the prevention of occupational disease.[37]

The medical director of a large chemical firm concurs:

> I don't know of any evidence that chemicals in industry are known to cause chromosome breaks or abnormalities. We need to know a lot more about what the hazards are in relation to exposure and genetic abnormalities. We don't want to get involved in doing anything that gives us results where we don't really know what they mean. If you do get high levels of chromosome breaks, what do you tell the workers?

A medical officer at a major steel company articulates the negative assessment other corporate officials share: "I would never recommend using chromosomal aberrations as a monitoring tool."

Support for Monitoring. Supporters believe that genetic monitoring is already of considerable value in industrial surveillance. They believe that chromosomal studies of populations exposed to particular industrial chemicals should be conducted now, in order to assess exposure risks before clinical symptoms appear. While admitting that some previous studies have been flawed, they argue that certain techniques are already sufficiently developed for identifying exposure hazards in industry.

Proponents of monitoring also state that those who want to limit genetic monitoring to research purposes either misunderstand or misrepresent the value of these tests. They believe that critics of current chromosomal testing insist on unreasonable, interminable programs of research as a prerequisite for testing in the workplace.[38] Instead, advocates cite academic and government research scientists who support current industrial monitoring:

> [Chromosomal monitoring tests] can be integrated rapidly into a medical surveillance program for workers at risk and can provide the most advanced technology currently available for the identification of genetic damage leading to pre-malignant conditions in man or to permanent germ cell damage, or both, thereby providing an advanced warning system for hazardous chemicals.[39]

> [E]ven though [a genetic monitoring test] cannot predict clinical outcomes, this should not preclude its use as an indicator of exposure to potentially harmful agents.[40]

> The results [of chromosomal studies] cannot be interpreted as an indication of ill health risk or susceptibility but can be used to help determine guidelines for safe exposure levels.[41]

> Genetic monitoring . . . is very important, because this can be used not only to identify specific individuals who might be more sensitive within a given concentration range, but also to identify hazardous work conditions in work areas and to judge the effectiveness of control technology.[42]

Industry personnel usually oppose genetic monitoring in the workplace and labor people typically support it.[43] The safety dimension of susceptibility policies, which is seldom made explicit, helps explain this split. It also helps explain why monitoring has received such a different response from genetic screening.

Managers and others in industry tend to oppose genetic monitoring because they have an interest in avoiding tests that could point to inadequate government standards or new exposure problems. They also have an ideological commitment to the safety of current industrial working conditions—a commitment shared by politicians and scientists who oppose government regulation.[44] In contrast, advocates of genetic monitoring see it as a valuable tool for monitoring populations, not individuals. They believe that elevated levels of chromosome abnormalities in a working population can serve to indicate an exposure problem that confronts all workers. Therefore, labor tends to favor genetic monitoring as a method of detecting hazardous working conditions at low exposure levels. The same reasoning that leads employees and union officials to oppose genetic screening generally contributes to their support of monitoring techniques. Compared with genetic screening, monitoring is more consistent with labor's view of how workers at risk should be located: through identifying high-risk substances and workplaces.

In general, then, the use of genetic monitoring reflects and reinforces a concern for occupational health that targets work risks, a concern that so far has focused on large groups or populations of workers. But if genetic monitoring that identifies particular workers with an unusually high proportion of chromosome aberrations comes to be widely used in industry, the response will change. If labor representatives, for example, discover that industrial firms have not informed workers of their findings, or have denied jobs to workers on the basis of those findings, they may protest against monitoring for the same reasons they protest against screening.

Company Testing

Employers are more likely to pursue testing programs, heed their results, and publicize their findings when they perceive the tests to be in their company's interest. For example, between 1964 and 1977, Dow Chemical carried out cytogenetic studies of all its employees—approximately 3,000 workers. In 1977 company scientists found evidence suggesting detrimental health effects from two of the corporation's biggest selling chemicals.[45] Concerned with these findings, they urged the company to publicize the evidence and change the production process. As one company scientist explained: "We wanted them to tell the workers what we had found, reduce the levels of benzene to which workers were exposed, and inform the appropriate government agencies and the rest of the petrochemical industry."[46] Instead, Dow abruptly cut off the testing, explaining that the tests were not useful.[47] Company officials claimed that the cytogenetic data were ambiguous and difficult to evaluate. As the Dow medical director stated: "Continuing periodic medical examination tells us much more than cytogenetic studies about the health of our people. . . . Cytogenetics is a new and developing tool that someday will probably be useful in medical surveillance, but is not regarded as such today."[48] Another Dow official, in a confidential interview, commented on the problem of whether to explain the chromosomal monitoring results to employees: "We can envision a situation in which our workers become unjustifiably apprehensive that they will develop leukemia or other forms of cancer. . . . We may even be the target of litigation and unfavorable publicity." After cancelling its large-scale monitoring program, Dow studied chromosomal changes in sixty to seventy of its office workers on a research basis only. Company officials described this small monitoring project as very costly but useful for pointing to the methodological complexity and limitations of studying chromosomal abnormalities.

Officials at Dow and other companies argue that Dow's experience with monitoring lends support to their view that it should not be used except for research. They argue that this large-scale program revealed the shortcomings of the technology, the inadequate methods researchers used, and the difficulty of explaining equivocal positive findings. Yet Dow's experience suggests that monitoring tests can successfully point to exposure hazards: The company was clearly unwilling to pursue its cytogenetic tests when they seemed to indicate occupational health risks; and its testing of office workers to study variables that influence normal chromosome breakage may well have been an expensive use of medical resources in an attempt to justify avoiding chromosome studies of production workers.

The disparity in perspective toward genetic monitoring clearly is considerable. There are different assessments of the testing programs already undertaken, and different views of the purpose and value of chromosomal studies for industrial surveillance. Because these differences mirror the power dynamics and conflicting interests within industry at large, they are unlikely to be reconciled soon.

NOTES

Epigraphs: Bearn, Presentation at the National Academy of Sciences, Conference on Genetic Influences on Responses to the Environment (July 10, 1980); Mazzocchi (1980).

1. Jackall (1988), McCaffrey (1982), Kelman (1980), Mendeloff (1979), and Green and Waitzman (1981) discuss business opposition to regulation and provide evidence of managers' belief that health hazards have been overestimated.

2. Employers typically claim that their own and similar workplaces are safe. However, they usually can point to employers they deem to be unusual, who do not give enough attention to safety, allowing hazards to remain instead. Furthermore, in some cases employers publicly have defended their safety record while withholding or distorting information implicating chemicals in their workplaces, as discussed in U.S. House Subcommittee on Health and Safety (1987, 1981), Legator (1987), Brodeur (1985), and Epstein (1980). Epstein (1979) offers a detailed analysis of industry efforts to show that a small proportion of cancer cases is due to industrial chemicals. See also Peto and Schneiderman (1981) and Ashford (1976).

3. A major component of this debate is asbestos. In the proceedings of a conference focusing on estimating occupational cancer (Peto and Schneiderman, 1981), estimates of the proportion of cancer due to asbestos exposure ranged from lows of 1 percent and 2–3 percent (Peto, 1980) to a high of 13–18 percent (Bridbord et al., 1978).

4. Doll and Peto (1981). Davis et al. (1983:364) state, "Most investigators agree that it is difficult to estimate with accuracy what proportion of new cancers may be associated with industrial production, smoking, and living patterns."

5. *Journal of Occupational Medicine* (1986), Brodeur (1985), and Davis et al. (1981) discuss confounding effects involving smoking and occupation in the study of cancer causation.

6. Robinson (1987, 1984) and Davis and Rowland (1983), consider the concentration of blacks and other racial minorities employed in more hazardous jobs.

7. Doll and Peto (1981), for example, calculate cancer causation only on the basis of people age 65 and under. They argue that Medicare access, duplicative recordkeeping for individuals treated in multiple locations, diagnostic fads, and improved cancer diagnosis in the aged artificially make it appear that old people are dying of cancer much more frequently than in the past. Therefore, they argue, it is best to restrict attention to trends among younger people, since these are less subject to error. However, the weight of the factors Doll and Peto cite is reduced by the nearly thirty-year history of Medicare and the fact that problems of diagnosis and duplicate cancer records cross age groups.

8. Davis et al. (1981), Epstein and Swartz (1981), and Bridbord et al. (1978).

9. Davis et al. (1983) and Peto and Schneiderman (1981).

10. Labor organizations that have taken a stand on the proportion of cancer that is related to work have tended to favor higher estimates. For example, U.S. and British trade unions have joined the International Labor Organization and 17 U.S. federal agencies in endorsing the Bridbord et al. (1978) report by government scientists estimating occupation-related cases to be "up to 20 to 40 percent." Epstein and Swartz (1981), Davis et al. (1981), and Doll and Peto (1981) discuss controversy over this report, along with more general issues of cancer and work exposure.

11. See Harris and Associates (1980). This survey was the most expensive that the Harris organization ever had carried out. It was conducted for the world's largest insurance company, Marsh and McLennan.

12. William Alexander Newman Dorland, *Illustrated Medical Dictionary*, 25th ed., pp. 744 and 1505, Philadelphia: W. B. Saunders; cited in Reinhardt (1978:319) as the classical medical

definition of hypersusceptibility. "Susceptibility," "hypersusceptibility," and "high-risk" commonly are used interchangeably, as they are in this chapter. Each of these terms refers to a significant predisposition or special vulnerability to chemical agents. Levy and Wegman (1983: 134–135) define hypersusceptibility as "an unusually high response to some dose of a substance." "Sensitivity" is one type of special risk that refers to immediate, short-term, acute responses to chemicals. They refer to sensitivity as one form of susceptibility "characterized by an acquired, immunologically mediated sensitization to a substance." Indications of sensitivity include rashes, asthma, and other symptoms more akin to allergies than the type of longer term and chronic potential disease effects stemming from inborn susceptibility that are the focus of this chapter.

13. Ernest Dixon (Tabershaw Associates, former corporate medical director of Celanese), statement prepared for delivery to Hearings on Genetic Screening and the Handling of High-Risk Groups in the Workplace, U.S. Congress (1981:12).

14. Bernard D. Davis, Adele Lehman Professor of Bacterial Physiology, Harvard Medical School (National Academy of Sciences, 1980:233).

15. Other expressions of this approach include: Charles Reinhardt (Director, Haskell Laboratory for Toxicology and Industrial Medicine, E. I. duPont de Nemours and Company) (1978:319); H. E. Stokinger (Chief, Toxicology Branch, Laboratories and Criteria Development Department NIOSH) and L. D. Scheel (Toxicology Branch, NIOSH) (1973:564); and Arno Motulsky, Professor of Medicine and Genetics, University of Washington School of Medicine, Seattle (National Academy of Sciences, 1980:18).

16. H. E. Stokinger (Chief of the Toxicology Section, U.S. Public Health Service) and J. T. Mountain (Occupational Health Program, U.S. Public Health Service) (1967:538).

17. Stokinger and Mountain (1967:541).

18. The threshold limit values (TLVs) developed by the American Conference of Governmental Industrial Hygienists (ACGIH) are published in the ACGIH, *Documentation of the Threshold Limit Values*, 4th ed. (Cincinnati: ACGIH, 1980). In 1971, OSHA adopted 450 TLVs from the ACGIH as established federal standards. See Salter (1988) for insight into the heated debate over incorporation of ACGIH threshold limits in government standards.

19. Stokinger and Mountain (1967:537). See also Stokinger (1981).

20. Salter (1988), McCaffrey (1982), Englehardt and Kaplan (1987), Ashford (1976), Peto and Schneiderman (1981), Epstein (1979), and OSHA (1980) discuss the controversy over safe exposure levels for carcinogens.

21. See Samuels (1983) for a similar statement opposing the threshold concept of safe carcinogen exposure.

22. In its guidelines, the Office of Science and Technology Policy (1985) rejected the threshold concept for carcinogen safety. Peto and Schneiderman (1981) provide evidence of declining support for the "threshold" concept of safe carcinogen exposure levels.

23. Guthier (1986), Ruttenberg and Powers (1986), Daniels (1985), and Boden (1986) discuss employer costs and the economic feasibility of protecting vulnerable workers.

24. Academic scientist, personal interview. Similar expressions of the genetic explanation for responses to environmental chemicals appear in Kouri et al. (1984) and Omenn and Gelboin (1984). See Alexander et al. (1980) for a report on studies of brain tumors among petrochemical workers to which the scientist referred. Subsequent investigation uncovered a total of at least forty-nine brain tumor deaths in two Texas chemical plants (Dow and Union Carbide). Two industry reports of these brain tumor mortality studies are Austin and Schnatter (1983) and Bond et al. (1983).

25. Hillman (1981). Another textile industry official claims that "fewer than two percent of textile workers may be adversely affected by cotton dust" (Salmans, 1981).

26. Heterozygous means having different alleles (alternate forms of a gene) at a genetic locus. Homozygous refers to having the same alleles at a genetic locus. Sickle cell trait is an example of a heterozygous case and sickle cell anemia is an example of a homozygous case. Although exceptions exist, genetic screening typically identifies heterozygous carriers. These individuals appear normal but disease manifestation may be more likely than in "normal" individuals. Individuals who are homozygous for a particular abnormality usually are identified as individuals with a "disease" rather than carriers with a "trait" or "condition."

27. For example, other hemoglobins (D or G) can be mistaken for hemoglobin S, and other conditions (sickle beta thalassemia or hereditary persistence of fetal hemoglobin) can be mistaken for sickle cell anemia because of the presence of hemoglobin F.

28. Office of Technology Assessment (1983) offers details concerning the narrow applicability and limited predictiveness of current screening tests. The OTA report states "[T]he ability of the techniques to identify people who are predisposed to occupational illness has not been demonstrated" (p. 5) and "None of the genetic tests evaluated by OTA meets established scientific criteria for routine use in an occupational setting" (p. 9). Their critical assessment applies to both screening and monitoring. See also OTA (1990).

29. Hans Weill, Professor of Medicine, Tulane University (U.S. Congress, 1981: 24).

30. Ashford et al. (1984) considers disease risks that may stem from screening workers from jobs and replacing them with others.

31. For example, see Ruttenberg and Hudgins (1981).

32. Goffman's (1963) work offers insight into processes by which individuals become stigmatized. The process of casting occupational health problems in individuals terms is akin to the process of blaming victims of poverty rather than the social and economic forces creating inequality that Ryan (1971) examines.

33. U.S. Congress (1981:109–110) discusses one company (Southern Pacific Transportation Company) that includes this question in their preemployment medical history. Test results can disqualify workers from jobs in specific companies or from entire industries.

34. "Statement of the Chemical Manufacturers Association for the Health and Safety Subcommittee of the House Education and Labor Committee Hearing on the Withdrawal of the Department of Labor's Proposed Standard on Hazards Identification," May 27, 1981. See also Richard Lewis (Occupational Health Specialist, International Chemical Workers Union), Testimony before the Health and Safety Subcommittee of the House Committee on Education and Labor on the Worker's Right to Know and the Withdrawal of OSHA's Proposed Rule on Hazards Identification, Washington, D.C., July 14, 1981.

35. For discussion of employers' standards of evidence for proving and disproving chemical hazards, see Salter (1988), Englehardt and Kaplan (1987), Brodeur (1985), U.S. House Subcommittee on Health and Safety (1987), Field and Baram (1986), Epstein (1980, 1979), Nelkin (1984), McCaffrey (1982), and Rothstein (1984).

36. A good example of the company's argument about evidence of chemical safety is W. C. May (Manager, Labor Relations, Lederle Laboratories, American Cyanamid Company), letter to W. P. Siemers, Jr. (President, Local 143, International Chemical Workers Union), December 19, 1978. American Cyanamid's and other companies' fetal exclusion policies have been based on ambiguous scientific data.

37. Betty Dabney (biochemist and molecular geneticist, industry consultant) (1981:630–31).

38. A representative expression of this research-only view—and one that industrial monitoring advocates often cite to bolster their case—is Dabney (1981:630): "Clearly positive findings in an occupational group, confirmed by repeated samplings at more than one time, should be an indication for further clinical studies which might include formal reproductive epidemiology and prospective morbidity and mortality studies."

39. Jill D. Fabricant (Department of Preventive Medicine and Community Health, Division of Environmental Toxicology, University of Texas Medical School) and Marvin S. Legator (Director, Division of Environmental Toxicology, University of Texas Medical School) (1981: 624). See also Legator (1987, 1982).

40. Anthony V. Carrano (Section Leader for Cell Biology and Mutagenesis, Biomedical Sciences Division, Lawrence Livermore National Laboratory) (1982:307). See also Carrano and Natarajan (1988) and Carrano (1986).

41. Theodore Meinhardt (Industrywide Studies Branch, Division of Surveillance, Hazard Evaluations, and Field Studies, NIOSH) (1981).

42. David Brusick (Director, Department of Genetics and Cell Biology, Litton Bionetics), in National Academy of Sciences (1980).

43. Samuels (1983), Otten (1986), Legator (1987), Dixon, in U.S. Congress (1981), Venable (1978), Mazzocchi (1980), and U.S. Congress (1981).

44. Lynn (1986) offers evidence that scientists who work for industry tend to share employers' views on safety and regulation—and do so to a much greater extent than do scientists who work for universities or government. See also Englehardt and Kaplan (1987), Wertz and Fletcher (1989), Dabney (1981), and U.S. Congress (1981, 1982a, b).

45. Workers exposed to benzene and epichlorohydrin showed high rates of chromosome breakage. Earlier cytogenetic studies of vinyl chloride had revealed no significant problem.

46. Dante J. Picciano (geneticist, Dow Chemical Company), in Severo (February 5, 1980).

47. Legator (1987, 1981), Dabney (1981), Office of Technology Assessment (1990), Otten (1986), Epstein (1980), Severo (1980), and personal interviews with corporate officials, government officials, and academic scientists provide additional evidence regarding Dow's testing program. Both advocates and opponents of industrial monitoring refer to this case in their arguments for or against testing.

48. John Venable (Medical Director of Dow Chemical, Texas Division) (1978).

GLOSSARY

sickle cell trait   a genetic condition characterized by the presence of an abnormal molecule in the red blood cells. Health hazards from this heterozygous trait generally are considered minimal or nonexistent.

susceptibility   biological predisposition to harmful effects of a chemical or other agent.

TLV   threshold limit value: the maximum time-weighted average concentration to which a healthy worker may be exposed up to 8 hours a day for a normal 40-hour week, over a 40- to 50-year working lifetime without becoming ill.

tort   a private or civil wrong, independent of a contract, arising from a violation of a duty.

toxin   a substance that is poisonous, carcinogenic, mutagenic, or teratogenic.

REFERENCES

Alexander, Victor, Sanford S. Leffingwell, J. William Lloyd, Richard D. Waxweiler, and Richard L. Miller. 1980. "Brain cancer in petrochemical workers: A case series report." *American Journal of Industrial Medicine* 1, No. 1:115–23.

Ashford, Nicholas A. 1976. *Crisis in the Workplace: Occupational Disease and Injury*. Cambridge, Mass.: MIT Press.

Ashford, Nicholas A., Christine J. Spadafor, and Charles C. Caldart. 1984. "Human monitoring: Scientific, legal, and ethical considerations." *Harvard Environmental Law Review* 8, No. 2:263–363.

Austin, Susan G., and A. Robert Schnatter. 1983. "A case-control study of chemical exposures and brain tumors in petrochemical workers." *Journal of Occupational Medicine* 25, No. 4 (April): 313–20.

Boden, Leslie I. 1986. "Impact of workplace characteristics on costs and benefits of medical screening." *Journal of Occupational Medicine* 28, No. 8 (August): 751–56.

Bond, Gregory G., Ralph R. Cook, Peter C. Wight, and George H. Flores. 1983. "A case-control study of brain tumor mortality at a Texas chemical plant." *Journal of Occupational Medicine* 25, No. 5 (May):377–86.

Bridbord, K., P. Decoufle, J. F. Fraumeni, Jr., D. G. Hoel, R. N. Hoover, D. P. Rall, U. Saffioti, M. A. Schneiderman, and A. C. Upton. 1978. Estimates of the Fraction of Cancer in the United States Related to Occupational Factors. National Cancer Institute, National Institute of Environmental Health Sciences, and National Institute for Occupational Safety and Health (September 15). [Appears in Peto and Schneiderman, 1981.]

Brodeur, Paul. 1985. *Outrageous Misconduct: The Asbestos Industry on Trial*. New York: Pantheon Books.

Calabrese, Edward J. 1978. *Pollutants and High Risk Groups: The Biological Basis of Enhanced Susceptibility to Environmental and Occupational Pollutants*. New York: John Wiley and Sons.

Carrano, Anthony V. 1982. "Sister chromatid exchange as an indicator of human exposure." In Bryn A. Bridges, Byron E. Butterworth, and I. Bernard Weinstein (editors), *Indicators of Genotoxic Exposure, Banbury Report* 13. Cold Spring Harbor, N.Y.: Cold Spring Harbor Laboratory.

———. 1986. "Chromosomal alterations as markers of exposure and effect." *Journal of Occupational Medicine* 28, No. 10 (October):1112–16.

Carrano, Anthony V., and A. T. Natarajan. 1988. "Considerations for population monitoring using cytogenetic techniques." *Mutation Research* 204, No. 3 (March):379–406.

Dabney, Betty. 1981. "The role of human genetic monitoring in the workplace." *Journal of Occupational Medicine* 23, No. 9 (September):626–31.

Daniels, Norman. 1985. *Just Health Care*. New York: Cambridge University Press.

Davis, Devra Lee, Kenneth Bridbord, and Marvin Schneiderman. 1981. "Estimating cancer causes: Problems in methodology, production, and trends." In Richard Peto and Marvin Schneiderman (editors), *Quantification of Occupational Cancer, Banbury Report* 9. Cold Spring Harbor, N.Y.: Cold Spring Harbor Laboratory.

———. 1983. "Cancer prevention: Assessing causes, exposures, and recent trends in mortality for U.S. males, 1968–1978." *International Journal of Health Services* 13, No. 3:337–67.

Davis, Morris E., and Andrew S. Rowland. 1983. "Problems faced by minority workers." In Barry S. Levy and David H. Wegman (editors), *Occupational Health: Recognizing and Preventing Work-Related Disease*. Boston: Little, Brown.

Doll, Richard. 1981. "Relevance of epidemiology to policies for the prevention of cancer." *Journal of Occupational Medicine* 23, No. 9 (September):601–9.

Doll, Richard, and Richard Peto. 1981. "The causes of cancer: Quantitative estimates of avoidable risks of cancer in the United States today." *Journal of the National Cancer Institute* 66:1191–1308.

Englehardt, Tristan H., and Arthur L. Kaplan (editors). 1987. *Scientific Controversies: Case Studies in the Resolution and Closure of Disputes in Science and Technology*. New York: Cambridge University Press.

Epstein, Samuel S. 1979. *The Politics of Cancer.* New York: Anchor Books.

————. 1980. "2,4,5-T and the Dow track record." Statement on Behalf of the Interfaith Center on Corporate Responsibility, Press Release (May 9).

Epstein, Samuel S., and Joel B. Swartz. 1981. "Fallacies of lifestyle cancer theories." *Nature* 289, No. 5794 (January 15):127–30.

Fabricant, Jill D., and Marvin S. Legator. 1981. "Etiology, role, and detection of chromosomal aberrations in man." *Journal of Occupational Medicine* 23, No. 9 (September):617–25.

Field, Robert I., and Michael S. Baram. 1986. "Screening and monitoring data as evidence in legal proceedings." *Journal of Occupational Medicine* 28, No. 10 (October):946–50.

Goffman, Erving. 1963. *Stigma: Notes on the Management of Spoiled Identity.* Englewood Cliffs, N.J.: Prentice-Hall.

Green, Mark, and Norman Waitzman. 1981. *Business War on the Law: An Analysis of the Benefits of Federal Health/Safety Enforcement.* Washington, D.C.: Corporate Accountability Research Group.

Guthier, William E. 1986. "Medical screening and monitoring as noted by the insurance industry." *Journal of Occupational Medicine* 28, No. 8 (August):765–67.

Haldane, J. B. S. 1938. *Heredity and Politics.* London: George Allen and Unwin.

Handler, Philip. 1979. "Some comments on risk assessment." In *National Research Council Current Issues and Studies, Annual Report.* Washington, D.C.: National Academy of Sciences.

Harris, Louis and Associates. 1980. *Risk in a Complex Society.* New York: Marsh and McLennan Companies.

Hillman, Carol B. 1981. "An industrial view." *New York Times* (June 26).

Jackall, Robert. 1988. *Moral Mazes: The World of Corporate Managers.* New York: Oxford University Press.

Journal of Occupational Medicine. 1986. Medical Screening and Biological Monitoring for the Effects of Exposure in the Workplace (Conference held in Cincinnati July 10–13, 1984). Vol. 28, Nos. 8 and 10 (August and October).

Kelman, Steven. 1980. *Regulating America, Regulating Sweden: A Comparative Study of Occupational Safety and Health Policy.* Cambridge, Mass.: MIT Press.

Kouri, Richard E., Arthur S. Levine, Brenda K. Edwards, Theodore L. McLemore, Elliot S. Vesell, and Daniel W. Nebert. 1984. "Sources of interindividual variations in aryl hydrocarbon hydroxylase in mitogen-activated human lymphocytes." In Gilbert S. Omenn and Harry V. Gelboin (editors), *Genetic Variability in Responses to Chemical Exposure, Banbury Report 16.* Cold Spring Harbor, N.Y.: Cold Spring Harbor Laboratory.

Lavine, Mary P. 1982. "Industrial screening programs for workers: Ethical and policy problems." *Environment* 24, No. 5 (June):26–38.

Legator, Marvin S. 1981. "Statement by Dr. Marvin Legator, Director of the Division of Genetic Toxicology of the University of Texas Medical Branch (Galveston)." Delivered at the Dow Chemical Annual Meeting. Midland, Michigan (May 8).

————. 1982. "Approaches to worker safety: Recognizing the sensitive individual and genetic monitoring."*Annals of the American Conference of Governmental Industrial Hygienists* 3 (September):29–34.

————. 1987. "The successful experiment that failed." In Tristan H. Englehardt and Arthur L. Kaplan (editors), *Scientific Controversies: Case Studies in the Resolution and Closure of Disputes in Science and Technology.* New York: Cambridge University Press.

Levy, Barry S., and David H. Wegman. 1983. *Occupational Health: Recognizing and Preventing Work-Related Disease.* Boston: Little, Brown.

Lynn, Frances M. 1986. "The interplay of science and values in assessing and regulating environmental risks." *Science, Technology, and Human Values* 11, No. 2 (Spring):40–50.

Mazzocchi, Anthony. 1980. "Genetic confrontation in the workplace." Presented at the University of California, Berkeley (October 17).

McCaffrey, David P. 1982. *OSHA and the Politics of Health Regulation.* New York: Plenum Press.

Meinhardt, Theodore. 1981. "Cytogenetic studies of workers exposed to ethylene oxide." Cincinnati, Ohio: National Institute of Occupational Safety and Health, Division of Surveillance, Hazard Evaluations, and Field Studies, Industrywide Studies Branch.

Mendeloff, John. 1979. *Regulating Safety: An Economic and Political Analysis of Occupational Safety and Health Policy.* Cambridge, Mass.: MIT Press.

National Academy of Sciences, Institute of Medicine. 1980. Genetic Influences on Responses to the Environment: Conference Proceedings. Washington, D.C., July 10–11.

Nelkin, Dorothy (editor). 1984. *Controversy: Politics of Technical Decisions.* 2nd ed. Beverly Hills, Calif.: Sage.

Occupational Safety and Health Administration (OSHA), U.S. Department of Labor. 1980. "Identification, classification, and regulation of potential occupational carcinogens." *45 Federal Register 5001* (January 22).

Office of Science and Technology Policy, Executive Office of the President. 1985. "Chemical Carcinogens: A Review of the Science and Its Associated Principles." *50 Federal Register,* No. 50.

Office of Technology Assessment (OTA), U.S. Congress. 1983. *The Role of Genetic Testing in the Prevention of Occupational Disease.* Washington, D.C.: U.S. Government Printing Office.

———. 1990. *Genetics in the Workplace.* Washington, D.C.: U.S. Government Printing Office.

Omenn, Gilbert S., and Harry V. Gelboin (editors). 1984. *Genetic Variability in Responses to Chemical Exposure, Banbury Report 16.* Cold Spring Harbor, N.Y.: Cold Spring Harbor Laboratory.

Otten, Alan L. 1986. "Probing the cell: Genetic examination of workers is an issue of growing urgency." *Wall Street Journal* (February 24).

Percy, Constance, Edward Stanek, and Lynn Gloeckler. 1981. "Accuracy of cancer death statistics and its effect on cancer mortality statistics." *American Journal of Public Health* 71, No. 3 (March):242–50.

Peto, Richard. 1980. "Distorting the epidemiology of cancer: The need for a more balanced overview." *Nature* 284, No. 5754 (March 27):297–300.

Peto, Richard, and Marvin Schneiderman (editors). 1981. *Quantification of Occupational Cancer, Banbury Report 9.* Cold Spring Harbor, N.Y.: Cold Spring Harbor Laboratory.

Plumlee, Lawrence, Stanton Coerr, Herbert L. Needleman, and Roy Albert. 1979. "Role of high risk groups in the derivation of environmental health standards: Panel discussion." *Environmental Health Perspectives* 29 (April):155–59.

Rao, K. S., and B. A. Schwetz. 1981. "Protecting the unborn: Dow's experience." *Occupational Health and Safety* (March):53–61.

Reinhardt, Charles. 1978. "Chemical hypersusceptibility." *Journal of Occupational Medicine* 20, No. 5 (May):319–22.

Robinson, James C. 1984. "Racial inequality and the probability of occupation-related injury or illness." *Milbank Memorial Fund Quarterly/Health and Society* 62, No. 4:567–90.

———. 1987. "Trends in racial inequality and exposure to work-related hazards, 1968–1986." *Milbank Memorial Fund Quarterly/Health and Society* 65, Supplement 2:404–20.

Rothstein, Mark A. 1984. *Medical Screening of Workers.* Washington, D.C.: Bureau of National Affairs.

————. 1986. "Discriminatory aspects of medical screening." *Journal of Occupational Medicine* 28, No. 10 (October):924–29.

Ruttenberg, Ruth, and Randall Hudgins. 1981. *Occupational Safety and Health in the Chemical Industry*, 2nd ed. New York: Council on Economic Priorities.

Ruttenberg, Ruth, and Marilyn Powers. 1986. "Economics of notification and medical screening for high-risk workers." *Journal of Occupational Medicine* 28, No. 8 (August):757–64.

Ryan, William. 1971. *Blaming the Victim.* New York: Random House.

Salmans, Sandra. 1981. "Cotton industry upset by ruling." *New York Times* (June 18).

Salter, Liora. 1988. *Mandated Science: Science and Scientists in the Making of Standards.* Boston: Klewer Academic Publishers.

Samuels, Sheldon W. 1983. "Genetic testing of workers." A Technical Report to the Executive Council of the AFL-CIO Industrial Union Department (June 22).

Severo, Richard. 1980. "The genetic barrier: Job benefit or job bias?" *New York Times* (February 3–6).

Sontag, Susan. 1977. *Illness as Metaphor.* New York: Farrar, Straus, and Giroux.

Stallones, R. A., and T. Downs. 1979. "A critical review of K. Bridbord et al." Report prepared for the American Industrial Health Council.

Stokinger, Herbert E. 1981. "Genetic screening of employees: Resistance and responsibility." *Dangerous Properties of Industrial Materials Report* 1 No. 7 (September-October):7–11.

Stokinger, Herbert E., and John T. Mountain. 1967. "Progress in detecting the worker hypersusceptible to industrial chemicals." *Journal of Occupational Medicine* 9, No. 11 (November): 537–42.

Stokinger, H. E., and L. D. Scheel. 1973. "Hypersusceptibility and genetic problems in occupational medicine: A consensus report." *Journal of Occupational Medicine* 15, No. 7 (July): 564–73.

Swartz, Joel B., and Samuel S. Epstein. 1981. "Problems in assessing risk from occupational and environmental exposures to carcinogens." In *Quantification of Occupational Cancer, Banbury Report* 9. Cold Spring Harbor, N.Y.: Cold Spring Harbor Laboratory.

U.S. Congress, House, Committee on Science and Technology, Subcommittee on Investigations and Oversight. 1981. *Hearings: Genetic Screening and the Handling of High-Risk Groups in the Workplace.* Washington, D.C.: U.S. Government Printing Office (October 14 and 15).

————. 1982a. *Hearings: Genetic Screening of Workers.* Washington, D.C.: U.S. Government Printing Office (June 22).

————. 1982b. *Hearings: Genetic Screening in the Workplace.* Washington, D.C.: U.S. Government Printing Office (October 6).

U.S. House Subcommittee on Health and Safety, Committee on Education and Labor. 1981. Hearings: OSHA Oversight Hearings on Proposed Rules on Hazards Identification. Washington, D.C.: U.S. Government Printing Office (April 7, 28; May 19, 27; July 8, 14, 21; and October 6).

————. 1987. *Hearings: High Risk Occupational Disease Notification and Prevention Act of 1987.* Washington, D.C.: U.S. Government Printing Office (March 17, 24, 26, 31; and April 8).

Venable, John. 1978. Statement Issued by Medical Director of Dow Chemical, Texas Division (June 12).

Wertz, Dorothy C., and John C. Fletcher. 1989. "An international survey of attitudes of medical geneticists toward mass screening and access to results." *Public Health Reports* 104, No. 1: 35–44.

Wright, Michael, D. Jack Kilian, Nicholas A. Ashford, and Paul Kotin. 1979. "Role of the knowledge of high risk groups in occupational health policies and practices: Panel discussion." *Environmental Health Perspectives* 29 (April):143–53.

QUESTIONS

1. Describe the three types of research practices often used in studies that demonstrate that only a small percentage (1–5 percent) of cancers can be attributed to exposures to workplace hazards. What criticisms does Draper offer for each of these practices?

2. What is the difference between genetic screening and genetic monitoring? Why does labor support genetic monitoring, while employers and management oppose it?

Part 5

Toxins in the Community

Toxic substances are transported from factories, mines, and other sources to residential neighborhoods by two primary routes: in the form of consumer products, and in the form of waste by-products. In industrial societies, consumers use many products containing hazardous ingredients. There was a time when people hung arsenic-laden wallpaper in their houses, and painted their walls with lead-laden paint. Potentially dangerous chemicals can be still be found in an array of common household items, from varnishes and paint thinners to cleaning fluids. And many people continue to be douse their homes, lawns, and gardens with pesticides.

Many types of economic activity, from mining to manufacturing to medical services, generate hazardous waste materials. While sometimes treated on-site, these materials frequently leave the boundaries of a company's property. Some waste is hauled away by truck or rail to off-site disposal facilities. Other departure routes are not always so visible: air currents may carry small particulate matter across county, state, or national lines, and chemicals may migrate for miles through soils or underground aquifers. As these substances travel off-site, they expose residents of nearby communities, and sometimes quite distant communities, to potential health hazards.

The four readings included in the present section provide insights into the diverse range of environmental hazards encountered by community residents in industrial societies. They also address the challenges contested knowledge presents to efforts to ameliorate community-based environmental hazards. Chapter 13 by Phillimore, Moffatt, Hudson, and Downey, "Pollution, Politics and Uncertainty: Environmental Epidemiology in North-East England," overviews efforts to identify the causes of poor health among residents living along a highly industrialized section of the River Tees. While some epidemiological findings are highly suggestive of the causal role of industrial air pollution, other findings present equivocal or contradictory evidence. Local officials and corporations exploited the uncertainties of this scientific knowledge base, selectively highlighting those findings that suggested non-industrial causes of poor public health, such as poverty and smoking.

Chapter 14 by Berney, "Round and Round It Goes: The Epidemiology of Childhood Lead Poisoning, 1950–1990," discusses an environmental contaminant found in many U.S. communities. Lead has long been known to be a poison. In high enough doses, it kills. At lower concentrations, it can result in neurological impairments and other health consequences. During the middle part of the twentieth century, concern about lead exposure was focused on the consumption of lead-containing paint and plaster by children residing in lower-income housing. Community screening programs established during the 1960s and 1970s revealed childhood exposure levels were more extensive than previously thought. Attention shifted to airborne lead as a source of contamination, with a particular focus on the possible role of exhaust and evaporation from leaded gasoline. After use of unleaded gasoline became commonplace, marked decreases in lead blood levels were noted.

In chapter 15, "Lead Contamination in the 1990s and Beyond: A Follow-Up," Widener recounts the most recent developments in the battle against lead contamination. Overall blood lead levels in children continues to decline. A substantial number of children remain at risk for lead exposure, however. Furthermore, risk of exposure remains highest among low-income and/or minority children. Parent and community organizations are taking a more active role in protecting children against lead poisoning.

In chapter 16, "Suffering, Legitimacy, and Healing: The Bhopal Case, Critical Events," Das reminds us that sometimes communities are subjected to environmental hazards whose deadly impacts are sudden and massive. The leak of a vaporous cloud of methyl isocyanate (MIC) from the Union Carbide plant in Bhopal, India, some time during the night of December 2–3, 1984, is one of the world's worst industrial accidents. Several thousand people died as a result of their exposure to MIC, and as many as 300,000 were injured. Victims of environmental hazards face difficulties in securing compensation for their suffering. In Bhopal, the legal system worked to silence the voice of the victims. The Indian government unilaterally declared itself the legal representative of those harmed by the gas leak. It then reached a settlement with Union Carbide that went against the wishes of the very victims the government claimed to represent.

Chapter Thirteen

Pollution, Politics, and Uncertainty
Environmental Epidemiology in North-East England

Peter Phillimore, Suzanne Moffatt, Eve Hudson,
and Dawn Downey

Introduction

The gradient in mortality of women under 75 years of age points to factors unevenly distributed across Teesside. Air pollution from industry exhibits this uneven distribution in a way that alternative potential confounding influences do not. . . . Differential proximity to industrial air pollution over a long period currently offers a more convincing explanation for the observed variations in lung cancer mortality in women than any alternative explanation. (Pless-Mulloli et al. 1998, 196)

There is a theory in the report that air pollution in the past may have contributed to high incidences of lung cancer in women. On this question, however, no cause has been established in the report, and ICI has no evidence that our activities have or are causing ill-health [*sic*]. ICI believes this situation requires further careful research to allay concerns. (ICI 1995)

These quotations illustrate graphically how contentious environmental epidemiology can prove to be, particularly when the activities of major corporations are implicated.[1,2] Arguments about the causes of poor public health in Teesside, a conurbation of over 400,000 in north-east England, have lasted for most of this century and show no signs of diminishing. A critical issue is the role of air pollution emitted from the various steel, coking, and chemical plants which line the River Tees. Though greatly reduced in scale by comparison with the 1970s, Teesside remains one of the dominant centres of both steel and chemical production in western Europe. This chapter examines recent twists in this long-running controversy about industrial pollution and public health, interweaving epidemiological and social-political aspects of a vigorously contested story. Our aim is not to reveal a bedrock of epidemiological fact blurred behind distracting political misrepresentations. Such a goal is all but illusory, particularly given the uncertainties which surround interpretation in environmental epidemiology (Taubes 1995). It is this very uncertainty that concerns us, for it is regularly used to support powerful interests. Indeed, the inherent caution of medical science assists industry and public authorities alike in promoting a rhetoric of reassurance (Draper 1991; Nash and Kirsch 1988). In this rhetoric, credible contrary evidence is played down.

The evaluation of environmental "risk-factors" is commonly acknowledged to pose major methodological challenges for epidemiology (Lippmann and Lioy 1985). An inevitable mismatch between the ambitious aims of environmental epidemiology and inherent limitations in its methodology compounds the uncertainties with which it grapples. The consequences extend beyond public health, given the many groups with a stake in the outcome of epidemiological studies (Little 1998; Taubes 1995). Most importantly here, these limitations and the disagreements they provoke offer a loop-hole for political and corporate interests. Contradictory findings from different studies permit authorities to make reassuring claims about public safety, based on selective readings of the available evidence. No matter how carefully designed or conducted, studies in environmental epidemiology often permit radically different inferences to be drawn about particular risks. What investigators conclude, and what interested parties claim their researches demonstrate, are not necessarily one and the same thing. For the social scientist, an important task is to explore the motivations and power relations that help to determine which interpretations become authoritative. In this chapter we ask: how did industrial and local political interests manage to present an account of safety and risk in the Teesside environment that effectively distorted complicated epidemiological conclusions, and minimized the threat of local controversy?

We begin by discussing recent approaches to risk in the social sciences which inform our approach. We then outline the historical background to concerns about industrial pollution and health in Teesside, before summarizing the recent epidemiological evidence and the puzzles it raises. The remaining sections of this chapter examine the politics of air pollution in Teesside, particularly in relation to this epidemiological evidence, and reflect on the implications of efforts to promote an official narrative of reassurance about the Teesside air.

The Sociology of Risk

For a long time social scientists had little to say about the problems of environmental pollution and its attendant risks. It is safe to say, however, that pollution and risk are now among the salient concerns of both sociologists and anthropologists. Credit for this must go above all to Mary Douglas and Ulrich Beck. They offer perhaps the most exhaustive, certainly the most cited, work on risk and society (Douglas and Wildavsky 1982; Douglas 1992; Beck 1992). For both theorists, risk is an organizing concept with extensive explanatory reach, though their approaches to the concept differ radically.

Douglas was the first to recognize "the idea of 'risk' as a central cultural construct of our time" (Lock 1998, 10). In association with Aaron Wildavsky, she posed the question of why every society magnifies certain risks or pollution threats and minimizes others (Douglas and Wildavsky 1982; Douglas 1985). For Douglas, the risks that particular societies "select" from the myriad number of dangers in the world is ultimately a consequence of cultural preoccupations that have little to do with the inherent danger of any one risk. In theory, this transcultural approach to the construction of risk offers a chance to escape the invidious dichotomising of "rational calculation" versus "irrational obsession," for it emphasizes that there is no language of risk that is neutral, or independent of culture. We are all subject to cultural bias. When applied to industrial pollution in North America, however, Douglas and Wildavsky do not so much explain as explain away the concerns of individuals and environmental groups about toxins and their effects. By seeing environmental groups as akin to secular sects they effectively diminish any claims such groups may make. Paradoxically, they thus reproduce the polarity of the rational versus the irrational which their approach held the promise of overcoming. As Nash and Kirsch, two anthropological critics, have observed (1988, 158–59).[3]

> Douglas and Wildavsky trivialize apprehension about present-day dangers. The book serves as an argument for corporations that wish to minimize the real and growing danger of chemical toxins in the environment.

This charge could not be levelled at Beck. Like Douglas, he highlights the pervasiveness of contemporary discourses surrounding technology and risk. Unlike her, he takes much more seriously public concerns about these risks. In *Risk Society* (1992) and subsequent work (1996a, 1996b), Beck makes the shift from older to newer forms of risk one of the fundamental transformations defining the character of the world today. His thesis places environmental pollution, its associated risks, and the systems or apparatus through which we know such risks, at the centre of theoretical understanding of the modern world. We now live in what he calls the "risk society." A consistent thread running through Beck's writing concerns what he regards as the incalculability of many, if not most, risks (Beck 1992, 170–71). This has led him to explore the social, legal, and political consequences of such inherent uncertainty. In this chapter, we emphasize not so much the incalculability of pollution risks as the scope within many epidemiological studies for alternative ways of construing evidence and measuring risk. But the latitude that this gives corporations or government institutions to re-present evidence in ways convenient to themselves links up with Beck's arguments, and provides our main theme (see also Draper 1991; Latowsky 1998; MacGill 1987; Nash and Kirsch 1988; Paigen 1982).

One consequence of the growing public unease about environmental risks, theorized so differently by Douglas and Beck, has been manifest in increasingly frequent challenges to scientific claims and expertise. The concepts of "popular epidemiology" (Brown 1992) and "citizen science" (Irwin 1995) have emerged in this context. Both describe citizen challenges to official epidemiology, and encompass various levels of resistance to official silence on, or rejection of, public concerns about environmental health risks. These range from alternative models of causation to full-fledged research strategies (Irwin and Wynne 1996; Kroll-Smith and Floyd 1997; Whittaker 1998).[4] The Teesside story we tell here is only indirectly about citizens and their resistance to official versions of science. Our focus, rather, is on official efforts to minimize debate about pollution and health in Teesside and to defuse possible repercussions arising from epidemiological studies into this link. Popular concerns do not altogether disappear in this context however, as we will see; epidemiological reports filter through the media into local debate, and people often make their own assessments of health research (and official responses to such research), judging conclusions against the benchmark of their personal experience, and with their own ideas about credibility.

Teesside's Historical Background

The Teesside towns were built first around iron and later steel.[5] Teesside's association with chemical industries started after the First World War. At various periods in the twentieth century, but especially in the major recessions of the 1920s to 1930s and 1980s to 1990s, these towns have seen exceptionally high unemployment. This history of recurrent economic insecurity, and the poverty associated with it, are important for an understanding of the politics and epidemiology of pollution in Teesside.[6] Throughout this century, pollution from the industries along the banks of the River Tees has conjured up contradictory associations: on the one hand a sign of the particular risks to which its population was exposed, but more emphatically a visible demonstration of the area's economic vitality. This double vision is reflected in a pair of statements by the Medical Officer of Health for Eston, a district down-river from Middlesbrough. Written as far back as 1921 and 1925, at a time of recession, they reflect the two sides of economic inactivity:

> During the year the atmosphere has been very clear and the air pure, owing to the fact that the large industrial works have been standing most of the time. The absence of the clouds of black smoke which are emitted in normal times from the large chimneys have contributed to the favourable health statistics of the District. . . . While we all deplore the fact that the trade of the District is so bad, yet I am of the opinion that with greater care when the industries are in full swing much of the polluted smoke may be prevented. (Medical Officer of Health 1921)

> The health of the District is fairly satisfactory when one takes into consideration the privations suffered by many of the inhabitants through lack of work, as this has been one of the hardest hit districts in the kingdom. (Medical Officer of Health 1925)

As wider national and international concerns about environmental pollution have grown in recent decades, Teesside's image as a place burdened by abnormally severe air pollution has proved hard to shake off. For instance, as far back as the 1950s the tobacco industry saw Teesside as a valuable natural laboratory. If a link between industrial air pollution and lung cancer or respiratory diseases could be established, it might help to deflect attention from the rising body of evidence establishing the role of smoking in these causes of death. The UK Tobacco Research Council accordingly funded studies in Teesside to examine environmental factors associated with lung cancer and bronchitis mortality (Wicken and Buck 1964). Three areas in and around the Tees conurbation and four rural districts were included in the study. A follow-up study continued this work through the 1960s (Dean and Lee 1977). Both studies argued that neither conventional indicators of social class (based on men's occupations) nor smoking habits accounted for the variations in mortality reported, and concluded that air pollution contributed to high mortality from these two causes. Yet there were important differences in their emphasis. The earlier study, based on deaths during the years 1952–1962, concluded that lung cancer mortality was more strongly associated with smoking habits than with air pollution, identifying bronchitis as the cause of death more clearly associated with pollution (Wicken and Buck 1964). By contrast, the later study, based on death rates in 1963–72, concluded that lung cancer variations showed a more evident influence of industrial pollution (Dean and Lee 1977).

Recent Epidemiological Evidence

The latest phase of concern with environmental health in Teesside coincided with new evidence that mortality under 65 years was unexpectedly high in the late 1970s and early 1980s in poorer neighbourhoods (Townsend, Phillimore, and Beattie 1988; Phillimore and Morris 1991). Was poverty in Teesside simply more extensive than anywhere else in England's Northern Region (a region dominated by older industries), and did death rates at the time reflect those contemporary circumstances? Or were there additional influences—current or historical—at work?

Renewed examination in the 1990s of the impact of industrial air pollution was a response to that question.[7] Researchers from three universities were supplemented by two local government officers with specialist knowledge of air quality monitoring and by two public health specialists.[8] Funding was obtained from various local government and health service sources. Industry was not approached for funding. The relationship with industry, however, was not wholly an arms-length one. Industry kept a watchful eye over the course of the research. In return for access to stack emissions data from all the main industries on Teesside, a consortium representing industry interests (known at the time as the Teesside Clients Association) was given periodic briefings on the progress of the research and a preview of the findings one week prior to publication.

This new work focused on a comparison solely involving the poorest neighbourhoods in Teesside and Sunderland, a city 40 kilometres to the north. Sunderland was included for several reasons. In the first place, earlier research had identified localities there with poverty as severe as in Teesside, yet with mortality that was significantly lower among people under 65 years of age (Phillimore and Morris 1991). Moreover, Sunderland was also a heavy industrial centre. Its economy had been based around coal mining and shipbuilding. Like steel and chemicals, both of these industries have been associated with a heavy toll on employees' health. Yet in contrast with steel and chemicals, they have not produced pollution extending beyond the bounds of the workplace into neighbouring residential areas. A further reason for including Sunderland in the study was to guard against the possibility (admittedly slim) that Teesside-wide pollution might obscure any relationship between residential proximity to industry and human health. Within Teesside the selected neighbourhoods were differentiated by their proximity to the major industries, and were grouped into zones accordingly. The research design presupposed the interlinking of health, socio-economic, and air quality data.[9]

There can, of course, be no completely disinterested summary of such research, and much hinges on the weight given to "positive" and "negative" findings (compare Bhopal et al. [1998] with Pless-Mulloli et al. [1998]). The most persuasive evidence for a link between industrial air pollution and health was in relation to death from lung cancer in particular, and, secondarily, from other respiratory diseases (emphysema, bronchitis, pneumonia) (Bhopal et al. 1998; Pless-Mulloli et al. 1998; TEES 1995). Gradients were also more striking for women than men. Lung cancer death rates among women in Sunderland were well in excess of national rates; nevertheless, across Teesside they were higher still, rising with proximity to the main industries. In the areas closest to industry, death rates among women under 65 years of age from 1981 to 1991 were virtually four times the national level (and more than twice as high as in the Sunderland area chosen for comparison). Less steeply, this gradient between zones was apparent in the decade from 65 to 75 years. The same applied to respiratory diseases (in areas closest to industry, death rates among women under 65 from respiratory causes were three times the national level). Among men under 65, differences for lung cancer and respiratory diseases also showed the same trend, though in these cases the gradient across areas was much less pronounced. For women under 65, lung cancer and respiratory diseases together accounted for 1:4 deaths in areas closest to industry, 1:5 deaths in areas of intermediate distance, and approaching 1:6 deaths in areas of Teesside furthest from industry or in Sunderland. If these deaths were excluded from the comparison, differences between areas diminished greatly (TEES 1995).

Overall, however, the evidence produced a number of puzzles and inconsistencies, and did not all point in the same direction, whether we consider mortality on its own or morbidity as well. For example, evidence on asthma and chronic bronchitis morbidity showed no clear variations associated with proximity to industry. Nor was there a temporal association between pollution peaks and a subsequent increase in visits to a family doctor. Evidence on asthma severity, and on the exacerbation of asthma or bronchitis by air pollution, was inconsistent, with different datasets indicating divergent patterns (Bhopal et al. 1998). Further, there was no evidence of differentials related to proximity to industry in indicators from around the time of birth (foetal abnormalities, stillbirths, sex ratios). We summarize below key questions arising from these findings, concentrating on the mortality evidence specifically.

(a) How are we to explain why the gradient across zones from female lung cancer and respiratory disease mortality was not paralleled to the same extent among men?

(b) How are we to explain why the pattern of variation among women for these causes of death disappeared or even reversed over 75 years?

(c) How are we to explain why deaths from cardiovascular causes did not reveal any signs of being associated with proximity to emission sources?

Some of these puzzles can be addressed. Pless-Mulloli et al. (1998) have argued that the weaker gradient among men is plausibly explained. For men in all areas, Teesside and Sunderland, employment until the 1980s was commonly found in heavy industries which all necessitated regular exposure to polluted air. It would be hard to argue that exposure to pollution was greater in some of these industries than in others. For men dying in middle age through the 1980s, any residential exposure was likely to have been overlaid, therefore, by a heavier and more uniform occupational exposure. The particularly heavy toll of respiratory deaths among men in Sunderland, for example, surely reflects the heavy health costs of coal mining. By contrast, female employment patterns indicate no comparable workplace exposure to pollution to outweigh variations in residential exposure. Thus, the argument is not that men's mortality is unaffected by living close to industry, but that it is more difficult to distinguish than in the case of women, given men's greater occupational exposure.

Harder to explain is the apparent reversal of the gradient at older ages, in the case of women particularly. It is well documented that health inequalities in general are reduced at older ages (over 75 for instance). But they do not reverse. One possibility may be that the effects of exposure have not only augmented the total incidence of lung cancer or respiratory disease in areas close to industry, but have also brought it forward in the life-course, shifting the burden to younger ages. Yet such an argument can only be speculative. Equally problematic, the absence of confirmatory gradients for certain other cancers where links to air pollutants have been reported, or in relation to cardiovascular diseases such as congestive heart failure, robs the Teesside evidence of a consistency that epidemiologists look for if they are to be convinced that a link is causal.

Nonetheless, we do not consider that these complexities and ambiguities are a reason to label the research inconclusive.[10] As is evident from the quotation which started this chapter, our judgment is that it is hard to overlook the sharp increase in lung cancer mortality among women under 75 in areas close to the main industrial sites. Indeed, it would be irresponsible to do so. The same applies to deaths from respiratory diseases among women, even though the gradient observed is slightly less steep than for lung cancer. Spanning a period of more than a decade, these data are not easily dismissed as a short-lived anomaly, and we would argue that no other factors convincingly explain the unequal geographical distribution of deaths from these causes. Consequently, we doubt very much whether it is plausible to account for this mortality excess without reference to a significant causal role for industrial air pollution, as we have argued in recent epidemiological publications (Bhopal et al. 1998; Pless-Mulloli et al. 1998). It will become plain, nevertheless, that a different reading has prevailed in Teesside politics.

Putting the Air Pollution Debate to Sleep?

One year after initial publication of this epidemiological evidence, chemical industry executives expressed satisfaction that the research had created few repercussions for them:[11]

Since the TEES study findings were published, it's easier to breathe out there.

We were expecting the equivalent of World War Three.

I thought the publicity was well balanced, positive for Teesside, the best publicity for many a year. It put the air pollution debate to sleep.

How was industry able to draw these conclusions so confidently, given the evidence summarized above? Official responses to the publication of the study highlighted for us the corporate and local government pressures to re-present the research. In the reading of the study that swiftly emerged, Teesside residents and other interested parties were assured that, whatever health problems the area experienced, they were unlikely to be attributable to industrial air pollution. The official desire to head off potential controversy was hinted at even prior to publication. When industry representatives were given advanced briefing on the findings, one offered a succinct response to the study: " 'Industry causes lung cancer!' Would you be happy with that headline?"[12]

The ICI press release quoted at the start of this chapter summarizes the early corporate response: to focus attention on the inconclusive and uncertain language of the study and call for more research. A press release from Her Majesty's Inspectorate of Pollution (HMIP), at the time the UK pollution regulatory authority, used similar language to the same end (HMIP 1995):

> HMIP also notes that the report can not [sic] explain the high rates of lung cancer in women living in residential areas close to current industrial complexes and there is a need for more research in this area.

Local government and health authorities funding the research went a step further. Their joint press statement started to reinterpret the findings, and cast doubt on any plausible link between industrial air pollution and mortality. It re-emphasized instead the adverse effects of smoking, despite evidence that smoking levels, however high, were very similar across the populations under consideration (Statement by Tees Health, Middlesbrough and Langbaurgh Councils, 6 December 1995):

> The results do reveal, however, an increase in deaths from lung cancer among women living in the areas closest to industrial sites. It should be noted that these areas do not correspond well to the major sites of industry 20–30 years ago when the disease would be developing and tobacco is by far the most potent cause of lung cancer. Rates of smoking are still rising in young women and this needs to change if we want to see reducing rates of lung cancer in women.

The implication that the research had overlooked the gradual migration of industry downriver through this century was in our view unfounded. Nonetheless by suggesting that the "wrong" areas may have been chosen, the statement highlighted an official willingness to challenge inconvenient evidence, and by implication to question the competence of the researchers (perhaps surprisingly considering that staff from these organisations had been co-opted as members of the research team).

A different emphasis was taken by Middlesbrough Borough Council acting independently. Their statement chose to highlight the issue of poverty (statement by Middlesbrough Council 6 December 1995):

The report says that current air quality on Teesside is good, but also strongly reaffirms the link between poverty, smoking and ill-health. Said Middlesbrough Council's chair of Public Protection and Trading Standards:

> Action must also be taken at Government level to reduce the yawning gap between health in some of Teesside's most disadvantaged communities and the rest of the UK. That assistance should not be confined to advice on lifestyle, helpful as that might be. It must come in the form of measures to alleviate the hardship and poverty which is literally killing people.

Poverty has proved to be a less contentious issue than pollution in Teesside politics. By magnifying the well-recognized links between unemployment or poverty and health as an explanation for unequal health patterns, any role for pollution in this equation is effectively weakened.

The carefully worded responses to the study by local and regional influentials were predictably reflected in local news reports. The *Northern Echo* quoted the chair of the Middlesbrough Public Protection Committee as saying that the report "helps us nail the myth about local air quality" (7 December 1995). Even so, the initial local news coverage showed uncertainty as to which pieces of evidence to highlight, as the following newspaper headlines show:[13]

> Air "link" to cancer: report warns women.

> No evidence of asthma link to air pollution.

> Illness: it's not the Teesside air. Poor health cannot be blamed on pollution, says new study.

In the weeks after publication of the TEES Report there were a number of official and unofficial meetings at which the evidence for and against a link between industrial air pollution and ill-health was summarized. These summary accounts diverged sharply in the emphasis they placed on different aspects of the evidence, depending on whether the story was being told by academic or local government members of the research team. Each appeared to be reacting to and counteracting the other's version of the evidence. Not surprisingly, one consequence seems to have been to compound public confusion as to what the study signified. In this process, the following statement by a local government officer indicates official intentions:

> Fortunately our past experiences with the local media helped us to prepare for the public launch in ensuring that a balanced view was represented to the public and the "bad news" was not unduly emphasised. Our views as officers are that the research findings need to be minimized. So we were pleased with the low key media response. The fact that people don't know much is good news, we're pumping millions into the area, we don't need more bad news about Grangetown. For the council it's a double edged sword. We need to breathe confidence into the area and attract industry, we need jobs. . . . I see my role in relation to the research . . . [to] play it down . . . focus attention on crime not health . . . pump money into lifestyles, better housing . . . not to address lung cancer findings at this stage.

In their study of the political repercussions following an epidemiological investigation in Pittsfield, Massachusetts, which centred on the environmental impact of

General Electric's operations, Nash and Kirsch entitle one section "The Denouement: Another Reprieve" (1988, 165). The same verdict might be applied to the present case. Yet ironically, just when it seemed that the findings of the epidemiological study were unlikely to have any further impact in the local arena, national concerns attracted renewed attention to the research. As a result of ICI's being criticized by the Environment Agency, the new national regulatory authority, for contravening discharge standards, one of us was contacted by the *Sunday Times* to explain the evidence regarding the effects of industrial air pollution on the health of local communities. Under the headline, "Revealed: the Chemical Giants Polluting Britain," reference was made to death rates from lung cancer being four times the national level among women in Teesside who lived in areas close to industry. Suzanne Moffatt was quoted as saying: "We concluded that the most plausible explanation was exposure to industrial pollution" (*Sunday Times* 1 July 1997). This newspaper report produced an exasperated reaction from one local government officer:

> It was diabolical and has severely strained relations between the Local Authority and the TEES researchers.

Meanwhile, local responses to publication of the research findings were muted.[14] To the limited extent that local residents of the selected neighbourhoods were aware of the work (Hudson et al. 1998), reactions were marked by a mixture of puzzlement at the terminology and the cautious qualifications of the conclusions.[15] This combined with weary cynicism about either the integrity of researchers themselves or the commitment of local government to act on the findings. The comments below reflect this mixture of reactions.

> The majority of people around here won't have took any notice. It is an everyday thing, it goes over people's heads, something is always going on. People have more to worry about, they look in the paper for what is of interest to them, somebody is stabbed or the house is burnt down. They are more concerned about where the next loaf of bread is coming from.

> There's lots of leaks from industry. Nothing ever materializes, people lose faith, and they're fed up, just resigned to living here and getting on with it. Unless something is done people won't take any notice.

> Take your houses, you can run a cloth across the washing line and window sills every day and it's thick with black dust, what is that doing to us? But no one is interested.

> I wasn't interested because nothing ever gets done.

> Do you genuinely believe yourself what you have told us, you have been brain washed to read that, look at the whacking great black clouds, the budgies round here have gas masks.

Although staff in local government and industry alike expressed relief that the research generated so little reaction, several comments suggest that they were surprised that no environmentalist organization stepped in to use the new evidence. But as one industry executive remarked:

If the Green Party were not interested and did not have an angle on it, local people won't have a handle on it.

The image of Teesside as a dirty polluted place is just not true!

The continued sensitivity in Teesside surrounding industrial air pollution was brought sharply into focus in a recent leaflet circulated to local residents and businesses by Teesside local government (from which the quotation above is taken). Entitled *Air Quality Today* (1996), the leaflet has two photographs on its cover, comparing Middlesbrough in the 1960s with the 1990s:

Smokey old Teesside?

Fact: Teesside used to be one of the most polluted places in Britain. In the 1960's it suffered from some of the worst air pollution in the country, due partly to domestic coalburning.

Fact: it used to be—but not any more!

The latest analytical techniques available have shown that our air is as good as other towns and cities in the country—and in many cases a lot better. Take airborne particle pollution for example, which is the most significant local pollutant nowadays. National statistics show that in 1994 Middlesbrough had the lowest reported levels . . .

Our biggest air quality problem on Teesside is one of perception.

So, do we ever get poor air quality?

Yes. On about a dozen occasions through the year. Sometimes these occur in the summer when we get *photochemical smog* drifting into Teesside from as far away as Europe.

Despite great care being taken with industrial processes, there are occasions when locally-produced short-term emissions of pollutants do occur. . . . Thankfully these episodes are quite rare. You might also be interested to know two things:

1 Peak pollution levels occur on 5th November.[16]

2 Nowadays, much of our local air pollution comes from road traffic.

Do you get annoyed when people talk down Teesside? . . . If we have wrong perceptions of our area, *we cannot be surprised if others do as well.* A wrong perception could mean people do not invest in our area . . .

The way forward involves everyone getting together to challenge wrong perceptions. (Extract, original emphasis)

Ironically, considering its defensive tone, this leaflet was part of a policy of increased openness and improved provision of information to the public.[17] Yet two assumptions about the public are paramount: that popular perceptions are misguided, and that these can be corrected by greater exposure to "the facts" (cf. Belsham 1991; Nash and Kirsch 1988). Behind the rhetoric, we can discern several strategies employed in official efforts to deflect attention from industrial pollution and to promote an image of Teesside as a "clean" environment (cf. Burgess 1990; Burgess and Harrison 1993). One strategy is to acknowledge that Teesside has had a history of heavy pollution, making the contrast between past and present to highlight the improvements that have been made, particularly the reductions in sulphur dioxide and smoke. A second strategy, and one which is becoming more important, is to emphasize traffic as a far greater pollution burden than industry. A third approach is to place stress on the contribution to poor air quality of air pollution from outside the region—whether imported from the power stations of South Yorkshire or from across the North Sea. This is particularly ironic, given that long-distance pollution effects are plagued by even greater uncertainties than local ones (cf. Irwin 1995, 55–56). A final strategy brings together local government and industry in a joint approach to educate the public regarding the "real" dangers of pollution and the "effective" mitigation efforts of industry. The public objective is to improve air quality and increase general awareness of the improvement. A less public consequence, however, is to bind local authorities more closely to the industries' interests and their own gloss on these improvements.

These messages fail to acknowledge, however, that air pollution has never been evenly distributed across Teesside. Despite major improvements in air quality overall, certain localities continue to be disproportionately affected by pollution, just as they were in the past when emissions were greater.[18] Recent developments in methods of air quality monitoring have helped to obscure this crucial detail. Today, modern measuring equipment permits continuous monitoring of a range of pollutants, and the emphasis has shifted away from the older pollutants such as sulphur dioxide, smoke and ferric oxide, to ozone, small particles (PM_{10}), volatile organic compounds (VOCs) such as benzene, and polycyclic aromatic hydrocarbons (PAHs), few of which were measured before 1990. Cost, however, means that monitoring is centralized in one or two locations, situated at some distance from the most important sources of pollution. Gains in the range of pollutants covered and the temporal sensitivity which allows real-time monitoring must be set against the loss in geographical coverage of the 1960s and 1970s, when less sensitive equipment nonetheless allowed comparisons to be made between the different pollution burdens of different areas within Teesside. Paradoxically, it has become harder than it used to be to tell how much heavier pollution is in some areas than others within Teesside, even if it is now easier to compare how Middlesbrough (standing for Teesside as a whole) fares in relation to other British cities with similar data. It has thereby become easier for the local authorities in Teesside to publicize "good news" about the Teesside air, partly because some of the potential "bad news" is no longer collected. In areas close to industry, this reassurance is not wholly convincing as it flies in the face of personal observation, as comments such as these illustrate:

> We tend to get a lot of dust in the house blowing through windows. I think it comes from nearby works such as ICI and British Steel.

> A lot more could be done to make the local atmosphere cleaner. We notice that the outpouring from ICI, Enron and BSC [British Steel Corporation] tends to increase at night-time.

Conclusion

One of Beck's fundamental arguments is that the character of risks today demands that we step outside frameworks of social analysis based on class and its inequalities.

> Reduced to a formula: *poverty is hierarchic, smog is democratic* . . . risk societies are *not* exactly class societies; their risk positions cannot be understood as class positions, or their conflicts as class conflicts. (Beck 1992, 36, original emphasis)

At the same time, he contends, this argument "does not exclude risks from often being distributed in a stratified or class-specific way" (Beck 1992, 35). The qualification is important, for without it Beck's claims are too sweeping and distinctly premature. Certainly in Teesside, pollution risks reinforce rather than cut across class inequalities. In the past, exposure to pollution has been as unequally distributed as wealth, and affluent areas have been far less affected than poor ones. This remains true today. Moreover, there is now credible recent evidence for an effect on mortality from this exposure in poor neighbourhoods, intensifying the inequalities in health associated with other more widely acknowledged effects of poverty. We have sought to show how this potentially important knowledge has been handled in official accounts of Teesside's public health problems. In conclusion, we draw together three main observations.

First, any evidence pointing to a link between industrial pollution and ill-health is a sensitive matter in a place like Teesside (cf. Nash and Kirsch 1988). In retrospect, it is clear how determined the main industries and local government in Teesside were to keep control over the pollution "story." A leaked draft and an advanced briefing of industry representatives certainly assisted them in exercising control over how the research was perceived. More important, however, was the inclusion of local government staff on the research team to keep track of the way the research findings were pointing and to influence interpretation. Even without these opportunities to shape the public perception of the study, we doubt whether industry and local government in Teesside would have allowed a mere research team free rein to present unchallenged an independent account of environmental health risks in the area. In any field of writing, from science to the novel, the link between authorship and authority—the authority to determine what the text signifies—is subject to challenge. Epidemiological research is no exception. Nonetheless, a distinctive feature of events in Teesside was the extent to which efforts were made to rival the researchers' interpretation with an alternative account from the first moment.

A second observation concerns the substance of this rival interpretation. Official accounts repeatedly emphasized the importance of the evidence on poverty, deprivation, and individual lifestyles (particularly smoking) as health risks. The effect of air pollution was thereby overshadowed, if not altogether discounted. One tacit and rarely stated implication of this tendency to focus upon poverty and lifestyle has been to treat pollution as if it had nothing to do with poverty. There is no sign in official pronouncements of exposure to pollution itself being an aspect of poverty. To claim as much would be to re-politicize issues of poverty and social exclusion. In the Teesside context it is scarcely accidental that such a connection should be studiously overlooked (see note 4). Such sleight of hand allows variations in mortality from lung cancer or respiratory diseases which elsewhere might be expected to arouse enormous concern to be played down and absorbed within wider pronouncements about the need to reduce smoking.

A final theme is the nagging matter of uncertainty and inconclusiveness in environmental epidemiology itself. The scope for uncertainty is magnified by the well-recognized difficulty of establishing reliable evidence in this field (Lippmann and Lioy 1985). It is in the nature of epidemiological research design that there will always be pieces missing from the jigsaw, unrecognized or underestimated confounding factors or biases. Some of these inherent problems are sharply thrown into relief if we consider one relevant factor: time. The concept of the "long-term" is relevant in three senses here, each of which makes judgements about health effects harder: the long duration of most exposures to pollution; the long time lag between cumulative exposures and medical symptoms; and the chronic nature of ill-health once symptoms manifest themselves. These long time frames quite reasonably militate against confident claims about causation in epidemiological studies (Scott 1988; Adam 1993, 1996; Das 1995), and help to ensure that such claims are more than usually hedged around with caution and qualification. Yet as we have seen, caution may swiftly be interpreted as inconclusiveness *for political reasons*, as evidence of adverse health effects is played down or disregarded by corporations under scrutiny and other supporting institutions. At the very least, the charge that certainty or "proof" is lacking buys time for these corporations. In this way, epidemiological science finds itself giving unwitting legitimacy to corporate public relations efforts intended to exonerate business.

NOTES

1. ICI's statement came in a "General Position Statement" released on the day of publication of a report entitled *Health, Illness, and the Environment in Teesside and Sunderland* (TEES 1995), which was based on a four-year epidemiological study. Pless-Mulloli et al. (1998) derives from this research (as also does Bhopal et al. [1998]).

2. Two of the present authors (PP and SM) shared in the design and conduct of the research mentioned in note 1, and consequently have a stake in the interpretation of that work. As social scientists, however, PP and SM have been as interested in the political "noise" surrounding this work as in the epidemiology itself: hence a commentary such as this one (see also Phillimore 1998; Phillimore and Moffatt 1999a, 1999b). These sociological interests had earlier led PP and SM to set up a separate research project to trace the impact of the epidemiological study in Teesside. EH and DD undertook this research. Taking the publication of the epidemiological report (TEES 1995) as a starting point, EH and DD looked at the repercussions of this new environmental health evidence among certain of the groups and interests in Teesside most directly affected—policy makers and the public at large (Hudson et al. 1998).

3. For another anthropological critic of Douglas and Wildavsky, see Kaprow (1985). For a more sympathetic recent assessment, see Boholm (1996).

4. One difference between the USA and Britain would appear to be the importance attached in the former to issues of equity concerning environmental quality. The notion of rights to a clean environment, and the application of the principle of equity to such considerations, have yet to emerge strongly in local environmental campaigns in Britain, especially where the affected groups are primarily working class. In these contexts, where the activities

of local industries are often at issue, the preservation of jobs often clashes directly with concerns about pollution and its risks (Phillimore and Moffatt 1999a, 1999b).

5. The Teesside towns are Middlesbrough, Stockton, Thornaby, Billingham, and Redcar. Even now the conurbation remains a patchwork of distinct towns, with the name Teesside rarely being a badge of local affiliation.

6. However, Teesside's story is not altogether typical of England's north-east. The 1950s and 1960s were times of economic growth and long-term optimism that Teesside's future development in oil-related and chemical production would galvanize the region and offset the waning of more traditional industries (Beynon, Hudson, and Sadler 1994; Hudson 1989). The economic collapse of the 1980s was not so much the culmination of a steady process as a more abrupt seizure in an area not seen as seriously in decline.

7. The same question was also being asked by some local doctors and health workers, and received attention in national television documentaries and newspaper articles (e.g., Ghazi 1991). Funding for this research reflected, in part, the widespread awareness of the issue in north-east England.

8. Having local government officers and health authority personnel on the research team brought valuable expertise, but it also created unavoidable dilemmas. Almost inevitably, given the sensitivity of such research, membership of a research group and local government responsibilities created conflicts of interest.

9. Details of a complex study design are not appropriate here: see Bhopal et al. (1998), Pless-Mulloli et al. (1998). A note on smoking is appropriate, however, in a study of the effects of air pollution. By selecting neighbourhoods with closely matched socioeconomic characteristics, we assumed that we were likely to include populations with similar smoking habits. A survey in a sample of these neighbourhoods provided subsequent empirical confirmation that our assumption was justified, permitting us to discount smoking as a significant factor in mortality *variations* observed.

10. "We" in this paragraph refers to the two of us who were part of the epidemiological research team (PP and SM).

11. These quotations, and other comments by individuals in the sections following, come from interviews conducted by EH and DD (Hudson et al. 1998).

12. This occasion was not the first opportunity that industry figures had to learn about the main epidemiological findings. Several weeks earlier, industrialists had also received a leaked draft of the TEES Report, passed on anonymously by "a concerned doctor."

13. The first headline comes from the *Evening Gazette* in Teesside (6 December 1995); the latter two from the *Northern Echo* (7 December 1995).

14. It might be anticipated that local residents would include employees of the firms whose emissions were under scrutiny. After the redundancies of the 1980s, this was less the case than it had been in earlier decades (Phillimore and Moffatt 1999a, 1999b).

15. Belsham's (1991) discussion of local scepticism about epidemiological research on cancers and air pollution in a poor area of Philadelphia shows parallels here.

16. November 5 is "bonfire night" or "Guy Fawkes night" in England, with numerous large fires and firework displays all over the country.

17. For a parallel to this piece of public relations, see Irwin's (1995, 52) discussion of an advertisement by the U.K. Meat and Livestock Commission in 1990 as the BSE crisis loomed.

18. For example, the vicinity of chemical and steel operations (including coking) in South Bank and Grangetown, on the south side of the River Tees, has throughout this century been particularly severely affected. In the 1970s, a report by Teesside County Borough (1974) noted, with regard to smoke:

A marked decrease in smoke pollution is revealed over this period in Teesside . . . the only location which showed no significant improvement in all Teesside was again in South Bank.

For sulphur dioxide, it stated:

> Again a general trend to decrease in Teesside but two sites with a significant increase in pollution were . . . South Bank having the highest monthly average, and Normanby . . . having the greatest increase.

And for ferric oxide:

> There is no real evidence of a general improvement trend in Teesside in this pollutant and again the sites with the three highest monthly averages were in the Eston District in South Bank. . . . Furthermore one of the sites in South Bank showed a significant increase in pollution.

REFERENCES

Adam, Barbara. 1993. Time and environmental crisis: an exploration with special reference to pollution. *Innovation* 6:399–413.

———. 1996. Re-vision: The centrality of time for an ecological social science perspective. In S. Lash, B. Szerszynski and B. Wynne, eds., *Risk, environment, and modernity: Towards a new ecology*, pp. 84–103. London: Sage.

Air Quality Today. 1996. Leaflet. Middlesbrough, UK.

Beck, Ulrich. 1992. *Risk society: Towards a new modernity*. London: Sage.

———. 1996a. World risk society as cosmopolitan society? Ecological questions in a framework of manufactured uncertainties. *Theory, Culture, and Society* 13(4):1–32.

———. 1996b. Risk society and the provident state. In S. Lash, B. Szerszynski, and B. Wynne, eds., *Risk, environment, and modernity: Towards a new ecology*, pp. 27–43. London: Sage.

Belsham, Martha. 1991. Cancer, control, and causality: Talking about cancer in a working-class community. *American Ethnologist* 18:152–72.

Beynon, Hugh, Ray Hudson, and David Sadler. 1994. *A place called Teesside: A locality in a global economy*. Edinburgh: Edinburgh University Press.

Bhopal, Raj, Suzanne Moffatt, Tanja Pless-Mulloli, Peter Phillimore, Chris Foy, Chris Dunn, and Jacqui Tate. 1998. Does living near a constellation of petrochemical, steel, and other industries impair health? *Occupational and Environmental Medicine* 55:812–22.

Boholm, Asa. 1996. Risk perception and social anthropology: Critique of cultural theory. *Ethnos* 61:64–84.

Briggs, Asa. 1968. *Victorian cities*. Harmondsworth: Penguin.

Brown, Phil. 1992. Popular epidemiology and toxic waste contamination: Lay and professional ways of knowing. *Journal of Health and Social Behaviour* 33:267–81.

Burgess, Jacquelin. 1990. The production and consumption of environmental meanings in the mass media: A research agenda for the 1990s. *Transactions, Institute of British Geographers* 15:139–61.

Burgess, Jacquelin, and Carolyn Harrison. 1993. The circulation of claims in the cultural politics of environmental change. In A. Hansen, ed., *The mass media and environmental issues*, pp. 198–221. Leicester: Leicester University Press.

Das, Veena. 1995. Suffering, legitimacy, and healing: The Bhopal case. In V. Das, ed., *Critical events: An anthropological perspective on contemporary India*, pp. 137–74. Delhi: Oxford University Press.

Dean, G., and P. Lee. 1977. *Report on a second retrospective mortality study in north-east England*. London: Tobacco Research Council.

Douglas, Mary. 1985. *Risk acceptability according to the social sciences*. London: Routledge.

———. 1992. *Risk and blame: Essays in cultural theory*. London: Routledge.

Douglas, Mary, and Aaron Wildavsky. 1982. *Risk and culture: An essay on the selection of technological and environmental dangers*. Berkeley: University of California Press.

Draper, Elaine. 1991. *Risky business: Genetic testing and exclusionary practices in the hazardous workplace*. Cambridge: Cambridge University Press.

Ghazi, Polly. 1991. Breath of blighted life. *Observer Magazine*, 3 March.

Gladstone, Frank. 1976. *The politics of planning*. London: Temple Smith.

Her Majesty's Inspectorate of Pollution (HMIP). 1995. News release. London, 12 December.

Hudson, Eve, Dawn Downey, Suzanne Moffatt, and Peter Phillimore. 1998. *Putting the air pollution debate to sleep? The impact of a health study in Teesside*. Joint Working Paper No. 1, Departments of Social Policy and Epidemiology and Public Health, University of Newcastle-upon-Tyne.

Hudson, Ray. 1989. *Wrecking a region*. London: Pion.

ICI. 1995. General Position Statement. TEES Health Study. ICI Teesside, UK, 6 December.

Irwin, Alan. 1995. *Citizen science: A study of people, expertise, and sustainable development*. London: Routledge.

Irwin, Alan, and Brian Wynne, eds. 1996. *Misunderstanding science? The public reconstruction of science and technology*. Cambridge: Cambridge University Press.

Kaprow, Miriam. 1985. Manufacturing danger: Fear and pollution in industrial society. *American Anthropologist* 87:342–56.

Kroll-Smith, Steve, and Hugh Floyd. 1997. *Bodies in protest: Environmental illness and the struggle over medical knowledge*. New York: New York University Press.

Latowsky, G. 1998. Health effects of PCB exposure among Native American populations and community response to contaminant research. *Epidemiology* 9(4):S108.

Lippmann, Morton, and Paul Lioy. 1985. Critical issues in air pollution epidemiology. *Environmental Health Perspectives* 62:243–58.

Little, Miles. 1998. Assignments of meaning in epidemiology. *Social Science and Medicine* 47(9): 1135–45.

Lock, Margaret. 1998. Breast cancer: Reading the omens. *Anthropology Today* 14(4):7–16.

MacGill, Sally. 1987. *Sellafield's cancer-link controversy: The politics of anxiety*. London: Pion.

Medical Officer of Health. 1921. Annual report, Eston Urban District, UK.

———. 1925. Annual report, Eston Urban District, UK.

Nash, June, and Max Kirsch. 1988. The discourse of medical science in the construction of consensus between corporation and community. *Medical Anthropology Quarterly* 14:158–71.

Paigen, Beverley. 1982. Controversy at Love Canal: The ethical dimension of scientific conflict. *The Hastings Center Report* 12:29–37.

Phillimore, Peter. 1998. Uncertainty, reassurance, and pollution: The politics of epidemiology in Teesside. *Health and Place* 4(3):203–12.

Phillimore, Peter, and Suzanne Moffatt. 1999. Narratives of insecurity in Teesside: Environmental politics and health risks. In J. Vail, J. Wheelock, and M. Hill, eds., *Insecure times*, pp. 137–53. London: Routledge.

———. 1999b. " 'Industry causes lung cancer.' Would you be happy with that headline? Environmental health and local politics." In S. Allan, B. Adam, and C. Carter. eds., *Environmental risks and the media*, pp. 105–16 London: Routledge.

Phillimore, Peter, and David Morris. 1991. Discrepant legacies: Premature mortality in two industrial towns. *Social Science & Medicine* 33:139–52.

Pless-Mulloli, Tanja, Peter Phillimore, Suzanne Moffatt, Raj Bhopal, Chris Foy, Chris Dunn, and Jacqui Tate. 1998. Lung cancer, proximity to industry and poverty in northeast England. *Environmental Health Perspectives* 106(4): 189–96.

Scott, Wilbur. 1988. Competing paradigms in the assessment of latent disorders: The case of Agent Orange. *Social Problems* 35:145–60.

Taubes, Gary. 1995. Epidemiology faces its limits. *Science* 269, 14 July.

Teesside County Borough. 1974. *A report on the prevailing environmental conditions in the District of Eston and proposals for its improvement.* Middlesbrough: Teesside County Borough Council.

Teesside Environmental Epidemiology Study (TEES). 1995. *Health, illness, and the environment in Teesside and Sunderland: A report.* Newcastle: University of Newcastle-upon-Tyne.

Townsend, Peter, Peter Phillimore, and Alastair Beattie. 1988. *Health and deprivation: Inequality and the north.* London: Routledge.

Whittaker, Andrea. 1998. Talk about cancer: Environment and health in Oceanpoint. *Health & Place* 4(4):313–25.

Wicken, A., and S. Buck. 1964. *Report on a study of environmental factors associated with lung cancer and bronchitis mortality in areas of North-East England.* London: Tobacco Research Council.

Newspapers

Evening Gazette. Middlesbrough, 6 December 1995.
Northern Echo. Darlington, 7 December 1995.
Sunday Times. London, 1 July 1997.

QUESTIONS

1. What types of evidence presented in this chapter suggest there is a link between industrial air pollution and poor public health in Teesside?

2. In what ways did local officials and corporations exploit the uncertainties in the scientific knowledge base to point the finger at nonindustrial causes of poor public health in Teesside?

Chapter Fourteen

Round and Round It Goes
The Epidemiology of Childhood Lead Poisoning, 1950–1990

Barbara Berney

"Lead is toxic wherever it is found, and it is found everywhere." The 1988 report to Congress on lead poisoning in children by the Agency for Toxic Substances and Disease Registry (1988) thus neatly summarized the last 25 years of epidemiological (and toxicological) studies of lead.

Lead has been a known poison for thousands of years. The ancient Greeks described some of the classical signs and symptoms of lead poisoning: colic, constipation, pallor, and palsy (Lin-fu 1980). Some historians suggest that lead acetate used by the Romans to process wine contributed to the fall of the Empire (Mack 1973). Despite its known toxicity, lead use in the United States increased enormously from the industrial revolution through the 1970s, especially after World War II. Between 1940 and 1977, the annual consumption of lead in the United States almost doubled. In the 1980s, largely as a result of regulation of lead in gasoline, lead use in the United States leveled off and began to decrease.

In this chapter I explore the interaction of epidemiology and social forces in the continuing evolution of knowledge about the effects of low-level lead exposure, the extent of the population's exposure, and the sources of that exposure. I will concentrate on the effects of lead on the central nervous system (CNS) of children.

In the United States, the problem of lead poisoning and low-level lead exposure has been pursued with some consistency over the last 25 to 30 years. Research done in the 1950s and 1960s was generally confined to studies of the symptoms, diagnosis, and treatment of acute lead poisoning in low-income children suspected of eating the paint and plaster in their deteriorated housing. However, the questions raised by the research have expanded rapidly to broader political and epidemiological questions: What is the proper definition of an "adverse health effect"? What levels of exposure cause those effects? Is there a threshold below which no effects occur? Who is exposed to how much lead, and from what sources? What are the pathways of exposure?

Increasing attention to the epidemiology of lead coincided with what Vandenbrouche (1990) calls the "second wave of vocational epidemiology," which began in the 1950s. "Vocational epidemiology is epidemiology based on a profound, personally felt vocation to improve the fate of mankind by fighting the environmental and societal causes of disease" (Vandenbrouche 1990). Increasing interest in lead also coincided with epidemiology's transition to the study of noninfectious diseases. At the turn of the century, public health made a conservative turn away from the environment toward infected and infectious individuals as the source of disease. By the 1950s the trend was turning back toward the environment (Ozonoff 1988).

The problem of lead poisoning was perfectly suited to these developments in epidemiology. Lead poisoning had an environmental cause, exacerbated by social problems associated with poverty including slum housing, racial discrimination, and malnutrition. Its identified victims were children—poor black children—who continue to be at the greatest risk for lead poisoning even within the currently expanded definition of the population at risk. Not only did concerns about lead poisoning fit the emerging epidemiological agenda; they also were in tune with the emerging political agenda of the movement for civil rights and social justice. The problem of lead poisoning provided an opportunity for a generation of public health professionals to combine their professional, personal, and political goals. These public health professionals worked with community activists to establish and carry out screening programs. They presented the results of their research at public hearings, testifying for legislation and regulations aimed at preventing exposure to lead, and they argued in scientific and political arenas with the lead industry for increasingly stringent regulation of lead.

Lead epidemiology can be seen as a series of interactive "rounds" in which case finding or screening increased awareness of the disease and expanded the defined populations at risk. The increased numbers of children at risk led to more intensive study of the effects of lead and a lowering of the lowest observed effects level (LOEL). Lowering the LOEL, by definition, increased the population at risk. Because exposure of the enlarged population at risk could not be attributed only to deteriorating lead paint and pica (the tendency to eat nonfood items), epidemiologists began to look for additional potential sources of lead in the environment and considered normal behavior patterns, such as hand-to-mouth contact in toddlers. Augmented sources of lead, in turn, meant a larger population at risk of exposure, and so on. Each step in this process carried with it political implications related to both the cause and prevention of exposure. I have chosen as my topic the last 30 years of rounds.

The medical literature of the first half of the twentieth century contains dozens of articles on childhood lead poisoning. In fact, by 1934 nine countries and Queensland, Australia, had decided, based on the existing literature, to ban or restrict the use of leaded paint (Rabin 1989). Most of this literature consists of case reports or reports on series of cases of children hospitalized for acute lead poisoning. Several articles describe the effects of lead poisoning, especially the neurological effects (Rabin 1989). However, after the introduction of tetra-ethyl lead as a gasoline additive in the early 1920s, lead research in the United States was dominated by the lead industry, particularly by Robert Kehoe. Kehoe was "the nation's most vocal and influential scientist working on lead hazards" (Graebner 1988) from the 1920s to the mid-1960s. From 1925 to 1958, he was also the medical director of Ethyl Corporation—manufacturer of tetra-ethyl lead—and director of the Kettering Laboratory, which received funding from the lead and automobile industry (Graebner 1988).

Round One: Midcentury to the Mid-1960s

Until the mid- to late 1960s, childhood lead poisoning was viewed as an acute disease leading to encephalopathy and was diagnosed in its early clinical stages only by the suspicious and informed physician. Lead levels in diagnosed children generally exceeded 80 micrograms per deciliter (μg/dL), and were often well above 100 μg/dL. The epidemiology of this period, which consisted largely of case finding and case summaries, accurately identified the population at greatest risk; described the symptoms, methods for diagnosis, and causes of lead poisoning; and suggested most of the neurological sequelae that were to be studied for the next 30 years. Researchers consistently found that prolonged and repeated exposure was more often associated with significant neurological damage than even large single exposures. Therefore, they frequently reiterated the need for eliminating lead from the environment of children in general and poisoned children in particular (Mellins and Jenkins 1955; Eidsvold, Mustalish, and Novick 1974; Jacobziner 1966).

In the 1950s, a few clinicians in children's hospitals and health departments in several eastern cities became concerned with the cases of childhood lead poisoning that they saw and treated. Their interest resulted in expanded case finding and in research based on case summaries of the long-term neurological sequelae of exposure to lead.

Extensive field studies were carried out in Baltimore, Chicago, and New York City (Blanksma et al. 1969; Specter and Guinee 1970; Guinee 1971). The Baltimore City Health Department, in cooperation with the department of pediatrics at Johns Hopkins and the University of Maryland medical schools, had the nation's largest program of case finding and research on the treatment and evaluation of lead-poisoned children. Julian Chisolm, one of its leaders, has been a prominent researcher on childhood lead poisoning for more than 30 years. As early as 1951, the Baltimore Health Department (1971) completed a study of 293 lead cases that occurred in children from 1931 to 1951. The results showed that most cases occurred in two-year-old children and that incidence was greatly increased during the summer months. The children "lived in old rented properties and ate paint flakes or chewed on windowsills" (Eidsvold, Mustalish, and Novick 1974). These demographics and case characteristics were confirmed in other cities with expanded case-finding programs and in the results of larger studies over the next 30 years. Thus the population at greatest risk, the source of the poison, and the seasonal variation of the poisoning were all identified in the early 1950s (Eidsvold, Mustalish, and Novick 1974).

Although this information was available in respected medical journals for more than a decade, few cities took action to prevent exposure of children to lead paint. The exception was Baltimore: presented with evidence from its health department, the city attempted primary prevention through legislation and succeeded, in 1951, in passing one of the country's earliest lead paint laws. The law prohibited the use of paint with "any lead pigment" on interior surfaces, but, like many other lead paint laws, it was rarely enforced. In 1958, after recording its highest-ever number of lead-poisoning cases and deaths—133 cases and 10 deaths—the city passed another ordinance requiring that all lead-containing paint be labeled with a warning against its use on interior surfaces, furniture, toys, windowsills, or any place used for the care of children. This law also proved ineffective in preventing lead poisoning. Twenty-six years later, in 1984, 75 percent of Baltimore's housing was believed to contain lead-based paint. It was projected that some 6,000 Baltimore children had suffered lead-paint poisoning (Agency for Toxic Substances and Disease Registry 1988).

Although the clinical case finding done in the 1950s and early 1960s was based on identifying overt symptoms of lead poisoning, investigators began looking for neu-rological sequelae of lead exposure. Their studies included children with relatively mild symptoms of poisoning and their collected evidence suggested that asympto-matic levels of exposure might cause CNS effects. These findings led clinicians and public health officials to suggest that the cut-off between normal and elevated blood lead levels was too high. The Baltimore group consistently suggested that the upper limit of normal be lowered: first, in the 1950s, from 60 to 80 μg/dL down to 50 μg/dL; then, in the 1960s, from 50 or 60 down to 40 μg/dL (Lin-fu 1972).

The definition of "normal" blood lead levels was a political as well as a scientific controversy. During the late 1920s, and in the 1930s and 1940s, Robert Kehoe, sup-ported by the lead industry, wrote extensively on the presence of lead in the environ-ment and on its uptake and excretion by humans. He concluded that lead occurred naturally, including in human tissues and excreta. He further argued that the body did not store lead as a result of exposure to "naturally" occurring amounts of lead, but rather that it established an equilibrium between lead intake and elimination, and that "beyond the point of equilibrium absorption did not occur" (Graebner 1988). He argued further that these facts meant "there was no necessary relation between lead absorption and lead intoxication—no necessary connection between lead concentration in feces, urine, or tissues and lead poisoning" (Graebner 1988).

To refute these widely accepted arguments, much research carried out during the 1950s, 1960s, and 1970s was designed to show the following:

1. Lead in the environment was a result of human use of lead in industry.
2. Lead accumulated in the human body in proportion to the amount of lead found in the environment.
3. Lead was absorbed by the body from the environment.
4. Such absorption, measured in feces, urine, blood, and other tissues, was an indication of exposure and poisoning.

Additional controversy resulted from the fact that falling "hazardous" levels not only threatened the lead industry, but also increased the need for public health programs to find and treat poisoned children and to eliminate lead exposure. Re-sources required for such programs could be immense. For example, in 1957, Balti-more mounted the first large-scale screening for lead paint, one of the few programs ever to look first for lead in paint and only secondarily for lead in children. The program aimed to "assess the prevalence of lead paint in Baltimore homes . . . with a view to its possible removal as a preventive measure" (Baltimore Health Department 1971). Mass screening of paint was possible because a rapid screening test had just been developed by the city's Bureau of Laboratories. Of 667 dwelling units tested in 1957, 70 percent had lead in excess of 1 percent.

In 1961, after testing thousands of dwellings and children, the health department suggested extensive removal of lead paint from housing in the areas where childhood poisoning was the highest. The department encountered tremendous opposition from landlords, who did not want to bear the expense of removing lead paint from rental units. After calculating that 100 person-years of sanitarian time would be required to enforce removal, the city rapidly abandoned the notion of preventive removal (removing paint before a child is poisoned in the dwelling unit) as too expensive (Schucker et al. 1965). Although some states and cities have legislation requiring lead removal prior to poisoning, it is hardly ever enforced. This means that, since 1961, virtually no jurisdiction has used primary prevention to eliminate lead paint poisoning. Instead, public health departments have relied on screening children's blood as the warning system for poisoning. Only in the last few years have cost–benefit studies been developed to show that the cost of damage done by lead paint outweighs the cost of its removal (Florini, Krumbhaar, and Silbergeld 1990; Szabo and Pollack 1987). These studies relied on 20 years of research that showed adverse health effects at blood lead levels from 10 to 15 µg/dL, or even less. Several programs to eliminate lead paint from dwellings have been proposed recently (Needleman 1989; Florini, Krumbhaar, and Silbergeld 1990).

Several researchers working in the 1950s studied the neurological sequelae of lead poisoning. Mellins and Jenkins (1955) studied the mental and emotional development of the children in a Chicago cohort. Their findings suggested that symptoms of CNS involvement preceded hospitalization. Furthermore, symptoms and effects that were noted either while the children were hospitalized for poisoning or while they were undergoing follow-up examinations suggested virtually all the subtle damage reported in much later studies of low-level exposure: speech problems, especially in the naming of objects and conceptualization, which would "limit the symbolic processes so necessary to mature verbal behavior" (Mellins and Jenkins 1955); problems with visual motor coordination, especially fine motor coordination; distractibility; and short attention span. The authors concluded that improved housing and the elimination of lead paint were "essential to prevention." The Chicago study and one by Smith (1954) came to virtually identical conclusions. These studies considered 50 to 60 µg/dL as normal blood lead levels and 70 to 80 µg/dL as levels indicative of frank poisoning.

In 1959, Byers, who worked at the Children's Hospital in Boston, and whose studies of childhood lead poisoning spanned 40 years at the time of his death a decade ago, published his classic review article on lead. He cited the known findings, but his work also suggested that lead poisoning might be a much bigger problem than previously indicated. He observed that some poisoning might not be attributable to pica, but rather to the normal mouthing behavior of children in environments where paint contained very high levels of lead. (This observation had been made in 1904, but had been lost or forgotten [Rabin 1989].) Byers noted that intact, as well as peeling, paint could represent a hazard. He reported that some researchers had noted that poisoning might occur at blood levels below 60 µg/dL, even at levels as low as 40 µg/dL, although he concluded that levels above 60 µg/dL were generally agreed to be pathological. He further reported the presence of lead in umbilical cord blood and in infants less than six months old. He suggested as well that chronic exposure or reexposure to lead after treatment appeared to result in greater risk of retardation than single, or short-term, high-dose exposure that was properly, adequately, and quickly treated (Byers 1959).

The field epidemiology and case finding of the 1950s laid the foundations for later work by indicating that increasingly intensive case finding or screening would uncover more cases and prevent death; by accurately describing the population at greatest risk; and by providing provocative data on the neurological sequelae of the disease that no doubt formed the basis of the cross-sectional and prospective studies of low-level exposure reported from the mid-1970s into the 1990s.

In the 1940s titanium oxide began to replace lead as the pigment of choice for white paint; and, in 1955, the American National Standards Institute adopted a voluntary limit of 1 percent lead for paint. Although many people believed that these changes would eliminate the problem of lead paint poisoning, this was not to be. Paint stocks with a lead content greater than 1 percent continued to be produced. Furthermore, a 1 percent limit was not adequate to protect the health of children exposed to it. Moreover, although changes in the production of new paint represented effective toxic use reduction, they did nothing to remove highly leaded paint from existing housing. Thus, although interest in the problem ebbed at least until the mid-1960s, the problem persisted.

Round Two: The Mid-1960s through the 1970s

In the 1960s lead poisoning was characterized as epidemic in scope and was named a national health problem. Lead paint in deteriorated housing continued to be seen as the major source of lead exposure. Politically active health professionals, in cooperation with community groups, began screening programs, which were later expanded and sponsored by city agencies and which provided evidence of a widespread problem. Civil rights and progressive political groups challenged the medical and governmental establishments to do something to abate the problem. In 1970, the surgeon general issued a formal statement on lead poisoning (U.S. Department of Health, Education, and Welfare 1970). The following year Congress passed the Lead Paint Poisoning Prevention Act, which eventually provided funds for greatly expanded screening programs that offered further evidence of an even more widespread problem.

Lead was a perfect issue for the emerging social and political movements of the 1960s. Lead poisoning highlighted many problems of concern: the lack of preventive health care, the focus of medicine and public health on the individual to the exclusion of the environment, the lack of community services in low-income neighborhoods, and the relation between poverty, racial discrimination, and health. Because the population at greatest risk was poor and black, lead poisoning could be defined in terms of race and class.

Lead poisoning resulted from bad housing conditions. Blacks were forced into bad housing by discrimination. The housing was allowed to deteriorate by gouging land-lords and city housing officials, who did little to eliminate the problem or protect the health of the children affected. Lead poisoning could be used to tie the emerging environmental movement (and the reemerging consciousness of the environment in public health) to civil rights issues.

Lead poisoning was preventable, but, if neglected, it could permanently disable or kill its victims. Prevention required a low-tech, community-level environmental in-tervention, whereas cure required painful, expensive treatment in a hospital. Preven-tion could be carried out by low-skilled members of the community who needed work. Cure involved overworked doctors and crowded hospitals. Well-designed pre-vention programs required concerned community workers to canvas door to door and raise people's consciousness about the connection between bad housing and children's health and to inform parents about available services.

Lead carried tremendous symbolic power: It was a poison. It was deceiving—hidden in sweet-tasting paint and plaster on the very walls of homes. Its victims were innocent and already disadvantaged. It could be used as a symbol of what was wrong with society: the indifference of landlords, government officials (especially health and housing officials), and industry (paint and gasoline makers). Articles appearing in popular magazines, such as *Time, Reader's Digest, Saturday Review*, and *Good House-keeping*, during the 1960s and early 1970s made these points and used these symbols (*Time* 1969; *Scientific American* 1969; Block 1970; Craig 1971; Remsburg and Remsburg 1972; *Parents' Magazine* 1973).

After the War on Poverty was declared in 1964, Medicaid and community health centers made resources available for addressing the lead problem. Doctors and com-munity activists, who were often employed by hospitals and new federally funded neighborhood health centers that served inner-city communities, began advocating and organizing screening programs for children with symptomatic lead paint poison-ing in Chicago, New York, Philadelphia, and other cities. They were assisted by progressive organizations of scientists and professionals that were formed during this period to focus attention on the social and political implications of scientific and health policy and research, among them Science for the People, Scientists for Public Information, Center for Science in the Public Interest, Medical Committee for Hu-man Rights, and Physicians for Social Reform.

The issue of lead poisoning appealed to many staff members of health centers and hospitals in low-income communities. Many staffers were newly graduated physicians who chose to work in federal jobs in lieu of military service in Vietnam. Others were conscientious objectors to the war who had found alternative service jobs in health care. Still others were drawn to health care, especially in the inner cities, because they saw it as a means of pushing for social change and having a professional career at the same time. Departments of community and social medicine were established or expanded at many medical schools in the 1960s (David Rosner

1990, Benjamin Siegel 1990: personal communications). These departments provided a base for people concerned with environmental and community health issues and programs.

The push for expanded lead screening grew from the inference that if intensified case finding led to the identification of larger numbers of cases, then screening—more systematic and widespread case finding—would identify even more cases. The lead belts—old, deteriorated housing inhabited by children—had been defined. Organizers and selected health professionals in lead-belt communities first put together volunteer-based screening programs. Then, when those programs found high numbers and rates of cases, they went to local governments to demand funds for expanded, well-organized screening programs.

In the mid-1960s, in Chicago, after failing to convince the city council to establish a screening program, a group of health activists started one themselves, using volunteer professionals and concerned members of the community (Quentin Young 1990: personal communication). Shortly after this program began, two children were admitted to the hospital with acute lead poisoning and died. The newspapers picked up the story and, in 1965, responding to pressure from the Citizens' Committee to End Lead Poisoning, the American Friends Service Committee, and the Medical Committee for Human Rights, the Chicago Board of Health began the first large-scale screening for lead poisoning in the United States (Quentin Young 1990, Jane S. Lin-fu 1990: personal communications; Lin-fu 1979). This screening program found that from 5 to 15 percent of the children screened had excess body burdens of lead, defined as blood lead levels in excess of 50 μg/dL (Jane S. Lin-fu 1990: personal communication; Lin-fu 1979). The Chicago findings encouraged community and professional groups in other communities to pursue similar efforts.

The screening programs that began in the 1960s resulted in the "discovery" of large numbers of cases. In New York, the number of reported cases grew from 116 in 1958 to 700 in 1968–1969. The screenings carried out in several large cities during the years 1967–1970 showed that from 25 to 45 percent of one- to six-year-old children living in high-risk areas had blood lead levels (pbB) exceeding 40 μg/dL (considered at the time to be the upper limit of normal).

> Most of these children had no symptoms of lead poisoning. Suddenly, undue lead absorption unassociated with overt clinical evidence of toxicity gained recognition as a phenomenon which required careful investigation because of the enormous number of young children involved. (Lin-fu 1979)

The data provided indisputable and overwhelming evidence that the lead-poisoning problem was immense.

Increasing evidence about the average lead levels in urban populations compared with levels in geographically remote and preindustrial populations defeated Kehoe's model of naturally occurring lead in the environment. Accumulating evidence both of adverse health effects at ever-decreasing blood lead levels and of lead being stored in the body from all sources similarly challenged Kehoe's argument that the human body established a natural equilibrium between lead intake and elimination. These changes in the definition of "natural" forced down the definition of the upper limit of normal blood lead levels.

Evidence on lead levels was carefully accumulated by researchers who wished to show that existing levels in urban populations were a result of increasing industrial use and pollution and not, as the lead industry argued, "naturally occurring" (Patterson 1965; Shapiro, Grandjean, and Van Neilsen 1980). Studies of levels of lead at various depths in the arctic ice showed them increasing with time, especially after World War II. Studies on the remains of ancient Nubians and Peruvians demonstrated levels of lead in bone and teeth 100 times lower than those found in current urban populations (Shapiro, Grandjean, and Van Neilsen 1980; Ericson, Shirahata, and Patterson 1979). These studies were constantly cited in articles about what blood lead level should be considered "normal" (Agency for Toxic Substances and Disease Registry 1988; Lin-fu 1980, 1985; Cohen, Bowers, and Lepow 1973; Environmental Protection Agency 1977, 1986).

The combination of the number of children found to have elevated blood lead levels and the evidence that elevated levels were not "normal" led to greatly increased interest in exploring the biological and behavioral effects of low-level exposure, and in determining whether or not these effects were adverse and at what level.

The 1970s: The Surgeon General's Statement, the Lead Paint Poisoning Prevention Act, and a Decade of Screening

In 1970, the surgeon general issued a statement that shifted the focus in lead poisoning from case finding and treatment of overt lead poisoning to its prevention through mass screening of young children and the termination of hazardous exposure for children with evidence of undue lead absorption. It defined "undue exposure" as a blood lead level of 40 µg/dL at a time when 45 percent of the children screened in New York City had blood leads above this level (U.S. Department of Health, Education, and Welfare 1970). The burden on local health departments was immense. The justification for choosing this blood lead level, then considered to be asymptomatic, was that time was needed to remove a child from leaded surroundings after "undue exposure" had been noted and before poisoning occurred.

The agitation, epidemiology, and publicity of the late 1960s led to congressional hearings on lead poisoning and the passage of the Lead Paint Poisoning Prevention Act of 1970. This act marked the beginning of two decades of often ambivalent government investigation and regulation of lead. Under its provisions, the Centers for Disease Control (CDC) funded the screening of close to four million children from 1972 to 1981 (Lin-fu 1985). Thus, the 1970s became the decade of screening.

The early 1970s also brought dramatic changes in screening techniques. In 1973, erythrocyte protoporphyrin transformed testing for lead poisoning. The new tests were much cheaper in terms of equipment, consumable supplies, and skill required to take and process blood samples. They used finger stick rather than venipuncture techniques. Results of the tests could be obtained on site in a few minutes, thus limiting loss to follow-up. Because the procedure did not test directly for lead, it was not subject to environmental contamination.

Screening Results Expand the Population at Risk

Data from the screening programs of the late 1960s through the 1970s had a profound effect on the understanding of who was affected by undue lead absorption. Surveys in the early 1970s indicated that the problem was not confined to large urban slums and areas east of the Mississippi River. Fourteen cities with populations ranging from 10,000 to 150,000 were found to have problems comparable to those in large cities (Lin-fu 1979). A Department of Health, Education, and Welfare (DHEW) survey of 52 communities throughout the nation revealed that undue lead absorption among children was geographically widespread, occurring in cities of every size and in rural areas as well (Lin-fu 1979; Cohen, Bowers, and Lepow 1973). "The clearly defined borders of lead belts began to disappear when screening extended beyond them" (Lin-fu 1979). Although poor black children in the inner cities were still at highest risk for poisoning, excessive lead absorption affected urban middle-class and even rural children of every race, making it perhaps the largest preventable childhood health problem in the nation.

In the early 1970s, biological and epidemiological findings on lead absorption and enzyme and CNS effects led to greatly increased concern about low-level exposure in children. King et al. (1972) suggested that children might absorb lead from the gut more efficiently than adults. A year earlier, King (1971) had estimated the "maximum safe daily dose," assuming absorption of 10 percent of ingested lead. By 1974, at least one study had confirmed that children absorbed close to 50 percent of the lead they ingested (Alexander 1974). These new findings represented a quintupling of the absorption rate for children, and dramatically decreased the amount of lead that children needed to consume in order to raise blood lead levels to the level of concern. These findings were enough to change the thinking about the importance of pica in cases of undue lead absorption (Lin-fu 1973; Sayre, Charney, and Vostal 1974). The suggestion that normal hand-to-mouth activity in normal ambient environments could cause undue exposure was confirmed, which meant that many more children were at risk.

During the same period several studies found that inhibition of amino levulinic acid dehydratase (ALAD), an enzyme involved in hemoglobin synthesis, showed a continuous dose-response to blood lead levels ranging from 5 to 95 µg/dL (Hernberg et al. 1970; Hernberg and Nikkanen 1970; Millar et al. 1970; Secchi, Erba, and Cambiaghi 1974). These findings indicated that lead might affect hemoglobin production at blood levels as low as 5 µg/dL. Retrospective studies suggested that mental retardation and learning disabilities occurred in children previously considered asymptomatic (de la Burde and Choate 1972, 1975; Perino and Ernhart 1974; Rummo 1974). In 1973 an article appeared in the *Journal of the American Medical Association* comparing blood lead levels in rural and urban populations and suggesting that 40 µg/dL might be too high for a definition of "undue absorption" (Cohen, Bowers, and Lepow 1973). The author referred to other studies of screened children and articles on hyperactivity in children with low-level lead exposure that had appeared in *Lancet* the previous year (David, Clark, and Voeller 1972). Although these findings were disputed and challenged, they increased the pressure for additional research on CNS effects at low exposure levels. . . .

The Overlapping Debate: Where Does the Lead Come From? (1960–1988)

Lead in Air

Concern with low-level exposure and growing awareness of widespread exposure led researchers to examine sources of lead other than paint. Lead in air became an important subject of research and debate. In examining the relation of lead in air to lead poisoning, several questions had to be addressed: Did lead in the air cause exposure? That is, was exposure to air with higher concentrations of lead associated with higher blood lead levels? If air lead contributed to the lead body burden, did it cause poisoning? How much of the lead body burden could be attributed to air lead? Where did the lead in air come from? These questions were complicated by the fact that much of the lead in air was not inhaled or absorbed directly, but rather was deposited on dust and soil and then ingested or inhaled.

Lead in Air and Gasoline

The initial battles over the health effects of lead in gasoline were fought when tetra-ethyl lead was first added to gasoline as an antiknock additive in the early 1920s. Leaded gasoline was actually banned in several cities while studies of its potential effects were carried out. By 1925, the public health forces had been soundly defeated by the lead industry, ensuring the "accumulation of tons of lead dust on every New York City street" and the streets of every city and town across America, just as Yale physiologist Yandell Henderson had predicted in 1925 (Graebner 1988).

In 1958, the Ethyl Corporation, manufacturer of tetra-ethyl lead, asked the surgeon general for advice on increasing the concentration of lead in gasoline. A committee, established by the surgeon general, reported back that "the proposed increase in lead apparently would pose no health hazard" (Graebner 1988), but asked for additional research on atmospheric lead. The surgeon general commissioned a study, managed by the Public Health Service, but conducted with the cooperation of the automobile industry, gasoline producers, and Kettering Laboratory. The resulting "Tri-City Study" concluded that levels of airborne lead were lower in 1961 and 1962 than they had been 25 years earlier. At the 1966 Senate hearings on air pollution, Clair Patterson of the California Institute of Technology accused Robert Kehoe of conducting a whitewash in his analysis of the data, and pointed out that lead levels in U.S. cities were 100 times higher than they had been in the mid-1930s. He further challenged the role of industry in public health research. "It is not just a mistake for public health agencies to cooperate and collaborate with industries in investigating and deciding whether public health is endangered—it is a direct abrogation and violation of the duties and responsibilities of those public health organizations" (Graebner 1988).

In 1971, the year after Congress passed the Clean Air Act and the Lead Paint Poisoning Prevention Act, the Environmental Protection Agency (EPA) put together a working seminar on lead. That year the working group received a position paper that reviewed the available research in order to determine "the contribution of atmospheric lead to the endangerment of public health" (Engel 1971). It stated that nonoccupational exposure to air lead might increase body burden, but that evidence from available studies was uneven. However, it concluded that settled lead in dustfall in the streets and soil "is sufficient to produce poisoning. . . ." The report also noted that heavy traffic increased lead dustfall significantly, suggesting that gasoline may be an important source of environmental lead contamination. Although it acknowledged that lead paint was the source of poisoning in most children, it noted that lead is accumulated from all sources and stored in the body. Therefore, the paper suggested, air lead could push children over the edge from a nonpoisoned to a poisoned state. Over the next 10 years, evidence accumulated to show that some 50 percent of children's blood lead could be attributed to lead in gasoline.

In 1971, Dr. Lin-fu of DHEW wrote to Irwin Billick, program manager for Lead-Based Paint Poisoning Prevention Research at the U.S. Housing and Urban Development Department (HUD). In her memorandum, Dr. Lin-fu outlined the nature and extent of lead-based paint poisoning in the United States by reference to several papers and screening data from several cities. She asked HUD to concentrate its efforts on finding methods to remove lead from residential housing as required by the act (Lin-fu 1971). While examining data collected by screening programs in New York City from 1970 to 1976, Billick noticed that blood lead levels were dropping and he turned to falling air lead levels for an explanation of these data. He obtained data from a single air-monitoring station in New York City and noted that blood lead levels tracked air lead levels very closely (Billick, Curran, and Shier 1979, 1982).

There were two main sources for lead in air: point sources, which were usually lead smelters, and mobile sources, which were cars that burned gasoline containing tetra-ethyl lead. A series of lead smelter studies in the 1970s looked at lead levels of children living at various distances from smelters (and at CNS effects of lead in those children). These studies showed that subjects who lived closer to the smelters had higher blood lead levels than children living farther away. The studies also measured air lead and lead in dust, soil, paint, water, and other media. They demonstrated that leaded air emissions from point sources contributed to lead in air, soil, and dust and to lead body burden and that lead was absorbed both from the air and from dust (Yankel, von Lindern, and Walter 1977; Landrigan et al. 1975, 1976). In its 1977 *Air Quality Criteria Document*, the EPA states: "The conclusion to be drawn from (these studies) is that people who live in the vicinity of a major industrial source of lead are exposed to abnormally high lead concentrations" (Environmental Protection Agency 1977).

Several studies of the relation between lead isotopes in the environment and in blood were carried out to determine the sources of lead and the amount of body burden that could be attributed to each source (Manton 1977; Garibaldi et al. 1975).

Manton determined that 7 to 41 percent of blood lead came from air (through gasoline). The isotope lead experiments showed that gasoline was responsible for 90 to 95 percent of lead in air (Agency for Toxic Substances and Disease Registry 1988). The combined effect of these studies was to defeat Kehoe's position that exposure to lead in air would not necessarily lead to either increased absorption or a greater lead body burden.

From 1975 to 1984, gasoline lead consumption fell by 73 percent because of EPA regulation of lead in gasoline and the introduction of catalytic converters that required the use of unleaded gasoline; lead levels in air fell by a similar amount over the same period (Agency for Toxic Substances and Disease Registry 1988).

From 1976 to 1980, the National Center for Health Statistics carried out the second National Health and Nutrition Examination Survey (NHANES II), collecting data on a stratified random sample of the U.S. population. Almost 10,000 blood lead determinations formed part of the data collected. The results were shocking.

Median blood lead levels for the U.S. population as a whole were 13 µg/dL. The median level in children (aged six months to five years) was 15 µg/dL; in black children it was 20 µg/dL. More important, the data collected for NHANES II exhibited significant time trends. From 1976 to 1980, blood lead levels dropped 37 percent, from 14.6 to 9.2 µg/dL. Regression models controlling for a large number of confounding variables showed that this reduction was almost entirely the result of decreased use of lead in gasoline. Similar time trends were observed in data from lead-poisoning screening programs (Environmental Protection Agency 1986; Billick et al. 1979, 1982; Annest et al. 1983; Schwartz, Janney, and Pitcher 1984). . . .

Using these studies, which were early examples of the use of large data bases for epidemiological studies, the EPA was able to determine the effects of gasoline lead on lead in blood and to predict the number and demographic characteristics of children and fetuses (and adults) who were at risk at different allowable levels of gasoline lead. The predictions were straightforward. They did not rely on complex and questionable models for tracking lead in air, nor on extrapolation either from animals to humans or from high doses to low doses, nor on highly uncertain mathematical models (Environmental Protection Agency 1985, 1986).

This research also showed that while poor children in inner cities were no doubt at the highest risk of lead poisoning, all children were exposed to lead from gasoline. Combined with a sophisticated cost–benefit analysis showing that the benefits of removing lead in gas outweighed the costs by more than 5 to 1 (Environmental Protection Agency 1985), these studies greatly facilitated government actions to lower allowable levels of lead in gasoline and provided the basis for defending those actions. . . .

Round Three: The 1980s

In 1970 Congress passed not only the Lead Paint Poisoning Prevention Act, but also the Clean Air Act, the Clean Water Act, and the Occupational Safety and Health Act. By 1980, the National Institute of Occupational Health and Safety (NIOSH) had

produced a number of evaluations of the lead literature and at least two recommendations for setting standards. The Occupational Safety and Health Administration
(OSHA) had promulgated an occupational standard for lead, lowering the allowable
levels by 75 percent. The EPA had produced an air-quality criteria document for lead,
set an ambient air standard for it, begun regulating lead in gasoline, and published
documents related to regulating lead in drinking water. The Consumer Product
Safety Commission had limited lead in paint to 0.06 percent. None of these agencies
even existed before 1970. All of these activities required extensive review and evaluation of the epidemiology and other health literature on lead.

Several publications summarized the research on neurobehavioral effects of lead
at the turn of the decade. In 1979 the Office of Maternal and Child Health of DHEW
sponsored a conference entitled "Management of Increased Lead Absorption in
Children: Clinical, Social and Environmental Aspects" and published its adapted
proceedings in 1982 (Chisolm and O'Hara 1982). In 1980, Herbert Needleman edited
Low Level Lead Exposure: Clinical Implications of Current Research and Michael Rutter
published a careful review of the literature on the CNS effects of low-level lead
exposure.

These reviews and compilations of the literature noted several problems and data
gaps. In addition to the difficulty of controlling confounders, studies completed
before 1980 shared a number of problems in demonstrating neurobehavioral effects
(NBEs):

1. They lacked a single accepted indicator of dose or lead exposure.
2. The age, duration, and intensity of exposure that might cause NBEs was not
   known.
3. Outcome measures were not standard and were often not very sensitive.
4. Retrospective or cross-sectional studies could not demonstrate that NBEs occurred after exposure.

Several prospective studies carried out or reported in the 1980s were designed to
remedy these problems. Longitudinal prospective studies were conducted, using birth
cohorts whose blood lead levels were tracked from birth (from cord blood) or even
before (using maternal blood lead levels as a marker for prenatal exposure). Neurobehavioral functioning was also tracked from birth, thereby enabling researchers to
show that lead exposure preceded neurobehavioral deficits. Because blood lead was
tested every three to six months from birth, a fairly reliable history of exposure was
obtained, greatly reducing the risk of misclassification by exposure.

Repeated testing also gave some indication of the importance of timing, duration,
and dose of exposure. Several of these studies used the same standardized measures
of neurobehavioral functioning: the Bayley Mental Development Index, Stanford-
Binet IQ, the Wechsler Preschool and Primary Scale of Intelligence (WPPSI), and the
McCarthy Scales. Using multiple reliable, validated tests reduced the risk of misclassification by outcome, thus enhancing the studies' ability to find differences between
the exposed and unexposed groups.

Confounding was controlled in two ways. First, information was collected on
many confounding variables such as socioeconomic status (SES), home environment,

parental IQ, perinatal disease, parenting practices, trauma, and family size. These covariates were controlled by statistical techniques. Second, studies were designed to limit the domain, so that confounders would not vary greatly across the population included in the study. For example, Needleman and Bellinger (Bellinger et al. 1984; 1985; 1986a, b; 1987a, b; 1989) and their colleagues studied a cohort of white upper- and upper-middle-class infants in Boston. Dietrich and Bornschein (Bornschein et al. 1989; Dietrich et al. 1986; 1987a, b) and their colleagues studied a cohort of inner-city children in Cincinnati. Control of confounding through limitation of the study domain, rather than through matching, was desirable because potential confounders were not only numerous, but were also difficult to define and to measure. Limiting variability reduced the possibility of errors caused by poor measurement of confounders such as SES.

Prospective studies also allowed researchers to address the issue of the persistence of lead's effects. By 1989, children had been followed in several studies for approximately five years (Bellinger et al. 1987b; Ernhart and Morrow-Tlucak 1987). In 1990, Needleman published an 11-year follow-up study of the Chelsea/Somerville cohort he first reported on in 1979 (Needleman et al. 1990). Needleman found that effects such as school dropout rates and rank in class persisted into young adulthood. The birth cohort studies showed mixed results. Exposure at birth was inversely related to neurobehavioral performance up to 24 months of age (Bellinger et al. 1984; 1985; 1986a, b; 1987a, b; 1988; 1989) in some studies and up to six months in others (Dietrich et al. 1986; 1987a; Bornschein et al. 1989). In the Boston and Cincinnati studies, postnatal exposure was associated with longer-lasting deficits. In reviewing these studies, the EPA (1989) pointed out that, because of the low power of the particular studies involved, the positive results were more telling than the negative ones.

The prospective studies of birth cohorts in Boston, Cleveland, Cincinnati, and Port Pirie, Australia, were all of low-level exposure. The highest exposure levels were 25 µg/dL. Effects were observed in groups with exposure levels below 10 µg/dL. The medical community and the federal government concluded that 10 to 15 µg/dL was a level at which adverse NBEs occurred. In 1991, the CDC lowered the "level of concern" to 10 µg/dL to reflect this new information.

By the late 1980s, the epidemiological studies of lead and neurobehavioral deficits had finally met the requirements, summarized by Hill in 1965, for showing that an association is causal: the association was *strong* at high levels of exposure and was *consistent*, even at low levels of exposure. Deficits of four to six points on various intelligence scales were observed at exposures in the range of 10 to 25 µg/dL. The effects of lead were often, although not always, *specific*. (Hill was careful to point out that specificity should not be stressed too much, as even bacteria may cause more than one effect; streptococcus, for example, can cause sore throats, heart disease, and skin infections.) Although the NBEs of lead, as measured by standard development and IQ tests, may not always be the same, its biochemical effects were well documented and specific. The *temporal* relation between lead exposure and NBEs has been demonstrated: exposure precedes effect. A *dose–response* relation was evident: extremely high exposures cause encephalopathy and death, lower doses cause severe retardation, and lesser doses lead to school problems, small but significant shifts in

IQ, and other measures of CNS function. Huge numbers of in vitro and animal studies demonstrated not only the *biological plausibility* of the observed effects, but also many of their physiological mechanisms. Finally, the evidence was *coherent*.

Although research on the health effects of lead continued apace in the 1980s, other aspects of lead-poisoning prevention changed substantially. Following Ronald Reagan's election in 1980, federal programs in lead-poisoning prevention were cut back. The CDC's Lead Poisoning Prevention Program, which had distributed $89 million in the decade preceding 1981, was subsumed, along with many other categorical programs, into maternal and child health block grants given to the states. The total amount of the block grants was less than the amounts previously provided for all the programs they replaced. States made their own programming decisions and the reporting, data collection, and federal technical assistance aspects of the individual programs were lost. Most states, in fact, continued to do some lead screening, but federal programs were dismantled. In 1982 HUD ended its lead research program.

Social activism decreased notably in the 1980s around all issues, not just lead, partly in response to decreases in funding for community organizations and public information and partly in response to growing conservatism. Political energy was concentrated in defending public programs eroded both by the Reagan administration and by increasingly cash-starved state and local governments.

Summary and Conclusion

In the 40 years from 1950 to 1990, lead epidemiology and public policy based on it made enormous strides. Exposure levels that caused concern in the medical and public health community fell from 80 to 10 µg/dL of blood. In the space of 20 years, beginning in 1970, first the surgeon general and then the CDC lowered the official "level of concern" from 50 or 60 µg/dL. The public health community has turned its attention from the prevention of poisoning that results in encephalopathy, mental retardation, and death to the reduction of exposure to avoid subtle neurobehavioral deficits that are detectable only in fairly large epidemiological studies.

Numerous advances in technology and analysis have facilitated progress in lead epidemiology. In order to show that intellectual deficits were related to lead exposure rather than to such confounding variables as parental education, parental IQ, income, or parents' age at time of birth, researchers performed extensive regression analyses of fairly large sample populations, controlling for as many as 39 confounding variables. These analyses would have been virtually impossible but for the development of computer software programs that became available beginning in the late 1960s and early 1970s. Handling of large data bases, such as NHANES II, also required access to computer hardware and software not generally available earlier. . . .

Lead is an example of how changing science policy actually redefined adverse health effects. Until the 1970s, only clinical manifestations of disease such as encephalopathy, frank anemia, wrist drop, or kidney damage were considered adverse health effects of lead exposure. As the ability to study more subtle changes improved, accompanied by a better understanding of their significance, researchers came to recognize that interference with heme synthesis and the production of other proteins, as well as subtle changes in neurobehavioral functioning, were adverse effects. The change in the definition of adverse effects reflected a policy change as well as a change in the ability to observe and measure biological changes and to understand their biological importance. These policy changes were facilitated by a combination of advances in scientific understanding of the mechanisms and natural history of disease and attention to risk factors that accompanied the rise of chronic disease epidemiology. Policy shifted also partly in response to the growing demands of the environmental and occupational health movements for a more preventive approach to exposure to environmental toxins.

The population at risk to the effects of lead has certainly affected the degree of interest in and attention to the problem of lead exposure. Lead epidemiology captured the public eye and the interest of scientists in the 1960s in part because the poor black children who constituted lead's most obvious victims were at the center of a growing movement for civil rights, economic justice, and social change. Some of the scientists and public health professionals who participated in this movement were attracted to the study of lead because of its political implications. Professionals working on lead were drawn into the political debates surrounding the implications of their work. Lead epidemiology rose on a crest of vocational epidemiology and has been carried forward by several groups of researchers committed to studying effects at ever-decreasing levels of exposure.

The differential exposure of poor and black children to lead has also affected the epidemiology of low-level lead exposure because being poor and black are considered confounders of intellectual achievement and appropriate school behavior—the outcomes of interest in studies of low-level lead exposure. Thus, much of the task of the last 20 years of lead epidemiology has been to show that lead exposure and absorption cause—rather than result from—neurobehavioral problems. The counterargument—that neurobehavioral problems cause children to eat more lead—implied that poor, stupid black children, whose parents neither care for them properly nor keep their homes clean, eat more lead than "normal" children. When stated so baldly, the underlying racial and class prejudice of the argument becomes apparent.

Attention has increasingly focused on the effects of low-level lead exposure in part because of overwhelming evidence—especially data from NHANES II—of the size of the population exposed. The number of children (and adults) exposed to lead at levels that resulted in blood lead levels above 25, 15, and 10 μg/dL was so great that even relatively small shifts in such measures as IQ carried enormous social costs in terms of the number of children who would fall into the below-normal category or who might fail to achieve brilliance (Needleman 1990).

Lead has come to be recognized as a ubiquitous toxin presenting a hazard in soil, dust, and air as well as in more traditional sources such as paint. As the country grew more conservative in the 1980s and public concern with civil rights and social justice diminished, interest in the lead problem and in regulation to abate it was enhanced by the discovery of the importance of gasoline lead, which affected all children, not just those living in old housing who were disproportionately black and poor. Although action to remove lead from gasoline was not swift, it was complete and effective.

The falling observed effects level, combined with evidence of significant dispersion of lead in the environment and its bioavailability from practically any source—paint, air, soil, dust, ceramic glaze, water—has expanded the population at risk to include some 17 percent of American children under six years, making lead the health problem that affects the largest number of American children. Lead in food has been largely eliminated (Mushak and Crocetti 1990); the EPA has lowered allowable levels of lead in drinking water, but existing lead paint lingers, exposing millions of children—mostly poor and disproportionately black—annually. Regulatory efforts to eliminate lead paint from the environment of young children have been a dismal failure. Massachusetts, with one of the best state laws on deleading, deleaded less than 0.5 percent of its leaded housing stock from 1982 to 1986 (Mushak and Crocetti 1990). At that rate complete deleading would take 800 years. Although deleading homes is technically more difficult and expensive than deleading gasoline, certainly the public policy failure is related to the class and race of the affected population as well as to the difficulty and cost of the task. It remains to be seen whether 25 years of intensive study resulting in the increased understanding of the dangers of lead will result in effective demands to eliminate exposure and protect our most vulnerable citizens.

REFERENCES

Agency for Toxic Substances and Disease Registry. 1988. The Nature and Extent of Lead Poisoning in Children in the United States: A Report to Congress. Document no. 99-2966. Atlanta: U.S. Department of Health and Human Services/Public Health Service.

Alexander, F. W. 1974. The Uptake of Lead by Children in Different Environments. *Environmental Health Perspectives* 7:155–70.

Annest, J. L., J. L. Pirkle, D. Makuc, et al. 1983. Chronological Trends in Blood Lead Levels between 1976 and 1980. *New England Journal of Medicine* 308:1373–77.

Baltimore Health Department. 1971. *Baltimore Health News: Chronology of Lead Poisoning Control. Baltimore 1931–71.* (December):34–40.

Bellinger, D., A. Leviton, H. L. Needleman, et al. 1986a. Low-Level Lead Exposure and Infant Development in the First Year. *Neurobehavioral Toxicology and Teratology* 8:151–61.

Bellinger, D., A. Leviton, M. Rabinowitz, et al. 1986b. Correlates of Low-Level Lead Exposure in Urban Children at 2 Years of Age. *Pediatrics* 77:826–33.

Bellinger, D., A. Leviton, C. Waternaux, et al. 1985. A Longitudinal Study of the Developmental Toxicity of Low-Level Lead Exposure in the Prenatal and Early Postnatal Periods. In *International Conference: Heavy Metals in the Environment*, ed. T. D. Lekkas (vol. 1, 32–4). Edinburgh, U.K.: CEP Consultants.

Bellinger, D., A. Leviton, C. Waternaux, et al. 1987a. Longitudinal Analyses of Prenatal and Postnatal Lead Exposure and Early Cognitive Development. *New England Journal of Medicine* 316:1037–43.

Bellinger, D., A. Leviton, C. Waternaux, et al. 1988. Low-level Lead Exposure, Social Class, and Infant Development. *Neurotoxicology and Teratology* 10(6):497–503.

Bellinger, D., A. Leviton, C. Waternaux, et al. 1989. Low-level Lead Exposure and Early

Development in Socioeconomically Advantaged Urban Infants. In *Lead Exposure and Child Development: An International Assessment,* eds. M. A. Smith, L. D. Grant, and A. I. Sors (International Workshop on Effects of Lead Exposure on Neurobehavioral Development, September 1986, Edinburgh). Lancaster, U.K.: Kluwer.

Bellinger, D., J. Sloman, A. Leviton, et al. 1987b. Low-Level Lead Exposure and Child Development: Assessment at Age 5 of a Cohort Followed from Birth. In *International Conference: Heavy Metals in the Environment,* eds. S. D. Lindberg and T. C. Hutchinson (vol. 1, 49–53). Edinburgh, U.K.: CEP Consultants.

Bellinger, D. C., H. L. Needleman, A. Leviton, et al. 1984. Early Sensory-Motor Development and Prenatal Exposure to Lead. *Neurobehavioral Toxicology and Teratology* 6:387–402.

Billick, I. H., A. S. Curran, and D. R. Shier. 1979. Analysis of Pediatric Blood Lead Levels in New York City for 1970–1976. *Environmental Health Perspectives* 31:183–90.

———. 1982. *Predictions of Pediatric Blood Lead Levels from Gasoline Consumption.* Washington: U.S. Department of Housing and Urban Development.

Blanksma, L. A., H. K. Sachs, E. F. Murray, and M. J. O'Connell. 1969. Incidence of High Blood Lead in Chicago Children. *Pediatrics* 44:661–67.

Block, J. L. 1970. My Family Is Dying. *Reader's Digest* 96 (April):171–72.

Bornschein, R. L., J. Grote, T. Mitchell, et al. 1989. Effects of Prenatal Lead Exposure on Infant Size at Birth. In *Lead Exposure and Child Development: An International Assessment,* eds. M. A. Smith, L. D. Grant, and A. I. Sors (International Workshop on Effects of Lead Exposure on Neurobehavioral Development, September 1986, Edinburgh). Lancaster, U.K.: Kluwer.

Byers, R. K. 1959. Lead Poisoning: Review of the Literature and Report on Forty-Five Cases. *Pediatrics* 23:585–603.

Byers, R. K., and E. E. Lord. 1943. Late Effects of Lead Poisoning on Mental Development. *American Journal of Diseases of Children* 66:471.

Charney, E., B. Kessler, M. Farfel, and D. Jackson. 1983. Childhood Lead Poisoning: A Controlled Trial of the Effect of Dust-Control Measures on Blood Lead Levels. *New England Journal of Medicine* 309:1089–93.

Chisolm, J. J., and D. M. O'Hara. 1982. *Lead Absorption in Children: Management, Clinical, and Environmental Aspects.* Baltimore: Urban & Schwarzenberg.

Cohen, C. J., and W. E. Ahrens. 1959. Chronic Lead Poisoning: A Review of Several Years' Experience at the Children's Hospital, District of Columbia. *Journal of Pediatrics* 54:271–84.

Cohen, C. J., G. N. Bowers, and M. L. Lepow. 1973. Epidemiology of Lead Poisoning. A Comparison between Urban and Rural Children. *Journal of the American Medical Association* 226:1430–33.

Craig, P. P. 1971. Lead, the Inexcusable Pollutant. *Saturday Review* 54 (Oct. 2):68.

David, O., J. Clark, and K. Voeller. 1972. Lead and Hyperactivity. *Lancet* 2(783):900–903.

de la Burde, B., and M. S. Choate, Jr. 1972. Does Asymptomatic Lead Exposure in Children Have Latent Sequelae? *Journal of Pediatrics* 81:1088–91.

———. 1975. Early Asymptomatic Lead Exposure and Development at School Age. *Journal of Pediatrics* 87:638–42.

Dietrich, K. N., K. M. Krafft, et al. 1986. Early Effects of Fetal Lead Exposure: Neurobehavioral Findings at 6 Months. *International Journal of Biosocial Research* 8: 151–68.

———. 1987a. The Neurobehavioral Effects of Early Lead Exposure. In *Toxic Substances and Mental Retardation: Neurobehavioral Toxicology and Teratology,* ed. S. F. Schroeder, 71–95. Washington: American Association on Mental Deficiency (monograph no. 8).

———. 1987b. Low-level Fetal Lead Exposure Effect on Neurobehavioral Development in Early Infancy. *Pediatrics* 80: 721–30.

Eidsvold, G., H. Mustalish, and L. Novick. 1974. The New York City Department of Health: Lessons in Lead Poisoning Control Program. *American Journal of Public Health* 64: 956–62.

Engel, E. E. 1971. Health Hazards of Environmental Lead: A Position Paper. Paper presented at the Working Seminar on Lead to the Assistant Administrator for Research and Monitoring, Environmental Protection Agency, May 4.

Environmental Protection Agency. 1977. *Air Quality Criteria for Lead.* Pub. no. EPA-600/8-77-017. Washington: Office of Research and Development.

———. 1985. *Costs and Benefits of Reducing Lead in Gasoline. Final Regulatory Impact Analysis.* Pub. no. EPA-230-05-85-006. Washington: Office of Policy Analysis.

———. 1986. *Air Quality Criteria for Lead.* Pub. no. EPA-600-8-86-028aF. Research Triangle Park, N.C.: Environmental Criteria and Assessment Office.

———. 1989. *Supplement to the 1986 Air Quality Criteria for Lead* (Vol. 1, addendum A1-A67). Pub. no. EPA-600/8-89/049A. Washington: Office of Health and Environmental Assessment.

Ericson, J. E., H. Shirahata, and C. C. Patterson. 1979. Skeletal Concentrations of Lead in Ancient Peruvians. *New England Journal of Medicine* 300:946–51.

Ernhart, C. B., B. Landa, and N. B. Schell. 1981. Subclinical Levels of Lead and Developmental Deficit: A Multivariate Follow-up Reassessment. *Pediatrics* 67: 911–19.

Ernhart, C. B., and M. Morro-Tlucak. 1987. Low Level Lead Exposure in the Prenatal and Early Preschool Years as Related to Intelligence Just Prior to School Entry. In *International Conference: Heavy Metals in the Environment,* eds. S. E. Lindberg and T. C. Hutchinson (vol. 1, 150–52). Edinburgh, U.K.: CEP Consultants.

Facchetti, S. 1985. Isotope Lead Experiment—An Update. Paper presented at a workshop on Lead Environmental Health: The Current Issues, Durham, N.C., Duke University, April/May.

Florini, K. L., G. D. Krumbhaar, Jr., and E. K. Silbergeld. 1990. *Legacy of Lead: America's Continuing Epidemic of Childhood Lead Poisoning.* Washington: Environmental Defense Fund.

Garibaldi, P., S. Facchetti, A. Quagliardi, et al. 1975. Petrols Additivated with Isotopically Differentiated Lead: Proposal of an Experiment to Estimate the Incidence of Traffic on the Environment of Pollution by Lead—First Experimental Results. In *Proceedings of an International Symposium: Recent Advances in the Assessment of the Health Effects of Environmental Pollution* (vol. 3, 1287–99). Luxemburg: Commission of the European Communities.

Graebner, W. 1988. Private Power, Private Knowledge, and Public Health: Science, Engineering, and Lead Poisoning, 1900–1970. In *The Health and Safety of Workers,* ed. R. Bayer, 1–15. New York: Oxford University Press.

Guinee, V. F. 1971. Lead Poisoning in New York City. *Transactions of the New York Academy of Sciences* 33: 539–51.

Harvey, P. G., M. W. Hamlin, et al. 1984. Blood Lead, Behavior and Intelligence Test Performance in Pre-school Children. *Science of the Total Environment* 40: 45–60.

Hernberg, S., and J. Nikkanen. 1970. Enzyme Inhibition by Lead under Normal Urban Conditions. *Lancet* 1: 63–64.

Hernberg, S., J. Nikkanen, G. Mellen, and H. Lilius. 1970. Delta-Amino-Levulinic Acid Dehydratase as a Measure of Lead Exposure. *Archives of Environmental Health* 21: 140–45.

Hill, A. B. 1965. The Environment and Disease: Association or Causation? *Proceedings of the Royal Society of Medicine; Section of Occupational Medicine* (Jan. 14): 295–300.

Jacobziner, H. 1966. Lead Poisoning in Childhood. *Clinical Pediatrics* 5:277–86.

King, B. G. 1971. Maximum Daily Intake of Lead without Excessive Body Lead-burden in Children. *American Journal of Diseases of Children* 122:337–40.

King, B. G., A. F. Schaplowsky, E. B. McCabe, et al. 1972. Occupational Health and Child Lead Poisoning: Mutual Interests and Special Problems. *American Journal of Public Health* 62 (August): 1056–59.

Kotok, D., R. Kotok, et al. 1977. Cognitive Evaluation of Children with Elevated Blood Lead Levels. *American Journal of Diseases of Children* 131:791–93.

Landrigan, P. J., E. L. Baker, R. G. Feldman, et al. 1976. Increased Lead Absorption with Anemia and Slowed Nerve Conduction in Children near a Lead Smelter. *Journal of Pediatrics* 89:904–10.

Landrigan, P. J., S. H. Gehlbach, B. F. Rosenblum, et al. 1975. Epidemic Lead Absorption near an Ore Smelter: The Role of Particulate Lead. *New England Journal of Medicine* 292:123–29.

Lin-fu, J. S. 1971. Memorandum of May 5 to Dr. Irwin H. Billick: Information Pertinent to Title III—Federal Demonstration and Research Program of PL 91-695, The Lead-Based Paint Poisoning Prevention Act. Washington: Department of Health, Education and Welfare.

———. 1972. Undue Absorption of Lead among Children—A New Look at an Old Problem. *New England Journal of Medicine* 286:702–10.

———. 1973. Vulnerability of Children to Lead Exposure and Toxicity. *New England Journal of Medicine* 289:1229–33, 1289–93.

———. 1979. Lead Poisoning in Children. What Price Shall We Pay? *Children Today* 8:9–13, 36.

———. 1980. Lead Poisoning and Undue Lead Exposure in Children: History and Current Status. In *Low Level Lead Exposures. The Clinical Implications of Current Research*, ed. H. L. Needleman, 5–156. New York: Raven Press.

———. 1985. Historical Perspective on Health Effects of Lead. In *Dietary and Environmental Lead: Human Health Effects*, ed. Mahaffey. New York: Elsevier.

Mack, R. B. 1973. Lead in History. *Clinical Toxicology Bulletin* 3:37–44.

Manton, W. I. 1977. Sources of Lead in Blood: Identification by Stable Isotopes. *Archives of Environmental Health* 32:149–59.

Mellins, R. B., and C. D. Jenkins. 1955. Epidemiological and Psychological Study of Lead Poisoning in Children. *Journal of the American Medical Association* 158(1):15–20.

Milar, C. R., S. R. Schroeder, P. Mushak, et al. 1980. Contributions of the Caregiving Environment to Increased Lead Burden of Children. *American Journal of Mental Deficiency* 84:339–44.

Millar, J. A., R. L. C. Cummings, V. Battistini, et al. 1970. Lead and Delta-Amino-Levulinic Acid Dehydratase Levels in Mentally Retarded Children and Lead-Poisoned Suckling Rats. *Lancet* 2:695–98.

Mushak, P., and A. F. Crocetti. 1989. Determination of Numbers of Lead Exposed American Children as a Function of Lead Source: Integrated Summary of a Report to the U.S. Congress on Childhood Lead Poisoning. *Environmental Research* 50(2):210–29.

———. 1990. Methods for Reducing Lead Exposure in Young Children and Other Risk Groups: An Integrated Summary of a Report to the U.S. Congress on Childhood Lead Poisoning. *Environmental Health Perspectives* 89 (Nov.):125–35.

National Center for Health Statistics. 1984. Blood Lead Levels for Persons Ages 6 Months to 74 Years: U.S. 1976–1980. Data from the Second National Health and Nutrition Examination Survey (NHANES II), series 11, no. 233. Washington.

Needleman, H. L. Ed. 1980. *Low Level Lead Exposure: Clinical Implications of Current Research*. New York: Raven Press.

———. 1989. The Persistent Threat of Lead: A Singular Opportunity. *American Journal of Public Health* 79:643–45.

————. 1990. Low Level Lead Exposure: A Continuing Problem. *Pediatric Annals* 19(3):208–14.

Needleman, H. L., C. Gunnoe, A. Leviton, et al. 1979. Deficits in Psychologic and Classroom Performance of Children with Elevated Dentine Lead Levels. *New England Journal of Medicine* 300:689–95.

Needleman, H. L., A. Schell, D. Bellinger, et al. 1990. The Long-Term Effects of Exposure to Low Doses of Lead in Childhood. An 11-Year Follow-up Report. *New England Journal of Medicine* 322(2):83–88.

Ozonoff, D. 1988. Failed Warnings: Asbestos-related Disease and Industrial Medicine. In *The Health and Safety of Workers*, ed. R. Bayer, 139–220. New York: Oxford University Press.

Parents' Magazine. 1973. Does Lead Poisoning Threaten Your Child? 48 (August):59–60.

Patterson, C. C. 1965. Contaminated and Natural Lead Environments of Man. *Archives of Environmental Health* 11:344–60.

Perino, J., and C. B. Ernhart. 1974. The Relation of Subclinical Lead Level to Cognitive and Sensorimotor Impairment in Black Preschoolers. *Journal of Learning Disorders* 7:26–30.

Perlstein, M. A., and R. Attala. 1966. Neurologic Sequelae of Plumbism in Children. *Clinical Pediatrics* 5:292–98.

Rabin, R. 1989. Warnings Unheeded: A History of Child Lead Poisoning. *American Journal of Public Health* (December):1668–74.

Rabinowitz, W. B., and H. Needleman. 1983. Petrol Lead Sales and Umbilical Cord Blood Lead Levels in Boston, Massachusetts. *Lancet* 1:63.

Remsburg, C., and B. Remsburg. 1972. Youngest Victims. *Good Housekeeping* 48 (August):59–60.

Rummo, J. H. 1974. Intellectual and Behavioral Effects of Lead Poisoning in Children. Unpublished Ph.D. dissertation, University of North Carolina, Chapel Hill.

Rutter, M. 1980. Raised Lead Levels and Impaired Cognitive/Behavioral Functioning. *Developmental Medicine and Child Neurology* 42 (suppl.):1–26.

Sayre, J. W., E. Charney, and J. Vostal. 1974. House and Hand Dust as a Potential Source of Childhood Lead Exposure. *American Journal of Diseases of Children* 127:167–70.

Schucker, G. W., E. H. Vail, E. B. Kelley, and E. Kaplan. 1965. Prevention of Lead Paint Poisoning among Baltimore Children. *Public Health Reports* 80:11:969–74.

Schwartz, J., A. Janney, and H. Pitcher. 1984. *The Relationship Between Gasoline Lead and Blood Lead.* Washington: Environmental Protection Agency, Office of Policy Analysis.

Scientific American. 1969. Silent Epidemic. 220 (May):54.

Secchi, G. C., L. Erba, and G. Cambiaghi. 1974. Delta Amino Levulinic Acid Dehydratase Activity of Erythrocytes and Liver Tissue in Man. Relationship to Lead Exposure. *Archives of Environmental Health* 28:130–32.

Shapiro, I. M., P. Grandjean, and O. Van Neilsen. 1980. Lead Levels in Bones and Teeth of Children in Ancient Nubia: Evidence of Both Minimal Lead Exposure and Lead Poisoning. In *Low Level Lead Exposure: Clinical Implications of Current Research*, ed. H. L. Needleman. New York: Raven Press.

Silbergeld, E. K., and A. M. Goldberg. 1973. A Lead-induced Behavioral Disorder. *Life Sciences* 13:1275–83.

————. 1974a. Hyperactivity: A Lead-induced Behavior Disorder. *Environmental Health Perspectives.* 7:227–32.

————. 1974b. Lead-induced Behavioral Dysfunction: An Animal Model of Hyperactivity. *Experimental Neurology* 42:146–57.

Smith, H. D. 1954. Lead Poisoning in Children. *American Journal of Nursing* 54(6):736–38.

Smith, M. 1985. Recent Work on Low Level Lead Exposure and Its Impact on Behavior,

Intelligence and Learning: A Review. *Journal of the American Academy of Child Psychiatry* 24:24–32.

Specter, M. J., and V. F. Guinee. 1970. Epidemiology of Lead Poisoning in New York City—1970. Paper presented at the American Public Health Association, 98th Annual Meeting, Houston, Texas, October 26.

Szabo, N. B., and S. Pollack. 1987. *A Silent and Costly Epidemic: The Medical and Educational Costs of Childhood Lead Poisoning in Massachusetts.* Boston: Conservation Law Foundation of New England.

Thurston, D. L., J. N. Middlekamp, and E. Mason. 1955. The Late Effects of Lead Poisoning. *Journal of Pediatrics* 47:413–23.

Time. 1969. Deadly Lead in Children. 93 (April 4):42.

U.S. Department of Health, Education, and Welfare. 1970. Medical Aspects of Lead Poisoning. Statement of the Surgeon General, November 8.

Vandenbrouche, J. P. 1990. Epidemiology in Transition: A Historical Hypothesis. *Epidemiology* 1(2):164–67.

Yankel, A. J., I. H. von Lindern, and S. D. Walter. 1977. The Silver Valley Lead Study: The Relationship between Childhood Blood Lead Levels and Environmental Exposure. *Journal of the Air Pollution Control Association* 27:763–76.

QUESTIONS

1. What is the "lead belt?" How and why did the definition of its boundaries change during the early 1970s?

2. What evidence from 1960s and 1970s investigations suggested air-borne lead from gasoline was an important contributor to the total amount of ambient lead, and lead in the blood, of many individuals?

Chapter Fifteen

Lead Contamination in the 1990s and Beyond
A Follow-Up

Patricia Widener

The number of American children poisoned each year by lead has dropped considerably since the 1970s, due in large part to the government's initiative to eliminate lead in gasoline, paint, and soldered food and drink cans. However, lead has remained an important health issue in the 1990s. Unable to be broken down by the environment, lead persists in the homes and play areas of children. It continues to be ingested through hand-to-mouth behavior and contact with contaminated soil, paint chips, and dust.

This report is a follow-up to the preceding chapter by Barbara Berney, "Round and Round It Goes: The Epidemiology of Childhood Lead Poisoning, 1950–1990." It will examine the politics and science of lead contamination and how this tenacious, preventable childhood epidemic has been refashioned by the government and health care profession in the 1990s. Specifically, it will look at the dramatic decline in the number of children exposed to lead, alongside the socio-demographic factors that place a glaringly disproportionate number of some children at greater risk. This chapter will also look at the government's response to lead contamination in the home, particularly its efforts to reduce the prevalence of lead-based paint in rental housing. The government's screening policy is also discussed. Not surprisingly, it continues to be hotly debated, dividing medical professionals into either the universal or targeted screening camps. And finally, this chapter will examine the emergence of citizen action groups—the grassroots mobilization of parents, community workers, and medical professionals to educate and assist each other and the nearly one million children affected in the United States.

The Good News about Lead

From 1976 to 1991, according to comparisons of the first National Health and Nutrition Examination Survey (NHANES) and Phase I of the third NHANES (1988–1991), the mean blood lead level of people aged one to 74 plummeted 78 percent, from 12.6 µg/dL to 2.8 µg/dL (Pirkle et al. 1994).[1] By race, the percentage of children with blood lead levels at or above 10 µg/dL fell from 85 percent to 5.5 percent for non-Hispanic white children aged one to five years, and from 97.7 percent to 20.6 percent for non-Hispanic African American children of the same age. Continuing the steady decline, Phase II of the third NHANES (October 1991–September 1994) showed a decrease of 51 percent from Phase I to Phase II in the prevalence of blood lead levels at or greater than 10 µg/dL in the U.S. population (CDC 1997). Quite clearly, remarkable gains have been achieved in the overall reduction of lead levels in children.

The Bad News about Lead

Despite the striking decline in lead poisoning, nearly one million children between the ages of one and five had elevated blood lead levels during the 1991–1994 period of NHANES Phase II (CDC 1997). Stark evidence exists that children most at risk are male, minorities, urban residents and from low-income families (Brody et al. 1994). Elevated blood lead levels were 2.5 times higher in one-to two-year-old non-Hispanic African American children than non-Hispanic white children, and two times as high in Mexican American children. Children from low-income families were four times more likely to have elevated blood lead levels than children from high-income families. By race/ethnicity and residence, non-Hispanic African American children who live in urban areas of one million people or more are seven times more likely to have blood lead levels at or above 10 µg/dL than non-Hispanic white children living in non-urban areas. Mexican American children in urban areas also had elevated blood lead levels. Brody et al. summarize their data:

> Blood lead levels were consistently higher for younger children than for older children, for older adults than for younger adults, for males than for females, for blacks than for whites, and for central-city residents than for non-central-city residents. Other correlates of higher blood lead levels included low income, low educational attainment, and residence in the Northeast region of the United States. (Brody et al. 1994, 281)

In 1998 the figures have not improved. Approximately 890,000 children in the United States or 2.7 percent, have elevated blood lead levels, while 8 percent of children of low-income families, 22 percent African American children and 14 percent Mexican American children, have blood lead levels that exceed the recommended level of toxicity set by the Center for Disease Control and Prevention (CDC) (National Safety Council 1998).[2] Goldman and Carra coined the term "sad injustice" to capture this marked inequality in the distribution of lead poisoning (1994).

In studies of exposure pathways, race is a major factor in predicting the degree of lead exposure. Lanphear, Weitzman, and Eberly (1996) found that interior lead exposures for African American children and exterior lead exposures for white children in Rochester, New York, were major contributors. For African American children, who lived in poorly maintained rental homes, lead-contaminated dust and the condition and lead content of painted surfaces, such as windowsills, were the greatest sources of lead exposure. These children were also more likely to suck on windowsills and to use a bottle. White children, on the other hand, were exposed to lead from the ingestion of lead-contaminated soil. White children spent more time out of doors than did African American children and were more likely to suck their fingers. Lanphear, Weitzman, and Eberly (1996) suggest that as the degree of lead-contaminated house dust declines, soil ingestion becomes an important contact of lead for children. Moreover, although paint in the homes of white children had a higher lead content, the painted surfaces were usually in better condition. According to them, lead-based paint tastes sweet and may be one reason children suck on painted surfaces.

In Massachusetts, a study was conducted in 1991 following the passing of a law that required yearly lead poisoning screening of children between nine months and four years (Sargent et al. 1995). In a follow-up on the screening, Sargent et al. concluded that the incidences of lead poisoning in Massachusetts are linked with socio-demographics and housing conditions. They found that female-headed households with children under 18 had a "strong association" with lead poisoning and that children living in areas with a high number of children living in poverty were 8.9 times more likely to have lead poisoning than children in areas with a low poverty rate. When analyzing their data, seven variables were linked with lead poisoning: percentage of female-headed households, percentage of the population who were African American, income, percentage of children under 5 living in conditions of poverty, percentage of rental housing, percentage of housing built before 1950, and screening rates.

Commenting on the environmental justice issue raised in studies of lead and social class, Silbergeld writes:

> These differences have little or nothing to do with biology and everything to do with the coincidence of environmental lead sources with other conditions of economic disadvantage in American society in the later 20th century. As noted by many commentators, lead poisoning is a reflection of the problem of environmental racism, in which disadvantaged and politically powerless groups are more likely to live in areas with high levels of lead in their environments. (Silbergeld 1996, 50)

Weintraub (1997) attributes some of the racial disparity to racial segregation. Segregation and its concentration of poverty may decrease access to health care and increase lead intake through inadequate nutrition and pica behavior, or the ingestion of non-food items.

In addition to the voices of epidemiologists and social researchers, legal voices also began to speak out on the racial disparity. Advising plaintiff lawyers on landlord liability in lead-paint poisoning cases, Roisman (1995) argued that lead poisoning is a result of environmental racism. He advised trial lawyers to contact the National Association for the Advancement of Colored People (NAACP) and the Legal Services Corp to build stronger cases against lead poisoning incidents.

Finally, late in the twentieth century, another at-risk group of children was identified. Whelan et al. (1997) found the children of construction workers to be six times more likely to have elevated blood lead levels compared with a control group, which was selected from within the same neighborhood as each construction worker. The study revealed that many of these workers wore their street clothes at work and laundered them at home. Dust lead levels in the workers' automobiles and homes were significantly higher than in those of the control groups, although there was no difference between lead-based paint and the lead level in the drinking water between the groups. They also found that although some employers monitor workplace lead levels, they do not provide measures for home prevention. And medical personnel who screen workers are not recommending lead screening for the workers' children.

The Government's Response

In 1992, Congress passed the most comprehensive federal legislation to curb elevated blood lead levels. The Residential Lead-Based Paint Hazards Reduction Act, or Title X, took effect in 1996 (Goldman and Carra 1994; Binder et al. 1996). According to the Department of Housing and Urban Development (HUD), sellers, agents, and landlords are required to warn homebuyers and tenants of the existence of lead-based paint and its hazards in pre-1978 housing. The buyer or tenant must be made aware of the location of such paint in the home, paint hazards, and the condition of painted surfaces. Sellers must also provide a 10-day paint inspection and risk assessment period for homebuyers. Recognizing that lead occurs in privately owned rental housing and that lead removal expense may concern landlords, Congress established a 40-member Task Force on Lead-Based Paint Hazard Reduction and Financing to recommend hazard control, suggest financial alternatives for lead removal, and address liability and insurance concerns (Most 1996).

A year later, the Environmental Protection Agency (EPA) and the HUD agreed to work together to enforce the regulation, especially in regard to multifamily units of high-risk populations (HUD). They intend to issue graduated responses to landlords ranging from warning letters to fines. Moreover, in conjunction with the National Institute of Environmental Health Sciences, the HUD is conducting its own nationwide study of 1,000 homes to look at the prevalence of lead-based paint and lead-contaminated dust and soil. Results are not expected until mid-1999.

Despite federal, state, and local regulations prohibiting the use of lead-based paint and recommending and/or requiring the abatement of lead-based paints from homes, lead poisoning continues. Due to the persistent, yet pervasive, characteristics of the problem, plaintiff lawyers are beginning to hold landlords, who have not adhered to disclosure and abatement regulations, accountable (Keenan and Hurley 1998). Previously, landlords and the lead industry successfully refuted the hazards of lead on human health and development. Now, with valid evidence that lead is in fact toxic to children, defendants target the plaintiffs' experts for attack, calling into question their expertise, knowledge, and reliability (Roisman 1995). In spite of this, accusing landlords of negligence in removing hazardous materials from homes with children has met with some success in the courts.

Who Is to Be Screened? From Universal to Targeted

Although Berney (Chapter 14) described the 1970s as the decade of screening, screening has remained a key contested issue in the lead debate. In 1991, the CDC recommended universal screening of children. Despite the call for increased screening to insure early diagnosis and treatment of childhood lead poisoning, the majority of children were not being screened. Some six years later, "universal" screening remained a dream of health care workers. Recognizing the futility of the universal approach to screening and aware of studies that continued to underscore how certain subgroups in the population were at greater risk of lead exposure, late in 1997 the CDC altered its universal screening recommendation. Today, the agency advocates systematic targeted screening of high-risk children (Tips, Falk, and Jackson, 1998). This change in screening strategy, which is supported by the American Academy of Pediatrics (AAP 1998), is to aid states and communities in screening children who are at greatest risk of exposure and limit screening children who are not.

It should come as no surprise, given the protracted history of this contested issue, to learn that there is some disagreement in the medical community regarding the appropriateness of a stratified screening strategy. Manheimer and Silbergeld, for example, argue that

> targeting children in certain neighborhoods will not only miss the large number of exposed children not living in those neighborhoods but will also further ghettoize a disease to which all children are susceptible. Using "race" and poverty as a marker for lead poisoning reinforces the stereotype that the problem exists only among poor, inner-city people of color, creating a false sense of security among higher socioeconomic status groups and generating little political pressure to tackle the problem, which is seen as afflicting less politically powerful groups. The result is a further delay of much-needed control efforts. (1998, 44)

Whelan et al. (1997) would quite likely agree, arguing that children of construction workers, regardless of neighborhood demographics, are at great risk of lead exposure and would probably be missed in a targeted screening.

In an odd, but perhaps expected, twist to the screening debate, anti-social and delinquent behavior in children might be early expressions of lead poisoning (Needleman et al. 1996) and should therefore serve as flags to possible lead exposure. In a four-year study of 301 boys at public schools, Needleman et al. concluded that lead exposure is associated with anti-social and delinquent behavior. Looking at bone lead burden and behavioral reports from parents, teachers, and the individual students, and controlling for nine social and economic variables, they found that children who were considered asymptomatic for lead poisoning, but actually had elevated bone lead levels, were judged as more aggressive by their parents and teachers and were shown to have higher delinquent scores. They suggest that increased lead toxicity, by affecting motor skills, could hinder social development skills:

> Lead exposure in association with reduced verbal competence, increased rates of reading disabilities, frustration, and increased academic failure. Reduced verbal skills could

interfere with the use of internal language to mediate behavior and to delay immediate responding. (Needleman et al. 1996, 368)

A potential snare in the association of lead contamination, anti-social behavior, and academic failure is the possibility that lead screening might become a means of predicting future "problem children." In any screening program, as Draper (Chapter 12) notes, the strategy of screening can shift from one of diagnosis to prognosis. In short, lead screening strictly among the poor and disadvantaged could be used to predict and thus single out and control children who are expected to evidence problems sometime in the future.

Cost is also a contested issue in screening. Manheimer and Silbergeld claim that medical professionals and HMOs who oppose the universal screening recommendation may be against it because such screening entails more time with no monetary compensation.

> Lead poisoning is predominantly—although not exclusively—a disease of poor children, whose limited buying power leaves both the pharmaceutical and medical supply industries and medical care providers unenthusiastic. The consequences are already visible. The manufacturer of the newest lead chelator[3] has recently stopped production. Industrial research on new detection and treatment methods is at a low level. Because the potential return on investment in research and development of the pharmaceutical and medical supply industries is uncertain in the context of a changing policy on screening, it is even less likely that new agents and devices will be developed. (Manheimer and Silbergeld 1998, 46)

Although they acknowledge that only a small percentage of children are currently being screened, Manheimer and Silbergeld argue that CDC-endorsed targeted screening will only stabilize or reduce the level of screening. Tips et al. of the CDC responded:

> Chasing the distant and retreating mirage of universal screening is a dubious mission. ... It is wasteful of all our capital—the time and goodwill of parents and health care providers as well as health care dollars for our children—to screen every child irrespective of lead exposure of plausible risk. (Tips, Falk, and Jackson 1998, 51)

Studies support the CDC's conclusion that universal screening is an illusory and budget-busting goal. Bar-on and Boyle (1994) queried primary care pediatricians and subspecialists in Virginia and found that less than 12 percent screened all of their patients for lead toxicity and 25 percent never screened any of their patients. Their results do not bode well for universal screening. Only 13.5 percent of the primary care physicians and 5.6 percent of the subspecialists screened all of their patients. And, although primary care physicians were more knowledgeable about childhood lead poisoning, both groups lacked sufficient knowledge to educate and assist parents and children. Less than 75 percent of the pediatricians answered correctly half of the questions pertaining to lead poisoning. Some were unaware of the lead level currently thought to be toxic, lead sources, screening, effects, and treatment of lead poisoning. In essence, Bar-on and Boyle argue that before screening becomes a routine part of a pediatrician's examination, they must first become more knowledgeable on all aspects of lead poisoning.

In another study of physician practice, Fairbrother et al. (1996) reported that physicians in low-income, inner-city neighborhoods in New York who submitted large volumes of Medicaid billing claims were not immunizing or screening children, despite opportunities to do so. Only 26 percent of the children were up to date on their immunization, compared with 49 percent for the city. And opportunities to immunize were missed in 84 percent of the eligible cases. Children who were not immunized were also not likely to be screened for lead and tuberculosis. Only 20.4 percent of the children who visited the physician offices, mostly for sick care and follow-up visits, had been screened.

> It is clear that the quality of pediatric primary care is suboptimal in these inner-city storefront[4] physician offices. These physicians appear to be providing mostly episodic sick care to these children. . . . Seeing children episodically at the time of illness generates more income under current reimbursement policies, and that is clearly what physicians are doing. (Fairbrother et al. 1996, 789)

Citing figures from the General Accounting Office (GAO), the Alliance to End Childhood Lead Poisoning reports that 81 percent of Medicaid children between one to five years of age were not screened, despite a federal policy that requires such screening of all Medicaid-enrolled children. Moreover, the Alliance cited a GAO estimate that 65 percent of children on Medicaid with lead poisoning are never diagnosed (Alliance).

In Massachusetts, where a law has been passed requiring annual screening of all children nine months to four years, 58 percent of the children were screened (Sargent et al. 1995). This high, yet not exhaustive, figure is from a state that has been quite pro-active in combating lead poisoning and treating those affected. Approximately 34 percent of the screened children and 71 percent of the lead-poisoned children identified during this study were from communities with more than 25 percent of the children at or under five living in poverty. Since at-risk communities are least physician-served, Sargent et al. suggest combining physician in-office screening with door-to-door screening by community nurses in high-risk neighborhoods.

In a national telephone survey in 1994, approximately 24 percent of children aged six and under had been tested for lead poisoning and lead paint testing was conducted in only 9 percent of the housing units (Binder et al. 1996). However, the percentage of children most likely to have been screened were from low-income families, lived in rental housing and/or were residing in the northeast, possibly due to government-supported initiatives to provide testing for high-risk children.

New Voices: Parents and Communities

In the fray of conflicting opinions and contested issues, new voices were heard in the 1990s. Communities and parents of at-risk children began to seek information for themselves on the effects of lead and to speak out on its continued pervasiveness. Community health care workers sought ways of educating and counseling parents and children, while parents sought ways of altering their children's behavior to reduce lead exposures. Informal networks of information exchanges among parents, community workers, and medical professionals led to the formation of coalitions aligned in the struggle to eradicate lead poisoning among children.

The Alliance was formed in 1990 by leaders in pediatrics, public health, environmental protection, education, civil rights, and children's welfare to disseminate information among policy makers, health professionals, the press, and parents (Alliance). And although located in the United States, the Alliance has spread its education and information exchange campaign across the country's border. In 1994, it organized an international prevention conference that brought together 250 diverse organizations and/or representatives from 37 countries, and it continues advocating the accelerated phase-out of leaded gasoline in other countries. The first step in this international campaign has been the production of the document *Myths and Realities of Phasing Out Leaded Gasoline*, which has been translated into Spanish and Russian. According to the Alliance, only about 12 countries have succeeded in eliminating the use of leaded gasoline.

Despite the voluminous amount of information on lead poisoning, its sources, effects, and treatments, parents seemed the least informed in the early 1990s. Jackson (1996) writes a harrowing account of the discovery and treatment of lead poisoning in her two daughters and the eventual formation of Parents Against Lead (PAL) in 1992 to support and educate parents on the dangers of lead exposure. Jackson and other parents of lead-poisoned children were shocked to discover the amount of research and information available through state governments that was not readily accessible to parents. Through their investigation, they discovered that each state had different lead-information programs, with Massachusetts and California being the most progressive, and Louisiana and Oklahoma having no government-sponsored lead program at the time.

> The reason for this lack of information has a great deal to do with funding trends for lead-information and -abatement programs. In the late 1970s and early 1980s, programs were in place to help both children with lead poisoning and their families. When funds for these programs dried up, these efforts collapsed—and so did public awareness and outreach. The problem of lead poisoning persisted, but neither parents nor their advocates could turn to specific programs for help. (Jackson 1996, 58)

In 1994, the PAL was expanded to encompass a national network of parents who called themselves the United Parents Against Lead (UPAL). These activists learned that organized, united efforts were able to make a change. One of their first challenges was to ensure that parents received the same information on safe lead abatement procedures that landlords received. The UPAL also devised a way to educate children through songs and rhymes on washing their hands and running tap water before drinking it (Jackson 1996).

Studies indicate that educating parents on the risks of lead poisoning and the behaviors of children associated with elevated lead levels is an effective tool in lead prevention. In one study, for example, low-income parents with a high school or less education were told to wash their children's hands before meals and when they went to be bed. They were also asked to keep their children's fingernails short and to provide a well-balanced diet.[5] Finally, parents were told how to remove peeling paint and decrease lead-contaminated dust. After four months, the blood lead level in the children on average had declined, but after one year the blood lead levels had increased slightly, but were still at the level that existed prior to counseling (Kimbrough et al. 1994).

Community-level involvement in the drive to eliminate lead toxicity also expanded in the 1990s. In Baltimore, the Coalition to End Childhood Lead Poisoning (the Coalition) sponsored a Parent Lead Forum in 1999 to train parents on how to educate communities on lead. The Coalition is also associated with the Community Lead Education and Reduction Corps (CLEARCorps), an AmeriCorps Program, which works to reduce lead in urban areas across the United States.

Communities in urban areas also became places to build temporary lead-safe shelters to protect children who were undergoing chelation treatment,[6] and to protect all family members during lead abatement and remodeling work when lead-tainted dust is in the air. When children received chelaton to treat elevated lead levels, they may actually increase their lead level if exposed during treatment (Davoli 1997; Jackson 1996). Therefore, many of these children would have to be treated as inpatients. Lead-free shelters, in addition to reducing medical costs for families, provide a home-like environment for these children and their families, while the children are being treated.

A lead-safe, community-based shelter was established in Baltimore in 1989 to house families in need of temporary assistance (Farfel and Quinn 1994; Davoli 1997). Between 1989 and mid-1994, the shelter housed 33 families, including 88 children (Farfel and Quinn 1994). Farfel and Quinn predict that the need for such shelters will climb with increased focus on prevention and screening. The Kennedy Krieger Institute in Baltimore, which established the shelter through its Lead Poisoning Prevention Program, also opened a Community Lead Poisoning Prevention and Treatment Center in 1994 to provide medical treatment and community outreach and family education. In its first two years of operation, the program assisted nearly 400 children, finding lead-safe housing for more than 200 of them.

Summary

Lead hangs on. It persists in the environment, in the home, in medical and social discourses, and in the bodies of children. Although political, medical, and social advances in recent years have elevated the public's lead awareness and reduced lead toxicity in the environment and population, efforts to once and for all eliminate childhood lead poisoning have fallen short of success. Lead continues to invade the bodies of children. Hit hardest by this preventable epidemic are poor African American and Mexican American children, a vulnerable group requiring adults to speak for them.

Parents, communities, and medical activists have united and are loudly voicing their concerns, giving life to educational, resource, and work-based coalitions. Federal and state officials and medical professionals are responding. Together, they have reduced the risk of lead contamination dramatically. In 1994, after four years of opposition, the Senate passed legislation to gradually restrict the use of lead in consumer products such as toys, plumbing, and packaging material (Camia 1994).

At the close of the twentieth century, there are few indicators that the problem of childhood lead poisoning will be resolved any time soon. The debate on universal and targeted screening will continue and some health care workers will remain uninformed on the effects, symptoms, and treatment of this childhood illness. New sources of lead in common household objects are likely to be discovered and the actual danger level of lead may be shown to be below 10 µg/dL.

NOTES

1. The expression "µg/dL" denotes the number of micrograms of lead per deciliter of blood.

2. In 1991, the CDC revised the blood lead level from 25 µg/dL to 10 µg/dL (CDC 1991). At 10 µg/dL, health care professionals and policy makers deem lead toxic to the individual.

3. A chelating agent is recommended by the Food and Drug Administration for children with a blood lead level at or above 45 µg/dL to reduce elevated lead levels. However, when taken in a lead-contaminated environment, it can actually increase lead absorption (Davoli 1997).

4. Storefront physician's offices are often located in buildings that had previously been used as stores. A plate glass window that had formerly displayed merchandise is now boarded up or behind bars (Fairbrother et al. 1996).

5. Low calcium levels (Sargent et al. 1995; Lanphear et al. 1996) and iron deficiency (CDC 1991) may increase lead intake.

6. Chelation therapy involves either the injection or oral ingestion of a chelating agent, such as succimer, that binds to lead, thus enabling lead's excretion from the body. Treatment duration varies. Jackson's daughter received injections for more than eight months because of high on-going contamination in the home (Jackson 1996). If a child is not exposed to lead during treatment, oral medication may be administered for approximately 28 days (Davoli 1997).

REFERENCES

Alliance to End Childhood Lead Poisoning. http://www.aeclp.org/
American Academy of Pediatrics (AAP). 1998. "AAP Recommends Targeted Lead Screening, Universal Screening in High-Risk Areas." http://www.aap.org/advocacy
Bar-on, Miriam E., and Russell M. Boyle. 1994. "Are Pediatricians Ready for the New Guidelines on Lead Poisoning?" *Pediatrics* 93(2):178–82.
Berney, Barbara. 1993. "Round and Round It Goes: The Epidemiology of Childhood Lead Poisoning, 1950–1990" *The Milbank Quarterly* 71(1):3–39.
Binder, Sue, Thomas D. Matte, Marcie-jo Kresnow, Barbara Houston, and Jeffrey J. Sacks. 1996. "Lead Testing of Children and Homes: Results of a National Telephone Survey." *Public Health Reports* 111:342–46.
Brody, Debra J., James L. Pirkle, Rachel A. Kramer, Katherine M. Flegal, Thomas D. Matte,

Elaine W. Gunter, and Daniel C. Paschal. 1994. "Blood Lead Levels in the US Population." *Journal of the American Medical Association* 272(4):277–83.

Camia, Catalina. 1994. "Senate Bill Calls for Limits, Ban of Lead in Products." *Congressional Quarterly Weekly Report* 52:1384.

Center for Disease Control and Prevention (CDC). 1991. "Preventing Lead Poisoning in Young Children: A Statement by the CDC." Atlanta, GA: U.S. Dept of Health and Human Services, Public Health Service.

———. 1994. "Blood Lead Levels—United States, 1988–1991." *Morbidity and Mortality Weekly Report* 43(30):545–48.

———. 1997. "Update: Blood Lead Levels—United States, 1991–1994." *Morbidity and Mortality Weekly Report* 47(7):141–46.

Coalition to End Childhood Lead Poisoning. http//www.leadsafe.org/

Davoli, Cecilia T. 1997. "Use of Capitated Reimbursement to Provide Comprehensive Management of Childhood Lead Poisoning." *American Journal of Public Health* 87(12):2056–57.

Department of Housing and Urban Development (HUD). http//www.hud.gov/lea/

Fairbrother, Gerry, Stephen Friedman, Kimberly A. DuMont, and Katherine S. Lobach. 1996. "Makers for Primary Care: Missed Opportunities to Immunize and Screen for Lead and Tuberculosis by Private Physicians Serving Large Numbers of Inner-city Medicaid-eligible Children." *Pediatrics* 97(6):785–90.

Farfel, Mark R., and Ruth Quinn. 1994. "A Lead-Safe Family Shelter in an Urban Minority Community." *American Journal of Public Health* 84(8):1338–39.

Goldman, Lynn R., and Joseph Carra. 1994. "Childhood Lead Poisoning in 1994." *Journal of the American Medical Association* 227(4):315–16.

Jackson, Maurci. 1996. "Living with Lead: A Personal Account." *Forum for Applied Research and Public Policy* 11:55–58.

Keenan, Judy, and Erik K. Hurley. 1998. "Get the Lead Out: Proving Notice in Lead Paint Cases." *Trial* 34 (March): 21–37.

Kimbrough, Renate D., Maurice LeVois, and David R. Webb. 1994. "Management of Children with Slightly Elevated Blood Lead Levels." *Pediatrics* 93(2):188–91.

Lanphear, Bruce P., Michael Weitzman, and Shirley Eberly. 1996. "Racial Differences in Urban Children's Environmental Exposures to Lead." *American Journal of Public Health* 86(10): 1460–63.

Lanphear, Bruce P., Michael Weitzman, Nancy L. Winter, Shirley Eberly, Benjamin Yakir, Martin Tanner, Mary Emond, and Thomas D. Matte. 1996. "Lead-Contaminated House Dust and Urban Children's Blood Lead Levels." *American Journal of Public Health* 86(10): 1416–21.

Manheimer, Eric W., and Ellen K. Silbergeld. 1998. "Critique of CDC's Retreat from Recommending Universal Lead Screening for Children." *Public Health Reports* 113:38–46.

Most, Heidi. 1996. "Financing Lead Control: Weighing Safety and Cost." *Forum for Applied Research and Public Policy*. 11:65–67.

National Safety Council. 1998. "Childhood Lead Poisoning: Still a Threat." *Environmental Issues Bulletin*. 1(1):http://www.nsc.org/ehc/

Needleman, Herbert L., Julie A. Riess, Michael J. Tobin, Gretchen E. Biesecker, and Joel B. Greenhouse. 1996. "Bone Lead Levels and Delinquent Behavior." *Journal of the American Medical Association* 275(5):363–69.

Pirkle, James L., Debra J. Brody, Elaine W. Gunter, Rachel A. Kramer, Daniel C. Paschal, Katherine M. Flegal, and Thomas D. Matte. 1994. "The Decline in Blood Lead Levels in the United States." *Journal of the American Medical Association* 272(4):284–91.

Roisman, Anthony Z. 1995. "Getting the Lead Out." *Trial* 31 (January) 31:26–30.

Sargent, James D., Mary Jean Brown, Jean L. Freeman, Adrian Bailey, David Goodman, and Daniel H. Freeman, Jr. 1995. "Children Lead Poisoning in Massachusetts Communities: Its Association with Sociodemographic and Housing Characteristics." *American Journal of Public Health* 85(I4):5 28–34.

Silbergeld, Ellen K. 1996. "Getting Ahead of Lead." *Forum for Applied Research and Public Policy* 11:48–54.

Tips, Nancy M., Henry Falk, and Richard J. Jackson. 1998. "CDC's Lead Screening Guidance: A Systematic Approach to More Effective Screening." *Public Health Reports* 113:47–51.

Weintraub, Max. 1997. "Racism and Lead Poisoning." *American Journal of Public Health* 87(11): 1871–72.

Whelan, Elizabeth A., Greg M. Piacitelli, Barbara Gerwel, Teresa M. Schnorr, Charles A. Mueller, Janie Gittleman, and Thomas D. Matte. 1997. "Elevated Blood Lead Levels in Children of Construction Workers." *American Journal of Public Health* 87(8): 1352-55.

QUESTION

1. How is citizen mobilization around the lead issue changing the debate about children and lead?

Chapter Sixteen

Suffering, Legitimacy, and Healing
The Bhopal Case

Veena Das

The industrial disaster in Bhopal was acknowledged all over the world as the worst such event in history. Unlike the crises created by the largescale abduction of women or the occurrence of sati (which could be attributed to the ills of Hindu society needing correction by a progressive and civilizing state), the magnitude of the gas leak in Bhopal directly involved the scientific, legal, and administrative structures of modern society. Analysing it, we glimpse the ways in which theories of suffering come into being within the institutional frameworks of modernity. But before we get to Bhopal, let us consider certain general characteristics of chemical disasters.

One of the best-known studies of the legal and moral issues arising out of chemical disasters was made by Schuck (1987) in the Agent Orange case. This case represents the attempts of millions of Vietnam veterans to seek legal redress for grievous and debilitating illnesses which they believed were caused by their exposure to Agent Orange, a herbicide used by the United States Army to defoliate Vietnam's jungle cover. This herbicide was initially considered harmless to animals and human beings. Yet as veterans spread over different parts of the United States began to experience similar symptoms, including delayed onset of cancer, a pattern began to emerge in which it seemed probable that the impact of Agent Orange on human health had been severely underestimated.

Schuck argued that Agent Orange represented a new kind of case which required judges to incorporate vastly altered quantitative dimensions in their understanding of tort. The financial and personnel demands of the case were staggering. There were 1500 networks of law firms representing the plaintiffs. It was estimated that the defendants, i.e. the various companies responsible for the manufacture of this chemical, spent $100 million simply to prepare for the trial. The costs of litigation and administrative machinery were staggering enough, but the issue also called for a new kind of jurisprudence. Whereas traditionally the tort of personal injury cases "were essentially isolated disputes in which law's role was simply to allocate losses between putative injurers and victims according to a moral conception that Aristotle called corrective justice" (Schuck 1987), in the new cases of mass toxic tort there were large-scale economic and social interests involved. A relatively simple and comprehensible body of evidence was simply not available in this case, and courts found themselves weighing risks and benefits for which individual tort-oriented jurisprudence—which depends upon mechanical notions of causality—seemed to be ill suited. In general, there are scientific and legal complexities of a staggering nature in cases of mass toxic tort; in addition, there are other crises within society which get articulated in the course of such a legal dispute. In the Agent Orange case, for example, courts became sites on which the Vietnam war and its hardships were symbolically re-enacted. The same lawsuit could also be seen as a cathartic drama in which war veterans gave public expression to annoyance at their dismal treatment at the hands of their society: the court case became an occasion to comment upon the moral problems of contemporary society.

Other observers have similarly argued that chemical disasters and public response to them provide an opportunity to understand the construction of normality in a society. In an excellent paper on the subject, Reich (1982) observes:

> Chemical disasters appear by surprise. They represent an extraordinary event that disrupts the normal flow of social life. But paradoxically such crises in society create windows on normality. Through the windows of a chemical disaster, one can peer at political and social processes not usually accessible or visible.

One of the most important aspects of these political and social processes is that despite the pervasive uncertainty which surrounds such disasters—an uncertainty arising from the fact that the impact of toxic chemicals in the environment on human beings cannot be described in mechanistic terms—bureaucratic decisions are presented as if they were grounded in certainty. Toxic victims and their representatives often contest these bureaucratic definitions of illness and safety, and also try to expand the scope of the event from private suffering to the political issue of public policy. Reich's characterization of what happens in the case of chemical disasters, and especially the difficulties encountered by victims in bringing these issues into the domain of public consciousness, is lucid. However, the nature of the chemical disasters he analyses (contamination of cooking oil in Japan and animal feed in the USA) has meant that his emphasis is on understanding the transformation of private trouble into public issue. He also shows how bureaucratic rationality resists this transformation.

In the Bhopal case, however, the disaster was experienced right from the start as a *collective disaster*. Yet the inherent legal complexities of mass toxic tort cases, the power of the organized chemical industry (especially a multinational), the financial and organizational difficulties faced by victim groups, and the application of bureaucratic expert norms to the definition of disease, all came together in such a manner that the victims were denied their suffering. This denial did not occur through repression and censorship alone, but rather by talking about suffering in such a way that it came to be constituted purely as something verbal. Language came to be deployed as an end in itself, creating a discourse of which the function was to dissolve the concrete and existential reality of suffering victims.

The Event

Sometime at night on 2-3 December 1984, anything between 30 and 40 tonnes of methyl isocyanate (MIC), stored in a huge tank, escaped from the Union Carbide factory located in Bhopal, the industrial capital of Madhya Pradesh. The earliest descriptions of the gas were of a creeping, deadly, yellow, choking vapour which was intensely irritating to the eyes and lungs. Between 3 and 6 December about 2500 people died as a result of inhaling this deadly vapour.[1]

Within a month of this tragedy, suits were filed by many American lawyers in US courts on behalf of several victims. Earlier, within a week of the tragedy, several American lawyers had flown into Bhopal and obtained powers of attorney from victims in order to pursue cases against the Union Carbide Corporation (UCC) and its Indian subsidiary. These lawyers were described by many as "ambulance chasers," because their fees were said to be based on a contingency sharing of the obtained damages. In any case, in view of the difficulties of pursuing the case by such a large number of poor victims against a most powerful multinational, the Government of India passed the Bhopal Gas Leak Disaster (Processing of Claims) Act in March 1985. With this act the Government of India, in accordance of its *parens patriae* function, took upon itself responsibility for the conduct of the case, as well as for the provision of welfare to the victims.

The legal and litigational issues arising out of this case are important,[2] and they will surely become part of the ongoing debates on toxic torts.[3] Here I shall extract two kinds of texts from this complex web. The first is the medical discourse as presented in the courts; the second is the judicial discourse as embodied in two judicial texts, i.e. the judgement of the Supreme Court of India delivered on 22 December 1989 and 16 February 1991, on the constitutional validity of the Bhopal Act. These are texts in which we can see clearly the professional transformation of the suffering of the victims of Bhopal and the window it opens on the understanding of how suffering of victims can become an occasion for the legitimation of power, exercised in this case through the mediums of science and state in contemporary India.

The Litigational History

The major events in the litigational history of this episode which need to be kept in mind are the following:

1. The Bhopal Act was passed by the Government of India in March 1985, consolidating all claims arising out of the Bhopal disaster and making the government the only competent authority, on the basis of the *parens patriae* function of the state, to represent the victims.

2. The case was initially pursued in 1985 in a New York District Court, in which the Government of India represented that its judicial system was not competent to deal with the complex legal issues arising out of this disaster. The Government of India's strategy was to either force Carbide to submit to US laws on environmental protection, safety regulations, and liability of hazardous industry, and thereby secure compensation for the victims in accordance with law and standards of compensation in the USA (which Union Carbide seemed anxious to avoid); or to ensure that

Carbide, as a multinational entity, was brought under the jurisdiction of the Indian courts. This was important, since Carbide's strategy was to deny that any entity other than its Indian subsidiary, with far lower assets, was liable. Judge Keenan, in whose court this was adjudicated, declared faith in the Indian judiciary and bound Carbide to the jurisdiction of Indian courts.[4]

3. Hearings of the case began in 1986, in the district court of Bhopal. An application for interim relief was filed on behalf of several victim organizations in 1987, and the judge ruled that Carbide should pay Rs 350 crores as interim relief, which would be adjusted against the final settlement. Carbide appealed against this ruling in the High Court of Madhya Pradesh, where the quantum of interim relief was reduced to Rs 250 crores.

4. Carbide appealed to the Supreme Court against the Madhya Pradesh High Court order. It was while hearing the petition for interim relief that the Supreme Court, to the surprise and dismay of victims, made a judicial order on 15 February 1988, asking the Government of India and Carbide to settle at 470 million dollars. Part of the settlement deal fashioned by the court was immunity granted to Carbide and its subsidiaries against any criminal or civil liability arising from the Bhopal disaster at present and in the future.

5. There were widespread protests against the settlement, although it received the support of some important jurists.[5] It was believed by many that the courts had been pressurized or influenced by the Congress government, which was then in power, to issue the order, and that the government had made a private deal with Carbide. Regardless of the truth or otherwise of these allegations, the order was challenged on several legal grounds. One of these was a challenge to the constitutional validity of the Bhopal Act, since it took away from victims the right to be heard.[6]

6. The court gave its judgement on the Bhopal Act, holding it to be constitutionally valid, on 22 December 1989. The court was not required to pronounce on the settlement since that was to be heard by another bench; nevertheless, the court made several pronouncements defending the settlement.

7. Review petitions were filed in the Supreme Court, challenging the settlement on behalf of some victim groups. The Janata government, which came into power in January 1990, supported the review petitioners. The Supreme Court, in a judgement delivered in October 1991, upheld the settlement which it had itself fashioned. But it struck off the criminal immunities granted to Union Carbide, instructed the government to provide insurance cover to victims, and recognized the right of unborn children to reopen litigation in case further proof of the mutagenic impact of MIC on the human organism became available at a later time.

The particular text that I would like to analyse in some detail is the judgement upholding the Bhopal Act. This contains an explicit defence of the settlement, primarily through a discourse on suffering, supplemented by an analysis of the October judgement. To the extent that a judgement creates a master discourse in which the various voices are appropriated in a kind of monologic structure, it is important to see how these voices represent the suffering of the victims, to which references abound.

Before I come to the judgement, however, I will briefly comment on the various contentions regarding the medical evidence, for these have a bearing on my subsequent analysis.

The Victim's Body as Contested Site

There are certain similarities, as I said, between the Agent Orange and Bhopal cases. In the first the judge had fashioned an overall settlement because he felt that, despite the moving evidence of human suffering, legal uncertainties would create further hardship for the veterans and would be risky for the chemical companies. There were severe criticisms on the amount of this umbrella settlement, and Schuck (1987) notes that this amount was reliably learnt to be lower than the chemical companies had been willing to pay. This aspect of the case concerns us less at the moment than the anger felt by several veterans, not only at the amount of the settlement but also at the refusal of the chemical companies to admit guilt, of the government to admit its responsibilities, and at the fact that certain damning documents had been returned to the defendants without even unsealing them. This anger was very similar to feelings expressed by the victims of the Bhopal disaster after the settlement. In the fairness hearings after the settlement, there were in the Agent Orange case more than a thousand veterans who came to testify, expressing anger and anguish that they had not been given an opportunity to speak of their suffering in court. The words of the judge, as quoted in Schuck (1987):174, acknowledged this suffering:

> Broken-hearted young widows who have seen their strapping young husbands die of cancer, wives who must live with husbands wrecked with pain and in deep depression, mothers whose children suffer from multiple birth defects and require almost saint-like daily care, the strong men who have tears welling up in their eyes as they tell of fear that their families will be left without support because of their imminent death, the man whose mind is so clouded he must be prompted by his wife standing by with his defective child in her arms to go on with his speech, the veterans trying to control the rage that wells up within them, the crippled and diseased with running sores and green fungus growths, and the women who volunteered for field or Red Cross duty and now feel themselves rejected and sick with what they believe are Agent Orange related diseases.

But having acknowledge the reality of this suffering the judge nevertheless felt the court was not the right place to pronounce in a decisive way on whether guilt could be unambiguously laid at the door of the chemical industry or the government. As he remarked, he expected at least the lawyers to understand and acknowledge the limited power of the court in the matter of establishing proof: "It is likely that even if plaintiffs as a class could prove that they were injured by Agent Orange, no individual class member would be able to prove that his or her injuries were caused by Agent Orange." At the moment I shall not comment upon the standards of proof enunciated here, nor on whether such notions of causality are applicable in cases of mass toxic tort. What interests me here is the question of whether we are witnessing two alternative visions of a court of law in relation to the sufferings of the victims.

One view constructs the court of law as a public space in which a dramatic expression can be given to the suffering imposed upon vulnerable people, in this case veterans who were exposed to the toxic dioxin through Agent Orange. The public and dramatic re-enactment of private sufferings, in this view, would impose upon a whole society the necessity to recognize mutilated bodies and damaged minds of veterans and families, rather than hide these from public view, and to think seriously about the question of guilt. In addition, the court of law became an appropriate space to enact this, since modern society has vested it with the authority to pronounce the truth.

The second view was that pronounced by the judge on the strictly limited nature of modern courts of law. In this view, scientific innovations of the kind on which the prosperity of industrial society depends involve some risk taking.[7] While the suffering of the Vietnam veterans was real, it could not be decisively proved that it was caused by Agent Orange. Hence, chemical companies should respond to the suffering, but they could not be held guilty for it. In fact, in an extremely ambiguous statement, Judge Weinstein asserted that the public could assume for the first time that there was some merit to the claims of those exposed to Agent Orange "that they are suffering because of their war and post-war experiences." The suffering was real, but the cause of suffering was diffused and spread over a range of war and post-war experiences. Guilt could not be pronounced on any one party.

At first reading it seems that the Vietnam veterans were operating within a cosmology in which suffering, expiation, and guilt were interlinked. They were formulating a theodicy in which they had suffered because of the wilful actions of giant profit-making corporations which callously exposed human beings to health risks for the sake of financial gain. In contrast, the chemical companies and the judge who fashioned the settlement worked with a view of the world that was contingent, the argument here being that the very process of creating a prosperous society exposes some people to risks; chemical disasters were part of the risk of living in a modern world. Questions of guilt were irrelevant in such a world. What was necessary was "risk regulation."

Underlying these two views is a certain similarity: it is assumed by both sides to the dispute that courts of law are spaces in which a victim becomes a plaintiff by acquiring the means of proving that damage had been done to him.[8] But the very certainty demanded by the judges, within a context where the toxic hazards of chemicals were either not known or not revealed by the chemical industry, robbed the victims of the means by which the damage done to them could be *proved*. Victims were in effect being told to learn how to transform their sufferings into the language of science in order for it to be judicially recognized. But if both plaintiffs and defendants were obliged to speak *only* in the language of science, then surely it has to be conceded that the proceedings in the court were being conducted on two different registers: one, the register of scientific speech; the other, the anguished expression of victims in a case-by-case enumeration. If the second kind of evidence was finally to turn out not to be evidence at all, then why was this suffering exhibited as display? I shall argue later that this was allowed because, being in the nature of figures of speech, it performed an ornamental function which made the legal text appear proper to the occasion (see Witteveen, 1988). The stylistic function was similar to the use of metonymy in dreams, wherein the expression of a trauma is permitted but is so enmeshed within other images that it simply slides under.

The Construction of Medical Knowledge

How was medical knowledge constructed in the petitions before the Bhopal court, and what was the place it occupied in the judgements? Was the construction of this discourse deliberately such that it obstructed the representation of the suffering of the victims?

Recall here that, at the time of the leak, very little was known about methyl isocyanate. Diseases caused by a single inhalation of methyl isocyanate in dose-concentrations as high as this were simply unknown to medical science. During the earliest stages of the tragedy, Union Carbide officials repeatedly claimed that MIC was no more potent than tear gas, and that it caused only surface damage.[9] The large number of deaths within a few hours of the leak were attributed to the behaviour of the victims. Dr. Peter Halberg, one of the three doctors flown in by UCC as part of its relief operations within the first few days of the leak, stated that MIC created a heavy cloud which settled very close to the earth, killing children because of their immature lungs, the elderly because of their diminished lung capacity, and those who ran because their lungs expanded too rapidly. The survivors, said this report, included those who stood still and covered their faces with handkerchiefs (quoted in Dembo et al. 1987, 42). While it is true that any activity which led to increased inhalation of the poisonous gas would have led to greater damage, what is striking in the report is that it is not the character of methyl isocyanate that seems to be at issue but the behaviour of the people who tried to run when faced with this inexplicable phenomenon.

Between the period of the gas leak and 1989, several studies were conducted on methyl isocyanate. These were experimental studies conducted on animals, clinical studies on survivors, epidemiological surveys, and autopsy reports. The production of this knowledge was primarily through three sources. The first was the large number of animal experiments conducted by Carbide scientists. The second were clinical studies under the guidance of the Indian Council of Medical Research (ICMR), a body entrusted with sole responsibility for providing medical evidence to the Government of India (which was representing the victims). The third source was the experimental studies and small surveys on health, reproductive disorders, etc. conducted by certain independent doctors in pursuit of scientific interest and a concern for the victims. It can be seen that the production of this medical knowledge was hardly disinterested. The victim groups were not in a position to generate scientific knowledge themselves. They suspected that Carbide scientists were likely to set up the experiments so as to minimize the importance of the long-term impact of methyl isocyanate on human health. The ICMR scientists, whose work was simply not available to the victim groups until 1990—it was treated as confidential by administrative order—did not appear to be taking the suffering of the victims seriously. The reasons for this are complex and varied, but one of them is that the very process by which doctors become transformed from healers into certifying agencies seems to contribute to this desensitizing process.[10]

The first assumption in mass toxic tort cases, which is that objective scientific knowledge on the impact of toxic chemicals would be available to the judge, was therefore belied by the facts. Hazardous chemicals continue to be used by the chemical industry, and an assessment of their impact on human health only begins after an accident has occurred or after a pattern of disease establishes itself in repeated low-exposure cases. Victims are nevertheless expected to provide proof, linking exposure to the chemical with their disease pattern through a straightforward chain of cause and effect. Let us see how victim groups fared in this process.

On behalf of victim groups, the following medical facts were presented before the court.

First, it was submitted that, on the basis of experimental studies conducted after the Bhopal disaster, it had been established that, of all the known industrial isocyanates, MIC is the most potent pulmonary irritant and the second most potent sensory irritant. Its toxic hazard is related to the fact that the vapour pressure of MIC is exceedingly high: hence even a short duration of exposure is likely to result in severe effects.[11] Among the acute effects of MIC is immediate death due to profound hypoxia, resulting from blockage of airways by necrotic epithelial cells, mucous, fluid, and fibrin. Experimental research on animals as well as clinical observations on survivors have confirmed that there is a complex response to varying degrees of exposure to the gas. At higher dose levels, there were life-threatening manifestations of pulmonary damage among the survivors, reflected by clinical conditions of bronchi-constriction, pulmonary oedema, etc. At intermediate levels of exposure there was eye damage, including corneal opacity, and high risks of cataract. At relatively low doses, sensory and neurological responses were found.[12]

Second, although Carbide scientists tried to maintain that MIC had no long-term effects, it was submitted that other research seems to make it almost certain that there are several long-term effects, including chronic respiratory conditions, the development of corneal opacity, and an involvement of the immune system. It is not only likely that local immunity gets altered but also, as experimental research with animals has shown, there is persistent bone-marrow suppression, and a suppression of lympho-proliferative responses.[13]

Third, review petitioners stated that, while initially scientists had assumed that most symptoms caused by MIC exposure were reversible, on the basis of the epidemiological surveys it could be concluded that signs and symptoms in exposed populations had increased; that populations which had initially been asymptomatic were now showing signs of lung and eye disorder, besides chronic fatigue; that period-prevalence surveys of morbidity (as distinct from point-prevalence studies) revealed exposed populations were ill for longer periods of time in the year than unexposed populations; and that children's health had deteriorated. Apart from relying upon published experimental research, clinical reports, results of epidemiological surveys, and unpublished documentation in ICMR reports, the review petition was supported by an affidavit from Dr. Charles Mackenzie, Head of the Pathology Department of

Michigan State University, who had conducted epidemiological, clinical, and experimental studies.

Union Carbide contended on the other hand that the medical categorization done by the Madhya Pradesh government had shown that, out of the 359,793 claimants who appeared for medical examination, only 2574 suffered serious injury, only 40 suffered permanent disability, and only 10,534 suffered temporary disabilities. They claimed that the medical examination done by the doctors of the Madhya Pradesh government, and the scoring methods they used to allot points for assessing the nature of the injury, were more reliable than any results obtained by ongoing clinical observations or partial surveys. It was also alleged that many people had put in false claims. The petitioners on behalf of Union Carbide quoted from the first ICMR manual, which had stated that the symptoms reported by victims in 50 percent of the cases for certain disorders were out of proportion to the clinical, radiological, and pathophysiological features of the patients (Jain and Dave 1986).

The power of the medical profession to pronounce people as malingerers became starkly evident during the hearings. To take a few random examples, Union Carbide arranged the visit of one Dr. H. Weill in a consultative capacity one week after the disaster. After a week's stay Dr. Weill said that while he had not had an opportunity to examine injured persons, nor had direct access to lung function or pathological data, the anecdotal information collected by him enabled him to make an informed speculation: the symptoms reported by patients were highly exaggerated.

Carbide's main method of dealing with the medical evidence presented on behalf of victim groups was plain denial. I give only one example of this style of argumentation to show the casual manner in which the health of victims was treated. On the toxic hazard posed by MIC, review petitioners had stated the following:

> The petitioners wish to place before the court some of the expert studies which show that MIC causes and has caused long-term irreversible damage to the vital systems of the exposed population. These studies revealed that MIC is about the most potent sensory and pulmonary irritant known among industrial chemicals. A copy of one such study published in Environment Health Perspectives is annexed as Annexure D.

The response in the UCC petition to this was: "It is denied that MIC is the most potent sensory and pulmonary irritant among industrial chemicals." The counter-response to this by the review petitioners was: "it was stated that MIC was *about* the most potent pulmonary and sensory irritant." The paper on the basis of which this statement was made, in fact, appeared in a volume that reported work on MIC in a conference that had been sponsored by Carbide itself. This paper had concluded: "It is clear that MIC is a potent sensory irritant. To date only 2,4 toluene diisocyanate has been found to be more potent than MIC. . . . It can be seen that MIC is the most potent pulmonary irritant of the substances tested in this bioassay." One is constrained to conclude that there was a certain levity on the part of Carbide's lawyers in dealing with issues of such grave magnitude as the nature of the toxic hazard posed by methyl isocyanate.

Examples of this kind can be multiplied: selective quoting, omission of qualifications, and other stylistic devices were used to speak with an authoritative voice on matters that did not admit of any certainty. What is worse, while the scientific documentation quoted by Carbide, as defendant, was geared towards proving that the sufferings of the victims were highly exaggerated, the government, which was acting as *parens patriae*, also ended up by producing documentation in which the operative sentences showed victims exaggerating their symptoms. The subjective symptoms did not correspond with the known objective indices of disease said the government. In effect, victims were responsible for the fact that their disease was not understood by modern medicine.

The most offensive biological argument was perhaps the one which said that most of the victims were suffering from malnutrition or a previous disease, such as tuberculosis; hence it was not possible to distinguish between a disease caused by the inhalation of MIC from that which may have resulted from a combination of factors, such as a history of lung disease. This was like saying that because human beings were not like laboratory animals, the toxic insult to their bodies by the inhalation of methyl isocyanate—about which science did not possess definite knowledge—could not be decisively linked to the diseases encountered. One might even rephrase this to mean that those whose lives were already wasted by poverty and disease could scarcely claim just compensation merely on account of this further exposure to industrial disaster. This professional transformation of the experience of suffering, guilefully encoded in the language of science, ended up blaming the victim for his suffering.

The population of more than 300,000 which resided in the thirty-six wards near the factory, and which were identified as gas affected, went through total and abject suffering. Yet they did not possess the language now required to transform their horror into the legal, technical, and scientific forms that were being demanded as making sense in the political domain within which their grievances had to be addressed. In the medical text the causation by which links could be established between the inhalation of methyl isocyanate and the disease pattern of survivors was only possible through a slow and laborious process. Even then the links were likely to look only probable, given the complexity of the human organism and the fact that carcinogenic and mutagenic effects can take anything up to thirty years (or one generation) to manifest themselves. Meanwhile, the necessity to speak definitively in court probably led doctors to speak of survivors as malingerers—people who were filing false claims, exaggerating their illnesses, feigning symptoms. Even allowing for the fact that there is likely to have been a small percentage of false claims (given the strained economic circumstances of the affected populace), the callousness of this discourse is overwhelming. The suffering of survivors became an occasion for the exercise of power by the medical profession, with no promise of the healing that provides a delicate balance to this relationship.

The Judicial Discourse

Under the Bhopal Act the government used its *parens patriae* functions to represent the victims. Evoking this particular power of the state was an important judicial innovation since the costs of pursuing the case would have been too staggering for the victims. But we need to enquire a little more closely how the state constructed itself as a surrogate victim, as implied by its evocation of the *parens patriae* function.

In normal circumstances, *parens patriae* refers to the inherent power and authority of a legislature to provide protection to the person, and property of persons, who are *non sui juris*—such as minors, the insane, and the incompetent. The victims of the Bhopal disaster had been assimilated into this category of *non sui juris*. The circumstances were compelling enough to regard their lack of resources to pursue the case themselves as equivalent to their being judicially incompetent. The Bhopal Act was challenged on the grounds that, by its enactment, the government had constituted itself as a surrogate of the victims and taken away their right to be heard. Certain sections of the act, it was argued, amounted to a naked usurpation of power. The implication of this argument and the timing of the challenge was that, having usurped power in this manner, the government had used it to compromise the rights of victims by unilaterally arriving at a settlement and granting immunity to Carbide, against the expressed wishes of the victims. Thus, guardianship in this case amounted to a pretext by which the issue was sought to be resolved against the interests of the victims.

The counter-argument to this plea was presented by the Attorney-General, who had represented the government in February 1988, and who had defended the settlement with Union Carbide in court and outside. In defence of the Bhopal Act, he argued that the disaster had been treated as a national calamity by the Government of India. The government had a right, indeed a duty, to take care of its citizens in the exercise of its *parens patriae* jurisdiction or in a principle analogous to it. He reminded the court that they were not dealing with one or two cases, but with a large class of victims who, because they were poor and disabled, could not pursue the case against a powerful multinational. In the course of the arguments he maintained that:

> Rights are indispensably valuable possessions, but the right is something an individual can stand on, something which must be demanded or insisted upon without embarrassment or shame.

> When rights are curtailed—permissibility of such a measure can be examined only upon the strength, urgency, and pre-eminence of rights and the largest good of the largest number sought to be served by curtailment.

> If the contentions of the petitioners are entertained . . . rights may be theoretically upheld—but ends of justice would be sacrificed . . . the consent of victims should be based upon information and comprehension of collective welfare and individual good.

It needs to be pointed out here that the issue of government acting as surrogate for handicapped people who are unable to make their own decisions has been widely debated in recent years. It is generally agreed that the surrogate has to apply a best-interest standard in making decisions on behalf of people declared to be judicially incompetent.[14] Naturally, the interpretation of "best interest" has varied. According to one interpretation, the best-interest standard requires a surrogate to do what, from an objective standpoint, appears to promote a patient's good, without reference to the patient's actual supposed preferences. This view has been challenged on the grounds that a surrogate is acting in the best interests of the patient only if he tries to replicate the decisions that a patient would have made had she been capable. The evocation of objectively determined interests, as against subjectively expressed preferences, can act as a mask for exercising power in the name of doing good.

The same question now presented itself before the Supreme Court. In its earlier judicial order, the Supreme Court had claimed it acted in the best interests of the victims. But the victims were contesting this. How could the court defend the actions of the government, as well as its own actions promoting the settlement, when the victims themselves were protesting against them? It was in this context that the "suffering" and "agony" of the victims seem to have found their *raison d'être*: they allowed the judiciary to create a verbal discourse which legitimized the position of the government as guardian of the people and the judiciary as protector of the rule of law.

In their judgement the judges upheld the constitutional validity of the Bhopal Act. They declared: "Our Constitution makes it imperative for the State to secure to all its citizens the rights guaranteed by the Constitution and when the citizens are not in a position to assert and secure these rights, the State must come into the picture and protect and fight for the rights of the citizens." This is a principle which expands greatly the power of the state to act as guardian and deprive people of their normal constitutional rights, including the right to petition a court of law. Aware of this contradiction, the judges further stated:

> It is necessary for the state to ensure the fundamental rights in conjunction with the Directive Principles of State policy to effectively discharge its obligations and for this purpose, *if necessary, to deprive some rights and privileges of the individual victims in order to protect these rights better and to secure them further.* (Emphasis added)

But in what way had the meagre settlement served to secure the rights of the victims or led to the "greater good of greater numbers"—the utilitarian principle formulated by the Attorney-General and the judges at several times in the course of the argument?

The judgement said that the court had, in its earlier judicial order instructing both parties to settle, acted in the interest of the victims. "The basic consideration," the court recorded, "motivating the conclusion of the settlement was the compelling need for urgent relief, and the Court set out the law's delays duly considering that there was a compelling duty both judicial and humane, to secure immediate relief for the victims." Here was the first invocation of the suffering of the victims. Since there was a compelling need to provide immediate relief, the court had been moved by humane considerations to instruct the parties to settle. The court simply omitted to mention that the urgent need for immediate relief was precisely what the victims had asked for in their petition for interim relief, which had been upheld by the lower courts! Further, this immediate need for relief was baldly stated as the motive for settlement; there was nothing in the judicial order to instruct government to provide this immediate relief. Neither any procedure for the dispersal of money nor any time-table was formulated by the court. It would not be far off the mark, by any means, to conclude therefore that the suffering of the victims was more a verbal ploy than a condition which caused serious concern to either judiciary or government.

On the question of why the court had not found it fit to invite the opinion of the victims themselves (or certain representative organizations) about the proposed settlement before passing the judicial order, the judges noted that this had been because of a certain degree of unease and scepticism, expressed by the learned counsels of both parties, "at the prospects of success in view of their past experience of such negotiations when, as they stated, there had been uninformed and even irresponsible criticism of the attempts to settlement." So, we now discover that the first imperative to arrive at a meagre settlement was the "suffering" of the victims. The second imperative to do this with complete secrecy, and to present the victims with a *fait accompli*, was the irresponsibility and inability of the victims to understand the issues that affected their own lives.

In sum, the situation can be summarized in this manner: A multinational corporation was engaged in the production and storage of an extremely hazardous industrial chemical for which it had been given license to operate by the Indian government. Despite the known hazards of industrial isocyanates and diisocyanates, neither the multinational corporation nor its Indian subsidiary nor the Government of India had considered it important or necessary to enquire into the nature of hazard to the people posed by the manufacture and storage of this toxic material between the setting up of the factory and the spillage of gas. The people of Bhopal, and especially those staying around the factory, had not been warned of the dangers posed to them by these industrial activities, nor had any regulations been made and implemented about the placement of such factories. The result of all these activities, geared towards the "development" and "industrialization" of India, was that more than 300,000 people were suddenly, one night, blighted by a crippling disease, of which more than 2500 died horrible deaths. Yet the people declared incompetent and irresponsible were neither the multinational nor the government but the sufferers.

How had the court arrived at the particular sum of 470 million dollars as adequate compensation for the 300,000 victims known to have been affected? Were the victims right in alleging that there was no evidence of the application of a judicial mind behind this order? The fact is that medical examination of only half the claimants had been completed at the time that the court ordered settlement, and this through highly questionable procedures.[15] Similarly, medical folders which recorded the basis on which a victim's status was determined for purposes of compensation were not made available to them or to their representatives on grounds of medical secrecy. These arbitrary bureaucratic decisions could not be contested by the victims.

After the 22 December judgement, certain victim groups were able to secure some primary documentation on the procedures through which the extent of victimage was estimated. It was found that even the figures produced in the Supreme Court during a hearing by the Madhya Pradesh Council were arrived at after the settlement, and therefore could not have been the basis for the settlement. Further, the procedure by which the status of a claim was determined was based on a kind of scoring system: each symptom was given a score, and finally these scores were aggregated. Despite the fact that medical journals have constantly used the rhetoric of MIC being the most lethal gas known to man, the scoring system used the analogy of organ-by-organ injury rather than systemic damage as a result of toxic insult. To compound it all, patients who could not produce documentary evidence in the form of records of hospital admission, or proof of having been treated in the first few days of the gas disaster, were declared "uninjured" regardless of the state of their health at the time of examination.[16] It does not require any familiarity with mass disaster to know that the immediate task at the time of a disaster is to reach help rather than keep meticulous individual records. Yet all this incredible and appalling disregard of human and medical ethics has been justified on the grounds of bureaucratic and legal necessity. When repeatedly challenged on how the court had arrived at its particular sum for settlement, the court, in its own clarificatory orders, noted that the reasonableness of the sum was based not only upon independent quantification "but the idea of reasonableness for the present purpose was necessarily a broad and general estimate in the context of a settlement of the dispute and not on the basis of an accurate assessment of adjudication." In other words, whereas the victims were irresponsible and uninformed when they enquired about the basis on which the court had arrived at its estimate of victimage and the principles by which it determined compensation, the court itself had only been motivated by humane considerations in refusing to divulge medical information on the nature and extent of damage caused by MIC.

In its final transformation, then, the suffering of the victims was considered sufficient reason to justify the settlement and uphold the usurpation of power by the government. The principle on which this was defended was that the Bhopal Act could be upheld if it had proved to be in the interests of the victims (the proof of the pudding is in the eating, as the learned judges stated). Therefore, although the judges conceded that justice did not *appear* to have been done (since the victims were not given a hearing), they also argued that justice was nevertheless *done*—on the principle that a small harm can be tolerated for a larger good. The judges were concerned to signal their own humanity and protect the legitimacy of the judicial institution of which they were a part. The "suffering" of the victims was a useful narrative device which could be evoked to explain why victims had not been consulted; why their protests over the settlement could be redefined as the actions of irresponsible and ignorant people; to explain away the fact that the judges had not felt obliged to ask the government and its medical establishment to place for public scrutiny what it had accomplished by way of relief and help; to ignore the knowledge that had been generated on the impact of the deadly isocyanate on the health of people; and finally, to obfuscate the fact that they had completely failed to fix responsibility for the accident and had thus converted the issue of multinational liability into that of multinational charity.

In the course of the judgement it was evident that there was no lack of concern for the impact of a hazardous industry on *society in general*; it was only the interests of these particular victims that could not be fully protected. There were acute reflections on the dangers arising from hazardous technology. The judges stated that there were vital juristic principles which touched upon problems emerging "from the pursuit of such dangerous technologies for economic gains by multinationals." The learned judges also noted the need to evolve a national policy to protect national interests from the ultra-hazardous pursuit of economic gain. Only the sufferings of people came in the way of evolving these juristic principles:

> In the present case, the compulsions of the need for immediate relief to tens of thousands of suffering victims could not wait till these questions, vital though these may be, were resolved in due course of judicial proceedings; and the tremendous suffering of thousands of persons compelled this court to move into the direction of immediate relief, which, this Court thought, should not be subordinated to the uncertain principles of law.

In all these discussions at the judicial and bureaucratic levels, "victim" finally became a completely abstract category. The judges did not seem to have any clear idea on how many victims there were, nor the extent of havoc caused by MIC poisoning on their bodies and minds. In a moving speech I heard in Bhopal, protesting the settlement, one illiterate woman stated: "We only ask the judges for one thing— please come here and count us." In the judicial discourse, however, every reference to victims and their suffering only served to reify "suffering" while dissolving the real victims in order that they could be reconstituted into nothing more than verbal objects.

NOTES

1. See the report on the Bhopal disaster in the *Lancet*, 15 Dec. 1984.
2. For a clear and concise picture of these issues, see Baxi and Dhanda (1990). This book puts together, in one place, all the legal records necessary for an understanding of the issues.
3. See, especially, Royce and Callahan (1988); Rosenberg (1984); and Elliot (1988).
4. See Baxi and Paul (1988); and Baxi (1990).
5. On the inadequacy of the settlement, see for example the letter of the reputed tort lawyer Melvin Belli, characterizing the settlement as the most "unethical" in his fifty-year experience of dealing with tort law. *Times of India*, 7 March 1989. My own calculations show that, even if only half the claims were valid, the sum of 470 million dollars was just enough to sustain the affected households at the minimal income of Rs 700 p.m., and a minimal subsidy for risk to life, for a maximum period of ten years. I have not included costs of medical care in these calculations. Thus, without even going into questions of rights, the sum was not even adequate to ensure the survival of severely affected households for twenty years.
6. Writ Petition No. 268 of 1989, No. 164 of 1986, No. 198 of 1989, and No. 1551 of 1986.
7. Compare Beck's (1992) formulation that modern societies have made a transition from industrial societies to risk societies.
8. For a fine discussion of this distinction, see Lyotard (1988).
9. Thus L. D. Loya, medical officer, Union Carbide India Ltd., stated at the time of the disaster on 3 December 1984 that "The gas is non-poisonous. There is nothing to do except ask patients to put a wet towel over their eyes." Quoted in David Dembo et al. (1987).
10. I would hazard a guess that doctors in the local hospital, who were in daily contact with their patients, responded to their suffering quite differently from those to whom the patients were part of a population on whom scientific data were to be produced for litigational

purposes. The fact that the diseases defied cure, and the best that could be done for the patients was to provide symptomatic relief coupled with care, probably made these doctors defensive and anxious. But the most distant and arrogant in relation to patients were doctors whose job could be defined as that of certification.

11. See Alaris et al. (1987) for a comparison of the toxic hazards of methyl isocyanate, as compared to twenty-five other known isocyanates or diisocyanates.

12. See Gassert et al. (1987); for a description of the three patterns of toxicity, see Anderson et al. (1985).

13. The long-term effects are lucidly summarized in Anderson (1989). See also Verma (1989) for an account of how the long-term effects of methyl isocyanate were sought to be obscured by a clever manipulation of the suggestion that the deaths had been caused by hydrogen cyanide, which is much less lethal than methyl isocyanate, but in popular imagination is considered to be the most lethal poison since it is associated with cyanide deaths.

14. See *Deciding to Forego Life Sustaining Treatment: Ethical, Legal and Medical Issues in Treatment Decisions*. Report of the President's Commission for the study of ethical problems in medicine and biomedical and behavioral research (U.S. Government Printing Office, March 1983).

15. To mention only one fact, although clinical studies with survivors showed development of small airway and large airway obstructions which worsened with the passage of time, tests of pulmonary functions were done on a minuscule number of claimants. Mostly, it was assumed that survivors were exaggerating their symptoms.

16. For a reasoned critique of the procedures of evaluation of health status of victims by the Directorate of Claims, see Sathyamala, Vohra, and Satish (1989).

REFERENCES

Alaris, Y., et al. 1987. "Sensory and pulmonary irritation of methyl isocyanate in mice and pulmonary irritation and possible cyanide-like effects of methyl isocyanate in guinea pigs." *Environmental Health Perspective* 159–67.

Anderson, N., M. Kerr Muir, V. Mehra, and A. G. Salmon. 1985. "Exposure and response to methyl isocyanate: Results of a community based survey in Bhopal." *British Journal of Industrial Medicine* 45:469–75.

Anderson, N. 1989. "Long term effects of methyl isocyanate." *Lancet* (3 June): 1259.

Baxi, Upendra. 1990. "The Bhopal victims in the labyrinth of the law: An introduction." In *Valiant Victims and Lethal Litigation: The Bhopal Case*, ed. Upendra Baxi and Amita Dhanda, p. i–lxix. Delhi: Indian Law Institute.

Baxi, Upendra, and Amita Dhanda. 1990. *Valiant Victims and Lethal Litigation: The Bhopal Case.* Delhi: Indian Law Institute.

Baxi, Upendra, and T. Paul. 1988. *Mass Disasters and Multinational Liability: The Bhopal Case.* Delhi: Indian Law Institute.

Beck, Ulrich. 1992. "From industrial society to risk society: Questions of survival, social structure, and ecological enlightenment." In *Cultural Theory and Cultural Change*, ed. Mike Featherstone, pp. 97–123. London: Sage Publications.

Dembo, David, et al., eds. 1987. *Nothing to Lose but Our Lives.* New York: Council of International and Public Affairs; Delhi: Indian Law Institute.

Elliot, E. Donald. 1988. "The future of toxic torts: Of chemophobia, risk as a comparable injury, and hybrid compensation systems." *Houston Law Review* 25: 78–99.

Gassert, L., et al. 1987. "Long term pathology of lung, eyes, and other organs following acute exposure of rats to methyl isocyanate." *Lancet* 72:95–103.

Jain, S. K., and S. K. Dave. 1986. *Working Manual 1: On the Health Problems of Bhopal Gas Victims*. Delhi: ICMR and DST Centre for Visceral Mechanisms.

Lyotard, Jean-François. 1988. *The Differend: Phrases in Dispute*. Minneapolis: University of Minnesota Press.

Reich, Michael R. 1982. "Public and private responses to a chemical disaster in Japan: The case of Kanemi Yusho." *Law in Japan: An Annual* 15:102–29.

Rosenberg, D. 1984. "The causal connection in mass exposure cases: A public law vision of the tort system?" *Harvard Law Review* 97(4): 849–929.

Royce, M. D., and T. G. Callahan. 1988. "Isocyanates: An emerging toxic tort." *Environmental Law* 18(2): 293–319.

Sathyamala, C., N. Vohra, and K. Satish. 1989. *Against All Odds: The Health Status of the Bhopal Survivors*. Dehli.

Schuck, Peter H. 1987. *Agent Orange on Trial: Mass Toxic Disasters in the Courts*. Cambridge: Harvard University Press.

Verma, Daya R. 1989. "Hydrogen cyanide and Bhopal." *Lancet* (2 September): 567–68.

Witteveen, William J. 1988. The rhetorical labors of Hercules. Paper presented in the session on Hard Cases. International Association of Law and Semiotics, University of Pennsylvania.

QUESTIONS

1. What strategies did Union Carbide pursue to play down the extent and severity of health consequences stemming from MIC exposure?

2. How did the legal system work in this case to exclude the voices of victims from the very court case where questions of victim compensation were being decided?

Part 6

Living with Environments and Contested Diseases

Scientists and public health officials work with abstractions. People are grouped together on the basis of population characteristics that are of interest to us—their HIV status, the amount of fat in their diet, their socioeconomic status, or their proximity to a hazardous facility. It is sometimes easy to forget that real individuals stand behind these abstract conceptual labels. The concrete expression of "the AIDS crisis" or "a cancer cluster" is actual individuals, together with their families, friends, and communities, whose lives are ripped apart by devastating illness, and the associated emotional, physical, and financial toll.

The chapters in part 5 have provided some indication of the pervasive presence of environmental hazards in modern societies. For example, chemical toxins may be found in our homes and in our workplaces, in the food we eat, the water we drink, and the air we breathe. The number of individuals potentially threatened by these hazards is staggering. The readings included in the present section seek to move beyond the impersonal treatment of conceptual labels. In often compelling accounts, they record the voices of people living with illnesses they ascribe to exposures to environmental hazards.

In chapter 17, "Time," Steingraber intersperses discussions about her own, and a close friend's, cancer diagnosis with more general statistics about cancer incidence in the United States. She recounts the history of the development of cancer registries. She also highlights a number of problems with these registries, including their virtual exclusion of environmental causes of cancer, the lack of a national registry, limited availability of long-term data, and the inclusion of data on cancer mortality but not cancer morbidity. She presents evidence that environmental hazards have played a central role in increasing rates of certain kinds of cancers, such as non-Hodgkin's lymphoma.

In Balshem's chapter 18, "A Cancer Death," the author relays to the reader the voice of a widow grieving over the recent cancer death of her husband. The widow is also angry and embittered by the treatment her husband received at the hands of the medical professionals. A particular source of contention is the likely "cause" of the cancer listed in her husband's medical record. At first, the record lists only lifestyle factors, such as drinking and smoking. The wife fights to have her husband's exposure to environmental hazards included in the record. Balshem illustrates how emotions become the vehicle through which class and status differences between doctors and patients gain concrete expression.

Lawson's contribution, "Notes from a Human Canary," presents the perspectives of an environmentally ill person on the causes and consequences of multiple chemical sensitivity. Lawson recounts the multiple exposures to chemicals she has had over the course of her life. A particularly important source of exposures has been her work environments, which included at various times a department store, an art institute, and an office building. After suffering for years from severe headaches, Lawson finally consulted an allergist who diagnosed her with having several common food allergies. He also found she reacted to a number of synthetic chemicals. After two days of following the doctor's advice of eliminating her exposure to these substances, Lawson's headaches disappeared, never to return. She currently lives a highly structured life designed to minimize her contact with a wide range of problematic substances.

Chapter Seventeen

Time

Sandra Steingraber

Like a jury's verdict or an adoption decree, a cancer diagnosis is an authoritative pronouncement, one with the power to change your identity. It sends you into an unfamiliar country where all the rules of human conduct are alien. In this new territory, you disrobe in front of strangers who are allowed to touch you. You submit to bodily invasions. You agree to the removal of body parts. You agree to be poisoned. You have become a cancer patient.

Most of the traits and skills you bring with you from your native life are irrelevant, while strange new attributes suddenly matter. Beautiful hair is irrelevant. Prominent veins along the soft skin at the fold of your arm are highly prized. The ability to cook a delicious meal in thirty minutes is irrelevant. The ability to lie completely motionless on a hard platform for half an hour while your bones are scanned for signs of tumor is, conversely, quite useful.

Whether it happens at a hospital bedside, in a doctor's office, or on the phone, most of us remember the event of our diagnosis with a mixture of photographic recall and amnesia. We may be able to describe every word spoken, the arrangement of photographs on the doctor's desk, the exact color of the office draperies—but have no memory of how we got home that day. Or we may remember nothing that was said but everything about the bus ride. The scene I happen to remember most vividly—and this must have occurred weeks after my discharge from the hospital— is unlocking my door and discovering that my roommate had moved out. She did not want to live with a cancer patient. This was my redefining moment. Fifteen years later, the sight of a bare mattress can still cause me to burst into tears.

In 1995, an estimated 1.2 million people in the United States—thirty-four hundred people a day—were told they had cancer. Each of these diagnoses is a border crossing, the beginning of an unplanned and unchosen journey. There is a story behind each one.

These diagnoses also form a collective, statistical story. When all the diagnoses of years past and present are tallied, an ongoing narrative emerges that tells us how the incidence of cancer has been and is changing. Changes in cancer incidence, in turn, provide key clues about the possible causes of cancer. For example, if heredity is suspected as the main cause of a certain kind of cancer, we would not expect to see its incidence rise rapidly over the course of a few human generations because genes cannot increase their frequency in the population that quickly. Or if a particular environmental carcinogen is suspected, we can see if a rise in incidence corresponds to the introduction of such substances into the workplace or the general environment (taking into account the lag time between exposure and onset of disease). Such an association does not constitute absolute proof, but it gives us ground to launch additional inquiries.

The work of compiling statistics on cancer incidence is carried out at a network of cancer registries, which exist in the United States at both the state and the federal levels. Theoretically, for each new cancer diagnosis, a report is sent to a registry. How a diagnosed person has experienced, reacted to, coped with, remembered, or re-pressed this stunning event are aspects not included in this accounting, of course. What each report does contain is a coded description of the type of cancer; the stage to which it has advanced; and the geographic region, age, sex, and ethnicity of the newly diagnosed person.

This incoming information is then processed, analyzed, audited, graphed, and disseminated by teams of statisticians. In and of itself, a head count is not very useful. The prevalence of cancer is higher now than it was a century ago, in part because there are simply more people now. There are also proportionately more older people alive now than ever before, and the aged tend to get more cancers than the young. Between 1970 and 1990, for example, the U.S. population increased by 22 percent, and the number of people over sixty-five increased by 55 percent. To eliminate the effects of the changing size and age structure of the population, cancer registries standardize the data. One way of doing this is to calculate a cancer incidence rate, which is traditionally expressed as the number of new cases of cancer for every 100,000 people per year. For example, in 1982, 90 out of every 100,000 women living in the state of Massachusetts were diagnosed with breast cancer. By 1990, the inci-dence rose to 112 out of 100,000.

These numbers are also age-adjusted. That is, the data from all the differently aged people from any given year are weighted to match the age distribution of a particular census year. Thus standardized, the statistics from various years can be compared to each other. In this way, we know that the 24 percent rise in breast cancer in Massachusetts that occurred between 1982 and 1990 did not happen because the population of New England women was aging. Alternatively, cancer registry data can be made age-specific: the percentage of forty-five- to forty-nine-year-olds contracting breast cancer can, for example, be compared with the percentage from a decade ago.

I have often wondered about the daily lives of tumor registrars, those souls responsi-ble for keeping count of cancer's casualties. How strange it must be to monitor the thousands of cancer reports that flow into the registries every day in the form of paper files or electronic transfers. Surely I would want to pluck each one from the current and imagine the life behind the name. A seventy-five-year-old black woman from an urban area with advanced-stage breast cancer . . . or a forty-five-year-old white man from a farming community with chronic lymphocytic leukemia . . . or a seven-year-old girl with a brain tumor. I would long to sit down and talk with each one. "What has happened to you since your diagnosis? Are you getting good care? Are you surrounded by people who love you?"

As a group, tumor registrars seem like an affable lot—happy to converse about their work. Susan Gershman is the director of the Massachusetts Cancer Registry. Speaking to the public at a small, suburban library one Saturday afternoon, she was cool, well organized, and articulate as she stood in the tiny spotlight of the overhead projector, illuminated by her data. People in the audience took notes. Later, during the coffee and doughnut reception, she mentioned casually that her mother and father had both died of cancer when they were young, and I knew that she must bring a double perspective to her work.

Cancer registries publish their findings in thick annual volumes replete with tables and graphs, much like sports almanacs. My own reaction to these reports follows a particular evolution. At first examination, my eyes disassemble the data. In a graph displaying the age-adjusted rates for ovarian cancer, for example, I initially focus on the points rather than the lines that connect them. I wonder at the individual women whose lives are contained by the little black circles and gray squares that float in a white field of mathematical space. Gradually, as when I am looking at a picture that contains a hidden pattern, another way of seeing emerges from the page. Years of biological training kick in, and my eyes automatically begin to trace the slope of the lines, check the coordinates, imagine how the data might appear if displayed logarithmically.

In many ways, tracking the changing patterns of cancer incidence is not unlike tracking the patterns of ecological change. The statistical methods are certainly very similar—as are the vexing problems.

I once compiled old and current species inventories in order to monitor gradual changes in the composition of a Minnesota forest over several decades. During this time, some species became more common and others more rare. Sometimes I literally could not see the forest for the trees. The graphs constructed from my data showed clear trends often not apparent to me as I walked the deer paths that meandered among the pillars of the ancient canopy pines and through the green tangle of shrubs and saplings below. Without an exact count, I tended to overestimate the presence of rare plants because my delight at discovering them was more memorable than my efforts to note the existence of their more common neighbors. Perception can be misleading.

But I also had reasons to distrust parts of my data. To study time trends over half a century, one must rely on census counts conducted by many previous researchers, including some no longer living. If their system of coding and classifying differed significantly from mine, or if any one of us consistently misidentified certain species, then the changes indicated by my graphs were artifacts of our different techniques rather than reflections of a real biological shift. The seeming disappearance of a species that then suddenly reappeared in abundance five years later was a likely indication of a methodological snafu.

Cancer registry data are cursed with similar problems. We need these data because perception can mislead. It may seem to us that more and more people are getting brain tumors or that breast cancer is striking women at increasingly younger ages, but what do the numbers actually show? Perhaps people with cancer are now simply more outspoken than their predecessors. The numbers, on the other hand, can also deceive. Earlier detection, changes in the rate of misdiagnosis, and alterations in coding and classifying tumor types mean that apparent rises or falls in incidence rates can be artificial. How to quantify and correct such problems is a recurring question at tumor registrars' conferences and in publications such as *Cancer Registry News*.

Breast cancer incidence, for example, rose by nearly 25 percent in the United States between 1973 and 1991. During that time, the introduction of mammography changed the way many U.S. women were diagnosed with the disease, presumably because malignancies could be identified before being felt as a lump. How much of this rise can be explained by the increased use of mammograms? To answer this question, statisticians first look to see whether breast cancer incidence began to rise at the same time mammography became widely available. An internal audit of the data can also show whether groups of women with the highest rates of cancer are those receiving the most mammograms. And, since mammograms purportedly detect cancer earlier, statisticians can check whether the diagnosis of small breast tumors has been increasing faster than the diagnosis of large, advanced ones.

While still a matter of some debate, the most widely accepted estimate is that between 25 and 40 percent of the recent upsurge in breast cancer incidence is attributable to earlier detection, for which mammograms are partially responsible. Underlying this acceleration exists still a gradual, steady, and long-term increase in breast cancer incidence that has just recently begun to level off. This slow rise—between 1 and 2 percent each year since 1940—predates the introduction of mammograms as a common diagnostic tool. Moreover, the groups of women in whom breast cancer incidence is ascending most swiftly—blacks and the elderly—are among those least served by mammography. Between 1973 and 1991, the incidence of breast cancer in females over sixty-five in the United States rose nearly 40 percent, while the incidence of breast cancer in black females of all ages rose more than 30 percent. Therefore, the majority of the increase in breast cancer cannot be explained by mammograms.

This kind of analysis is possible only when many years of data are available. Unfortunately, many state cancer registries are new; they cannot look back across fifty years as I could with my tree inventories. The Illinois State Cancer Registry was created in 1985. My own diagnosis, which took place in 1979, is therefore not part of the collective story of cancer in Illinois. Unless I die from the disease, I will never be officially counted among those touched by cancer. The first year of reliable data in the Illinois State Cancer Registry is 1986. Moreover, like many state registries, Illinois is about five years behind in analyzing and publishing its data. Currently, therefore, Illinois residents have only a four-year picture of cancer incidence in their home state. Studying these time trends is like watching four minutes of a feature-length movie and trying to figure out the whole story.

Regional comparisons are often difficult because cancer registries in neighboring states can vary wildly in their length of operations. For example, Connecticut has the oldest functioning registry, one started in 1941. The Connecticut Tumor Registry provides one of the only truly long-term views of U.S. cancer incidence. Massachusetts, on the other hand, established its cancer registry in 1982. Nearby Vermont is one of ten states that had no cancer registry at all until 1992, when Congress established the National Program of Cancer Registries.

This patchwork of state-based registries is afflicted with another problem that we who count plants never have to worry about. People, unlike trees, move. Lifelong residents of one state, for example, may migrate to another upon retirement and become statistics in their new community. Without a comprehensive national cancer registry—which the United States does not have—state registries must rely on an elaborate system of data exchange. This is especially crucial for my elongated home state of Illinois, which shares a border with five other states. When faced with a serious health problem, many rural folk in the central and southern counties wind up being diagnosed across the Mississippi and Wabash Rivers because they would rather travel to cities in Iowa, Missouri, Indiana, or Kentucky than make the long trek north to Chicago. Illinois recently began trading registry data with its neighbors, thereby considerably boosting cancer incidence figures in its many east and west border counties.

Five state registries also contribute data to the federal cancer registry. The so-called SEER Program (Surveillance, Epidemiology, and End Results), overseen by the National Cancer Institute, does not attempt to record all cases of cancer in the country, but instead samples about 14 percent of the populace. SEER is a child of the War on Cancer as declared by President Richard Nixon and codified as the National Cancer Act of 1971. SEER has been collecting cancer diagnoses since 1973 and currently represents the states of Connecticut, Hawaii, Iowa, New Mexico, and Utah, as well as five specific metropolitan areas: Atlanta, Detroit, San Francisco—Oakland, Seattle, and Los Angeles. Everyone living in one of these states or cities who is diagnosed with cancer becomes a bit of data in the SEER Program registry, and their tumors stand in for all of ours.

Without a nationwide registry, no one can know exactly how many new cases of cancer are diagnosed in the United States every year. Instead, such numbers are estimated by applying rates from the SEER registry to the population projection for any particular year. To generate estimates before 1973, statisticians combine data from older individual state and city registries across the country. In this way, we now have reasonably reliable incidence figures going back to 1950.

Incidence data were not available to Rachel Carson when she first documented what she believed was the beginnings of a cancer epidemic. Instead, Carson focused on rising death rates from cancer. She was most disturbed by evidence that childhood cancer had jumped from the realm of medical rarity to the most common disease killer of American schoolchildren within a few decades.

Some researchers believe that mortality rates—which are also adjusted for age and population size—are still a more reliable indicator than incidence because they are less affected by changes in diagnostic technique. Death, after all, is certain and absolute. Moreover, causes of death, duly noted in all states of the union, have been tallied for far longer than tumors have been registered. We have a much deeper and wider view when we examine cancer trends over time using information gleaned from death certificates.

But mortality is also an imperfect measure of the prevalence of cancer. Not everyone diagnosed with cancer, thankfully, goes on to die from it. If treatment improves, mortality can decline even as incidence rises. This is certainly the case for childhood cancers, which, according to SEER data, jumped in incidence by 10.2 percent between 1973 and 1991 even as the death rate fell by almost 50 percent. Long-term trends show that childhood cancers have risen by one-third since 1950. Using mortality to measure the occurrence of cancer in children today would create a falsely rosy picture. Heroic measures may be saving more children from death, but every year more children are diagnosed with cancer than the year before. Increases are most apparent for leukemia and brain tumors. At present, eight thousand children are diagnosed with cancer each year in the United States; one in every four hundred Americans can expect to develop cancer before age fifteen.

Cancer among children provides a particularly intimate glimpse into the possible routes of exposure to contaminants in the general environment and their possible significance for rising cancer rates among adults. The lifestyle of toddlers has not changed much over the past half century. Young children do not smoke, drink alcohol, or hold stressful jobs. Children do, however, receive a greater dose of whatever chemicals are present in air, food, and water because, pound for pound, they breathe, eat, and drink more than adults. In proportion to their body weight, children drink 2.5 times more water, eat 3 to 4 times more food, and breathe 2 times more air. They are also affected by parental exposures before conception, as well as by exposures in the womb and in breast milk.

The night before Jeannie's death, I dreamed I traveled on a large boat with many other people. No shorelines were visible. Someone suggested I walk out onto the deck and get some sun. It's too hot, I said. But I walked out anyway and discovered the weather very pleasant. Someone suggested I go for a swim. Too dangerous, I said. But I dove in, and the water was cool and crystalline. Dolphins circled me protectively. Back in the boat, I asked. Where are we? And someone smiled and handed me a map.

Driving across the Charles River to the hospital the next morning, I took the dream as a sign that I had accepted what I understood now to be imminent. But by the time I crossed the river again that night, I knew I had not and never would.

I wanted time to stop. I wanted all the clocks unplugged and the calendars nailed flat to the walls. It was April. I wanted no leaves to emerge from the buds that blurred the outlines of the trees.

Time had become such a strange commodity in the preceding month. On the surface, it had seemed to speed up as the vague progression of Jeannie's various symptoms had suddenly accelerated. One day she found she could no longer type. A week later she could not turn doorknobs. The next week, buttons were impossible. Each loss was profound and irrevocable—the ability to write, to walk through a doorway, to undress.

But under the quick surface, in the deep water at the center of every hour and every moment, time was slowing down. Each meal, each conversation, each walk from one room to another unfolded with such deliberateness that an afternoon spent in Jeannie's apartment was the equivalent of a week.

"You understand this is a terminal event." A doctor's voice on the magnetic tape of my answering machine. The dazed drive to the intensive care unit. Each heartbeat visible as data on a video screen. Slow drippings in tubes. An endless night. A blue-black dawn. A nurse's voice, as though from a distant room: "Okay. These are her last breaths now."

The whole concept of time was unbearable. I wanted to be back in Illinois in the middle of winter. I wanted to walk across frozen fields. No ocean. No leaves. No boats. She was gone.

All types combined, the incidence of cancer in the United States rose 49.3 percent between 1950 and 1991. This is the longest reliable view we have available. If lung cancer is excluded, overall incidence still rose by 35 percent. Or, to express these figures in another way: at midcentury a cancer diagnosis was the expected fate of about 25 percent of Americans—a ratio Carson found so shocking that it inspired the title of one of her chapters—while today, about 40 percent of us (38.3 percent of women and 48.2 percent of men) will contract the disease sometime within our lifespans. Cancer is now the second leading cause of death overall, and the leading cause of death among Americans aged thirty-five to sixty-four.

More of the overall upsurge has occurred in the past two decades than in the previous two, and increases in cancer incidence are seen in all age groups—from infants to the elderly. If we exclude cancer of the lung and restrict our view to the period covered by SEER, overall incidence rose 20.6 percent between 1973 and 1991, while mortality declined 2.8 percent.

Adding lung cancer to the picture, overall cancer mortality *rose* by 6.9 percent from 1973 until 1991—a difference that testifies to the deadly nature of this disease. Happily, the decline in smoking is finally affecting the cancer death rate. In a recent study of cancer mortality rates from 1991 to 1995, researchers found a small but decisive decline in overall cancer mortality (about 3 percent) during this period. The single largest factor behind this decline is a decrease in lung cancer deaths.

One-fourth of all cancer deaths are from lung cancer. Because the fatality rate is so high, lung cancer incidence and lung cancer mortality are very nearly the same statistic, and, in the United States, both closely mirror historical patterns of cigarette consumption. (Among American women, who began smoking in large numbers later in the century than did men, lung cancer mortality is still rising.) Overall, approximately 87 percent of the deaths from lung cancer can be attributed to cigarette smoking.

This also means, of course, that 13 percent of all lung cancer deaths occur among people who do not smoke. Thus, although smoking dominates the lung cancer picture, additional mysteries need sleuthing here. And, while smoking remains the largest single known preventable cause of cancer, the majority of cancers cannot be traced back to cigarettes. Indeed, many of the cancers now exhibiting swift rates of increase—cancers of the brain, bone marrow, lymph nodes, skin, and testicles, for example—are not related to smoking. Testicular cancer is now the most common cancer to strike men in their twenties and thirties. Among young men both here and in Europe, it has doubled in frequency during the past two decades. These increases cannot be attributed to improved diagnostic practices. Brain cancer rates have risen particularly among the elderly. Between 1973 and 1991, brain cancers among all Americans rose 25 percent. Those over sixty-five suffered a 54 percent rise.

Mortality and incidence do not always track each other. No cancers are increasing in mortality while decreasing in incidence, but several cancers have increased in incidence even as their death rates have declined due to more effective treatments. According to SEER data, these include cancers of the ovary, testicle, colon and rectum, bladder, and thyroid. There are eight cancers whose incidence and mortality are both on the decline: those of the stomach, pancreas, larynx, mouth and pharynx, cervix, and uterus, as well as Hodgkin's disease and leukemia. Stomach cancer has been declining for decades, probably owing to improvements in food handling and the increased consumption of fresh foods made possible when refrigeration replaced more toxic methods of food preservation, such as smoking, salting, and pickling. Pap smears have been credited with bringing down the incidence of cervical cancer because precancerous lesions can be detected and cut out before they are transformed into invasive tumors.

However, these modest gains are swamped by the cancers that show both increasing incidence and increasing mortality: cancers of the brain, liver, breast, kidney, prostate, esophagus, skin (melanoma), bone marrow (multiple myeloma), and lymph (non-Hodgkin's lymphoma) have all escalated over the past twenty years and show long-term increases that can be traced back at least forty years. In recent years, breast cancer mortality among white women has begun to slow down, declining 6.8 percent from 1989 to 1993. However, the death rate is still higher than it was when Rachel Carson died of the disease in 1964, and it is still rising for black women. Moreover, breast cancer incidence rates are still rising for localized disease even as they are falling for more advanced-stage diagnoses (a shift probably indicating that breast cancer is being detected and treated earlier); the proportion of women developing the disease at all remains at the highest level ever recorded.

"Explanations for these increases do not exist," according to Philip Landrigan, a pediatrician and leading public health researcher (Landrigan 1992). Medical literature is accustomed to summations more temperate and indirect, but this one has been echoed again and again in recent research papers on trends in cancer rates. In a 1995 assessment of the situation, a research team at the National Cancer Institute similarly concluded, "Some trends remain unexplained . . . and may reflect changing exposures to carcinogens yet to be identified and clarified" (Zahm and Devesa 1995, 177).

Clarification about carcinogens, Landrigan believes, requires an environmental line of inquiry:

> The possible contribution to recent cancer trends of the substantial worldwide increases in chemical production that have occurred since World War II (and the resulting increases in human exposure to toxic chemicals in the environment) has not been adequately assessed. It needs to be systematically evaluated (Landrigan 1992, 945).

I have read the preceding two sentences many times. Most of my life spans the time between Carson's call for a systematic evaluation of the contribution of toxic chemicals to increased human cancers and Landrigan's repetition of this call. Both give one pause.

I am struck also by the symmetry between Landrigan's recommended course of action and an observation made thirty years earlier by two senior scientists at the National Cancer Institute, Wilhelm Hueper and W. C. Conway: "Cancers of all types and all causes display even under already existing conditions, all the characteristics of an epidemic in slow motion" (1964, 17). This unfolding crisis, they asserted, was being fueled by "increasing contamination of the human environment with chemical and physical carcinogens and with chemicals supporting and potentiating their action" (1964, 158). And yet the possible relationship between cancer and what Hueper and Conway called "the growing chemicalization of the human economy" has not been pursued in any systematic, exhaustive way (1964, 158).

The environment, it seems, keeps falling off the cancer screen. The circumstances surrounding the birth of the Illinois State Cancer Registry is a case in point. The registry came into being when the Illinois Health and Hazardous Substances Registry Act was signed by the governor in September 1984. As implied by its name, this state law was intended to "monitor the health effects among the citizens of Illinois related to exposures to hazardous substances in the work place and in the environment" (Illinois State Cancer Registry 1989). Accordingly, the registry system was to collect information not only on the incidence of cancer among the Illinois populace but also on their "exposure to hazardous substances, including hazardous nuclear material," thus prompting public health studies that would relate "measurable health outcomes to environmental data to help identify contributing factors in the occurrence of disease" (1989).

The cancer registry was funded. The hazardous substances registry was not.

Like a thriving child with a stillborn twin, the Illinois State Cancer Registry dutifully acquires information on health outcomes, but this activity now goes on independently of any attempt to correlate health with exposure to hazardous substances.

Two months pass before I visit the cemetery. It is June. Four days of stormy weather have pelted the last of New England's rhododendron blossoms into the grass. The just-awakening roses, however, are luminescent in the streaming rain. In fact, their buds seem to be opening before my eyes.

Time still seems speeded up, as in an old movie when the wind tears the pages from the calendar and the characters leap forward into another season. Cars drive too fast. People walk too fast. Food even seems to cook too fast. I have learned to avoid quickness—like dashing out to the post office before it closes—because sudden movements seem to rush time forward even faster. I am hoping an afternoon in the cemetery will slow the world down again.

I realize immediately that I have no idea where her gravesite is. When last here, I had noticed nothing but the flower-swathed casket and the mound of dirt draped in green plastic. There was some kind of old, severely pruned tree nearby, but I can't recall the species. In my mind, I can see the round bull's eyes of its sawed-off limbs and the humped roots that had pushed away the hurricane fence behind it. I scan the fence line. A row of old basswoods runs along the far side. Finally, I see it: nearly at the end and standing exactly as in my memory. So, it is a basswood. The tree leads me to the rectangle of earth I am looking for. At the top is a plastic plaque coated with wet petals, seed coats, leaf bits, and stems. JeanMarie Marshall. 1958–1995. Finally, everything seems still enough.

Studying cancer time trends is like ascending a glacial moraine in central Illinois. The rise is gradual, steady, and real. What seems imperceptible from the ground— percentage changes that unfold over miles or over decades—is plainly revealed by graphs of the data. In regard to cancer incidence in the United States, we are, in fact, walking on a sloping landscape.

The failure to evaluate systematically the relationship between rising cancer rates and rising exposures to environmental carcinogens is beginning to receive attention. The National Cancer Advisory Board stated bluntly in its fall 1994 report to Congress that a lack of appreciation for environmental and food source contaminants has frustrated cancer prevention efforts. Further asserting that the government has a responsibility to identify and prevent environmental hazards, the board called for a coordinated investigation of industrial chemicals and pesticides as causes of cancer.

Recent analyses of cancer registry data have made the need for such a coordinated investigation more urgent. Using data unavailable to Rachel Carson, the public health researcher Devra Davis and her colleagues have analyzed U.S. cancer patterns in a novel and revealing way. Rather than simply look at changes in cancer rates over calendar time, Davis grouped people according to year of birth, as well as year of diagnosis, and explored how cancer has affected successive generations. Because data on nonwhites in the early years of the SEER Program are unreliable, she restricted her view to U.S. whites and separated cancers generally believed to be associated with smoking from those not known to be so associated.

Davis found that cancer not tied to smoking has increased steadily down the generations. U.S. white women born in the 1940s have had 30 percent more non-smoking-related cancers than did women of their grandmothers' generation (women born between 1888 and 1897). Among men, the differences were even starker. White men born in the 1940s have had more than twice as much non-tobacco-related cancer than their grandfathers did at the same age. "What this is telling us," Davis says, "is that there is something going on here in addition to smoking, and we need to figure out what that is" (1994, 437).

The grandparents of those born in the 1940s are mostly all dead now. Of all the worries they carried for their baby-boom grandchildren—those riotous offspring who opened the original generation gap—cancer, as I recall, was not high on their list. I say this as a life-long observer of this birth cohort. At one point in my childhood, it seemed that the entire generation born in the decade before mine might die young. By eleven, we all wore metal bracelets engraved with the names of those officially missing in action in Vietnam. About the rest, we heard various dire predictions from adults: perhaps they would all be felled by police truncheons or end up crazed and deafened from rock and roll. But I never heard anyone's grandmother predict that those born in the 1940s would surely undergo chemotherapy regimens in record numbers or that a cancer diagnosis would become as significant a generational marker as patchouli oil.

Nothing slows time down as much as waiting for lab reports. This time I am the patient. In the interior waiting room, dressed in a wraparound smock identical to the ones worn by every other human being who has entered this room, I try to conjure Jeannie out of thin air. Of the ample supply of magazines provided us here, she would choose *Vanity Fair*. Of this, I feel certain. During these moments of waiting, which celebrity interview would she, in her unflagging attempt to bring me up to snuff on popular culture, read aloud to me? And when I drifted into anxious thinking, what clever thing would she say to keep me from floating off too far?

Last summer she waited with me for hours at the ultrasound clinic.

"They had a hard time seeing what they wanted to see," I reported back to her as we finally walked out the door. "And then one of the technicians looked at the image in the monitor and whistled."

She laughed. "You know that ranks right up there with 'Hey, nice tits!' "

My name is called and I follow the doctor down the corridor to her office. Like a defendant studying the faces of the jurors as they file back into the courtroom, I try to read her expression.

It seems my situation today is mostly good, but a little bit ambiguous. The specialists have conferred and would like to recommend I undergo a new type of test, which the doctor explains in clear detail.

"I know this isn't what you wanted to hear," she says, with genuine compassion. "But you don't need me to be your best friend right now."

Time lurches forward again. Where is she?

The rise in cancer incidence over calendar time is one line of evidence that implicates environmental factors. The increase in cancer incidence among successive generations is another. A third line of evidence comes from a close consideration of the cancers that exhibit particularly rapid rates of increase. If we restrict our view to these cancers, what patterns emerge? Who gets these cancers and what do we know about their possible causes?

After lung cancer in women, the three cancers ascending most swiftly in the United States are melanoma of the skin, non-Hodgkin's lymphoma, and multiple myeloma. These are not the most prevalent cancers—breast cancer remains the most frequently diagnosed cancer in women, for example—but these are the ones galloping forward at the fastest rate.

Melanoma accounts for only 5 percent of all skin cancers, but it is the most dangerous kind, accounting for 75 percent of skin cancer deaths. The U.S. incidence of melanoma rose nearly 350 percent between 1950 and 1991, and mortality rose by 157 percent. Between 1982 and 1989 alone, melanoma incidence jumped 83 percent. Each year, about 4 percent more people contract melanoma than the year before, and the average age at diagnosis is going down. The more common basal cell and squamous cell skin cancers are also on the rise. But because they rarely spread to other parts of the body and are seldom life-threatening, these skin cancers are not even included in cancer registry data. Only melanoma diagnoses are recorded in U.S. registries.

A melanoma is a cancer that begins in a melanocyte, a cell type that surely serves as the excuse for more wars, social strife, injustice, and oppression than any other human tissue. Melanocytes are the pigment-producing cells of the skin. Those who ponder the origins of racism would do well to consider the humble biology of the melanocyte. Comprising only about 8 percent of all skin cells, the melanocytes appear in microscopic cross section as dark, delicate shrubbery. They are surrounded by Langerhans cells, which migrate up from the bone marrow and play a role in immunity, and by keratinocytes, layers of flat stepping stones that comprise 90 percent of our epidermis and produce a waterproofing protein. The melanocytes' slender branches extend between and around the keratinocytes and deliver to them the molecules of melanin they cannot produce on their own. Once inside the keratin-ocytes, the blackish-brown granules float to the surface and form a sunlight-absorbing cloak that lies over the fragile chromosomes inside the nuclei. Exposure to ultraviolet radiation—that high-energy wavelength lying just below violet in the spectrum of visible light—stimulates the melanocytes to make more melanin and in darker shades. More grains of melanin are dispatched to the keratinocytes. Therefore, we tan.

Everyone, regardless of race, has approximately the same number of melanocytes. Differences in skin color represent differences in the amount of melanin produced. Not everyone, however, has an equal chance of contracting melanoma. Incidence among whites is ten times higher than among blacks. Among white men, the disease more often originates in a melanocyte located somewhere on the trunk of the body; among white women, on the lower leg. When a melanocyte becomes cancerous—multiplying out of control and, if undetected, seeding itself in deeper and more distant parts of the body—its pigment-producing activities do not stop. The dark interior spaces of the body, where rays of sunlight never penetrate, thus become filled with black tumors that go on crazily producing molecules of light-shielding melanin with no companion cells to receive them.

Melanomas are clearly associated with exposure to ultraviolet radiation (albeit in a complicated way that is a matter of some debate), and here is where individual behavior and changes in the global environment come together. Basal and squamous cell cancers, which arise from keratinocytes, appear to increase in proportion to one's cumulative lifetime exposure to sunlight. Melanomas, by contrast, are thought to be initiated by acute exposures, such as a bad sunburn in childhood. In essence, the cells designed to protect us from the chromosome-breaking effects of the sun are themselves damaged by an overdose of the very element they strive to shield us from. Decades later, another insult of some kind causes wild cell divisions within the damaged melanocyte to commence. A melanoma forms. A borderline is crossed. This second event may be more sunlight, but it may also include exposure to certain chemicals. Excess rates of melanoma are found in rubber and plastics workers, as well as in those employed in electronics and metal industries.

The accelerating incidence of melanoma means exposure to ultraviolet radiation is probably increasing. This could be happening for two reasons. First, more people are spending more time in the sun. Second, the sunlight to which we are exposed contains more ultraviolet rays. Since the 1974 discovery that earth's ultraviolet-shielding ozone layer is thinning, a growing group of physicians and climatologists have come to believe both forces are at work, especially in raising the risk for future melanomas. The U.S. Environmental Protection Agency (EPA) projects that tens of thousands of additional fatal skin cancers will result from the 5 percent loss of ozone that has already occurred in the stratosphere above North America. Individual behavior also plays a role. Melanomas have been on the rise for many decades—since Coco Chanel first popularized the suntan in the 1930s, according to some researchers. However, the *worldwide* increase in melanoma incidence points to a role for ecological factors. A study published in the *Journal of the American Academy of Dermatology* observes:

> Because of the worldwide increase in melanoma incidence, global factors need to be considered as potentially involved. Stratospheric ozone depletion, allowing more intense UV light to reach the earth's surface, may, in part, be responsible. (Rigel et al. 1996, 839)

Ultraviolet—or UV—light is a strange energy. It is responsible for creating the very layer of ozone that defends us from it. As UV rays stream into the stratosphere, they cleave oxygen molecules in half, creating free atoms of oxygen, which then react with intact oxygen pairs to form little triangles called ozone. Ozone in turn can absorb UV rays, preventing their further passage down miles of deep air to the earth's surface. The ozone layer intercepts some, but not all, of the UV rays beamed at us from the sun.

Chemicals responsible for destroying the layer of ozone that resides twelve to thirty miles above us include the now notorious chlorofluorocarbons (CFCs). They are unlikely culprits. Not carcinogenic or even toxic, CFCs belong to a big family of synthetic, organic, chlorinated chemicals whose other members—DDT and PCBs, for example—certainly are. But however harmless at ground level, a CFC molecule behaves quite differently when wind currents sweep it into the ozone layer. Ultraviolet rays split the CFC molecule apart, releasing a chlorine atom that quickly reacts with a molecule of ozone. The triangle of oxygens falls apart as the chlorine temporarily binds with one member of the triad. An unstable union, the chlorine atom soon shakes free of its oxygen partner and goes on to react with and destroy other ozone molecules. Before the chlorine is finally enveloped by a raindrop and redeposited on the earth's surface, it may break apart some 100,000 ozone molecules.

Less ozone allows more UV rays to beam their way through the atmosphere and down to us. Some will be halted by the veil of melanin spread across our skin for that purpose. But if its absorptive capacity is exceeded—which happens easily to the fair-skinned among us—some rays will penetrate further inside the skin cells until they are absorbed by the DNA strands themselves. If this occurs inside a melanocyte, the resulting genetic damage can place this cell on the pathway toward melanoma. In this way, noncarcinogenic CFCs contribute to rising cancer incidence by intensifying incoming sunlight, thereby making it more carcinogenic.

Non-Hodgkin's lymphoma strikes at another tissue designed to protect us from harmful invasions: the knobby lymph nodes clustered in our throats, armpits, groins, and elsewhere. Our tonsils, the most accessible example, represent a constellation of lymph nodes wrapped in a mucous membrane.

The watery fluid that fills the microscopic spaces between all of our cells is, for all intents, lymph. It does not receive that name, however, until it flows from those spaces, like rainwater from a field, into the creekbeds called lymphatic vessels. The origin of all this fluid is the bloodstream, and when held within that system, it is known as plasma. Each day, about three quarts of blood plasma leak out of the capillaries, swirl around freely, and then drain into the lymph vessels. Eventually, lymph becomes plasma again when it is poured back into blood just at the point where the jugular vein joins the subclavian in their return to the heart. Several tasks are accomplished during the ceaseless transformation of lymph to plasma and plasma to lymph. The identification and destruction of foreign substances is one of them. Lymph nodes, scattered along the lymph vessels at various intervals, are honey-combed with a diverse array of cell types specialized for immune response. As the fluid is channeled through the nodes' intricate meshwork, alien life forms are trapped and killed. Lymph nodes can also send immune-responsive cells forth to circulate in other territories of the body.

Because the lymph system also serves as a highway for runaway cancer cells of all kinds, lymph nodes are a significant feature in the cancer landscape. Breast cancers very often spread to nearby lymph nodes, for example. Breast cancer patients are quickly categorized as node positive or node negative, a distinction that depends on whether breast cancer cells, shed from the orginal tumor, have lodged themselves in the lymph nodes beaded between the arm and the trunk of the body. Their presence there indicates the disease has likely dispersed to other, more distant locations.

As a way of measuring the extent of this cancer diaspora, node-positive women are further classified by the number of nodes containing breast cancer cells: 1 to 4 is one kind of identity; 11 to 17 is quite another. "How many nodes?" is very often the first question women in breast cancer support groups ask each other.

But a lymphoma is a different condition. In this case, the tumors derive from lymph tissue itself, not from immigrant cells that have floated in from someplace else. Lymphoma can arise inside a node, or, because lymph tissue is diffused through-out the body, it can originate almost anywhere elsewhere—in the spleen, for example, or even in the skin. Non-Hodgkin's lymphoma (NHL) is therefore a collection of diseases, in contrast to the very specific and highly curable lymphoma called Hodg-kin's disease.

While the incidence of Hodgkin's disease has declined modestly over the past two decades, non-Hodgkin's lymphoma has shot up—approximately tripling in incidence since 1950. This increase is evident in both sexes and within all age groups except the very young. Non-Hodgkin's lymphoma is also far less curable than Hodgkin's disease. Jackie Kennedy Onassis was killed by one of its most malignant incarnations.

AIDS has contributed to some, but not all, of the increase in non-Hodgkin's lymphoma. A small but significant percentage of AIDS patients are diagnosed with lymphoma, which for many causes death. However, the steady upward momentum of non-Hodgkin's lymphoma incidence in the United States was already under way decades before the AIDS epidemic sank its teeth in.

Lymphomas do seem to be consistently associated with exposure to synthetic chemicals, especially a class of pesticides known as phenoxy herbicides. These synthetic chemicals were born in 1942 as part of a never-implemented plan by the U.S. military to destroy rice fields in Japan. The most famous phenoxy is a mixture of two chemicals, 2,4,5-trichlorophenoxyacetic acid (2,4,5-T) and 2,4-dichlorophenoxyacetic acid (2,4-D). This combination is called Agent Orange, and it was used between 1962 and 1970 by U.S. troops to clear brush, destroy crops, and defoliate rainforests in Vietnam. The military career of phenoxy herbicides was thus revived.

Linked to miscarriages and contaminated with dioxin, 2,4,5-T was eventually outlawed. By contrast, 2,4-D went on to become one of the most popular weed killers in lawns, gardens, and golf courses, as well as in farm fields and timber stands. It has been marketed under a schizophrenic collection of trade names: Ded-Weed, Lawn-Keep, Weedone, Plantgard, Miracle, Demise.

Evidence for an association between phenoxy herbicides and non-Hodgkin's lymphoma comes from several corners. Vietnam veterans have high rates of non-Hodgkin's lymphoma. So do farmers in Canada, Kansas, and Nebraska who use 2,4-D. Studies show that the risk of lymphoma to farmers rises with the number of days per year of use, the number of acres sprayed, and the length of time they wear their "application garments" before changing clothes. In Sweden, exposure to phenoxy herbicides was shown to raise one's risk of contracting lymphomas sixfold. In a comprehensive review of the topic, the National Cancer Institute scientists Sheila Hoar Zahm and Aaron Blair concluded:

> NHL is associated with pesticide use, particularly phenoxy herbicides. Exposure to phenoxy herbicides is widespread in the agricultural and general populations. The use has increased dramatically preceding and during the time period in which the incidence of NHL has increased, which could explain at least part of the rising incidence. (Zahm and Blair 1992, 5487s)

Similarly, an 812-page study of herbicide-exposed Vietnam veterans conducted by the Institute of Medicine offered the following terse opinion:

> Evidence is sufficient to conclude that there is positive association. That is, a positive association has been observed between herbicides and the outcome [non-Hodgkin's lymphoma] in studies in which chance, bias, and confounding could be ruled out with reasonable confidence.

Dogs also acquire lymphoma. One recent study showed that pet dogs living in households whose lawns were treated with 2,4-D were significantly more likely to be diagnosed with canine lymphoma than dogs whose owners did not use weed killers.

Risk rose with number of applications: the incidence of lymphoma doubled among pet dogs whose owners applied lawn chemicals at least four times per year.

From jungle warfare to suburban dandelions: our ongoing war against plants is now waged on a domestic grid of tiny battlefields. One in ten single-family American households now uses commercial lawn care services, and one in five applies the chemicals themselves. The evidence linking phenoxy compounds to non-Hodgkin's lymphoma is preliminary. No one knows exactly how traces of weed killer find their way into our extracellular fluid as it is funneled back and forth between blood and lymph. Absorption through the skin is considered the most likely route of exposure. No one has explicated the exact mechanism by which these chemicals might alter the cells inside the far-flung network of nodes, canals, and lymph tissues-at-large and thereby set the stage for a lymphoma. No one knows whether phenoxys require interactions with other agents to work their damage nor what proportion of the current rise in lymphoma might be attributed to phenoxy exposure.

Most of us are probably far less exposed to phenoxy herbicides than soldiers, farmers, or even our own beloved dogs who use our lawns for their bedrooms. Nevertheless, the presence of disease in these specific groups is a clue to which we, as readers of a complicated mystery, need to pay attention when trying to determine why non-Hodgkin's lymphoma casts an ever-longer shadow among us all.

Bone marrow is the mother of lymphocytes, the immune cells that inhabit the lymph nodes. Multiple myeloma is cancer of the cells inside the bone marrow that give rise to a particular type of lymphocyte called plasma cells. Its main symptom is horrible pain. As the tumors grow, blood, lymph, and bone marrow are filled with an excess of abnormal plasma cells, which then churn out an excess of abnormal antibodies. The bones themselves, riddled with lesions, begin to fracture. Calcium spills into the bloodstream. Although multiple myeloma is thought to begin with a single mutation in a single cell, tumors created by aggregations of plasma cells are usually diffusely present throughout the bone marrow by the time of diagnosis. The skull is often severely affected.

As with non-Hodgkin's lymphoma, the incidence of multiple myeloma in the United States has approximately tripled since 1950, and the mortality rate is not far behind. As a sign of its rise from obscurity, some cancer newsletters now run announcements for multiple myeloma support groups. (In San Francisco, the third Saturday of each month at the Women's Cancer Resource Center is multiple myeloma day.)

There is much less to say about myeloma than about the other contenders for Most Swiftly Moving Cancer. Much less is known. It tends to stalk the elderly, and blacks are at higher risks than whites—but for unclear reasons. Because the presence of abnormal antibodies in the blood and urine provides a definitive diagnosis, the registration of multiple myeloma is considered to be particularly accurate.

Exposure to ionizing radiation is recognized as one probable cause. U.S. radiologists exhibit excess rates of multiple myeloma, as do survivors of the 1945 atomic bomb blasts in Japan. Some evidence suggests that workers in the nuclear industry also have increased risks for myeloma.

In contrast to solar radiation, such as UV light, ionizing radiation has energy sufficient to penetrate the skin's surface, stream through the soft tissues, and in some cases, enter the bones themselves. Released when atoms are split, ionizing radiation is so called because it alters the molecules through which it passes, knocking away their electrons and creating electrically charged particles, or ions. Because of this property, ionizing radiation is classified as a known human carcinogen at any exposure level. When the atomic modifications induced by radiation involve molecules of DNA, tucked as they are into the nucleus of every cell, cancer-inducing mutations can result. Alternatively, radiation can create ions of surrounding atoms, which then bind with DNA to create mutations. In either case, our chromosomes are equipped with DNA repair mechanisms designed to detect and correct such problems, but it is a system that can be overwhelmed and overpowered. Multiple myeloma is one of the most recent cancers to be linked with exposure to radiation; the lag time between exposure and diagnosis is much longer than that for other cancers caused by irradiation of the bone marrow.

Multiple myeloma is also associated with exposure to a variety of chemicals—metals, rubber, paint, industrial solvents, and petroleum. Farmers and agricultural workers exposed to pesticides and herbicides have higher rates of multiple myeloma than the general population. Multiple myeloma is on the rise in all major industrialized countries. But the parallel increase among both sexes argues against a purely occupational cause. According to one researcher who has examined multinational mortality trends, the patterns of multiple myeloma among generational cohorts suggest a general environmental exposure of some kind, common to all industrialized countries, which would have begun increasing at the turn of the twentieth century.

Other researchers urge an investigation of one very specific industrial chemical: benzene. Consisting of a simple ring of six carbon atoms, benzene is used as a solvent in which other petrochemicals are dissolved, as an additive to gasoline, and as a raw material for the creation of synthetic materials including certain foams, plastics, and pesticides. It is a ubiquitous pollutant of outdoor and indoor air and a common contaminant of drinking water. Benzene can pass through the waterproofed layer of our skin and thus seep into blood upon direct contact; it also evaporates quickly and can be easily inhaled.

Benzene is a suspect in myeloma because it is a known offender in a related crime, namely, leukemia. A proven bone marrow toxin, benzene alters the cells of the marrow that give rise to leukocytes, or white blood cells. Could the same toxin also preside over alterations in the marrow's production of plasma cells? According to the U.S. Agency for Toxic Substances and Disease Registry, "Although this is plausible, no scientific proof of a causal relationship exists." The question thus becomes, Is anyone looking? (1992, 7).

Bone marrow. Lymph nodes. Skin. From the body's dark tunnels to its sunlit surface, cancers of all kinds are presenting themselves with increasing frequency. Melanoma, lymphoma, and multiple myeloma are simply traveling at especially high velocities.

*

A month before her death, Jeannie initiated a massive housecleaning project. She reorganized all her files, returned books, gave away clothing. Waiting for me on her kitchen table one morning was a stack of medical papers, department of public health reports, press releases, and newspaper clippings. They were her collection of articles about the cluster of cancer cases in southeastern Massachusetts, where she grew up.

"I thought you might want them for your research."

"You don't want to keep these?"

"You take them."

Eighteen months after Jeannie's death, I finally read them—prompted by the release of a new study confirming the patterns documented by the previous ones. Jeannie's cancer is not included in any of these these studies, which concern sharply rising leukemia rates in five neighboring towns during the 1980s and their possible relationship to documented radioactive releases at the Pilgrim nuclear power plant— the result of a fuel rod problem—ten years earlier. While no firm cause-and-effect relationship has been established, meteorological data indicate that coastal winds may have trapped the airborne radioactive isotopes and recycled them within a five-town area. "Individuals with the highest potential for exposure to Pilgrim emissions . . . had almost four times the risk of leukemia as compared with those having the lowest potential for exposure."

Although one of the towns is her own, Jeannie's cancer was far too rare for the case-control comparisons made here. Her cancer has no known cause, and cancer registries do not track its incidence. I will not find her here.

REFERENCES

Agency for Toxic Substances and Disease Registry. 1992. *Case Studies in Environmental Medicine: Benzene Toxicity*. Atlanta: U.S. Department of Health and Human Services, 1–32.

Davis, Debra L. 1994. "Decreasing Cardiovascular Disease and Increasing Cancer among Whites in the United States from 1973 through 1987: Good News and Bad News." *Journal of the American Medical Association* 271: 431–37.

Hueper, W. C., and W. D. Conway, 1964. *Chemical Carcinogenesis and Cancers*. Springfield, IL: Charles Thomas.

Illinois State Cancer Registry. 1989. *Cancer Incidence in Illinois by County, 1985–1987*. Springfield, IL: Illinois Department of Public Health.

Landrigan, Philip. 1992. "Commentary: Environmental Diseases—A Preventable Epidemic." *American Journal of Public Health* 82: 941–43.

Rigel, D. S., et al. 1996. "The Incidence of Malignant Melanoma in the United States: Issues as We Approach the 21st Century. *Journal of the American Academy of Dermatology* 34: 839–47.

Zahm, S. H., and A. Blair. 1992. "Pesticides and Non-Hodgkin's Lymphoma." *Cancer Research* 52: 5485s–88s.

Zahm, S. H., and S. S. Devesa. 1995. "Childhood Cancer: Overview of Incidence Trends and Environmental Carcinogens." *Environmental Health Planning* 103 (supp. 6): 177–84.

QUESTIONS

1. What are cancer registries, and what types of statistics do they collect?

2. Explain the similar problems that confront efforts to track patterns of cancer incidence and patterns of ecological change.

3. What three types of evidence implicate environmental factors as causes of cancer?

Chapter Eighteen

A Cancer Death

Martha Balshem

Have we reached such a point in our "health-conscious" society that every individual who suffers an illness classified as "preventable" must bear the burden of responsibility for that illness? Why isn't it possible to just get sick without it also being your fault? . . . We seem to view raising a cheeseburger to one's lips as the moral equivalent of holding a gun to one's head.

—Paul R. Marantz, "Blaming the Victim: The Negative Consequence of Preventive Medicine," *American Journal of Public Health*, October 1990

Like, they'll say he smoked and drank and that's why he got cancer. And I'm trying to say all the research isn't right.

—Jennifer, the patient's wife

Explosions in chemical plants surrounding Tannerstown are common. But are they related to incidences of cancer? The medical case history I will discuss involves the death of a Tannerstown man with metastatic pancreatic cancer. The conflicts involved are common in clinical oncology and may be experienced between physicians and patients of any social background. But in this case, as in the community setting, the issues are presented in an elaborately insightful and unusually outspoken way by a Tannerstowner, the wife of the patient.

This medical case history is more than the story of a cancer death. It is also the story of a dispute between a Tannerstown woman and her husband's physicians. This dispute was probably not experienced coherently as such by any one of the many physicians involved in the case. In the eyes of the patient's wife, however, the dispute is clear; it concerns control of decisions, information, and the official record of her husband's illness and death. The climax of the dispute occurs over the issue of what is legitimately recognized, in the medical record, as the cause of his cancer and his death.

In constructing my narrative of this case, I have drawn on three texts: the retrospective narrative of the patient's wife; the written medical record, obtained, with permission, from the hospital in which the patient died; and general reflections, again retrospective, from one of the attending physicians on the case. This multiplicity of texts complicates the story and renders the characters as too complex to fit into easy analytical categories. In the end, only a multiplicity of readings will fully represent my view of the case.

Presentation of the Case

John, the name I will give to the patient in this case, died at the age of forty-two. The direct cause of his death was pneumonia; the antecedent cause was pancreatic cancer, metastatic to both lungs and to numerous other areas of the body. The main character in the story to follow is his wife, whom I shall call Jennifer.

I first met Jennifer four years after John's death, when she agreed to be interviewed for my study on local perceptions of Project CAN-DO (a five-year Cancer-education program [1983–1988] in seven Philadelphia neighborhoods). Over the next several years, I visited her on two more occasions. Between these visits and telephone conversations, I have spoken with her for more than ten hours.

When we first spoke, Jennifer lived in Tannerstown, where her family had lived for three generations. When she and John were first married, they had left the neighborhood. But several years later, her parents purchased a bar in the next neighborhood and moved to the apartment above the bar. Her mother, she told me, wanted her close by and offered her the house in Tannerstown. So Jennifer and John moved into the house that Jennifer had grown up in, and they raised their children there. Jennifer found that a lot of people her age had moved away from Tannerstown for a time but were now returning. Prices for housing elsewhere were sky high, and there was something missing in those other neighborhoods, anyway.

> I think most of Tannerstown is family. . . . My aunt lives here. Oh, all my relatives lived here, but eventually they died off. My cousin lived on B Street. He just moved. He just moved this year, but he lived there while my kids were growing up. It's—the girl down the street, I know her mother-in-law. Lives a little ways away. So it's—the girl over there is cousins with the girl over here. [*Laughs*] But I like that. You know, I'll tell you—I like it. They really help me with my kids. If I want to go out or something, my kids know they can't party here because my neighbor has my key and if it gets too loud in here she's allowed to walk right in. . . . I don't even tell her. She just—as soon as she hears some banging, she runs in and says, "Don't you ruin this house! Your mother—" They are those type of people. . . . And even my, when they lived here, um, they shared everything. They said they remember the parties in their neighborhood. Just neighborhood parties, for a Friday night out. You didn't have money. So they said—he was the first one with a TV. So they used to come in here and watch TV. You know, it was that type of thing, years ago, see. And a lot of them remember my mom and dad, so they were really nice to me when I moved here.

At the time of his illness, John was employed at a metal-working plant located not far from Tannerstown. Photographs and family memories show a quiet family man, slightly built, with short blond hair and a kind expression.

Onset of Illness

Jennifer was an informant with an agenda. I began our first interview with my standard introductory questions. But when we got to the subject of cancer treatment, she began to tell her story.[1] (My own comments are bracketed and identified with my initials, MB.)

> I feel some of the hospitals are more educated than others. I didn't realize it at the time ... but I think they should teach people because I know when my husband got sick, I knew nothing. I didn't even know where to go. And you know, if you hit the right doctor, I think you have—not a better shot—I think timing is what it is—but, uh, maybe you're more pleased, let's put it that way, with the results. And I hit a lot of wrong hospitals.

At the onset of his terminal illness, John was without a primary-care physician.

> He just really didn't have a doctor. Well, he had one, but he died, which is ironic because the guy would have known him well enough, I think, to know that it wasn't his nerves. The new doctor felt it was his nerves. Yeah—and you get that all the time. [MB: Holy crow! Like sometimes if they don't know what it is, they might say that.] Well, I could see them believing it. Now I can. I thought it was his nerves. [MB: So he died of cancer?] Yeah. Yeah. I thought it was his nerves. My neighbor next door, being older, said she thought it was cancer. And she kept saying to me, "How's your husband? How's your husband?" I'd say, "Oh, God, he's being a pain in the ass this summer. I don't know what's wrong with him! He's not eating. He's not doing this—" You know. And of course, nobody wants to say an opinion, and I guess with age, you get to know or recognize things. Maybe she's seen people. I don't know. And then when everything came out, she said, "I had a feeling it was serious." She wasn't sure why, she said, but she'd known us like twelve years, and she said she saw him losing the weight so fast, and I didn't. [MB: Because you saw him every day.] I guess, yeah. And then when his friends saw him—they hadn't seen him for a while and we'd bump into them, people would look and say, "What the hell is wrong with you?" And then I'd start thinking, you know, as more people and more people approached you and said it, you know. And one day I—I had to really step back and look and I thought, "He is sick."

In late September, John developed diarrhea and began to vomit constantly. He was admitted to Hospital A, a fairly good community hospital, apparently as an emergency patient, and was put on intravenous therapy for rehydration. The physician they had gone to, who had originally diagnosed the problem as nerves, now wished the patient to have tests. But this physician did not have privileges at Hospital A, so John was transferred to Hospital B, a weaker institution. The following is from the Hospital B discharge summary:

> This is a 42 year old Caucasian gentleman who was admitted with weight loss of about 10 lbs. since July of this year. He had abdominal pain preceded by right loin pain since

June of this year. He had been vomiting for the last five days prior to admission and had diarrhea for the last two weeks. He was well until June of this year when he was ... afflicted with pain in the right kidney region. He also had a muscle pull and trauma to the right lower chest and attributed this pain to that trauma. However, the pain continued and it moved to midabdominal region. He consulted the doctor at his place of work who was treating him for muscle pain. Then he went to his family physician who treated him with antacids, Ativan and Librax but this was to no avail. ... PAST MEDICAL HISTORY: Positive for seizure disorder [seven years ago] for which he had extensive work-up but no treatment was prescribed. It was ascribed to possible alcohol intake. SOCIAL HISTORY: Smoked 1½ packs for the last 22 years and had not taken a drink in the last 3 or 4 months although he used to drink heavily in the past. ... PHYSICAL EXAMINATION: ... Diagnostic impression on admission was dehydration, vomiting, diarrhea, abdominal pain, weight loss, rule out malignancy, and rule out peptic ulcer disease. LAB DATA: ... of note was the presence of questionable antigen CEA [carcinoembryonic antigen]. On 10/5 it was 22.4 and on 10/11 it was 31.7 and it was repeated and report was pending at the time of discharge. Urinalysis was normal on two occasions. Barium enema normal. Dorsal spine x-ray—no bony abnormality in the dorsal spine. Ultrasound of abdomen and kidney normal, limited visualization of the pancreas which was unremarkable. This pancreatic study was done twice and each time was reported as normal, no suggestion of any pancreatitis.

Ultrasounds and x-rays taken at Hospital B were read at Hospital C, a teaching hospital with better facilities. The relevance of this will be seen below.

From the time of John's admission, physicians at Hospital B considered the possibility that he might have cancer. The note to "rule out malignancy" indicates that they had planned further tests to rule out or possibly establish, if only through a process of elimination, a diagnosis of cancer. Also noted is the elevation of John's CEA level. The presence of this antigen in the bloodstream signals the possible presence of a malignancy. Usually, medical care in the United States would at that point involve open discussion with the patient regarding the probability of cancer. Appropriately, Jennifer's view of John's stay at Hospital B focuses on this issue of diagnosis. The questions that linger in her mind, years later, concern control of information. How much did they know? Did they diagnose him as having cancer? Did they do so and not tell her? Or not tell John? Or did John know and not tell her?

> So he went to [Hospital B], and of course the tests take so long. He was in there about three weeks. Apparently—I don't know the whole story. Apparently, they must have said something to him, and he just wasn't saying anything to me. [MB: So they diagnosed him at (Hospital B)?] Nothing definite. Nothing said to me, but something must have been said to him. When I was talking to this doctor later, I said to him, "He has cancer." And he said, "I told him that." And of course I wasn't going to argue with my husband. He was sick then. And I didn't believe him. I said, "You couldn't have told him," I said, "because I think we're close enough—he would have told me." And he said, "I don't know what to tell you."

After three weeks, Jennifer decided to have her husband transferred to Hospital D, a small suburban hospital with limited resources, where he could be treated by her cousin's doctor. Their original primary physician questioned this: "He said—well, first he said, you know, 'Why are you transferring him?' I said, 'Well,' I said, 'I just think he has something,' I said, 'and you—' Plus, I was very upset."

The unfinished sentence marks a point at which Jennifer hesitates to verbalize, even now, the criticism she would level at this physician. Later, she describes her motive in seeking care from her cousin's physician: her cousin had described him as being easy to communicate with. This implies a finishing of her sentence: "I just think he has something," *and you are either not seeing it, or you are shutting me out of that knowledge.*

Another point of concern to Jennifer was the length of time it had taken for the results of the Hospital C radiology studies to be made available. To Jennifer's knowledge, the results were still not available after one week. It is not clear whether these results had in fact been known earlier to the Hospital B physicians, or to John, but they were not made known to Jennifer, who was trying to make informed decisions about her husband's care.

> I was really upset because it was like over a week and they didn't get the x-rays. That's why I really had him transferred ... over a week and they, nobody was getting any results back. So I just felt they were sort of pushing me off. You know, when we decided to transfer to [my cousin's] doctor, they told me to bring x-rays, I called [Hospital C] myself, told them the story and they said to me, "It usually takes us 24 hours to get x-rays together." Now, they could have had them in 24 hours. He said, "But—" I had like an 11 o'clock appointment. I called them at 9—he said, "I'll have them ready for you in an hour. Come down and get them." That's what I said to these other doctors: "Why do you do that—like when somebody's so sick?" I said, "Just—" I said, "Over a week you're telling me" ... I said, "I would have picked them up for you," I said, "if that was the problem."

Jennifer's memory of her complaints to the Hospital B physicians is intense. "I said" is repeated after each phrase, as the emotional experience of having said these things is relived.

When she went to pick up John's studies from the radiology department at Hospital C, Jennifer met a person whom she identifies only as the "blood doctor."

> That blood doctor, he was helping me get some of the x-rays together and when I was leaving he said to me, "Don't worry too much." He said, "I really don't think it's cancer." Well, then I thought they were all nuts. What the hell are they talking about? I still didn't think it was—and, just not being educated enough, I guess, I said—um, I just looked at him and thought he was nuts, although he said, "I think it's something with his liver."

The discharge summary for Hospital B shows a concern with the issue of liability, documenting that the patient had willingly and knowingly assumed full responsibility for the discharge:

> Consultation with hematologist was also requested for elevated sedimentation rate and CEA and serum haptoglobin level was ordered along with the ultrasound of the abdomen, the latter two tests are pending at the time of discharge. However, the patient and the patient's family was insistent that the patient be discharged and followed as an outpatient because of financial strains and domestic reasons and therefore, the patient

after being well informed of the consequences of his leaving the hospital and assuming the responsibility so that he has unfailing follow-up as an outpatient by the family physician, hematologist and myself, the patient was discharged.

The radiology report from Hospital C, dated the day after the patient's discharge, includes the following recommendation: "Ultrasound examination of the pancreas or specific CT [computerized tomography] examinations of the pancreas as indicated."

Diagnosis

At Hospital D, the diagnosis of cancer was quickly made clear to Jennifer.

> Yeah—well, as soon as [Hospital D] saw him, two days, they said, "It's cancer." I said, "What?" I said, "These doctors had him for three, four weeks," I said, "and you're telling me in two days." Of course it could have advanced by then, but he said—I said, "How do you know when they didn't know?" They said, "Well, the symptoms are all there."

John underwent two surgical procedures at Hospital D: removal of a mass blocking the lower bowel; and partial removal of a mass encircling and invading the bones of the spine, which caused pain and muscular weakness in his legs.[2] Hospital D physicians assumed that the bowel was the primary site of John's cancer, and his treatment was consistent with this diagnosis.

By this time, Jennifer herself felt in need of medical care. She asked her husband's new primary physician, her cousin's physician, for medication. He refused her request, thereby gaining her respect, because to Jennifer, as to most Tannerstowners, toughing it out without drugs is a positive value.

> I liked him. He wouldn't give me any medication so I liked him. Cause the one time, you know, I said to him, "You'd better give me something." And he said, "Get out of here," he said. "I've been watching you every day." He said, "You're driving too far. I just don't want you on medication." Because I said, "I can't eat." And he said, "That's your nerves." I said, "That's what they thought my husband had!" And he said, "Now don't go getting paranoid!" He said to me, "Really, it's your nerves, now."

Despite her personal satisfaction with this physician, she questioned the quality of care that her husband was receiving at Hospital D.

> Now, [Hospital D] isn't too educated as far as medicine goes. He was zapped out there. He was hallucinating there. And they weren't—[MB: That's awful.] Oh, *was* it awful! My cousin and her husband came once and they walked out—they said—you know! So of course, then I thought, "God! Is this how he's going to be like the rest of his life? On this medication and stuff?" . . . I mean, I couldn't even stand seeing him then. He even said, when he came out of it, "I realized I was doing that stuff," and he said, "but I just couldn't stop myself," he said, "from doing it." One time he was collecting eggs from the ceiling, putting them in baskets. And my kids saw him like that. And my kids said—my one boy said to me, "I'm not going any more."

John was being treated for pain with large doses of methadone, Demerol, and Vistaril. He was hallucinating, and moreover, his pain was not well controlled.

With the diagnosis of cancer in the open, Jennifer faced a decision about one more transfer. Hospital D had sent her husband's tests to Hospital F, a major teaching hospital with an excellent oncology program, for review. Jennifer arranged to have her husband transferred to Hospital F, seeking state-of-the-art care for John. At the time of this transfer, she again had contact with the blood doctor at Hospital C, regarding transfer of her husband's records. He urged her to have John admitted to Hospital E, an institution that Jennifer had not considered. Hospital E is a strong community hospital with a good oncology program.

> He said, because he just felt [Hospital E] was the best hospital then, not [Hospital F]— and that they would try harder, and they would work day and night with him. [MB: So— so then he was in (Hospital E)?] No, I just didn't know what to do at that point. I just already had arrangements for [Hospital F], so I said—I said, these doctors think I'm nuts the way it is—so I said I—[MB: Well, what are you supposed to do? I mean, you have to go by what they're telling you.] Yeah, but in a way now, I wasn't happy—he told me I wasn't going to be happy with [Hospital F].

The State of the Art

In Jennifer's estimation, John received superior medical care at Hospital F. At the time of his admission, he was described as cachectic—that is, he appeared to be wasting away. He was still heavily medicated, in pain, and unable to walk.

> He got to [Hospital F] and they said to me, "How long has he been like that?" I said, "Since his operation." They said, "It's the wrong medicine." So they brought their specialists in and the next day I couldn't *believe* the next day—then I thought a miracle was going to happen. The medication—they were more educated. As I'm saying, each hospital is *so* different. And it's terrible. Why don't they all get together and have meetings once a month and tell each other?

John's pain medication was changed to Dilaudid, with an almost immediate improvement in pain control. A social worker's note two days after admission gives us a rare glimpse of John himself:

> Patient alert, oriented and articulate at time of visit. He appears mildly anxious and is somewhat focused on details of his medical history. . . . He states he and wife feel much more hopeful since change in medical care and definitive diagnosis of problems and that prior to that he was "hanging in there." Denied depression, anger, etc., but states he feels his problem would have been diagnosed much sooner if long-time family doctor was still alive. Patient described numerous incidents in which he diagnosed and "cured" family medical problems, indirectly expressing wish for same now.

This professional judgment entails a subtext of criticism. The patient is "somewhat focused" on the details of a medical history that would almost undoubtedly command anyone's focus; depression and anger are not simply absent, they are denied; and the social worker puts quotes around John's claim that his family doctor cured many illnesses, indicating her distance from what she sees as folk understandings and his naive faith that some doctors are better than others.

Five days after admission, Hospital F physicians had settled on the lung or the pancreas as likely primary sites for John's cancer, with metastases to the intestine and bone. Oat cell carcinoma of the lung was chosen as a working diagnosis. Appropriate chemotherapy was begun, along with intensive rehabilitation therapy to help John regain his ability to walk. After a nineteen-day stay, John was discharged "with a considerable decrease in his pain and a considerable increase in his ability to ambulate, and in a overall much improved status." Outpatient chemotherapy and radiation therapy were scheduled.

Notes from the social worker indicate that both the patient and his wife were coping well, although at one point it is noted that Jennifer "expresses fear about asking physicians for medical information, especially concerning lung involvement, as this threatens her hopefulness he will be okay."

Again, the social worker's note is loaded with interpretations. Who suggested that Jennifer's motivation is to protect her hopefulness—Jennifer, or the social worker? According to Jennifer, her difficulties in communicating with John's physicians about lung cancer are rooted more in the realities of physician-patient communication patterns than in her own internal psychological state. In any case, Jennifer was, as the blood doctor had predicted, unhappy with Hospital F. Her unhappiness was not with her husband's medical care but with the management of medical knowledge by Hospital F staff.

> I wish I would have tried [Hospital E]. If I had it to do over again, I would have tried— if I was more educated and knew what hospitals did what. And I do think a study hospital—although [Hospital F] is a study hospital. But, uh, my idea there was—they were too smart. They were too smart. They didn't want to listen to people. . . . They got him, off the medication . . . That was the only thing I liked at [Hospital F]. The thing I didn't like there was, like I said, I think they're too educated.

"They were too smart"—"they're too educated." These are said with a low, intense tone, and an air of grim and final judgment.

Concerned about the Wrong Things

John spent the holidays at home and experienced some transient benefit from chemotherapy. After about six weeks, however, his general condition had worsened dramatically, with an increase in pain. He was readmitted to Hospital F for terminal care.

At this point, with the grave nature of John's illness clear and the best possible medical care being pursued, the issue of making sense of it came to a climax for Jennifer. Throughout John's illness, she had struggled with issues relating to the control of medical information. Now, she focused on the specific matter of the inclusion and exclusion of information on John's medical chart. Her underlying concern was control of the legitimate medical statement of the cause of John's illness and impending death.

At issue was the following section of John's medical history, taken at the first time of his first admission to Hospital F and now appearing in typewritten form in his chart:

PAST MEDICAL HISTORY: Remarkable for the questionable seizure, may have been alcohol related, no known etiology and no further seizures [for the past seven years]. . . . SOCIAL HISTORY: Patient is married, fairly extensive history of alcohol abuse, none in the last several months. Also there is a 22 year history of cigarette smoking (1½ packs).

Through a staff indiscretion and a Tannerstown connection, Jennifer was told that John's hospital chart identified him as an alcoholic. Medical charts are by law open to the patient and next of kin, but medical practice at Hospital F, as at most hospitals, discourages patients and their families from seeing them. Jennifer, however, was aware of her rights and was determined to voice her challenge.

So I waited a couple days, and the one—the therapist had left it. So I was looking through it and the doctor comes running back, and of course—they can't stop you. But already I'd found the page, you know. So the doctor said, "Oh, I forgot the book." He went to take it. I said. "Oh, what's this on here?" And he, like, looked at me, you know, and he said, "Oh, that's not important." I said, "It is important to *me!*" I said, "Now, you're a study hospital." I said, "You're going to take this report," I said, "saying he's an alcoholic and he smokes and this is what causes cancer." And I said, "Then you wonder why we get upset because the statistics are wrong!"

Jennifer angrily foresees her husband's case being counted as evidence that smoking and alcohol cause cancer. She sees the chart as representing an indictment of her husband's lifestyle and a statement that this lifestyle has caused his illness. Her dialogue with the physician continues:

I said—then I had told him, I said, "That girl thought he was an alcoholic." I said. "She talked to me fifteen minutes," I said, "and diagnosed that from what I said." I said, "She should have asked me how much *did* he drink. She never asked," I said. "And what he was drinking then was," I said, "maybe a case of beer a week." Which comes out to four cans a day. I said, "It's just that the doctor felt the combination might have triggered something off because they couldn't find nothing else." As a matter of fact, he stopped drinking after that. And I said, "I think some of the liquor *kills* these germs in our body if you don't overdo it." And I said, "I drink more than he does. Do I look like an alcoholic?" I was really upset. And he just kept saying to me, "You're concerned about the wrong things."

Jennifer's interest was not solely academic. She wanted the description of her husband as an alcoholic to be taken off his chart.

I said, "I want it off his records." But they never took it off. [MB: They didn't take it off?] No. Then he—then everything was happening too fast, and then I was just tired of arguing with everybody. So—'cause he said, "Oh, I'll have an input put in there." But they never showed it to me.

Jennifer attributes her husband's cancer not to his lifestyle but to his environment. The debate between Jennifer and her husband's physician is the debate between lifestyle and environmental theories of cancer causation.

See, ironically, he got hurt right before this all happened, and I really believed the hurt had caused this. They all said, "No, it doesn't." . . . [MB: He had had like a bad accident or something like that?] Just in work. Just wrenched his back. Just—and they were treating him for muscle spasm, a muscle pull. . . . He was never sick, until that accident. And then he was wearing a thing around the back and he just kept getting worse and worse and worse. [MB: And it was right after that that he started to lose weight.] This happened in July. By October he was in the hospital.

Jennifer's point is not only that her husband's cancer is connected to his accident but that the power inequity between herself and her husband's physicians caused her observations regarding possible causality to be dismissed. At this point in the interview, I asked her directly about attribution of blame:

[MB: When you were talking about your husband and the doctors, um, I thought that— you know, it was as if they were treating you—that it was his fault somehow? Or you know, that they were saying, "Well, it was because he smoked or because he drank" or something like that, you know?] He didn't actually come out and say that. No, it wasn't said. But when you find out that it's on the report—I'm saying to them, "Let's get *all* the facts together. Not just what you want. I want you to know he did get hurt before this." They didn't want to hear that. You know. "Well, how much does he smoke?" I said, "You know, I'll give you all that information, but take the whole surrounding of what it is! It's not just from the smoking." [MB: I see what you're saying. So it was like they had their ideas and they only wanted the information that filled that in.] 'Cause they're smart. They're smarter than us. They know they are. And they feel this other stuff isn't important, but maybe that's why they aren't getting ahead. [MB: So they never asked those questions—so they don't see that there's a pattern or whatever.] Right. And when I say, "Well, this happened"—they go, "Well, that has nothing to do with it."

Jennifer wanted her husband's chart to include the following information: that he had suffered an accident directly before becoming ill; that he had worked in a chemical plant for many years; and that he lived in a highly polluted neighborhood. She sees these factors as additive, along with lifestyle factors and a measure of fate, in total cancer risk.

Of course, this is a bad neighborhood. He died two years after that explosion, which my neighbor across the street died a year after and they told him it was lung cancer. . . . I really do—I feel it's—now, we all eat the same food at this table. He gets it—I don't. But I think he might have too much in his system. I think we all get it and our system fights it off somehow. Now, when things happen to you—he smoked—which—he got more cancer than I did. He worked at a place that had chemicals so he had more. He lives in a neighborhood that has chemicals. . . . A combination is what I'm telling you. If you probably weigh everything out—I don't think you'll ever beat it. I do believe your time is when it is, too.

Jennifer's challenge left a faint trace on John's medical record. Next to the typewritten words "extensive history of alcohol abuse," there is a marginal notation, handwritten and initialed by the attending physician, that reads: "(less than 1 case of beer/week)."

Death

Upon readmission, John was placed on narcotics for pain relief. He began to have trouble breathing and started to hallucinate. Four days later, a family conference was held, attended by the resident, the attending physician, the social worker, and Jennifer. All agreed that pain control was of primary importance. The resident's note states that Jennifer "is aware of critical nature" and that she requests an autopsy at time of death. The social worker's note states that Jennifer would like to bring John home.

The day after this conference, a morphine drip was started. John began to experience acute respiratory distress, with the type of agitation that accompanies this state. The next day, Haldol, a major tranquilizer, was added to his drip. On that day Jennifer, according to the progress report, became "quite agitated" and wanted to take John home. Hospital staff persuaded her to leave him in the hospital for a few more days, to allow the palliative medications to go into action. The next day's notes read:

> [Early a.m.] Wife stayed all nite, as did her sister. Patient-wife are much calmer today. In a.m., plans will be made for homecare.
>
> [9:15 a.m.] No pulse or respiration. Patient expired. No autopsy.

Autopsy

After enduring this long struggle, Jennifer faced one more. She asked that an autopsy be performed. After so much confusion about his diagnosis and anger at the physicians for shutting her out, she felt no sense of resolution at John's death. She reports facing the disapproval of the medical staff:

> I had an autopsy done, and they were very upset that I was having the autopsy done, which I was upset—and I said, "Well, I read a few books and they told me this is why youse aren't getting ahead; that enough people aren't autopsying." They felt he had lung cancer. I didn't feel he had lung cancer. I said he never had symptoms. [Hospital D] told me he had stomach cancer. [Hospital; F] said, "No, he doesn't. It's lung cancer." I said, "No, it isn't. He never was coughing. He was very athletic. He never got out of breath. He was never spitting up blood." He said to me, "Oh, he had the signs. Youse just didn't look for them." [MB: Well, what was it? I mean, what did the autopsy say?] The pancreas. A different type of cancer. They'd never seen it before. I have the autopsy upstairs. Then I was upset about that after the autopsy. I called the doctor up and he said to me—instead of them telling us—he said to me, "Well, what do you want to know" I said, "I want to know what you know. I don't know!" He said, "Well, ask me, and I'll tell you." Now, this was at [Hospital F]. And I said, "I don't know what to ask you." I said, "I can read the report, that it was pancreas." I said, "But I just wanted to know everything."
>
> . . . Well, they don't know anything about it because it's an—they've never seen this type of cancer. That's why the chemo was hard because, uh—the chemo they were giving him was—one type—I don't know what kind. I mean, I have it all upstairs written—but I—like they said to me, "Why do you want an autopsy?" And I said, "I'm not suing." . . . I said, "I just don't feel that's what it was."

Through the autopsy, Jennifer hoped to make sense of her long and troubling struggle regarding medical claims to legitimate knowledge. The question is with her still: did she make the right decisions? The suggestion that it was a difficult case, and the fact that it was misdiagnosed at least twice, provided some explanation, in Jennifer's view, as to why no hospital and no doctor seemed to do well by the patient.[3] But this was small comfort, and the autopsy report lay upstairs, a still-open book at which I was invited to look to help her make sense of the death.

Jennifer's View

Jennifer succeeded in finding the best medical care for her husband. Her perceptions that he was receiving suboptimal care at Hospitals B and D were correct. Through challenging medical authority at these institutions and choosing to have John treated at Hospital F, she found state-of-the-art care for him. But even the state of the art had no place for her voice.

Jennifer's ultimate struggle concerned the exclusion of her own voice from her husband's medical record, the authoritative text on his illness and death. Her struggle reached a climax when she challenged the implication that her husband was an alcoholic and that his cancer was attributable to his drinking and smoking. Her challenge was dismissed. She fought this dismissal because it framed her as an observer, and not an actor, in the drama of her husband's death. She also related the dismissal of her views to wider issues concerning medical authority, control, and social class.

The mechanisms of medical dominance are clear in this confrontation. The physician's response to Jennifer—that she is "concerned about the wrong things"—is a veiled moral judgment. It is the same judgment faced by women who resist cesarean sections and voice disappointment at missing the experience of a vaginal delivery (Irwin and Jordan 1987). Medical dominance dictates not only the accepted standard of medical care but the accepted boundaries of patient opinion and emotional response. By the doctrine of personal responsibility for health, patients and their families are culpable if they fail to stay within proscribed behavioral boundaries. When Jennifer does not do so, she is cast as troublesome and described as agitated. As John is judged responsible for his cancer through his poor lifestyles choices, so Jennifer is judged responsible for having illegitimate concerns. Within this framework, however, John's physician is humane and makes a note in the chart. With that note, Jennifer's voice is heard—and placed in the margins. The typed social history, citing alcohol abuse, remains very much the legitimate record. The note beside it is small, handwritten, an afterthought. Despite Jennifer's challenge, John's chart implies that his death was attributable to his lifestyle and that his lifestyle was recklessly unhealthy. Jennifer resists, but her resistance does not become fact in the official statement of her husband's social history. In Scott's terms, domination is not hegemonic, but it remains a social fact (1985:330).

Thus, the power of medical dominance does not determine Jennifer's thoughts and feelings. The "social sphere" in which she speaks freely appears, as Scott says, small only if we view her from within the medical system. This view of Jennifer is radically partial and shuts out her own construction of the events. John's physicians see each confrontation with Jennifer as a moment of trouble in their management of John's case. Jennifer herself constructs these confrontations as moments within her continual struggle to be an effective agent in her own management of the case. With regard first to medical decisions and then to the form of John's final medical record, Jennifer struggles to be included in the dialogue and for her voice to be considered legitimate. Like most people who are the primary support of very sick patients, Jennifer accepts this conflict with medical authority as a major part of her responsibility in caring for John. Although the constraining forces surrounding her are enormous, she persists in being concerned with her own agenda. She accepts the burden of living with the conflict between her own resistance and the social fact of medical dominance. She continues to charge that her voice is being denied.

There are two primary messages that Jennifer wants to document through John's medical record. First, she wants the medical record to say that John's cancer should not be attributed solely to his smoking and drinking. She wants the record to reflect "the whole surrounding of what it is." To Jennifer, John's record states that smoking and alcohol caused his cancer. Shut out of the record, to her eye, are his accident, his industrial exposure, and the fact that he lived in a highly polluted neighborhood. All these factors do appear in the record—as misdiagnosis, name of employer, and address, respectively. But smoking and alcohol take center stage, and occupational and community exposures are not mentioned in any discharge summary or professional correspondence.

In point of fact, alcohol and tobacco use, as factors that have wide and profound impacts on health, should be noted in any medical history. Both are risk factors for pancreatic cancer, and each is relevant to some of the working diagnoses used in John's case (alcohol to gastroenteritis and nonsignificantly to bowel cancer; smoking to both of these and to lung cancer). But alcohol and tobacco are routinely included in medical histories not only because they are relevant to health. They are also lifestyle factors, and therefore they fit the paradigmatic stance in scientific medicine that translates "social history" into a consideration of the ways in which people bear personal responsibility for their health or illness. It is a model that tends to minimize or discount the social, cultural, and environmental contexts of disease. Jennifer's objection speaks directly to this point. She does not object to the inclusion of smoking and alcohol-use history per se. She objects to the exclusion of other factors that she sees as related and to the ease with which the medical staff labels her husband an alcoholic. In both inclusions and exclusions to the medical record, Jennifer sees wider social and political issues at work. In a sense, the basic issue is her right even to enter into a debate about cancer causality. But the factors that she is told have "nothing to do with it" have everything to do with it for her. She fights for their inclusion in the record.

The other message that Jennifer wants included in John's official record is that the physicians who treated him were not infallible and that their claim to a monopoly on legitimacy is unwarranted. Her specific focus here is the issue of diagnosis—an issue on which there was much confusion in John's case, as there often is in cases of pancreatic cancer. At Hospital F, Jennifer challenged the diagnosis of lung cancer. She was aware, as most people are, that medical science sees a strong link between lung cancer and smoking, and she read the diagnosis of lung cancer as a judgment on John. Her challenge to that diagnosis is a challenge to that judgment. When her challenge was rebuffed, she became motivated to order an autopsy. John's physicians denied that her opinions were legitimate; she looked to the autopsy report to deny the legitimacy of *their* opinions. Through the autopsy, she sought to finish a tension, finally to say effectively: *You are not always right, and I am not always wrong; I should have had a voice.*

In one sense, the autopsy bolsters Jennifer in this feeling, by establishing that the primary site of John's cancer was not the lung, but the pancreas. Hospital F physicians had considered pancreatic cancer as a possible working diagnosis, and complications from metastases to the lungs were the proximal cause of John's death. But to Jennifer, the misdiagnosis demonstrates that it was illegitimate for the physicians to shut her out and blame John for his disease. She was not at the time aware that smoking is also a risk factor for pancreatic cancer—nor does this now make a difference in her reading of the social process behind John's diagnosis. In another sense, however, the autopsy only provided another occasion for Jennifer to be rebuffed. Years later, one of Jennifer's clearest memories is the refrain of the pathologist: "Well, what do you want to know?" To the end, and beyond, her sense of defiance is both bolstered and dismissed. She is left with an enduring conflict.

At root, what Jennifer sought to challenge was the medical profession's control of the meaning of her husband's death. Despite her efforts, his life slipped from her grasp. Then, the final medical judgment on his life remained beyond her control, as well. Expropriating the meaning of his death, medical science expropriated her ability to feel closure, to experience herself as having been an agent for good in her husband's life, and to construct their life together as having been a worthy one and not the cause of his death. In the physician's eyes, the medical record is objective. To Jennifer, it privileges an apolitical frame for a political statement. John's medical record denies politically sensitive assertions about cancer causation, refuses to acknowledge her analysis of medicine as a social system, and constructs her as an agitated wife. Jennifer steadfastly stuck to her interpretations, but this was not without its costs. The autopsy did not bring closure—it served to bring the conflict into clearer focus. Years later, she is unable to put this conflict to rest.

After John's death, Jennifer continued to live in Tannerstown. A few years later, her father died, leaving his bar to her. Although she knew nothing about running a bar, she decided to try to make a go of it. She and her mother traded domiciles, her mother moving back to the Tannerstown house, and Jennifer and her three children moving to the apartment above the bar. To her great delight, she has made a success of the business, and it has renewed her sense of control over her life. But she has not put John's death to rest and still seeks to legitimize her claim that he was not to blame for his cancer. Her agreement to work with me speaks to her continuing effort. Ironically, now that her insights are written down, they gain some measure of legitimacy. Her voice is finally, in a sense, added to the medical record.

The Physician's View

Up to this point in my description of John's case, Jennifer is the only character I have drawn clearly. She is represented by her own voice, which is vivid and intelligent. In contrast, the physicians involved are represented indirectly, by Jennifer's recollections and by the text of John's medical record.

So often, this is the case in medical anthropology. Patients and their families, and lay people in general, are often more accessible, and perhaps less intimidating, to anthropologists—most of whom prefer, in any case, to listen to the voices of the less powerful.

This preference can simplify our view of the medical encounter. It is easy to connect lay critiques of medicine to wider conceptual frameworks dealing with power, politics, economics, and conflict and to trace the labyrinthine manifestations of medical authority through various contexts and cases. Such an analysis can stand, *quod erat demonstrandum*, as an analysis of medicine as a social system. But in such a rendering, both patients and physicians appear only as caricatures. The patient is a hapless victim, and the physician is an embodiment of all-powerful forces. The medical encounter is a series of crude power moves, in which the physician wields a sledgehammer, and the patient cringes or fights back.

In Jennifer's story, we see nothing of the physician's consciousness of Jennifer's challenge to medical authority or of the physician's experience of his or her own power in the climactic confrontation with Jennifer. Nor should we expect to. To begin to understand the physician's experience, we must turn to the physician.

A number of years after John's death, I was able, using information from his medical record from Hospital F, to locate one of the attending physicians on his case. This physician graciously agreed to an interview. I originally wrote the section that follows without giving this physician a name: I realized on a later draft that this was refusing him the humanity that I had automatically thought to assign to John and Jennifer. I have therefore assigned him the pseudonym Dr. Hughes.

When I described John's case to Dr. Hughes, he remembered John but did not remember the confrontation with Jennifer or any details about John's social history. After a brief discussion of the case, I presented Jennifer's critique to Dr. Hughes, point by point. I began with Jennifer's assertion that physicians put in the medical record only that which affirms the views they already hold. The tape recorder begins a few words into Dr. Hughes's answer:

So I think it is true that—that most physicians look for what they—you know, want to expect. And ninety-nine percent of the time we're probably correct in making those assumptions, uh—in this instance, somebody saying that they drink, or a—a family member recounting that a patient drinks a—a—case of beer a week, generally speaking people underreport alcohol intake, and then there are studies to show that. And usually by a factor of two. That is, one-half to two-thirds. So, if somebody is saying they're drinking a, uh, case of beer a week, it's probably two and three cases a week. And that's a lot of beer. . . . As far as, uh, doctors including the work history and exposure history and the environment, I would think most doctors don't because they're not educated in that area; very few doctors have had sufficient epidemiology training to—or public health training—to look or even question that. Unless the patient would bring it out and would talk about some pretty gross exposure. But, ah—or unless a diagnosis would come forward that would—for instance, uh, a neurologic problem that might lead to a suspicion of lead—intoxication which might then lead to very specific questions about working with batteries or working with, uh, lining, uh, uh, you know, brake linings and things like that. So unless one got, you know, some hints from the workup of the patient, we probably don't do a very adequate—environmental—uh, history.

Dr. Hughes accedes to Jennifer's criticism. Yes, he says, we do include the details we most expect to hear. He defends the expert authority of the physician: we are, he says, usually on the right track; and it is known through studies that patients tend to underreport alcohol use. But his defense is not emotional, and with regard to the slighting of environmental factors in medical histories, he simply agrees with Jennifer's point. Although he contradicts some of what Jennifer says, Dr. Hughes is not tied to or motivated by a need to prove her definitively wrong point by point. His tone toward me is patient and not at all defensive. He seems accustomed to admitting his failings and the failings of his profession.

Again, I asked Dr. Hughes about his confrontation with Jennifer. The physician's emotions are often complex, he answered, and each case is different. But then he spoke of the psychosocial issue that would have been, in his view, important for most oncologists in such a confrontation. In the following passage, he offers his interpretation:

You have to re—understand that, yeah, there are frequently confrontations like this. Where, ah, here you can give the person your advice, you can tell them that this is important . . . and, ah, yet she doesn't want to listen to that at all . . . I—I think what— what people, what—what physicians learn in their training, and what they by nature begin to do, is become authoritarian to a certain extent, I mean, you have to tell people rather positively—you *must* take this medicine and you *must* do this and, ah, you know, because this is what's *best* for you. Well, patients, in this modern age interpret that as being—you know, not having their own—authority, not having their own—being in control of their own destiny. And so there's *always* this kind of push and pull between people who, ah, who wanna be in control and those physicians who are trying to outline the best *treatment* for the patient. [MB: Do you think that—that those issues are, um, ah—really especially central in oncology?] Well, maybe more so, ah—obviously it's more serious 'cause a mistake is usually a—a—life threatening, you know, or *fatal* mistake. In fact if you *don't* do the right thing up front that's usually where the fa-fatalities occur. So, ah—[MB: You mean if the—if the *physician* doesn't do the right thing.] Yeah, and the—and the patient. See, if the patient delays, or decides that he— he wants to continue to seek opinions until he finds an opinion that coincides with *his* desires, that may be in fact a *fatal* mistake for him. [MB: Uh-huh. Wow.] So, ah, so, oncologists probably more than most specialists have to become much more, um— [MB: Forceful.] controlling, [MB: Clear, and—] Yeah. And—and—and, uh—and we frequently hear that, you know, nobody di-discussed this with—well, I *pretty* well—I know a *lot* of oncologists, and they usually go out of their way to explain what's going on, and we usually don't use euphemisms like, ah, this could be a tumor or this could be a little lump. We usually talk, it's cancer and you've got to do this, and—and yet patients still seem to not hear what the doctor is saying.

To Dr. Hughes, the issue is not that the physician is too controlling, it is that the physician cannot control enough. He understands the needs of patients "in this modern age" to feel in control but sees this need as an adversary, "push and pull," to his own control of the treatment decision at stake. If the physician fails to control that treatment decision, the result is loss of life. Dr. Hughes stumbles over the word "fatalities," and puts heavy stress on the word "fatal," which is spoken twice. He stresses the idea that fatal mistakes are made in the beginning. Thus, he assumes an extraordinary burden: that of insuring that both physician and patient make the right decisions, at a time when the consequences of those decisions are not yet clear.

To Dr. Hughes, it is this consciousness of mortality that causes the physician to become "authoritarian." Where I choose "forceful," a kinder word, he insists on "controlling." And yet what strikes him, in the end, is that the physician is in the final analysis unable to control—partly, in his eyes, through being unable to communicate clearly enough with patients and their families.

My interview with Dr. Hughes then moved to a discussion of Jennifer's position. I opened the topic by describing Jennifer's frustration in trying to make the right decisions about John's care. Dr. Hughes cut into my description, the only instance in the interview in which he directly and aggressively interrupted me:

> [MB: And she did in fact make a lot of decisions and good ones.] But [MB: I mean, after all—] But why—why *should* she be the one making them? [MB: Um—] You see, I think she's assuming a lot of things that aren't—[MB: Yeah?] [*Pause*] Generally speaking, we feel the patient should make the decisions. Now, obviously family members should be present and should *hear* the information, but usually I—tend to reject—a family member—making the decisions for the patient. Because, ah, after all, it is still the patient that has to go *through* the treatment, or has to bear the results of, uh, of whatever the problem is. So in a sense, ah, that's very presumptuous on her part, that she should be telling her husband what—should happen, or telling the doctors how to treat her husband. [MB: Right. And of course I have no—*no* way of knowing anything, about the patient. I mean, there's a social worker's note here and there about things that he had said, but—] But I'm sure he was very reticent, and very quiet, and probably withdrawn, and—maybe depressed. And, ah, you know, in—in pediatrics it's—it's usual for the mother to answer for the child. And—and, you know, and decision making. But in adults, we generally reject that.

With some vehemence, Dr. Hughes insists that his moral responsibility and primary relationship is to the patient. He presents this as a central tenet of medical practice. Still, this passage must be read in the context of Dr. Hughes's earlier descriptions of the patient-physician relationship as fraught with oppositions and endangered by struggles for control. In rejecting family authority over the patient, the physician experiences a construction of himself as tied to the patient by bonds of common interest. The family is cast as among the many threats to physician control, as being outsiders to the primary patient-physician tie. Again, Dr. Hughes returns to the central nightmare of physician loss of control:

> It's a very complex specialty, it's one that takes, ah, a lot of time but also a lot of detail, ah, applied, eh, small errors or mistakes, uh, frequently are fatal. [*Pause*] They—they don't *appear* to be, right at that moment, but—but they probably add up to be—fatal for the patient, ah, in a short time, or long time.

The beginning of this passage is spoken slowly, word by word, until the words "frequently are fatal." This phrase is spoken in a rush, as one word, with a short pause following. In oncology practice, death is a constant companion. For Dr. Hughes, it is the only important foe. It is, to paraphrase his admonition to Jennifer, the right thing to be concerned about.

In medical oncology, more so than in some other areas of medicine, the concerns of families are an enormous emotional burden, one that is sometimes beyond bearing for the already emotionally exhausted physician. In a way, families and physicians may suffer from the same emotional burden: anger about the patient's cancer. This may be especially so in a case with a poor prognosis. But this anger, if it does indeed animate confrontations such as that between Jennifer and Dr. Hughes, is not felt as a commonality. Adversarial understandings of the medical encounter structure the interpretations of all participants and suggest a target of anger for each. Through such understandings, oncologists and patient's families may often be pitted against each other, in a complex of angry emotions that everyone is too overwhelmed to sort out.

Common angers notwithstanding, Jennifer and Dr. Hughes brought different issues to their confrontation with each other. Jennifer did not want John blamed for his cancer. Because this is her interpretation of the conflict between herself and the physician, she attributed to the physician the point of view that John should be blamed. But Jennifer and Dr. Hughes confront each other at cross-purposes. Dr. Hughes understands that medical records recreate and reaffirm the expectations of physicians, and he sees that this process can reify the view that cancer is caused by patient lifestyles. But to Dr. Hughes, this is not the important issue. It is not what he is defensive about, or the answer he ultimately comes to, or his interpretation of why he pushed Jennifer away. His central concern is in maintaining control, because to lose control is to be responsible for his patient's death.

In this chapter, I have presented a case study involving the cancer death of a Tannerstown resident. Two important questions arise. First, what practical lessons might clinicians and patients draw from John's case history? Second, how are the clinical and community contexts related? In developing answers to these questions, I will offer three separate readings of John's case.

My first reading highlights the commonality between my story of the community and my story of the clinic [told in another section of this book]. The commonality is clear. In both community and clinic, Tannerstowners voice a resistance to professional authorities that would blame them, openly or tacitly, for their cancer. John's case, and the confrontation between Jennifer and Dr. Hughes, may be read as a bit of lived, felt experience that illustrates the specific workings of wider social forces such as medical authority and class discrimination.

When they confront each other, Jennifer and Dr. Hughes both seek to forge a concrete response to the wider fields of power in which they are enmeshed and must act. In both of them, one sees the pain of people who are trying to do what they think is right and important, and to act concretely to defend what they value. Jennifer's struggle is about meaning and the official documentation of truth. To the confrontation with Dr. Hughes, she brings long experience that has taught her that people like her do not participate in the construction of truth. Now, she seeks to be a writer of truth, a writer of the story of her own husband's death. For his part, Dr. Hughes struggles to be an effective physician. His definition of effective action is narrowly constrained by his professional training and what it has taught him: a focus on the patient's body, a focus on life as in danger, a focus on his own responsibility to that body and that life, a focus on staying focused at all costs. Dr. Hughes defends himself against the undoing of focus that Jennifer demands of him. In the final analysis, of course, he holds more power than she does. His voice is recorded as real, and hers is not.

Virtually all patients experience such differences in power between themselves and their physicians. Our picture of life in Tannerstown, Jennifer and John's community, however, suggests that the issues of cancer causation and social-class power differences were linked for Jennifer prior to her confrontation with Dr. Hughes. Jennifer's strong sense of the issue of relative power is colored by the depth of this prior link. Interestingly, Jennifer does not talk explicitly about social class. Her dominant metaphors for social class are education and intelligence. She speaks of the physicians at Hospital F as being "too smart" and as "smarter than us." Thus, on both semantic and strategic levels, she casts a social-class struggle as a struggle about intelligence, knowledge, and truth. In one interview, she phrased it as follows:

> I was never good at asking questions the right way, or expressing myself the right way. And people like that—I guess we'll have a hard time, when you're in the medical field, if they're not with the right doctor that's going to sit down and say, "I know you don't understand, let's go over this." . . . Why do they treat them so different? An educated person comes in there, they *do* treat them different. . . . And it's a way of telling us. "Well, you're not smart enough to understand this." And I did feel that a lot, you know. And, uh, "I shouldn't be wasting my time cause you wouldn't understand it anyway." Yeah, I felt that way a lot.

"People like that" are people who are labeled in school as not intelligent; who are told that they cannot understand things as well; who have less formal education; who are told they are "not smart enough." In this language, we see the insidious and destructive confounding of school success and intelligence that is so common in our society. "People like that" clearly refers to Tannerstowners, or to working-class people in general. Elsewhere in that same interview, I asked Jennifer directly about social class. She drew a blank. She speaks of social class only in terms of a self-deprecating mystification.

What Dr. Hughes brings to the meeting, with respect to wider social forces that shape the interaction, is not class discrimination but medical authority. In this, Dr. Hughes serves as an interesting example. My judgment of Dr. Hughes is that he is not particularly prone to social-class prejudice. In fact, I read him as a person who freely enjoys fraternal relationships with people of various social backgrounds. In my few observations of him with patients, he appeared very open to their suggestions and not defensive about their critiques. But in this first reading of the case, that is beside the point. Dr. Hughes still represents those who are hated for their class power, and he is still committed to authority as a keystone of effective medical practice.

The force of collision between Tannerstown and the cancer science establishment—on a wider level, between the working class and the medical establishment—reverberates within the specific meeting between Jennifer and Dr. Hughes. Such a reading springs from a stance common in critical medical anthropology, in which prior insight into wider political and economic structures of power order perceptions of individual interactions. Such a stance orders the world very strongly and gives the analyst a firm commitment. This appeals to many academics in the United States, where we are fond of conceptual order, are comfortable with absolute heroes and villains, and are currently feeling bereft of a stable place from which to analyze. Thus, Jennifer is our hero, and Dr. Hughes, with perhaps some sympathetic caveats, is our villain. John's case, as noted above, is illustration and brightens up the narrative with a human-interest story. Human experience in the service of conceptual structure: this is indeed an alienated use of lived experience. It also offers us no practical suggestions and only a mechanistic tie between clinic and community.

My second reading of John's case will take us somewhat further. This second reading will be from the perspective of clinical medical anthropology, as defined by the work of Arthur Kleinman (Kleinman 1980, 1981; Kleinman, Eisenberg, and Good 1978). This reading will focus strongly on cognition within the clinic.

Kleinman's perspective—that of a physician-anthropologist widely respected in both anthropology and medicine—is that scientific medicine has great powers to make people well and that the improvement of patient compliance is a good goal for anthropologists who study the clinic. At the center of Kleinman's view is the concept of Explanatory Models (EMs)—that is, the cultural constructs of clinical reality that both patient and physician bring to the clinical encounter. This focus highlights cognitive disparities between patient and physician and defines communication as the basic problem. Kleinman argues for more culturally sensitive clinical practice. He distinguishes between disease (a physical phenomenon) and illness (a cultural phenomenon), like many medical anthropologists, and he exhorts anthropologists to teach physicians to treat both. This microsociological focus is strategic and programmatic, privileging a microsociological outcome: patient compliance.

Kleinman tells us that "generally speaking, the explanatory models of professional practitioners are oriented toward disease while those of the laity are oriented toward illness" (1980:73). This is certainly an accurate reading of the dissonance between Jennifer and Dr. Hughes. Jennifer openly insists on the relevance of social context and is concerned with the social construction of definitions, whereas Dr. Hughes insists on a concern with physical disease. Kleinman's model points us in the direction of these relatively opaque cognitions and suggests that if Jennifer and Dr. Hughes had negotiated an effective communication about what was important to each of them, Jennifer would not be left with so much unresolved emotional pain. This perspective has merit. If Dr. Hughes had stated that environment could have played a role in John's cancer—if he had, for instance, made the simple assertion that cancer causation is complex and physicians are often not sure how it develops in any given patient—it might have reduced Jennifer's suffering. Jennifer herself seems to believe this. Throughout, she represents herself as searching for communication. For instance, at one point, she changes hospitals in a search for a physician who can communicate better with her. At another point, she regrets that she never had an interview with the pathologist and seems to feel that in such a communication, she might have found what she wanted.

Thus, Kleinman offers an answer for one of our questions—that is, what concrete recommendations we might propose for reform of the clinic. Within the confines of the clinic, Kleinman's answer is a good one. With regard to our second question, however—that is, how to tie analyses of the clinic to analyses of wider social contexts—this answer needs amendment.

Kleinman considers the wider context, and states his wider ideal as follows:

> Anthropological medicine and psychiatry would seek to alter the power relationships within the health care system such that care became more patient-centered and the practitioner-patient relationship approximated a *negotiation* between relative equals in which the practitioner provided expert information, but the patient and family retained the responsibility for accepting or rejecting such advice. Again, this change would require preceding external alterations in power relationships in order to be effected. (Kleinman 1981:171; emphasis in the original)

But whether they negotiate a shared EM or not, Jennifer and Dr. Hughes will never meet as equals. As Kleinman acknowledges in this passage, deep change within the clinic needs change outside the clinic. Nothing Dr. Hughes does can change Jennifer's lifelong experiences. He cannot, through better communication, alter the fact of the chemical plant explosions in Tannerstown, which plays a significant role in Jennifer's EM about John's cancer. We cannot move in a linear fashion from a clinical perspective to the wider world. As critical medical anthropology can cause us to miss the specificity of the clinical encounter, so clinical medical anthropology does not provide a conceptual basis, pure and simple, for apprehending social reality outside the clinic, or for connecting clinical reality to wider contexts. This is recognized by many critical and clinical anthropologists, including Kleinman himself.

Community and clinic cannot be tied together in a simple way—either by using the clinic as illustration, or by formulating our goals within the clinic. Lessons, models, and perspectives from one context do not suffice for viewing the other. The analytical edifices we construct as academics are too transcendent, permanent, and consistent, and too rational and lucid across contexts, to fit real life. In the story I tell here, however, there is a thematic consonance between clinic and community— a consonance that neither of the above two readings can catch. What is missing is a recognition of emotion. In the case study, emotion shows us politics and social context as they are experienced on a personal scale. Emotion and cognition are not opposed: as Michelle Rosaldo says, "feeling is forever given shape through thought and ... thought is laden with emotional meaning" (1984:143). Still, without tracing emotions, we miss seeing how social class and medical authority are embodied in individual lives (compare Rosaldo 1984). In the present case history, in which cancer, social class, and medical authority are tied together with fear, anger, and moral passion, emotion illuminates a lot. Emotion, then, will be the focus of my third reading.

One key moment of dramatic emotion in Jennifer's story centers on her insistence that John's medical record not define him as an alcoholic. At this moment, she confronts Dr. Hughes and demands that John's drinking habits be fairly described. During one of our interviews, when Jennifer and I reviewed John's hospital record together, she saw Dr. Hughes's notation "less than 1 case of beer/week" for the first time. When she saw it, she leaned forward intently and I saw worry and effort flood her face.

[MB: Okay, page twenty-seven, it says "less than 1 case of beer per week." So, remember when you said, "I want that written down on the chart"?] That's where he added it? Where does it say—also, "there's a twenty-two-year history—" Here, now, where does it say there that—ah, for drinking, in here—[MB: It says—this is "SH," social history, so—] Oh! [MB: "Patient is married, fairly extensive history of alcohol abuse—" See, they're repeating the same thing.] Yeah. . . . I remember Dr. Hughes rushing in so fast cause he had to go on to his next meeting, wherever he was going, and somehow by instinct I felt, they're looking at me like, "well, you have nothing else to complain about. I know you're upset about him dying, so you're just picking on us." But it was my true feeling. My true feeling was, it wasn't true. If you're ever going to do anything, do the research right. . . . It didn't look like much of a big deal to them. Which I see it wasn't a big deal. They didn't even give it a paragraph. [MB: Right. Like a pinprick. It's not typed or anything.] Right, right, they just gave it a little—which I'm even surprised they did that. I always wondered if it was written in there. But let someone taste a beer—! It's almost exactly what I said, but that isn't correct. That didn't correct anything that I wanted it to correct.

Her words are included in the medical record—but the words do not correct anything. The words do not speak to her "true feeling," which was that John was not to blame for his cancer. They blame him, she says, for just tasting a beer—the smallest transgression, and they blame the patient. There is no emotional closure, no closure in spirit, from the fact that her few words were written in the margin. That does not correct the feelings she is left with.

Another powerful moment of emotion is Jennifer's description of the rudeness of the pathologist. In this moment, the pathologist resists talking to Jennifer about the autopsy. Jennifer has referred to this every time she has told me her story. It clearly continues to haunt her.

I was just very disappointed that I didn't have an interview. He called and said he was busy, he can answer me on the phone. . . . "Well, what do you want to know?" I didn't know. I said, "What *you* know." And he said, "Well, it took me years to go to school and I couldn't tell you that." That hurt.

Again, the injury is emotional, and so is the closure she needs. From the written autopsy report, she has what she wants to *know*: that she was right, John's primary disease was not lung cancer. This reads to her as a factual statement that he is not to blame for his disease. What she does not have is what she wants to *feel*, which is a sense of effectiveness as an agent for her husband, in his life and in his death. The rational fact in the autopsy does not provide that.

I would draw a third example from my interview with Dr. Hughes. The most emotional moment in this interview is when Dr. Hughes expresses the primacy of his own private relationship with his patient. When I say that Jennifer had made good decisions, he interrupts and asks, with emotional heat, "why *should* she be the one making them?" He expresses the view that his primary responsibility as John's physician is to communicate directly with John. He frames this as an ethical issue, trying his emotional vehemence to a "should." The stance he takes, however, is not a timeless ethical stance. Professional mores about open talk to cancer patients have changed through time in the United States, toward more open, direct communication with patients, and still differ from country to country, even within the realm of scientific medicine. But clearly, for Dr. Hughes, emotion and ethical sense are tied, perhaps specifically with the energy of anger, and deeply felt emotions are confounded with ideals to which he holds himself with regard to caring for and about his patients.

At first, this emotion-centered reading of John's case seems unconnected to my first reading, which was concerned with power and wider social forces. The focus on emotion, however, speaks to a central paradox in my first reading: how John's case can be about class and authority, and refer to class and authority, in cognitive and rational terms, so thinly. Emotion is the tie. The emotions expressed by both Jennifer and Dr. Hughes are the specific costs to them of life in the context of wider forces. Neither of them experiences social class and medical authority as such, and these rational structural edifices are not written into the case history verbatim. But many of the emotions they feel *are* class and authority in the form of lived experience. Disputes about class and authority are etched into their lives with pain, depression, exhaustion, anger, and all the rest. They are both transformed in the process.

The third reading also extends the second, by drawing us away from the confines of cognition and EMs. Medical authority is too big to fit into an EM. When we make it fit, we strip it of visible signs of its purpose. Dr. Hughes's EM, for instance, even if reformed as per Kleinman's suggestions, would still be rooted in a complex hope for patient compliance, because therein lies the best hope of cure. Likewise, working-class experience—in this case, growing up and living in Tannerstown—is too big to fit into an EM. When we make it fit, we strip it of the holistic perceptions and experiences of working-class people. Jennifer, for instance, still thinks about Tannerstown and still broods about whether air and water pollution pose health risks for her children. When rich and complex life processes are poured into EMs, they are reconstructed in the process and cleansed for the specific job at hand. "Medical authority" and "social class" become "barriers to communication." Again, this constitutes an alienated use of lived experience. The root problem in the clinic is not a problem of communication. It is a problem of power.

In the end, all three readings of John's case are best taken together. As suggested in the first reading, we see in the clinic what we saw in the community: working-class life experience and the workings of medical authority. As suggested by the second, failures of communications in the clinic are great cruelties, and we can stand in favor of Kleinman's suggestions for reform. But the third reading adds a sensibility through which we can make sense of community and clinic as a whole. In John's case, medical authority is sometimes muddled, and we are barely able to catch a glimpse of social class. Instead, we see the pain of unresolvable emotions. This speaks not to the weakness of class images but to the powerful and subtle forms through which they come into play.

NOTES

1. I present both Jennifer's story and the medical record more or less chronologically. Quotations from both documents are interspersed, but the line of Jennifer's story is preserved—that is, the quotations from the first interview with Jennifer are presented in the order in which they were originally said.

2. The first procedure was an exploratory laparotomy; a constricting lesion of the terminal ileum and ascending colon was resected. The second procedure was a decompressive laminectomy of the eighth and ninth thoracic vertebrae, with gross residual disease noted. The surgical pathology report from the laminectomy specified a diagnosis of a poorly differentiated carcinoma.

3. The autopsy report lists bilateral bronchopneumonia as the direct cause of death, with the antecedent cause listed as "carcinomatosis due to carcinoma of tail of pancreas." Metastasis was extensive. Jennifer's report that "[t]hey'd never seen [that kind of cancer] before" should not be taken to mean that Hospital F had never treated a case of pancreatic cancer; what this remark does refer to is unclear. Jennifer also raises the question of whether John's physicians prescribed the right chemotherapy for him. His chemotherapy program consisted of Cytoxan, Adriamycin, and 5-fluorouracil, a broad and aggressive treatment for an unknown primary tumor, thought most likely to be oat cell carcinoma of the lung. It is highly unlikely that a chemotherapy program designed for treatment of pancreatic cancer would have produced any more than the transient benefit that was produced by the program that was followed. It should be noted that the course of John's disease was not unusual. Pancreatic cancer is typically discovered late and has a very poor prognosis. In the early stages of the disease, there are often no symptoms. The first symptoms to appear—John's back pain and indigestion being common—are very frequently diagnosed as something else, such as happened in John's case. The earliest symptoms may be caused by metastatic tumors. In cancers of the tail of the pancreas, pain often does not appear until late in the course of the disease—usually too late, in fact, for effective treatment.

REFERENCES

Irwin, Susan, and Brigitte Jordan. 1987. "Knowledge, Practice, and Power: Court-Ordered Cesarean Sections." *Medical Anthropology Quarterly* 1: 319–34.

Kleinman, Arthur. 1980. *Patients and Healers in the Context of Culture.* Berkeley: University of California Press.

———. 1981. "The Meaning Context of Illness and Care: Reflections on a Central Theme in the Anthropology of Medicine." Pp. 161–76 in Everett Mendelsohn and Yehuda Elkana, eds. *Sciences and Cultures: Anthropological and Historical Studies of Science.* Dordrecht: D. Reidel.

Kleinman, Arthur, Leon Eisenberg, and Byron Good. 1978. "Culture, Illness, and Care: Clinical Lessons from Anthropological and Cross-cultural Research." *Annals of Internal Medicine* 88: 251–58.

Rosaldo, Michelle Z. 1984. "Toward an Anthropology of Self and Feeling." Pp. 137–57 in Richard Shweder and Robert LeVine, eds., *A Culture Theory: Essays on Mind, Self, and Emotion.* Cambridge: Cambridge University Press.

Scott, James C. 1985. *Weapons of the Weak: Everyday Forms of Peasant Resistance.* New Haven: Yale University Press.

QUESTIONS

1. Why was the attending physician so dismissive of Martha's attempts to influence her husband's diagnosis and treatment?

2. Explain how Balshem uses emotion to analyze the conflict surrounding John's cancer diagnosis. How does she relate these emotions to social class and medical authority?

Chapter Nineteen

Notes from a Human Canary

Lynn Lawson

How did I become chemically sensitive? I am now in my seventies. My first seventeen years were spent in a tiny farming town in southern Wisconsin, at a time when farmers fertilized fields with manure, not chemicals. Everyone ate organic food. There were no petroleum-based pesticides: fly-swatters or sticky pest strips caught flies. Housewives washed clothes with soap, not detergents. Synthetic fabrics, except for rayon, did not exist; there was no "need" for fabric softeners. Women wore cotton or, rarely silk stockings. Perfumes were made from flowers, medications mostly from plants. My teeth were poor; a local dentist put in the first of many fillings, probably silver/mercury amalgams. I remember playing, fascinated, with the mercury from a broken thermometer. A favorite hobby was developing and printing film. At twelve I had my first classical migraine (headache preceded by a visual disturbance called an "aura"—in my case flashing zigzag lights lasting up to a half-hour and gradually filling half of my vision). Except for the usual childhood diseases and an occasional headache of this type, I was reasonably healthy and remained so through four years of college in a small Wisconsin industrial city, where I was a chemistry major at the beginning of the chemical revolution.

In 1946 I moved to Hyde Park, on Chicago's South Side. There I spent many hours on the 55th Street promontory, excited by the dramatic contrast between the Loop skyscrapers on my left, the scrubber-less smokestacks of Gary and Hammond on my right, and ever-changing Lake Michigan in front of me. This was surely the best of both worlds, I thought—manmade and natural. Later, after I bought my first car, in 1953, driving with open windows through Gary and Hammond also excited me. Clichés flowed through my manipulated mind like pollutants through the Little Calumet River: "America's industrial might," "the industrial heartland," "city of the big shoulders." Though I hadn't yet heard of Ayn Rand, I was thinking like her. She is the conservative author who said during the environmental movement of the late 1960s, that we should go down on our knees daily and thank God for belching smokestacks (Rand and Schwartz 1998). Years later, I would remember that one of the smelly, belching smokestacks I passed on my way to work belonged to Argo, a major corn product company.

In the fall of 1947 I worked briefly for Marshall Field's department store in Chicago's Loop. As Christmas approached, I became very aware of the pungent odor from the store's elaborate Christmas decorations—probably fireproofing of some sort. From 1948 to 1954 I worked in an office surrounded by chemistry laboratories, where negative air pressure was carefully maintained in order to prevent radioactive and other substances from escaping. My teeth required more dental work—bridges, root canal work, and more "silver" fillings. My classical migraines recurred; since then I have had perhaps one a year. In addition, however, I began to have fairly severe aura-less headaches, about one or two a month.

From 1955 to 1956 I worked at the Art Institute of Chicago, where walls were usually painted before each new exhibit. The following year I moved to a suburb, where I have lived ever since. Just before and after my marriage in 1959, I painted, stained, or refinished over sixty pieces of furniture, often working in cramped, inadequately ventilated areas. (Who takes those label warnings seriously!) For a year and a half I worked for a major pharmaceutical company, where I walked from the train stop to my office building amid clouds of chemicals. The building was old and had windows that opened, letting in the fumes; my non-aura headaches worsened. Once, miserable with pain and nausea and thinking my condition caused by stress, I drank a beer "to relax," and felt much worse. Another time, after taking a mild tranquilizer "to relax," I drank a martini . . . and discovered that alcohol and tranquilizers don't mix. Both are poisonous.

In the summer of 1961 my husband and I bought our present home at a time when the spraying of elm trees with pesticides (to curb Dutch elm disease) was in full swing. The spraying was not always successful, and we often awoke to the sound of saws. That summer I found several dying robins in our yard. I brought one into our house, hoping to revive it, to no avail; the next year, after reading Rachel Carson's *Silent Spring*, I realized what had killed them. That same year I stopped smoking; I had been a pack-a-day smoker for sixteen years.

The sixties moved on. My own awareness of environmental issues was slowly awakening; not so my awareness of toxicity in humans, or of how potentially hazardous were even my own personal routines. Almost daily, I treated my continuing headaches with Fiorinal, a potent prescription drug containing phenacetin (now suspected of causing kidney damage). My then-internist also recommended buffered (aluminum-containing) aspirin, which I bought often in bottles of five hundred. Though I switched to hypoallergenic cosmetics (I had discovered a severe allergy to orrisroot, a common ingredient in cosmetics and potpourri), I continued to use perfume: after all, one is not completely dressed without a squirt of some highly touted scent. I welcomed the new synthetic fabrics that soon dominated the market: cotton/polyester sheets and easy-care, no-iron clothes washed in miracle-white detergents and no-cling fabric softeners all seemed the housewife's best friend. The miracle of modern chemistry had arrived.

In 1970, having become an ardent environmentalist, I felt fortunate to be able to teach several sections of a special freshman composition course on environment and population at a local university. My enthusiastic students wrote term papers on subjects ranging from the U.S. Bureau of Land Reclamation's depredations to the dangers from auto pollution to the harmful effects of pesticides. Yet I failed to connect my headaches with these issues. In the mid-1970s our house received its only (to our knowledge) indoor spraying of pesticides; in the mid-1980s we had several trees sprayed for cottony maple disease.

After I began full-time work in offices housed in new, tightly constructed buildings, with nearby copy machines, wall-to-wall carpeting, and sealed windows, my headaches worsened. Danger had moved indoors. Finally I went to a well-publicized Chicago headache clinic, where extensive testing failed to reveal anything organically wrong. There, with visits every three months, I was treated with a succession of prescription drugs, including Cafergot, Periactin, Migral, Midrin, Darvocet, Inderal, Decadron, and Miltown. None helped very much. Each time I went to the headache clinic, which was usually packed with sufferers, a nurse ushered me into an examining room, took my blood pressure, and asked a few questions. When the doctor finally came in, he would smile, shake my hand perfunctorily, look at what the nurse had written, and ask, "I've helped you, haven't I?" Foolishly, always the obedient patient, I would reply "Yes." Whereupon he would continue or change the prescription, smile again, and leave the room.

After seven years of this, I asked my present internist, Dr. William D. Kerr, Jr., if he would try to help me with my headaches. He prescribed Verapamil, a calcium channel blocker. It helped for a while, but had bad side effects. Finally, it stopped helping, my body apparently having developed a tolerance for it.

Working at a new job in an office with windows that opened, I was happier. However, I was increasingly bothered by odors of newsprint and cleaning solutions, and by other people's tobacco smoke, perfumes, shampoos, deodorants, and other cosmetic products. The office copier was in a small, unventilated room, and for a while the old stove in the building basement leaked carbon monoxide. I stopped walking down supermarket detergent aisles, because the smells had become obnoxious. I was becoming desperate, awakening early two or three days a week with a headache, nausea, and stiff neck. "Headache" is inadequate to describe my condition: it would be more accurate to call it a "sick-all-over-all-day-misery."

Most days I disciplined myself to get through the office routine somehow, but some days I would have to go home early, able to tell my supervisor only "I'm so sick." When the "headaches" were at their worst, I could not read, sleep, watch television, talk, or think. I was only able to lie motionless on a sofa with an icebag on my forehead. Many nights I tied the icebag to my head to help myself get to sleep. My husband would fill the icebag for me, even in the middle of the night.

I continued to take aspirin in large quantities. Sometimes, in desperation, I would take any other prescription or over-the-counter drug in our medicine cabinet that I thought might possibly help: tranquilizers or antihistamines or diuretics. For the nausea, I tried Coca-Cola syrup, purchased from a local pharmacist, and Emetrol, a nonprescription anti-nausea drug recommended by another pharmacist. Nothing helped very much.

Years of psychotherapy had given me valuable insights into my emotions, but did nothing for my sick headaches. Biofeedback and relaxation techniques, learned at the headache clinic and probably the only truly safe treatment the clinic offered me, nonetheless rarely had any effect. Two sessions with a hypnotist did not help. Fish-oil supplements, thought to help migraine sufferers, appeared to have little effect.

Plunging my head into ice-cold water and keeping it there until I could no longer stand *that* pain offered five minutes of relief. After reading that the plant feverfew had helped migraine sufferers in England, I planted some seedlings in our garden. Finally, unable to tolerate feeling so ill at work, I "retired" from a job that I had enjoyed. I thought that perhaps being away from professional stress would relieve my condition.

It didn't. Finally, in 1986, I went to Dr. Kerr and pleaded, in tears, "I'm desperate. Tell me again about your Dr. Randolph." Fifteen years before, he had told me about an allergist, Theron G. Randolph, M.D., who hospitalized people with "unmanaged" chronic illnesses in special "safe" environmental units, fasted them for four to five days, and then tested them for food and chemical sensitivities. With horror I had rejected the suggestion, for I knew that missing or delaying even one meal had given me excruciating headaches in the past. Would that I had taken that suggestion then! I could have saved myself fifteen years of pain.

Readiness Is All

This time I made an appointment with Dr. Randolph right away. I was ready. By then he no longer had an environmental unit, but his clinic sent an extensive questionnaire, covering my lifestyle as it related to foods and chemicals: What foods do you eat three or more times a week? Do you like/dislike the smell of gasoline and road-tarring compound? What medications have given you adverse side effects? Do you dislike perfumes? If so, which ones? And so on. At my first appointment, one of his nurse assistants took the most extensive medical history anyone had ever taken. Were you breast-fed? Where have you worked? Do you feel better when you travel? Do you feel better indoors or outdoors? The history took three hours and produced seven single-spaced pages. By then, I was beginning to think of possible connections between my health and my exposures, such as the corn product and pill-factory fumes of long ago. Dr. Randolph looked over the history and decided what I should be tested for.

The tests, sublingual and intradermal, given blind (I did not know what I was getting), revealed subtle but unmistakable reactions to several common foods, primarily corn and milk; several common chemicals, primarily phenol, formaldehyde, and ethanol; and several common mixtures of chemicals, primarily auto exhaust and tobacco smoke. The tobacco-smoke extract produced so much choking that the nurses gave me oxygen.

During my six days in the testing room, I heard others' stories and saw others' reactions. I saw a six-year-old girl become agitated and hyperactive after receiving a food or chemical and calm down after receiving a "neutralizing" dose. I heard a patient describe her uncontrollable tears while buying lunch in the building's grocery—tears that stopped as soon as she returned to the clinic's environmentally "safe" surroundings and filtered air. As I read materials in the testing room, especially Dr. Randolph's book *An Alternative Approach to Allergies: The New Field of Clinical*

Ecology Unravels the Environmental Causes of Mental and Physical Ills, (1989) I began to think that the mystery of my illness might now be solved . . . at last.

In the post-testing interview with Dr. Randolph, who had just turned eighty, he kindly but firmly warned me: "Your health is deteriorating and will continue to deteriorate until you find out what is causing this and do something about it." He knew what I had reacted to in the testing, but it was now up to me to find out what in my foods and surroundings I needed to avoid. Avoidance, he told me, is the "cure," not medications that only mask the symptoms. And though informally this condition is known as environmental illness, on the insurance form Dr. Randolph listed my major symptoms: migraine, rhinitis, gastroenteropathy, and chronic fatigue. Doctors and insurance companies tend not to accept the term "environmental illness," though "multiple chemical sensitivities" has gained broader acceptance.

The clinic sent me home with a prescription for aspirin in capsule form, without cornstarch, and elaborate instructions for starting the four-day rotation diet known as the "rotary diversified diet." Based on the biological food families, this diet forbids the person with suspected food allergies from repeating any food until three days have elapsed; the wait is one day if the foods are in the same biological family. For example, broccoli and cabbage are in the same family. If you have broccoli on Monday, you should not have it again until at least Friday, though you could have cabbage on Wednesday. This applies to every single food you eat. It is intended to keep you from developing new food allergies or reactivating old ones.

Sound bizarre? Probably, if you have never heard of it before. That night, determined to get better, I drew up a two-week schedule on a sheet of typing paper and tried to plan this weird diet with whatever foods I had on hand or could buy. The next day I started it, avoiding mainly milk and corn, plus other members of the grain family like wheat, rye, oats, barley, rice, and millet. What on earth does one eat for breakfast on this diet? Somehow I found enough to survive, though I had an extremely severe sick headache for two days, just as if I had been in the hospital environmental unit that I had declined to enter fifteen years earlier. My worst fears were realized, but I now knew that this reaction merely proved the point. As I knew by then from my reading, I was experiencing the symptoms of withdrawal from the foods to which I had become addicted. From that moment on, unbelievable though it may seem, I have never once awakened early with a stiff neck and a headache, my common pattern. Because those headaches ended so abruptly with my change of diet, I knew that they had been food-induced. Before I was properly diagnosed by Dr. Randolph, I used to have a glass of milk by my bed every night, for taking aspirin in the middle of the night. Those middle-of-the-night headaches, I now realized, occurred because of the long time lapse between dinner and breakfast: by five A.M. my body missed its "food fix." The sip of milk came too late.

The Detective Work

After those first two days were over, I began feeling better than I had for years. When I tested myself for corn, however, by eating a bowl of corn meal mush doused with corn syrup, several hours later I developed an excruciating two-day sick headache that confirmed for me my worst food allergy. I knew I had to avoid corn in all forms. Corn is the most common food allergy, possibly because, as corn syrup, corn starch, corn oil, and other less-obvious names, it is present in almost all processed foods and many medicines (just read the labels) Food allergies, according to Dr. Randolph and the many environmental physicians he has trained, are closely related to addictions. When we first react with dislike to a new food, we may still go on eating it for a variety of reasons, one being difficulty in avoiding it. After a while, we may become addicted to it and find it hard to give up. Tobacco is a good example of this phenomenon; those who choke on cigarettes at first try may eventually find themselves addicted.

During the first three weeks I was on the rotation diet, I lost eighteen previously hard-to-lose pounds. My meals consisted of only two or three foods each—dinner, for example, might consist entirely of turkey, sweet potato (without butter), and peas; lunch of almonds and cherries. Because at first I had a hard time finding enough "safe" foods to eat, the pounds melted off. They never have returned; now that I am no longer addicted to any food, it is easier to maintain my desired weight.

Actually, I got better in two stages—first from rotating my foods and avoiding those I was allergic to, second from avoiding chemicals. During the first year after I was diagnosed I continued to have headaches that usually came on during the day. As I became more sophisticated about chemicals, I could usually tell what exposure was causing the headache, though as a "delayed reactor"—one who does not react immediately to an exposure—I found the detective work somewhat tricky.

One night about a month after I was diagnosed, I smelled formaldehyde on my pillowcase. Formaldehyde, a suspected carcinogen, is in many, many products, including synthetic fabrics such as cotton/polyester sheets. I immediately put some old cotton sheets and pillowcases on my bed. A few days later. I detected the same smell in soap from a filling-station dispenser. Now I avoid using public soap dispensers and try to carry my own soap. And once again I hang my all-cotton sheets on lines in our basement after they have been washed in our washing machine with a plant-based liquid soap without fabric softeners. Fabric softeners from neighbors' dryers and others' clothes now smell nauseating to me.

One's sense of smell is invaluable in detecting possible chemical injury from modern synthetic products. As my husband and I gradually removed all or most smelly, suspect, petrochemically-based products from our home, we both began to be aware when someone who had used or was wearing such products entered our house or car. At first, such exposures were likely to bring on a headache the next day (now not so likely). Because of my sensitivity (my husband is aware of the smells but does not apparently react), I became homebound for a time. We stopped going out to public places and to restaurants, where there was no way I could stick to my rotation diet. Dr. Randolph had advised me to stay on the diet the rest of my life. I did observe it rigidly for about a year; now I follow a modified rotation, eating mostly organic (chemically less contaminated), whole, fresh, unprocessed foods. To my surprise, I really like the diet.

Cleaning Up

The greatest shock came shortly after I was diagnosed, when I got the results of a blood test that Dr. Randolph had prescribed. I was in high percentiles for two chemical solvents (toluene and tetrachloroethylene) and five pesticides, including a metabolite of DDT (DDE), heptachlor, and dieldrin—the last three restricted by the EPA since the early 1970s as carcinogenic. I had not known that I had these toxic chemicals in my body. I did not want them in my body. No one had asked for my consent to put them in my body. I had never worked for an exterminator. Our house had been sprayed only once. Where had they come from? Apparently the pesticides were from food and other common exposures, the toluene possibly from printer's ink and self-service gasoline, the tetrachloroethylene from dry-cleaned clothes left in their bags in my closet and from spending hours at a do-it-yourself dry-cleaner in order to save money. I was not alone, I discovered: the EPA's ongoing National Human Adipose Tissue Survey tells us that most Americans have modern toxic chemicals in their body fat and in their bloodstream.

I was determined to reduce my exposure to chemicals. I tried almost everything I read or heard about concerning health and the environment. My husband and I had our unvented gas kitchen stove removed and an electric one installed. We bought air filters that removed chemicals, one for our living room, one for our bedroom, and one for our car. We started driving our car with windows closed, the filter going, and the ventilating system on "recirculate." We tried to leave ample space between our car and other cars' exhaust pipes. My husband began reading our daily paper first, then putting it in a zippered nylon-mesh bag and "baking" it for forty minutes in our electric dryer vented to the outside, which outgassed chemicals in the paper and ink so that I could read it. (Heat activates molecules.) We gave away most of our synthetic clothes and bought only natural fabrics. We had clothes dry-cleaned as little as possible; when we did, we took them out of the bag immediately and aired them out in our basement for months before wearing them. I threw out all aerosol spray cans and other petroleum-based products (wishing that our community had a hazardous-waste disposal system). We switched to plant-based toothpaste, shampoos, deodorants, soaps, detergents, and other cleaning products. I put away my electric blanket (the heated wires give off fumes) and began sleeping under down, wool, or cotton, as pure as possible. Essentially, it was back to the thirties—to the products that I remember my mother using.

Plus a few new products. After reading about them, I bought a small negative-ion generator (positive ions are found in polluted areas, negative ions on unpolluted mountains and beaches), full-spectrum fluorescent tubes for our kitchen, and full-spectrum sun glasses (full-spectrum products maintain the natural balance of sunlight). I tried acupuncture, took vitamin and mineral supplements. (I rarely need prescription drugs and seldom have flu or colds, a welcome change from previous years.) In 1991 a dentist well-versed in nontoxic dentistry replaced my mercury amalgams with a composite to which I tested nonreactive. The only major thing not done is to change our gas forced-air heating system to electric in order to avoid exposure to combustion products (too expensive).

A Runaway Chemical Technology

A year after my diagnosis I was feeling much better. My headaches stopped almost completely, occurring only after unavoidable exposures. Which of these measures helped? I do not know—perhaps all of them. I finally do know, however, what caused my headaches for forty years: I had been slowly poisoned. Maintaining my health still requires considerable avoidance of certain places and situations. It also requires certain precautions, such as using a mask in theaters and traveling with my own bedding. The chemical revolution that I did not even dream of in my college chemistry lab has now taken place, and I will probably have to be careful for the rest of my life.

Samuel S. Epstein, MD, professor of occupational and environmental medicine at the University of Illinois Medical Center in Chicago, has written of a "runaway chemical technology": "With the dawn of the petrochemical era in the early 1950s, annual U.S. production of synthetic organic chemicals was about one billion pounds. . . . By the 1980s [it was] over four hundred billion pounds. The overwhelming majority of these industrial chemicals has never been adequately, if at all, tested for long-term toxic, carcinogenic, mutagenic, and teratogenic effects (Epstein and Briggs 1987, 10).

Actually, of course, it is we who are doing the testing. All of us, but especially those with environmental illness/multiple chemical sensitivities. We are the canaries in the modern chemical mines, the living environmental-impact statements, the not-so-distant early warning systems. Rachel Carson wrote, more than thirty-five years ago (a time when "he" was commonly assumed to mean both "he" and "she");

> The contamination of our world is not alone or a matter of mass spraying. Indeed, for most of us this is of less importance than the innumerable small-scale exposures to which we are subjected day by day, year after year. Like the constant dripping of water that in turn wears away the hardest stone, this birth-to-death contact with dangerous chemicals may in the end prove disastrous. Each of these recurrent exposures, no matter how slight, contributes to the progressive buildup of chemicals in our bodies and so to cumulative poisoning. . . . Lulled by the soft sell and hidden persuader, the average citizen is seldom aware of the deadly materials with which he is surrounding himself; indeed, he may not realize he is using them at all. (1962, 157)

NOTE

This article has been adapted from the introduction to *Staying Well in A Toxic World: Understanding Environmental Illness, Multiple Chemical Sensitivities, Chemical Injury, and Sick Building Syndrome*, by Lynn Lawson, M.A., available for $15.95 plus $3 s&h from Staying Well, P.O. Box 1732, Evanston, IL 60201. Grateful for her improved health, Lynn has devoted the past eight years to coordinating a support group for the chemically injured and publishing its newsletter, *CanaryNews*. She has also devoted herself to avoiding modern chemicals as much as possible, with the result that she is healthier and happier than she was for many years.

REFERENCES

Carson, Rachel. 1962. *Silent Spring*. New York: Houghton Mifflin.

Epstein, Samuel, and Shirley Briggs. 1987. "If Rachel Carson Were Writing Today: Silent Spring in Retrospect." *Environmental Law Reporter* 17: 10–18.

Randolph, Theron G., and Ralph W. Moss. 1989. *An Alternative Approach to Allergies: The New Field of Clinical Ecology Unravels the Environmental Causes of Mental and Physical Ills*. New York: Harper and Row.

Rand, Ayn, and Peter Schwartz, eds. 1998. *The Ayn Rand Column*. Los Angeles; Second Renaissance Press.

QUESTIONS

1. Illustrate how the author's account of the disappearance of her headaches constitutes an example of practical epistemology, as this term was defined by Kroll-Smith and Floyd in chapter 4.

2. Why would some physicians question Lawson's account of her illness? Does it seem plausible to you? Why?

Part 7

Citizen Responses to Contested Medicine

Sociologists and other scholars studying environmental health issues continue to be struck by the magnitude of citizen responses. Although the entire history of occupational and environmental diseases has been largely based on lay detection, it is only recently that this lay detection has garnered widespread public attention. Starting with Love Canal, residents' involvement with disease discovery and with health studies has been central to the entire environmental movement. Laypeople have been astute observers, hard-working researchers, devoted advocates, and effective publicists. They have been willing to take on professionals, institutions, and government agencies that have enormous social power and status. In many cases they have outsmarted people with much more education, and have outmaneuvered agencies with many more resources.

These citizen responses have largely been formulated by people who do not have the higher education and the political activism experience that we often associate with social movements concerned with knowledge issues. Rather, these toxic activists have been propelled by deep concern with actual and potential local health effects, topped off with a frequent bubble-bursting experience of corporate and government failure to understand their problems. These laypeople have formed the backbone of one of the most powerful social movements to develop in the last two decades.

Chapter 20, Stella Čapek's "Reframing Endometriosis: From 'Career Woman's Disease' to Environment/Body Connections," offers us an unfolding process of disease definition. She shows how many women have altered their ideas about endometriosis. Rather than view the disease as the inevitable outcome of emotionalism and of delayed fertility, women's health activists are examining possible routes of environmental etiology. Most interestingly, the support group formed to help women with this condition has wound up pursuing the alternative conceptualization of the disease. Thus, we see the links between social support and social discovery of illness.

Phil Brown's chapter 21, "Popular Epidemiology and Toxic Waste Contamination: Lay and Professional Ways of Knowing" develops the concept of popular epidemiology, based on the Woburn childhood leukemia cluster. This way of looking at community response to toxic waste contamination has been applied to a variety of sites involving citizen activism concerning toxics. Brown delineates a sequence of stages that typically occur in citizen responses to toxics. Through these stages, lay knowledge is shown to be especially valuable in the understanding of environmental health effects.

In chapter 22, "Environmental Movements and Expert Knowledge: Evidence for a

New Populism," Stephen Couch and Steve Kroll-Smith take on many of the same points as Brown. They use as examples Love Canal, multiple chemical sensitivity, and British regulation of 2,4,5-T herbicide. Couch and Kroll-Smith situate these examples in a theoretical framework that focuses on the transformation of scientific knowledge throughout all levels of society. Rather than viewing lay knowledge as primarily an alternative or complement to professional knowledge, they see it in the forefront of the general democratization of science, knowledge, and society.

Chapter Twenty

Reframing Endometriosis
From "Career Woman's Disease" to Environment/Body Connections

Stella M. Čapek

Introduction

In recent years an interesting shift has been taking place in the understanding and definition of endometriosis, a puzzling but common women's disease that affects over five million women in the United States and Canada alone (Nezhat 1987).[1] Endometriosis is being redefined, away from assumptions that locate causality in a woman's emotional state and career/fertility behavior, and toward a more holistic view that explores connections between the human body and a chemically toxic environment. This "reframing," or reinterpretation, is in large part due to the efforts of a grassroots self-help group called the Endometriosis Association (EA), founded in 1980 by women who had the disease.

Dissatisfied with the responses of the medical profession, the EA group came together as laypersons to pool their knowledge, to provide mutual support, and to seek new information about endometriosis. In doing so, they have successfully challenged the practices of mainstream medicine by raising the visibility of cutting-edge research, including a 1992 study of rhesus monkeys that found a statistically significant correlation between exposure to dioxin and presence and severity of endometriosis. Moreover, the EA has created opportunities for interaction among women with endometriosis as well as with researchers and physicians at conferences. In fostering these connections nationally and globally, the EA has become part of the data collection process itself. New findings are helping to remap prior understandings of endometriosis, shifting attention away from a woman's emotional state and life choices (typically viewed through a patriarchal lens) and toward the "social body" that produces an unhealthy environment in which women must function (Freund and McGuire 1995).

The EA is not the only group in recent years that is prodding the medical community into new research and understandings. Phil Brown (1992) coined the concept of lay or "popular epidemiology" to capture the reality of laypersons becoming involved in medical research on health issues (often because experts are seen as unresponsive or prone to misconceptualization). While the degree of lay involvement varies by situation, frequently laypersons collect their own data and have input into research design, cooperating with a few key experts to redirect research. Popular epidemiology affirms a bottom-up flow of knowledge, grounded in the practical, daily lived experiences of ordinary citizens. There are many examples of such persons making discoveries ahead of the medical establishment due to their practical knowledge and their different interests and objectives. There are also examples of lay groups bypassing the medical establishment altogether—after appropriating its language and tools—to affect social policy (Kroll-Smith and Floyd 1997).

It could be said that the EA represents part of a social movement in the late twentieth century to challenge top-down models of medical knowledge, working instead to incorporate lay experience of the body as a valid component of medical knowledge and research. The goal is not to do away with the medical establishment, but to redirect its priorities and, in some cases, to redesign its tools and to critique its deeply imbedded *social* assumptions about patients. Lay epidemiology is particularly likely to surface in "frontier" areas of medicine where the tools of the mainstream medical model are not sufficiently refined, and/or where vested interests distort the practice of medical science. Debates over toxic environments and health issues are one such example of a highly contested area of medicine; the EA has placed itself squarely at the heart of the debate by uncovering and pursuing links between endometriosis and such things as chemical exposure. The EA's successful efforts to redefine endometriosis in terms of a body/environment connection form the subject of this chapter.

To examine the key role of the EA in providing a new interpretive frame for endometriosis, I draw on qualitative materials such as EA documents (newsletters, books, pamphlets, and other publications) and participant-observation which I conducted in the dual role of sociologist and person diagnosed with endometriosis at the 15th Anniversary Endometriosis Association Conference in Milwaukee in 1995.[2] Theoretically, I draw on the concept of "framing" developed in the social movements literature (Snow and Benford 1988; Snow et al. 1986; Hunt, Benford, and Snow 1994). Snow and others have pointed out that people use interpretive frames to organize the flow of data coming their way on a daily basis. Frames are not only used to make sense of one's personal experience, but are used by social movement groups to construct a viable social identity and to influence public opinion. In order to work, frames must "resonate" with the experience of those who use them; by constituting reality in a certain way, they also attribute blame and prescribe an agenda for social change.

My chapter represents an initial effort to draw out the sociological dimensions of the "reframing" of endometriosis as an environmentally related illness. The case presented here offers striking evidence of the importance of lay knowledge in medical research. When lay experience is taken seriously, it challenges existing doctor-patient relationships, it redirects research in terms of topic and style, and it potentially yields new theories and treatments as well as altered popular conceptions of disease. Freud and McGuire (1995) have suggested that we must understand the individual human body as embedded in a "social body" constituted by particular cultural practices. A case study of the EA's efforts is intriguing precisely because one of its major goals has been to render visible the connection between women's bodies and the social body that structures not only medical practice, but human-environment relationships.

The ordering of my chapter is as follows. I begin with a brief description of endometriosis and theories about its etiology. Next, I consider some gender issues relating to the disease that have stood in the way of a timely or accurate diagnosis. I then examine the work of the Endometriosis Association (EA), particularly in the area of environment/body connections. Finally, I consider future directions relating to the project of understanding endometrosis.

Endometriosis: What Is It?

Endometriosis gets its name from the lining of the uterus, or "endometrium," which follows a monthly pattern of building up and breaking down in the menstrual cycle. For women with endometriosis, however,

> . . . tissue like the endometrium is found outside the uterus, in other areas of the body. In these locations outside the uterus, the endometrial tissue develops into what are called "nodules," "tumors," "lesions," "implants," or "growths." These growths can cause pain, infertility, and other problems. (EA 1987)

In most cases endometrial growths are found in the abdominal area, although some have been found elsewhere in the body, for example in the lungs. Although usually the growths are not malignant, they respond to the hormones of the menstrual cycle, causing buildup, breakdown, and bleeding. Since endometrial tissue outside the uterus has no way of leaving the body,

> The result is internal bleeding, degeneration of the blood and tissue shed from the growths, inflammation of the surrounding areas, and formation of scar tissue. Other complications, depending on the location of the growths, can be rupture of growths (which can spread the endometriosis to new areas), the formation of adhesions, intestinal bleeding or obstruction (if the growths are in the intestines), interference with bladder function (if the growths are on or in the bladder), and other problems. (EA 1987)

Common symptoms of endometriosis include pain before and during menstrual periods, pain during or after sexual activity, infertility, heavy or irregular bleeding, fatigue, painful bowel movements with periods, lower back pain with periods, diarrhea and/or constipation and other intestinal upset with periods (EA 1987). Some women with endometriosis have no symptoms, while others have severe, debilitating pain. The pain does not seem to be directly correlated with the amount of visible growths.

Diagnosis of endometriosis is usually accomplished through the technique of laparoscopy, a procedure in which a tube with a light in it is inserted into a small incision in the abdomen (EA 1987). Although a doctor can sometimes feel endometrial implants during a pelvic examination, laparoscopy is recommended to establish the exact location and extent of the growths, and to rule out other conditions with similar symptoms. EA research has documented a disturbing time lag of approximately ten years between onset of symptoms and diagnosis in the United States (Ballweg 1998b).[3]

The causes of endometriosis are still not well understood, but a number of theories concerning the etiology of the disease coexist in the medical literature. These include: (1) a "retrograde menstruation" theory, which assumes menstrual tissue backup through the fallopian tubes; (2) a theory that endometrial tissue is distributed from the uterus to other parts of the body through the lymph system or the blood; (3) a genetic predisposition theory; and (4) a theory that focuses on remnants of embryonic tissues in the adult body (EA 1987).[4] As the EA points out, none of these theories accounts for all cases of endometriosis, and new theories continue to be developed. An effective cure has been elusive to date.

Presently, the etiology of endometriosis is being revisioned. Given its common symptoms, and given the fact that endometriosis is assumed to be responsible for 30 to 40 percent of infertility cases among women (Denison 1997, 4), it is not surprising that it was conceptualized primarily as a reproductive or fertility disease. Now, endometriosis is coming to be understood more broadly as a complex autoimmune disease, of which some of the common symptoms may be "only the tip of the iceberg" (Ballweg 1991). Hand in hand with this concept is a new emphasis on environmental factors such as the chemical dioxin and PCBs (polychlorinated biphenyls) and PVCs (polyvinyl chlorides). The impetus for this new research direction did not initially come from the medical establishment, however. Rather, it came from the Endometriosis Association, which first made a connection between the disease and immunological problems. It was the EA that uncovered and pursued the link between endometriosis and dioxin exposure in the now well-known study of rhesus monkeys in 1992. A more detailed look at the new interpretive frame for endometriosis will appear in a later section. First, however, I will discuss some of the gender dimensions of endometriosis that have in the past prevented a scientific understanding of the disease, and in particular, have obscured its link to the "social body."

Gender Issues and Endometriosis

For many years, established medical assumptions about endometriosis have been infused with gender biases. For example, the disease was linked in the mind of the public and most doctors to a population of career-oriented women who had deferred childbearing—hence the name "career woman's disease." More precisely, according to the medical textbooks, it was assumed to affect "white, well-educated, perfectionistic, thin 'career women'" who were in their thirties or forties (Ballweg 1987). Doctors routinely suggested that the best cure for endometriosis was bearing a child fairly early in a woman's reproductive career, and this "prescription" was given to patients regardless of its appropriateness.[5] The other most commonly prescribed "cure" was hysterectomy. The medical establishment also assumed that much of the pain experienced by women with endometriosis was either part of the normal female physiological process, or psychological in origin.

The label "career woman's disease" not only carried—as it turns out—misinformation about the disease itself, but also implied a value judgment: a woman being punished for going against nature by putting her career before childbearing. The "career woman" label taps into an image of the working woman as selfish and destructive of the family, and ultimately herself—reminiscent of nineteenth- and early twentieth-century arguments that women would destroy their childbearing abilities if they pursued higher education. Neuroticism is often an additional component of this image of the "unnatural" career woman. Indeed, a psychologist claimed in a 1979 study to find psychological roots for endometriosis in the rejection of femininity, as reflected in negative self-esteem, hostility toward men, and fear of motherhood (Ballweg 1990).[6] A woman suffering from endometriosis was thus seen as bringing the disease on herself through her social choices and attitudes.

In addition, the painful symptoms of endometriosis, particularly those clustered around the menstrual period, were often dismissed by doctors as "all in your head." As Mary Lou Ballweg, co-founder and President/Executive Director of the EA, pointed out in 1987, an estimated 70 percent of women contacting the Association received such responses, with accompanying labels such as "psychosomatic," "hypochondriac," "frigid," and "You're just the nervous type" (Ballweg 1987, 9; see also Ballweg 1997). In fact, these doctor-patient interactions were so common that the EA developed a tongue-in-cheek and very popular cartoon character, "Joe with Endo," a male who, in the cartoon world, faces the typical responses given to females. By turning the tables and creating a male character, the absurdity of the medical advice is highlighted, as is the sexism of the medical profession (and the surrounding culture).[7] For women with endometriosis, this was a humorous protest against a medical "framing" of their disease that did not resonate with their lived experience. At the same time, they knew this was no laughing matter—these biases were especially wrenching because they affected the sphere of intimate relationships, women's relationships with their own bodies, and relationships to the larger society through financial and medical decisions.

Gender bias is found not only in the doctor-patient relationship and in general cultural responses to the woman with endometriosis, but also in the lack of research into women's diseases. This in turn contributes to frequent misdiagnosis. Although the incidence of endometriosis is now known to be quite high, it was not a research priority for many years. *Scientific American* recently called it "an epidemic ignored" (Denison 1997, 4). Due to the lack of research, it was easy for the "career woman's disease" myth to persist; in 1991 testimony offered before the U.S. Office of Research on Women's Health, National Institutes of Health, Mary Lou Ballweg summarized the situation as a "nightmare of misinformation, myths, taboos, lack of diagnosis, and problematic hit and miss treatments overlaid on a painful, chronic, stubborn disease" (Ballweg 1991).

Data-gathering and pressure on the medical establishment by the EA has challenged almost all of the previously held assumptions about endometriosis. As a result of their work, it is now known that endometriosis often affects very young adolescents, that it affects women of all races and income groups in various stages of their lives, that having a baby or a hysterectomy is not necessarily a cure, and that it is a global problem.[8]

The discovery of the onset of endometriosis in adolescence raises some particularly important gender issues. As Ballweg and others have pointed out, adolescence is a particularly sensitive time when such things as sexual identity are being worked out. If this experience is overlaid with cultural taboos about the body and with misinformation, the experience is much more emotionally traumatic. From a physical perspective, early diagnosis offers one of the most hopeful scenarios for dealing with endometriosis. Yet, as noted earlier, the EA has found an average lag time of ten years from onset to diagnosis, largely due to a lack of information among doctors as well as patients (Lemaire 1996). The problem has typically included a dismissal of symptoms that young women reported, as when they were told that the pain they experienced was "part of being a woman." Ballweg has pointed out, "If we could just get physicians and society to take this disease seriously and deal with it when the girl was fifteen . . . maybe she wouldn't be infertile when she was twenty-five" (Denison 1997, 7). Data for 1998 from the second major endometriosis survey undertaken by the EA recently confirmed a pattern of early onset of the disease as well as the time lag in diagnosis (Ballweg 1998b).

Problems of gender stereotyping and bias may be exacerbated by other cultural factors, which are coming to light as the EA creates an international network of women with endometriosis. Because of the isolating impact of universal taboos surrounding women's basic physical functions, as well as national and cultural differences, the creation of global networks has been particularly important as well as challenging (Denison 1997, 7). For example, in Algeria, it is forbidden for a young unmarried woman to go for a pelvic exam; her body literally belongs to a man who does not yet exist—her future husband (author's notes, 15th Anniversary Endometriosis Conference, 1995). This naturally makes diagnosis difficult and the work of a local EA group culturally precarious. In other cases, the gender bias is more subtle, but still affects a woman's life chances. A woman working with the Japanese endometriosis group recently reported a general feeling that ". . . a doctor more willingly treats the woman who wants to have a baby than the woman who does not" (EA 1994, 8).

The lack of an accurate medical picture of endometriosis has social policy implications as well. For example, Peggy Chumbley, an EA member of the San Francisco Bay Area Chapter, found out that endometriosis was not defined as "a valid illness" in the *Medical Book of Recognized Illnesses*, a document used for disability claims (Peterson and Grotberg 1997, 11). After being turned down a number of times in her request for Social Security disability, she enlisted the aid of a member of Congress and won recognition of the disease on the part of the Social Security Administration. While this lack of recognition is common to a number of less well-known diseases, EA members formed their organization in part because this problem was made worse by patriarchal gender relationships between doctor and patient that discouraged communication and also research; it was felt that neither the disease nor the female patient were taken seriously.

Women who actually experience endometriosis have long found the stereotypes to be demeaning and absurd, and very costly at both a personal and social level. Prior to recent research findings, however, they had little basis for credibly challenging this biased interpretive frame. It is not exaggeration to say that much of the recent research would not have seen the light of day without the active presence of the EA (see below). The research findings have permitted a reframing of endometriosis, one that locates causality not in the individual body of a woman, but rather in the "social body" organized around a particular set of social and cultural practices (Freud and McGuire 1995). This reframing questions not only past scientific "facts," but reallocates moral blame. Instead of pointing to bad personal choices made by the selfish career woman, it links a woman's body to "a new species of trouble" (Erikson 1991) stemming from a toxic environment. This new frame, then, does not stop with medical diagnosis, but rather broadens into the realm of social critique (Kroll-Smith and Floyd 1997).

"Reframing" Endometriosis: The Role of the Endometriosis Association

The EA was founded "to address the lack of information and research on the disease and lack of support available for those with it" (Ballweg 1991, 2). An important part of its work has been to challenge a biased cultural frame through which the disease has been interpreted. Since 1980, the EA, with its grassroots membership and a number of medical advisers, has had a major hand in redefining endometriosis, and in steering the mainstream medical profession in the direction of innovative new research and understandings. It is important to note that the EA goal has never been to reject medicine or science, but to rid it of its cultural biases. Ballweg made this clear in 1990:

> It's no wonder, with so much myth, taboo, and confusion surrounding endometriosis, that no major breakthroughs in understanding the disease, no new theories about its development, or new conceptual frameworks or treatment evolved from the 1920s to 1980. And so I make an appeal that may seem strange coming from a lay person, and that is an appeal to return to science, *true* science, on endometriosis. Let's look at the disease with new eyes, throw out the socioeconomic myths and taboos and assumptions about women, and start over. Women with endometriosis stand ready to be partners in the progress inevitable from this fresh start. (Ballweg 1990, 4)

The EA has worked toward its goal of better science by creating the first disease registry that collects data about people with endometriosis, by recruiting key doctors and researchers to the project of understanding the disease, by seeking out and supporting groundbreaking research, by organizing conferences that bring together researchers and laypersons, by providing public testimony and engaging in public education about endometriosis, and by creating international networks that promote the sharing of experience and information.[9] The new information created through this process has made it possible for the EA to begin to remap the disease; as noted in an earlier section, the data registry revealed, for example, that endometriosis afflicts women from all races and income groups, that it often starts in their teens and twenties, and that pregnancy does not cure the disease. Affirming the importance of lay knowledge in medical research, Mary Lou Ballweg recalled that her years of work on black/Hispanic issues, in the women's movement, and her own experience of seeking a diagnosis made her doubt that endometriosis was a white woman's disease.

A Broader Definition of Endometriosis

A first step in revisioning endometriosis was the connection made to a broader set of immunological problems observed in the data registry. In her 1991 testimony for research funding at a hearing called by the U.S. Office of Research on Women's Health, National Institutes of Health (an initiative that was itself a response to grassroots pressure), President/Executive Director of the EA Mary Lou Ballweg characterized endometriosis as a "chronic painful disease with many more symptoms than have been recognized in the past," symptoms that often occur all month long and reach moderate and severe levels (Ballweg 1991, 2). And in a dramatic departure from past models of the disease, she noted that:

> A picture of the disease is emerging that crosses specialities between reproductive health and immunology. Our data and the work of others has identified and has been following endometriosis as a "tip of the iceberg" disease, an immunological disease with many reproductive symptoms as well as many immunological symptoms, symptom as not yet widely known (Ballweg 1991, 2).

She went on to illustrate how endometriosis can manifest differently during various stages of the life cycle, beginning with a tendency toward allergies and autoimmune problems in young people. With this testimony, the EA made the case that a more holistic approach was needed to understand a disease that affected far more than just the reproductive system.

The Dioxin Connection

A major turning point in the medical understanding of endometriosis came in 1992, when research on a colony of rhesus monkeys revealed a link between endometriosis and the chemical dioxin. The EA had a key role in bringing this research to light and stimulating the medical community to do further research that is contributing to new understandings of the disease; the research has also generated controversy, since the emphasis on a toxic environment leads to a critique of the practices of the "social body" that produces it.

The dioxin connection was serendipitously discovered in early 1992, when EA President/Executive Director Mary Lou Ballweg found out that researchers at the University of Wisconsin-Madison had been studying the effects of the chemical dioxin on fertility in a colony of rhesus monkeys. The monkeys had trouble reproducing and two of the monkeys died due to severe endometriosis. Prior to that, no researcher had ever been able to create spontaneous endometriosis in a laboratory, and only mild endometriosis had ever been reported in animals in the wild. The results of this study were intriguing, but to Ballweg's dismay, the study had lost its funding, the research team had long been disbanded, and the monkeys were about to be sold. The EA Board made a decision to allocate emergency funds to the University so that laparoscopies could be performed on the remaining monkeys to determine whether they had endometriosis. Ballweg's lay experience with the disease taught her that it manifests very differently even in human populations, and might look different in monkeys; she therefore sought out the best experts she could find who could recognize the disease (Peterson and Grotberg 1997).[10] The completed laparoscopies revealed that 79 percent of the monkeys exposed to dioxin had endometrial growths on their abdominal organs, with a direct correlation between doses of dioxin and severity of the disease. These findings created a ripple in the medical science research community. They were reported in prestigious journals and as of 1999 have triggered dozens of additional studies on the link between dioxin and endometriosis by the Environmental Protection Agency, the National Institute of Environmental Health Sciences, and other investigators (Ballweg 1999; Denison 1997).

The dioxin study led the EA to pay close attention to the implications of a toxic environment for human bodies, and particularly for women. While this issue had been raised during the exploration of the immunological features of endometriosis, it now moved closer to center stage. The growing interest in the environment/body connection can be observed in the *Endometriosis Association Newsletter* as well as in conference agendas. This focus in turn led to new grassroots alliances, particularly with environmentally oriented groups that challenged the pervasive presence of chemicals in the environment.

The Health Care without Harm Campaign

As the dioxin findings surfaced, the emphasis on environmental factors made sense to many who had long been grappling with the mystery of endometriosis. The reframing of the disease in environmental terms corresponded better to their experienced reality and to the patterns that the EA was documenting. Mary Lou Ballweg, for example, observed that:

> One thing I and other people have been doing all along is looking for something that makes sense out of this pattern of immune dysfunction that you see in endometriosis. ... When I started looking at the literature on dioxin, PCBs, and other chemicals, it really started to fall into place because the immune abnormalities, the hormonal dysfunction, and certain other things that we noted in women's endometriosis is in the [dioxin] literature. I started to think we could really be on the right path. (Denison 1997, 5)

Among other things, the EA has noticed that the number of cases of endometriosis is sharply increasing over the course of the twentieth century. While only twenty-one cases were reported in the world literature prior to 1920, the current number is in the millions worldwide. As Denison (1997, 4) points out, "Although new surgical techniques have admittedly enhanced detection, the numbers could actually be higher, since endometriosis often goes undiagnosed for many years due to lack of public awareness." Ballweg observes that the rising number of cases parallels the increase in the use of hormonally active chemicals since World War II, particularly organochlorines, and notes that evidence from German, Canadian, and U.S. Air Force studies links endometriosis to exposure to PCBs and to radiation.

This focus on the environment—at a time when numerous other groups are calling attention to the impact of chemicals acting as endocrine disruptors on the human body—has led to new forms of activism in the EA. Joining the Health Care Without Harm campaign is an example of such activism. Learning that dioxin was produced at medical facilities, which incinerate PVCs (polyvinyl chlorides) in great quantities, an international coalition of groups from the environmental, medical, scientific, and labor communities, among others, formed to reduce the use of PVC, mercury, and other toxic chemicals in hospitals and to end incineration of substances which create dioxin in medical waste (EA 1997) (see Appendix for U.S. groups participating in Health Care Without Harm). The EA Board voted to endorse the EA's participation in this campaign by testifying before the Environmental Protection Agency, by contacting politicians, and by working with other organizations involved in a campaign to "Adopt a Hospital" in local communities. Participation in this campaign marked a new phase for the EA organization:

> For the first time in its history, the Association has been able to form alliances outside endo and women's health circles. Due to our groundbreaking work linking dioxin to the development of endometriosis, environmental organizations are working with us to educate about the dangers of hormonally-active chemicals and to reduce the dissemination of these chemicals in our environment. (EA Annual Report 1997, 9)

Thus the EA has found itself in the company of many other environmentally oriented groups in settings such as the Reproductive Toxins Conference in Washington, DC, organized by the National Environmental Law Center in 1996. A partial list of participants included Physicians for Social Responsibility, Greenpeace, the Santa Clara Center for Occupational Safety and Health, Communities for a Better Environment, the Sierra Club Legal Defense Fund, the National Wildlife Federation, Ralph Nader's Government Purchasing Project, the United Auto Workers, and the Center for Environmental Studies (Tremblay 1997). For the EA, this is a significant departure from past alliances, and brings in a new level of social critique, since reframing endometriosis as the partial outcome of a toxic environment implicates the society that produces the toxins.

The emphasis on environment/body connections has also enhanced the already existing goal of the EA to promote international networking and data gathering. Ballweg points out that since "[t]hese chemicals don't know any national boundaries" it is essential for women to work together globally (Denison 1997). For example, at the 15th Anniversary Conference of the EA, held in 1995 in Milwaukee, Wisconsin, the session "Endometriosis: A Worldwide Epidemic" included invited speakers from Japan, Brazil, the United Kingdom, Taiwan, Germany, New Zealand, and Algeria (EA 1995).[11] The environmental focus provides an additional connection between women across national boundaries, and the global approach itself has yielded important results that are reshaping knowledge about the disease, the frame through which it is interpreted, and, no less importantly, the experience of having it.

Understanding Endometriosis: The Future Agenda

The EA efforts to reframe endometriosis have yielded a number of important results. One result is the achievement of "medical respectability" for the claim of environment/body connections, and for a more holistic approach to the disease in general. This was first evident through the establishment of a research program at the Dartmouth Medical School following the showcasing of the EA's dioxin work at a May 1995 National Institute of Health Endometriosis 2000 Conference, which brought together endocrinologists and toxicologists to discuss the dioxin findings (Ballweg 1998a).[12] The conference was initiated in response to EA testimony about dioxin research at a U.S. Senate hearing. More recently, the Vanderbilt University School of Medicine has asked the EA to join it in establishing an Endometriosis Association research facility at their Nashville, Tennessee, campus, to be dedicated to addressing the mechanisms responsible for causing endometriosis (Ballweg 1998a).[13] This partnership with a nonprofit organization such as the EA indicates a participatory approach that incorporates the experience of patients themselves in the research process. Vanderbilt was impressed with the work done by the EA leading to major research breakthroughs in understanding the disease.

The emerging plan for future research is much more holistic than past efforts. At Vanderbilt, the research approach will be multi-disciplinary, and will facilitate the "synergy" that Ballweg called for in the research community in 1991. The EA has been dedicated to the holistic principle for a long time. For example, the program for the 15th Anniversary Conference of the EA in 1995 reveals a broad range of topics: dioxin research; immune therapy for endometriosis; connections between endometriosis and infertility, pain, nutrition, cancer, and genetic predisposition in families (the British OXEGENE project based at the University of Oxford, which collects international data on this topic, was represented at the conference); endometriosis and traditional Chinese medicine; the role of nurses; Hispanic women and endometriosis; black women and endometriosis; teens and endometriosis; husbands, partners, family and friends—and endometriosis; medical options; and surgical options. There were also workshops for Chapter/Support Group Leaders, affinity groups, updates from international EA groups, and numerous question and answer sessions. The dioxin research, in particular, has been highlighted at recent conferences.

All of this contributes to a sense that participants, both laypersons and medical experts, are actively contributing to an integrated larger picture of endometriosis. The VI World Congress on Endometriosis held in the summer of 1998 in Quebec, Canada, continued this trend, with broad incorporation of patients' perspectives (as Ballweg put it, "our *full* stories—not just what can be squeezed into a few minutes in the doctor's office) as well as "brainstorming sessions on key medical and scientific aspects" of endometriosis (Ballweg 1998c).[14]

This holistic approach makes sense from the point of view of a grassroots group such as the EA; one seeks answers anywhere and everywhere. The medical establishment, on the other hand, has often resisted holism, taking a much more specialized and therefore fragmented approach. The Vanderbilt project signals a shift away from such fragmentation. For example, the model includes not only a permanent multidisciplinary research group, but also promising researchers from other countries who will work for three years at a time under "New Investigator Awards." As the EA points out, these awards "will help advance research worldwide, provide cross-cultural input into the research process, develop potential lead investigators for endometriosis research groups around the world, and train new researchers in our field" (Ballweg 1998a).

The impact of the EA as a group of laypersons on the world of science and medicine is in many ways, then, a success story. As noted at the beginning of this chapter, the EA formed out of a dissatisfaction with the medical establishment, including researchers, who did not put priority on investigating endometriosis, and doctors, who were seen as not being well informed, and who, in addition, were seen as patronizing and sexist in their interactions with female patients with endometriosis. As a result of the EA's work since 1980, a significant shift has taken place. In addition to the changes reported above, there are new books on the shelf that serve as resources for laypersons, such as *Overcoming Endometriosis: New Help from the Endometriosis Association* (Ballweg and EA 1987) and *The Endometriosis Sourcebook* (Ballweg and EA 1995), both compiled by the EA. The EA has its own electronic web site, to counteract misinformation found elsewhere on the Internet. The EA has helped to transcend boundaries of race, class, gender, and nationality by exploding the myth of endometriosis as a white career woman's disease. By encouraging women to speak directly to each other and to engage in a direct, two-way dialogue with the medical profession, the EA has helped to undo the many levels of gender bias that have prevented the understanding of this disease. In doing so, the EA has amplified the culturally subdued voice of women in U.S. medicine, and has contributed to the sensitization of doctors—both male and female—trained by the medical system. Crucially, it has also brought the environment body connection to center stage in the framing of endometriosis.

At the same time, challenges remain. Apart from some specialists, most doctors still lack information on endometriosis, and do not have the skills to treat it comprehensively. Even with EA efforts, the "career woman's disease" myth still persists in many places, and endometriosis continues to be misdiagnosed by doctors (including gynecologists), as reflected in the lag time between onset and diagnosis.[15] There are many puzzles to the disease that are not yet well understood, such as the connection to other autoimmune diseases like lupus. In short, much remains yet to be mapped.

In the realm of social critique, as the EA focuses increasingly on environment/body connections, it joins other groups embroiled in the high-stakes controversy over the causal impact of toxic environments. It potentially gains new enemies, such as chemical manufacturers that are sources of pollution. At the very least, it can find itself in a contradictory position when a pharmaceutical corporation such as Zeneca provides funding for its conferences and at the same time manufactures suspected carcinogenic pesticides like acetachlor (Poe 1999).[16] While it has gained new allies, it has also been drawn into a broader "environmental justice" movement that faces efforts to delegitimize it by an active "countermovement" which plays down the significance of the body environment link.

In response, the EA concedes that much more study is needed. However, the environment/body connection is a frame that resonates with both personal experience and the data collected for many years by the EA, and the organization is dedicated to pursuing the environmental connection. Like a number of other successful grassroots groups (Čapek, in press), the EA has gained respectability under controversial circumstances by pursuing good science without abandoning grassroots efforts such as the Health Care Without Harm Campaign or the successful lobbying of Congress to allocate $5 million to endometriosis research in 1990–91. Niki Denison (1997, 7) describes the Endometriosis Association as an organization of "women around the world who refuse to remain complacent casualties of the twentieth century." This refusal links society, culture, and environment to the search for answers to the complex medical puzzle that is endometriosis.

APPENDIX: *Health Care Without Harm Participating Organizations*

1199, Occupational Safety & Health,
New York NY

Action for Women's Health,
Albuquerque NM

AFL-CIO Environmental Liaison,
Washington DC

American Indian Health,
Dearborn/Detroit MI

American Nurses Association,
Washington DC

American Public Health Assn., Cook Co.
Hospital, Chicago IL

Beth Israel Medical Center,
New York NY

Blue Ridge Environmental Defense
League, Wadesboro NC

Breast Cancer Action,
San Francisco CA

Breast Cancer Fund,
San Francisco CA

California Communities Against
Toxics, Rosamond CA

California Nurses Association,
Sacramento CA

Cathedral of Saint John the Divine,
New York NY

Catholic Healthcare West,
San Francisco CA

Center for Environmental Health,
San Francisco CA

Center for the Biology of Natural
Systems, Flushing NY

CGH Environmental Services,
Burlington VT

Chemical Impact Project,
Kentfield CA

Citizens Clearinghouse for Hazardous
Wastes (CCHW), Center for Health,
Environment and Justice,
Falls Church VA

Citizens Environmental Coalition,
Albany NY

Citizens Environmental Coalition,
Medina NY

Citizens for a Better Environment,
Chicago IL

Citizens for a Better Environment,
Milwaukee WI

Cleanup Coalition,
Baltimore MD

Committee of Interns and Residents,
New York NY

Commonweal/Jenifer Altman
  Foundation, Bolinas CA
Dept. of Environmental Health,
  Boston University School of Public
  Health, Boston MA
DES Cancer Network,
  Washington DC
Detroiters Working for Environmental
  Justice, Detroit MI
Earth Day Coalition,
  Cleveland OH
EarthSave,
  Louisville KY
EarthSave,
  Santa Cruz CA
Endometriosis Association,
  Milwaukee WI
Environmental Association for Great
  Lakes Education, Duluth MN
Environmental Stewardship
  Concepts, Richmond VA
Environmental Working Group,
  Washington DC
Farm-Verified Organic,
  Medina ND
First United Methodist Church,
  Santa Cruz CA
Fletcher Allen Health Care,
  Burlington VT
Gateway Greens,
  St. Louis MO
General Board of Church & Society,
  United Methodist Church,
  Washington DC
Government Purchasing Project,
  Washington DC
Grass Roots Environmental Organization
  (GREO), Flanders NJ
Great Lakes Center for Occupational &
  Environmental Safety & Health,
  Sustainable Hospitals Project,
  Chicago IL
Great Lakes Natural Resource Center
  NWF, Ann Arbor MI

Great Lakes United,
  Montreal Quebec
Greater Boston Physicians for Social
  Responsibility, Boston MA
Greater Cleveland Coalition for a
  Clean Environment, Cleveland OH
Greenaction,
  San Francisco CA
Greenpeace,
  Chicago IL
Hamtramck Environmental Action
  Team, Hamtramck MI
Human Action Community Organization,
  Harvey IL
Illinois Student Environmental Network,
  Champaign IL
Indigenous Environmental Network,
  Bemidji MN
Institute for Agriculture and Trade
  Policy, Minneapolis MN
Judith Helland Productions,
  New York NY
Kirschenmann Family Farms,
  Windsor ND
Learning Alliance,
  Jamaica Plain MA
Learning Disabilities Association,
  Pittsburgh PA
Legal Environmental Assistance
  Foundation, Tallahassee FL
Lone Star Chapter of the Sierra Club,
  Austin TX
Massachusetts Breast Cancer
  Coalition, Waltham MA
Massachusetts Nurses Association,
  Canton MA
Mercy Health Care Sacramento,
  Rancho Cordova CA
Methodist Federation for Social
  Action, Mason City IL
Mid-Michigan Environmental Action
  Council, E. Lansing MI
Minnesota Center for Environmental
  Advocacy, St. Paul MN

Mt. Sinai School of Medicine,
New York NY
Multinationals Resource Center,
Washington DC
National Environmental Law Center,
Boston MA
National Environmental Law Center,
Davis CA
National Medical Waste Resource
Center, Iowa City IA
National Women's Health Network,
Moretown VT
Natural Resources Defense Council,
San Francisco CA
New England Medical Center,
Boston MA
New York State Nurses Association,
Latham NY
North Carolina Waste Awareness &
Reduction Network, Durham NC
Ohio Network for the Chemically
Injured, Parma OH
Oil, Chemical and Atomic Workers
Union, Lakewood CO
Oncology Nursing Society,
Washington DC
Oregon Center for Environmental
Health, Portland OR
Physicians for Social Responsibility,
Atlanta GA
Physicians for Social Responsibility,
Cambridge MA
Physicians for Social Responsibility,
San Francisco CA
Physicians for Social Responsibility,
Washington DC
Reconstructionist Rabbinical
Association, Rockville MD
Reduce Recidivism by Industrial
Developments, Inc., Chicago IL
Safety & Environmental Programs,

Dartmouth Hitchcock Medical
Center, Lebanon NH
Save Our Country,
E. Liverpool OH
Science & Environmental Health
Network, Windsor ND
Sierra Club,
Olympia WA
South Bronx Clean Air Coalition,
Bronx NY
South Carolina Nurses Association,
Columbia SC
Southeast Michigan Sierra Club,
Detroit MI
Student Environmental Action
Coalition, Philadelphia PA
Students for a Healthy Hospital,
Ann Arbor MI
Toxics Reduction Project, Ecology
Center of Ann Arbor, MI
United Citizens and Neighbors,
Champaign IL
Vermont Public Interest Research
Group (VPIRG), Montpelier VT
Vietnam Veterans of America,
Michigan Chapter,
Saline MI
Washington Toxics Coalition,
Seattle WA
White Lung Association,
Baltimore MD
Women's Cancer Resource Center,
Berkeley CA
Women's Cancer Resource Center,
Minneapolis MN
Women's Community Cancer Project,
Women's Center, Cambridge MA
Women's Environment and Development
Organization, New York NY
Work on Waste USA,
Canton NY

Source: Endometriosis Association (Ballweg 1999). International organizations not included.

NOTES

1. The Endometriosis Association estimates that approximately 2 percent of women have the disease worldwide, which would bring the number to a staggering 89 million. It is not possible at present to scientifically prove such a claim, however, and the number usually cited includes the better documented number of five and a half million women in the United States and Canada with endometriosis (personal communication from Mary Lou Ballweg, January 21, 1999).

2. My thanks to the Hendrix College Faculty Travel Grants Committee for supporting my effort to integrate in a scholarly fashion the deeply personal and sociological elements that constitute this study.

3. A 1998 sampling of 4000 women with endometriosis confirmed the results of the first EA data registry from the early 1980s. Of these women, 47 percent reported that they had to see a doctor five or more times before being diagnosed or referred. The EA found that the overall delay between onset of symptoms and diagnosis was 9.28 years. While part of this time lag results from patient delay in reporting symptoms, on average it took doctors 4.6 years to diagnose endometriosis (Ballweg 1998b). It also found that the younger the patient, the greater the time lag, a pattern also found outside the United States.

4. The retrograde menstruation or transtubal migration theory posits that "during menstruation some of the menstrual tissue backs up through the fallopian tubes, implants in the abdomen, and grows. Some experts on endometriosis believe all women experience some menstrual tissue backup and that an immune system problem and/or hormonal problem allows this tissue to take root and grow in women who develop endometriosis" (EA 1987). The genetic theory "suggests that it may be carried in the genes of certain families or that certain families may have predisposing factors to endometriosis" (EA 1987). The embryonic tissue theory suggests that "remnants of tissue from when the woman was an embryo may later develop into endometriosis or that some adult tissues retain the ability they had in the embryo state to transform into reproductive tissue under certain circumstances" (EA 1987).

5. Mary Lou Ballweg has questioned the ethics of advising a person to have a child as a "prescription" for a medical condition. She has also asserted that "[t]o suggest to women, whether they are in their teens or older; single or coupled; heterosexual, gay, or celibate; in a relationship capable of sustaining a child or not; or even whether they *want* a child or not, that they have a child because they have a disease seems ludicrous even if it *did* cure the disease. That this "treatment" is "prescribed" so often, as is reported by thousands of women to the Association, is scandalous" (Ballweg 1990, 3).

6. The author of the study, Dr. Mary Lou Hollis, claimed that the woman with endometriosis ". . . has feelings of being limited and pressed by social and vocational aspects of her life space, is over sensitive and experiences pervasive and generalized resentment and hostility. She expresses this hostility in an indirect fashion toward men." Hollis also suggested that complaints of painful intercourse were "a nice protection for some women who do not want children" (EA 1990).

7. For example, Joe is told that his testicles will have to be removed (in an analogy to the usual advice given to females about hysterectomy). The doctor in the cartoon assumes that this makes good medical sense, and Joe needs to learn to live with it.

8. Hysterectomies do not always cure endometriosis, because when women receive hormone replacement therapy, the estrogen can reactivate the stray menstrual tissue and sufferers find themselves once again in pain (Denison 1997, 7). In many cases pregnancy leads to a temporary abatement of symptoms. However, symptoms often return later, sometimes even

after hysterectomy. In addition, surgery sometimes does not or cannot remove all traces of the disease.

9. The first disease registry was housed at the Medical College of Wisconsin, and in 1991 included over 3000 case histories with over 500 variables per case. Findings from the registry were published in medical journals as well as EA literature (Ballweg 1991). Presently, the second research registry is housed with the EA, and is accessible to researchers. The EA's commitment to speak and write for M.D.s and Ph.D.s as well as a lay audience is reflected in the fact that *The Endometriosis Sourcebook*'s (Ballweg and EA 1995) chapter titled "Endometriosis: The Patient Perspective" appeared originally in an medical textbook in 1992, while the chapter on "The Puzzle of Endometriosis" appeared in a medical textbook some years later (Ballweg 1999). The EA also undertook efforts such as the Society of Obstetricians and Gynaecologists of Canada project to help produce a consensus statement on endometriosis (Ballweg 1999).

10. The experts who were called in for the project included Dr. Dan C. Martin, Sherry Rier, Ph.D., and W. Paul Dmowski, M.D., Ph.D.

11. The speaker from Algeria was unable to attend due to visa problems, but the information on Algeria was presented by another speaker.

12. The Dartmouth program housed the Tracy H. Dickinson Research Chair of the Endometriosis Association and related research. Dr. Sherry Rier, who was a lead researcher on the dioxin study, held this chair.

13. The collaboration includes laboratory space, access to research equipment, a $2 million investment on the part of Vanderbilt in the facilities and programs, as well as the prestige of the Vanderbilt name (Ballweg 1998a).

14. Dr. Stanley Glasser, a world authority on endometrium and a participant in the World Congress, wrote that "directly and indirectly, the workshop and particularly the 'brainstorming sessions' had a much better pass-through effect; especially on those who might contribute to expanding, in all dimensions, the knowledge required to resolve the ontogeny and etiology of the disease" (Ballweg 1998c).

15. Data for 1998 revealed that it took gynecologists on average 4.1 years to diagnose endometriosis, reproductive endocrinologists 1.4 years, urgent care doctors 1.7 years, and family practitioners/general practitioners, 5.3 years (Ballweg 1998b).

16. Some have raised questions about the EA receiving money from pharmaceutical companies. Ballweg has responded with several points: first, that new drugs have made life tolerable for many women, and have been substitutes for hysterectomy; second, this money, particularly when it funds advertisements, gets the word out about the EA, serving a public education function for which the EA alone would have not have financial resources; third, the EA's operating budget (over 80 percent) is funded by members; corporate contributions are targeted for special projects, and these projects are proposed by and controlled by the EA. "While you're being pure, somebody else out there right now is screaming in pain or crawling to the bathroom. You know, I think that's part of what makes a difference—the Association is real women with real pain with a real disease. You become practical . . ." (EA 1990, 8).

REFERENCES

Ballweg, Mary Lou, 1987. "Why the Endometriosis Association Was Started." Pp. 1–10 in Mary Lou Ballweg and the Endometriosis Association (eds. J., *Overcoming Endometriosis-New Help from the Endometriosis Association.* New York: Congdon and Weed.

————. 1990. *They Changed World's View. 10 Years: Together We Make a Difference*, 1–14. Commemorative Booklet. Endometriosis Association.

————. 1991. Testimony for Research Funding, Task Force on Opportunities for Research on Women's Health, Office of Research on Women's Health, National Institutes of Health, June 12, 1991. *Endometriosis Association Newsletter* 12, 4:2.

————. 1997. Blaming the Victim: The Psychologizing of Endometriosis. *Obstetrics and Gynecology Clinics of North America* 24, 2: 441–53.

————. 1998a. Exciting News: Association Teams with Prestigious. Vanderbilt University School of Medicine to Create Dedicated Research Facility! *Endometriosis Association Newsletter* 18, 5–6: 1–2.

————. 1998b. New EA Research Shows Disease Is Starting Younger, Is More Severe. *Endometriosis Association Newsletter* 19, 1–2: 6–10.

————. 1998c. VI World Congress on Endometriosis Breaks New Ground with Patient Participation. *Endometriosis Association Newsletter* 19: 1–2.

————. 1999. Personal communication to the author, February 23, 1999.

Ballweg, Mary Lou, and the Endometriosis Association (EA), eds. 1987. *Overcoming Endometriosis: New Help from the Endometriosis Association*. New York: Congdon and Weed.

————. 1995. *The Endometriosis Sourcebook*. Chicago, IL: Contemporary Books.

Brown, Phil. 1992. Popular Epidemiology and Toxic Waste Contamination: Lay and Professional Ways of Knowing. *Journal of Health and Social Behavior* 33: 268–71.

Čapek, Stella M. (In press). Erasing Community: Institutional Failures and the Press Demise of Carver Terrace. *Research in Social Problems and Public Policy*, 7 (Special Issue on Institutional Failure in Environmental Management).

Denison, Niki. 1997. From a Candle to a Flame. *Endometriosis Association Newsletter* 18, 1:4–5, 7.

Endometriosis Association (EA). 1987. *Endometriosis Association Pamphlet*. Milwaukee, WI: Endometriosis Association, Inc.

————. 1990. *10 Years: Together We Make a Difference*. Commemorative Booklet. Endometriosis Association.

————. 1994. Association President Visits Growing Groups in Japan, Brazil, & Taiwan. *Endometriosis Association Newsletter* 15, 3:8.

————. 1995. *Final Program, 15th Anniversary Conference, Milwaukee, Wisconsin, November 3–5, 1995*.

————. 1997. Endometriosis Association Joins Health Care Without Harm Campaign: Effort to Push for Reduced PVC and Dioxins in Medical Waste. *Endometriosis Association Newsletter* 18, 1:1.

Endometriosis Association (EA) Annual Report. 1997. Annual Report, July 1, 1996–June 30, 1997. *Endometriosis Association Newsletter* 18, 5–6:9–10.

Erikson, Kai. 1991. A New Species of Trouble. Pp. 11–29 in Stephen Robert Couch and J. Stephen Kroll-Smith (eds.), *Communities at Risk: Collective Responses to Technological Hazards*. New York: Peter Lang.

Freund, Peter E. S., and Meredith McGuire. 1995. *Health, Illness, and the Social Body: A Critical Sociology*. Englewood Cliffs, NJ: Prentice Hall.

Hunt, Scott A., Robert D. Benford, and David A. Snow. 1994. Identity Fields: Framing Processes and the Social Construction of Movement Identities. Pp. 185–208 in Enrique Larana, Hank Johnston, and Joseph R. Gusfield (eds.) *New Social Movements: From Ideology to Identity*. Philadelphia: Temple University Press.

Institute for Optimum Nutrition (ION). n.d. *Endometriosis: The Hidden Epidemic*. Pamphlet put out by ION, London: Institute for Optimum Nutrition.

Kroll-Smith, Steve, and H. Hugh Floyd. 1997. *Bodies in Protest: Environmental Illness and the Struggle over Medical Knowledge.* New York: New York University Press.

Lemaire, Gail Schoen. 1996. Women's Experiences with Endometriosis: Survey Results. *Endometriosis Association Newsletter* 17, 5–6:4.

Nezhat, Camran. 1987. Foreword, Pp. xi–xv in Mary Lou Ballweg and the Endometriosis Association (eds.), *Overcoming Endometriosis: New Help from the Endometriosis Association.* New York: Congdon and Weed.

Peterson, Sheri, and Crystal Grotberg. 1997. EA Volunteers: Changing Lives One Person at a Time. *Endometriosis Association Newsletter* 18, 5–6:11.

Poe, Amy. 1999. Cancer Prevention or Drug Promotion? Journalists Mishandle the Tamoxifen Story. *Extra!* (January/February) 12, 1:15–16.

Snow, David A., and Robert D. Benford. 1988. Ideology, Frame Resonance, and Participant Mobilization. *International Social Movement Research* 2:197–217.

Snow, David A., E. Burke Rochford, Jr., Steven Worden, and Robert D. Benford. 1986. Frame Alignment Processes, Micromobilization and Movement Participation. *American Sociological Review* 56:464–81.

Tremblay, Lyse. 1997. "Reproductive Toxins Conference—Pollution Prevention Network." *Endometriosis Association Newsletter* 17, No. 5–6:13–15.

QUESTIONS

1. What does the author mean when she argues that established medial assumptions about endometriosis were gender-biased.

2. The Endometriosis Association advocated another way of framing this nascent disorder. What alternative did they propose?

Popular Epidemiology and Toxic Waste Contamination
Lay and Professional Ways of Knowing

Phil Brown

Medical sociology has long been concerned with differences between lay and professional ways of knowing (Fisher 1986; Roth 1963; Stimson and Webb 1975; Waitzkin 1989). Because of their different social backgrounds and roles in the medical encounter, clients and providers have divergent perspectives on problem definitions and solutions (Freidson 1970). Professionals generally concern themselves with disease processes, while laypeople focus on the personal experience of illness. For professionals, classes of disorders are central, while those who suffer the disorders dwell on the individual level (Zola 1973). From the professional perspective, symptoms and diseases universally affect all people, yet lay perceptions and experience exhibit great cultural variation. Similarly, lay explanatory approaches often utilize various causal models that run counter to scientific notions of etiology (Fisher 1986; Freidson 1970; Kleinman 1980). Medical professionals' work consists of multiple goals, among which patient care is only one; patients are centrally concerned with getting care (Strauss et al. 1964).

The study of these contrasting perspectives has centered on clinical interaction and institutional settings. Some scholars have examined lay-professional differences in occupational health (Smith 1981), community struggles over access and equity in health services (Waitzkin 1983), and genetic screening (Rothman 1986). Yet medical sociology has scarcely studied environmental and toxic waste issues.

Recently, lay perceptions of environmental health have manifested themselves in a burgeoning community activism. Following the landmark Love Canal case (Levine 1982), the childhood leukemia cluster in Woburn, Massachusetts, has drawn attention to the lay-professional gap. Woburn residents were startled beginning in 1972 to learn that their children were contracting leukemia at exceedingly high rates. Affected families and community activists attempted to confirm the existence of a leukemia cluster and to link it to industrial toxins that leached into their water supply. They pursued a long course of action that led to a major community health study, a civil suit against W. R. Grace Chemical Corporation and Beatrice Foods, and extensive national attention.

Building on a detailed study of the Woburn case and utilizing data from other toxic waste sites, this chapter discusses conflicts between lay and professional ways of knowing about environmental health risks. This discussion centers on the phenomenon of *popular epidemiology*, in which laypeople detect and act on environmental hazards and diseases. Popular epidemiology is but one variant of public participation in the pursuit of scientific knowledge, advocacy for health care, and public policy, as witnessed in such diverse cases as AIDS treatment, nuclear power development, and pollution control. The emphasis on ways of knowing makes sense because knowledge is often what is debated in struggles to win ownership of a social problem (Gusfield 1981, pp. 36–45).

In their popular epidemiological efforts, community activists repeatedly differ with scientists and government officials on matters of problem definition, study design, interpretation of findings, and policy applications. In examining the stages through which citizens become toxic waste activists, this chapter emphasizes lay-professional differences concerning quality of data, methods of analysis, traditionally accepted levels of measurement and statistical significance, and relations between scientific method and public policy.

Study Background

There were two sets of interviews with the Woburn litigants. The first set was with eight families; open-ended questions were asked dealing with individual experiences with the toxic waste crisis, including personal and family problems, coping styles, and mental health effects. These interviews were conducted in 1985 by a psychiatrist and reanalyzed in 1988 for an earlier phase of this research (Brown 1987). This reanalysis involved both the researcher and the psychiatrist rereading the interview material several times, and then discussing the most prominent themes. This process defined themes for discussion of the original, largely psychosocial, interviews. As well, it directed the creation of the interview schedule for the reinterview. For example, respondents in the original interviews expressed considerable anger at the corporations accused of contaminating the wells, and at the government officials investigating the disease cluster. This provided initial information on these important concerns, and directly yielded more specific reinterview questions.

The second set of interviews in 1988 (except for one family that did not wish to participate) comprised 20 open-ended questions on residents' perceptions of community activism, the litigation, government and corporate responsibility for toxics, and the relationship between lay and scientific approaches.[1] The first set of family interviews were taken down in writing. The second set of family interviews, as well as all interviews with other actors, were tape-recorded and transcribed.

Fourteen community activists, apart from the litigants, were also interviewed in 1990. In addition to basic personal data, respondents were asked 19 open-ended questions concerning toxic waste activism, knowledge about toxic wastes and their detection and remediation, attitudes toward corporate and governmental actors, and attitudes and participation in other environmental and political concerns. Between

1988 and 1991, the litigants' lawyer was interviewed, and other data were obtained from interviews with, formal presentations by, and official documents provided by state public health officials, federal environmental officials, and public health researchers. The interviews with public health officials sought responses to matters of lay-professional differences in methodology of, and interpretation of data from, both official and community health studies. Additional data came from legal documents, public meetings, and archival sources, and from research on other similar sites.

Material from all interviews, documents, meetings, and other sources was coded in two ways. First, codes were devised from prior knowledge gained from the first litigant interviews, from the themes that the litigant reinterviews questions and other interview questions were expected to tap, and from existing literature on toxic waste sites. Second, additional codes were identified after reading through the transcripts. In this second case, a number of codes were quickly apparent, such as the pride that citizens had in their nascent scientific abilities. The coding process therefore identified the beliefs and experiences of involved parties, enabling interpretations of those beliefs and experiences. In many instances, considerable congruence with other scholars' findings in case studies of toxic waste sites provided a degree of reliability.

In addition to this coding process, all data were examined in terms of their place in the historical/chronological development of the toxic waste crisis. While a clear line of unfolding events was previously apparent, the data culled from the detailed research allowed me to fill in fine-grained detail. This approach enabled me to create the stages model of popular epidemiology described in the next section. Here, too, other toxic waste studies offered support for the development of such a schema.

Lay Ways of Knowing

Popular Epidemiology

Traditional epidemiology studies the distribution of a disease or condition, and the factors that influence this distribution. These data are used to explain the etiology of the condition and to provide preventive, public health, and clinical practices to deal with the condition (Lilienfeld 1980, p. 4). A broader approach, seen in the risk-detection and solution-seeking activities of Woburn and other "contaminated communities" (Edelstein 1988), may be conceptualized as *popular epidemiology*.

Popular epidemiology is the process by which laypersons gather scientific data and other information, and also direct and marshal the knowledge and resources of experts in order to understand the epidemiology of disease. In some of its actions, popular epidemiology parallels scientific epidemiology, such as when laypeople conduct community health surveys. Yet popular epidemiology is more than public participation in traditional epidemiology, since it emphasizes social structural factors as part of the causal disease chain. Further, it involves social movements, utilizes political and judicial approaches to remedies, and challenges basic assumptions of traditional epidemiology, risk assessment, and public health regulation. In some cases, traditional epidemiology may reach similar conclusions to popular epidemiology. Yet scientists generally do not become political activists in order to implement their findings, despite exceptions such as Selikoff's work on asbestos diseases.

Popular epidemiology is similar to other lay advocacy for health care in that lay perspectives counter professional ones and a social movement guides this alternative perspective. Some lay health advocacy acts to obtain more resources for the prevention and treatment of already recognized diseases (e.g., sickle cell anemia, AIDS), while others seek to win government and medical recognition of unrecognized or underrecognized diseases (e.g., black lung, post-traumatic stress disorder). Still others seek to affirm the knowledge of yet-unknown etiological factors in already recognized diseases (e.g., DES and cervical cancer, asbestos and mesothelioma). Popular epidemiology is most similar to the latter approach, since original research is necessary both to document the prevalence of the disease and the putative causation.

From studying Woburn and other toxic cases (e.g. Couto 1985; Edelstein 1988; Krauss 1989; Levine 1982; Nash and Kirsch 1986), we observe a set of stages of citizen involvement. Participants do not necessarily complete a stage before beginning the next, but one stage usually occurs before the next begins:

1) A group of people in a contaminated community notice separately both health effects and pollutants.
2) These residents hypothesize something out of the ordinary, typically a connection between health effects and pollutants.
3) Community residents share information, creating a common perspective.
4) Community residents, now a more cohesive group, read about, ask around, and talk to government officials and scientific experts about the health effects and the putative contaminants.
5) Residents organize groups to pursue their investigation.
6) Government agencies conduct official studies in response to community groups' pressure. These studies usually find no association between contaminants and health effects.
7) Community groups bring in their own experts to conduct a health study and to investigate pollutant sources and pathways.
8) Community groups engage in litigation and confrontation.
9) Community groups press for corroboration of their findings by official experts and agencies.

Lay Observations of Health Effects and Pollutants. Many people who live at risk of toxic hazards have access to data otherwise inaccessible to scientists. Their experiential knowledge usually precedes official and scientific awareness, largely because it is so tangible. Knowledge of toxic hazards in communities and workplaces in the last two decades has often stemmed from lay observation (Edelstein 1988; Freudenberg 1984a; Frumkin and Kantrowitz 1987).

Although the first official action—closing Woburn's polluted wells—occurred in 1979, the Woburn water had a long history of problems. Residents had for decades complained about dishwasher discoloration, foul odor, and bad taste. Private and public laboratory assays had indicated the presence of organic compounds. The first lay detection efforts were begun earlier by Ann Anderson, whose son, Jimmy, had been diagnosed with acute lymphocytic leukemia in 1972.

Hypothesizing Connections. Anderson put together information during 1973–1974 about other cases by meetings with other Woburn victims in town and at the hospital where Jimmy spent much time. Anderson hypothesized that the alarming leukemia incidence was caused by a water-borne agent. In 1975 she asked state officials to test the water but was told that testing could not be done at an individual's initiative (DiPerna 1985, pp. 75–82). Anderson's hypothesis mirrored that of other communities, where people hypothesize that a higher than expected incidence of disease is due to toxics.

Creating a Common Perspective. Anderson sought to convince the family minister, Bruce Young, that the water was somehow responsible, although he at first supported her husband's wish to dissuade her. The creation of a common perspective was aided by a few significant events. In 1979, builders found 184 55-gallon drums in a vacant lot; they called the police, who in turn summoned the state Department of Environmental Quality Engineering (DEQE). When water samples were then taken from a number of municipal wells, Wells G and H showed high concentrations of organic compounds known to be animal carcinogens, especially trichloroethylene (TCE) and tetrachloroethylene (PCE). EPA recommends that the TCE be zero parts per billion and sets a maximum of 5 parts per billion; Well G had 40 times that concentration. As a result, the state closed both wells (Clapp 1987; DiPerna 1985, pp. 106–8).

In June 1979, just weeks after the state closed the wells, a DEQE engineer driving past the nearby Industri-Plex construction site thought he saw violations of the Wetlands Act. A resultant EPA study found dangerous levels of lead, arsenic, and chromium, yet EPA told neither the town officials nor the public. The public only learned of this months later, from the local newspaper. Reverend Young, initially distrustful of Anderson's theory, came to similar conclusions once the newspaper broke the story. Working with a few leukemia victims he placed an ad in the Woburn paper, asking people who knew of childhood leukemia cases to respond. Working with John Truman, Jimmy Anderson's doctor, Young and Anderson prepared a questionnaire and plotted the cases on a map. Six of the 12 cases were closely grouped in East Woburn.

Looking for Answers from Government and Science. The data convinced Dr. Truman, who called the Centers for Disease Control (CDC). The citizens persuaded the City Council in December 1979 to ask the CDC to investigate formally. Five days later, the Massachusetts Department of Public Health (DPH) reported on *adult* leukemia mortality for a five-year period, finding a significant elevation only for females. This report was cited to contradict the residents' belief in the existence of a childhood leukemia cluster.

Organizing a Community Group. In January 1980, Young, Anderson, and 20 others (both litigants and non-litigants) formed For a Cleaner Environment (FACE) to solidify and expand their efforts (DiPerna 1985, pp. 111–25). FACE pursued all subsequent negotiations with local, state, and federal agencies. It campaigned to attract media attention, and made connections with other toxic waste groups.

Community groups in contaminated communities provide many important functions. They galvanize community support, deal with government, work with professionals, engage in health studies, and provide social and emotional support. They are the primary information source for people in contaminated communities, and often the most—even the only—accurate source (Gibbs 1982; Edelstein 1988, p. 144). Through their organization, Woburn activists report pride in learning science, protecting and serving their community, guaranteeing democratic processes, and personal empowerment.

Official Studies Are Conducted by Experts. In May 1980 the CDC and the National Institute for Occupational Safety and Health sent Dr. John Cutler to collaborate with the DPH on further study. By then, the Woburn case had national visibility due to national newspaper and network television coverage. In June 1980 Senator Edward Kennedy asked Anderson and Young to testify at hearings on the Superfund, providing further important public exposure. Five days after Jimmy Anderson died, the CDC/DPH study was released in January 1981, stating that there were 12 cases of childhood leukemia in East Woburn, when 5.3 were expected. Yet the DPH argued that the case-control method (12 cases, 24 controls) failed to find characteristics (e.g., medical histories, parental occupation, environmental exposures) that differentiated victims from nonvictims. Lacking environmental data prior to 1979, no linkage could be made to the water supply (Parker and Rosen 1981). That report helped bolster community claims of a high leukemia rate, although the DPH argued that the data could not implicate the wells. Cutler and his colleagues argued that in addition to the absence of case-control differences and the lack of environmental water exposure data, the organic compounds in the wells were known as animal, but not human, carcinogens (Condon 1991; Cutler et al. 1986; Knorr 1991).

The government agencies and their scientific experts worked to maintain their "ownership" of the problem by denying the link with toxics, and by maintaining control of problem solution (Gusfield 1981, pp. 10–15). Activists struggled to solidify their claim to ownership of the problem, to redefine causal responsibility, and to take on political responsibility. While epidemiologists admit to the uncertainties of their work, their usual solution is to err on the side of rejecting environmental causation, whereas community residents make the opposite choice.

Activists Bring in Their Own Experts. The activists had no "court of appeals" for the scientific evidence necessary to make their case. It became FACE's mission to obtain the information themselves. The conjuncture of Jimmy Anderson's death and the DPH's failure to implicate the wells led the residents to question the nature of official studies. They received help when Anderson and Young presented the Woburn case to a seminar at the Harvard School of Public Health (HSPH). Marvin Zelen and Steven Lagakos of the Department of Biostatistics became interested. Working with FACE members, they designed a health study, focusing on birth defects and reproductive disorders widely considered to be environmentally related. The biostatisticians and activists teamed up in a prototypical collaboration between citizens and scientists (Lagakos 1987; Zelen 1987). The FACE/HSPH study was more than a "court of appeals," since it transformed the activists' search for credibility. They no longer had to seek scientific expertise from outside; now they were largely in control of scientific inquiry.

Sources of data for the Woburn health study included information on 20 cases of childhood leukemia (ages 19 and under) diagnosed between 1964 and 1983, the DEQE water model of Wells G and H, and the health survey. The survey collected data on adverse pregnancy outcomes and childhood disorders from 5,010 interviews, covering 57 percent of Woburn residences with telephone. The researchers trained 235 volunteers to conduct the survey, taking precautions to avoid bias (Lagakos, Wessen, and Zelen 1984).[2]

Litigation and Confrontation. During this period, the DEQE's hydrogeological investigations found that the bedrock in the affected area was shaped like a bowl, with Wells G and H in the deepest part. DEQE's March 1982 report thus determined that the contamination source was not the Industri-Plex site as had been believed, but rather facilities of W. R. Grace and Beatrice Foods. This led eight families of leukemia victims to file a $400 million suit in May 1982 against those corporations for waste disposal practices that led to water contamination and disease. A smaller company, Unifirst, was also sued but quickly settled before trial (Schlictmann 1987). In July 1986 a federal district court jury found that Grace had negligently dumped chemicals; Beatrice Foods was absolved. An $8 million out-of-court settlement with Grace was reached in September 1986. The families filed an appeal against Beatrice, based on suppression of evidence, but the Appeals Court rejected the appeal in July 1990, and in October 1990 the United States Supreme Court declined to hear the case (Brown 1987; Brown and Mikkelsen 1990; Neuffer 1988).

The trial was a separate but contiguous struggle over facts and science. Through consultant physicians, immunologists, epidemiologists, and hydrogeologists, the families accumulated further evidence of adverse health effects. The data were not used in the trial, which never got to the point of assessing the causal chain of pollution and illness. Nevertheless, the process made the residents more scientifically informed.

Pressing for Official Corroboration. In February 1984 the FACE/Harvard School of Public Health data were made public. Childhood leukemia was found to be significantly associated with exposure to water from Wells G and H. Children with leukemia received an average of 21.2 percent of their yearly water supply from the well, compared to 9.5 percent for children without leukemia. Controlling for risk factors in pregnancy, the investigators found that access to contaminated water was associated with perinatal deaths since 1970, eye/ear anomalies, and CNS/chromosomal/oral cleft anomalies. With regard to childhood disorders, water exposure was associated with kidney/urinary tract and lung/respiratory diseases (Lagakos et al. 1984). If only the children that were in-utero at the time of exposure were studied, the positive associations were even stronger (Lagakos 1987).

Due to lack of resources, this study would not have been possible without community involvement. Yet this lay involvement led professional and governmental groups—the DPH, the Centers for Disease Control, the American Cancer Society, the EPA, and even the Harvard School of Public Health Department of Epidemiology—to charge that the study was biased. The researchers conducted extensive analyses to demonstrate that the data were not biased, especially with regard to the use of community volunteers as interviewers.[2] Still, officials argued that interviewers and respondents knew the research questions, respondents had potential recall bias, and the water model measured only household supply rather than individual consumption (Condon 1991; Knorr 1991). Thus, although activists expected the results to bring scientific support, they saw only criticism.

Having laid out the stages of popular epidemiological involvement, the main lay-professional disputes will now be described. These stages include lay participation, standards of proof, constraints on professional practice, quality of official studies, and professional autonomy.

Professional Ways of Knowing

Sociologists of science note that despite the existence of competing paradigms and models in any science, there is nevertheless a mainstream canon of knowledge, interpretation, and casual reasoning (Aronowitz 1988; Dickson 1984). The discussion here of professional ways of knowing refers to that mainstream. This does not mean that there is a monolithic worldview; there are indeed alternative voices. Yet a dominant mainstream approach permits us to make generalizations.

Traditional science contains a narrowly circumscribed set of assumptions about causality, the political and public role of scientists, and corporate and governmental social responsibility (Ozonoff and Boden 1987). Political-economic approaches argue that scientific inquiry is tied to corporate, political, and foundation connections that direct research and interpretation toward support for the status quo (Aronowitz 1988; Dickson 1984). It is useful to draw on both the political-economic perspective, which provides a social context for science, and the ethnomethodological/constructivist perspective, which shows us the internal workings of the scientific community (e.g., Latour 1987). There is a dynamic relationship between these two approaches—social movement actions provide the impetus for new scientific paradigms, and those new paradigms in turn spawn further social movement action.

The critics of the Woburn health survey did not represent *all science*, since residents received help from scientists who supported community involvement and believed that contaminated communities fail to receive fair treatment. Often such scientists have worked without compensation. Some began as critics of mainstream approaches, while others became critical only during the investigation. Some believed in a different causal paradigm, and some were critical of prevailing canons of significance levels. Others simply believed they could conduct better studies than official agencies.

Lay pressure for a different scientific approach is not directed at "pure" science.

In environmental health, we are dealing with *combined* government/professional units, e.g., DPH, EPA, and CDC. The end goal for activists is mainly acceptance by *government agencies*, since they have the power to act. At the same time, activists seek to become "popular scientists" who can win the support of scientific experts for the sake of knowledge.

Popular Participation and the Critique of Value-Neutrality

Activists disagree that epidemiology is a value-neutral scientific enterprise conducted in a sociopolitical vacuum. Critics of the Woburn health study argued that the study was biased by the use of volunteer interviewers and by prior political goals. Those critics upheld the notion of a value-free science in which knowledge, theories, techniques, and applications are devoid of self-interest or bias. Such claims are disputed by the sociology of science, which maintains that scientific, knowledge is not absolute, but rather is the subject of debate among scientists (Latour 1987). Scientific knowledge is shaped by social forces such as media influence, economic interest, political pressure, and social movement activism (Aronowitz 1988; Dickson 1984). On a practical level, scientific endeavors are limited by financial and personnel resources (Goggin 1986; Nelkin 1985); lay involvement often supplies the labor power needed to document health hazards. Science is also limited in the method(s) it uses to identify problems worthy of study. As an academic and official enterprise, science does not take its direction from the lay public.

Toxic waste activists see themselves as correcting problems not dealt with by the established scientific community. The centrality of popular involvement is evident in the women's health and occupational health movements that have been major forces in pointing to often unidentified problems and working to abolish their causes. Among the hazards and diseases thus uncovered are DES, Agent Orange, asbestos, pesticides, unnecessary hysterectomies, sterilization abuse, and black lung (Berman 1977; Rodriguez-Trias 1984; Scott 1988; Smith 1981). In these examples and in Woburn, lay activists are not merely research assistants to sympathetic scientists, but often take the initiative in detecting disease, generating hypotheses, pressing for state action, and conceiving and overseeing scientific studies.

Standards of Proof

Many scientists and public health officials emphasize various problems in standards of proof, ranging from initial detection and investigation to final interpretation of data. Assessment of public health risks of toxic substances involves four steps. Hazard identification locates the existence and extent of toxics. Dose response analysis determines the quantitative effects of the substance. Exposure assessment examines human exposure to the substances. Risk characterization integrates the first three steps in order to estimate the numbers of people who will be affected and the seriousness of the effects. From the scientific point of view, there is considerable uncertainty about each of these steps (Upton, Kneip, and Toniolo 1989).

Scientists and officials focus on problems such as inadequate history of the site, lack of clarity about the contaminants' route, determination of appropriate water sampling locations, small numbers of cases, bias in self-reporting of symptoms, obtaining of appropriate control groups, lack of knowledge about characteristics and effects of certain chemicals, and unknown latency periods for carcinogens (Condon 1991; Knorr 1991). Epidemiologists usually do not choose the research questions they think are amenable to study based on clear hypotheses, firmer toxicological data, and adequate sample size. Rather, they respond to a crisis situation, engaging in "reactive epidemiology" (Anderson 1985). Traditional approaches also tend to look askance at innovative perspectives favored by activists, such as the importance of genetic mutations, immune disregulation markers, and non-fatal and non-serious health effects (e.g., rashes, persistent respiratory problems) (Gute 1991; Ozonoff and Boden 1987).

For public health officials, disputes over health studies arise from shortcomings in knowledge about toxic waste-induced disease. A DPH official involved in Woburn for over a decade reflected on the vast changes in knowledge, personnel, and attitudes over that period. At first, public health researchers knew little about investigating clusters; environmental epidemiology was a new field; the state had few qualified scientists; and officials did not know how to involve the public. The DPH was trying out new approaches as they proceeded, without clearly established protocols (Condon 1991).

Activists view scientists as too concerned with perfection in scientific study. Residents believe that there have been visible health effects, clear evidence of contamination, and strong indications that these two are related. From their point of view, the officials and scientists are hindering a proper study, or are hiding incriminating knowledge. Residents observe corporations denying their dumping of toxic waste and that such substances have health effects. When public health agencies fail to find adverse health effects, many people view them as supporting corporate polluters. While residents agree with officials that cluster studies and environmental health assessment are new areas, they believe the agencies should spend more effort on residents' perceptions of crucial matters.

The level of statistical significance required for intervention is a frequent source of contention. Many communities that wish to document hazards and disease are stymied by insufficient numbers of cases to achieve statistical significance. Some professionals who work with community groups adhere to accepted significance levels (Lagakos et al. 1984), while others argue that such levels are as inappropriate to environmental risk as to other issues of public health, such as bomb threats and epidemics (Paigen 1982). Ozonoff and Boden (1987) distinguish statistical significance from public health significance, since an increased disease rate may be of great public health significance even if statistical probabilities are not met. They believe that epidemiology should mirror clinical medicine more than laboratory science, by erring on the safe side of false positives.

Hill (1987) argues that even without statistical significance we may find a clear association based on strength of association; consistency across persons, places, circumstances, and time; specificity of the exposure site and population; temporality of the exposure and effect; biological plausibility of the effect; coherence with known facts of the agent and disease; and analogy to past experience with related substances.

Pointing to the above as well as to more "provable" experimental models and dose-response curves, Hill argues that there are no hard and fast rules for establishing causality. Given the potential dangers of many classes of materials, he believes that often it is wise to restrict a substance to avoid potential danger.

Epidemiologists prefer false negatives to false positives—i.e., they would prefer to claim falsely that an association between variables does not exist when it does than to claim an association when there is none. This burden of proof usually exceeds the level required to argue for intervention. As Couto (1985) observes:

> The degree of risk to human health does not need to beat statistically significant levels to require political action. The degree of risk does have to be such that a reasonable person would avoid it. Consequently, the important political test is not the findings of epidemiologists on the probability of nonrandomness of an incidence of illness but the likelihood that a reasonable person, including members of the community of calculation [epidemiologists], would take up residence with the community at risk and drink from and bathe in water from the Yellow Creek area or buy a house along Love Canal.

Indeed, these questions are presented to public health officials wherever dispute occurs between the citizen and official perceptions. Beverly Paigen (1982), who worked with laypeople in Love Canal, clearly believes that standards of evidence are value-laden:

> Before Love Canal, I also needed a 95 percent certainty before I was convinced of a result. But seeing this rigorously applied in a situation where the consequences of an error meant that pregnancies were resulting in miscarriages, stillbirths, and children with medical problems, I realized I was making a value judgment . . . whether to make errors on the side of protecting human health or on the side of conserving state resources.

This dispute suggests the need for a more interactive approach to the *process* of scientific knowledge-making. Applying Latour's (1987) "science in action" framework, the real meaning of epidemiological "fact" cannot be seen until the epidemiologist experiences the citizenry and the problem being studied. Conversely, the public has no clear sense of what epidemiology can or cannot do for them until they or their neighbors are part of a study sample. In addition, both parties' perceptions and actions are jointly produced by their connections with other components, such as media, civic groups, and politicians. Latour's method bids us to ask this question of epidemiological research: for *whose* standards, and by what version of *proof* is a "standard of proof" determined and employed?

Institutional Constraints on Professional Knowledge and Action

Professional knowledge formation is affected by various institutional constraints. Professionals rarely view public initiatives as worthy of their attention. Laypeople have fewer scientific and financial resources than government and corporations (Paigen 1982). Without an ongoing relationship with the community, professionals enter only as consultants at a single point, and are unlikely to understand the larger framework of lay claims-making.

University-based scientists, a potential source of aid, frequently consider applied community research to be outside the regular academic reward structure (Couto 1985). Further, universities' increasing dependency on corporate and governmental support has made scholars less willing to challenge established authority (Goggin 1986). Grant support from federal agencies and private foundations is less likely to fund scholars who urge community participation and who challenge scientific canons and government policy.

Scientists often ally themselves with citizen efforts because they see flaws in official responses. Challenging state authority sometimes leads them to be punished as whistle-blowers. When Beverly Paigen, a biologist at the New York State Department of Health (DOH), aided Love Canal residents' health studies, she was harassed by her superiors. The DOH withdrew a grant application she had written without telling her, and refused to process an already funded grant. She was told that due to the "sensitive nature" of her work, all grants had to go through a special review process. Her professional mail was opened and taped shut, and her office was entered and searched at night. Paigen's state tax was audited, and she saw in her file a clipping about her Love Canal work. Later, the state tax commissioner apologized to her. Two officials in the regional office of the Department of Environmental Conservation were demoted or transferred for raising questions about the state's investigation (Paigen 1982). Similar cases have been documented elsewhere (Freudenberg 1984a, p. 57).

Quality and Accessibility of Official Data

Massive complaints in Massachusetts about the state's response to lay concerns over excess cancer rates in 20 Massachusetts communities (including Woburn) led to state senate (Commonwealth of Massachusetts 1987) and university (Levy et al. 1986) investigations, which found that the DPH studies were poorly conceived and methodologically weak. Most lacked a clear hypothesis, failed to mention potential exposure routes, and as a result rarely defined the geographic or temporal limits of the population at risk. Methods were presented erratically and inconsistently, case definitions were weak, environmental data were rarely presented, and statistical tests were inappropriately used (Levy et al. 1986). Frequently, exposed groups were diluted with unexposed individuals, and comparison groups were likely to include exposed individuals (Ozonoff and Boden 1987). This situation is striking, since the damaging effects of the poor studies and nonresponsiveness to the community led to the resignation of the public health commissioner, Bailus Walker, then head of the American Public Health Association (Clapp 1987).

State agencies are often unhelpful. A survey of all 50 states' responses to lay cancer cluster reports found in an estimated 1,300–1,650 such reports in 1988, clearly a large number for agencies already short-staffed. Many state health departments discouraged informants, in some cases requesting extensive data before they would go further. Rather than deal specifically with the complaint, many health departments gave a routine response emphasizing the lifestyle causes of cancer, the fact that one in three Americans will develop some form of cancer, and that clusters occur at random (Greenberg and Wartenberg 1991).

Officials may withhold information on the basis that it will alarm the public (Levine 1982), that the public does not understand risks, or that it will harm the business climate (Ozonoff and Boden 1987). Many scientists oppose public disclosure on the grounds that laypersons are unable to make rational decisions (Krimsky 1984). Toxic waste activists often are called "anti-scientific" when in fact they may simply work at science in a nontraditional manner. Indeed, these activists express support for scientists as important sources of knowledge (Freudenberg 1984b). FACE activists report that they have become highly informed about scientific matters, and are proud of it.

A cardinal assumption of science is that its truth and validity are affirmed by widespread recognition of the findings through open access to data among members of the scientific community. Yet the Massachusetts cases were not even shared with all appropriate scientists. Local health officials typically heard of elevated cancer rates through the media, rather than from state health officials. The EPA began a secret investigation of the Woburn data, leaving out researchers and Woburn residents who had already been involved in many investigations. Formed in 1984, the study group's existence was only discovered in 1988 (Kennedy 1988). The EPA did not view this as secrecy, but merely an internal "tell us what you think of this thing" (Newman 1991).

Professionalism, Controversy, and Information Control

It is particularly ironic that epidemiology excludes the public, since the original "shoe-leather" work that founded the field is quite similar to popular epidemiology. Woburn residents' efforts are very reminiscent of John Snow's classic study of cholera in London in 1854, where that doctor closed the Broad Street pump to cut off contaminated water. Yet modern epidemiology has come far from its original shoe-leather origins, turning into a laboratory science with no room for lay input.

The combination of epidemiologic uncertainty and the political aspects of toxic waste contamination leads to scientific controversy. According to Latour (1987, p. 132), rather than a "diffusion" of ready-made science, we must study how "translations" by many parties of undecided controversies lead to a consensual reality. From the point of view of traditional epidemiologists, citizens' translations hinder consensual production of science. Yet, in fact, the scientific community is itself disunified on most issues of environmental epidemiology, and laypeople are partaking in the related consensual production.

In this struggle, citizens use controversy to demystify expertise and to transfer problems from the technical to the political arena (Nelkin 1985). This redefinition of the situation involves a lay approach to "cultural rationality" as opposed to the scientific establishment's "technical rationality" (Krimsky and Plough 1988). This form of struggle was described earlier, in reference to when residents ask officials whether they would live in and drink water from the contaminated community. We also may view gender differences as representative of differing rationalities. Women are the most frequent organizers of lay detection, partly because they are the chief health arrangers for their families, and partly because they are more concerned than men with local environmental issues (Blocker and Eckberg 1989; Levine 1982). From this perspective, women's cultural rationality is concerned with who would be willing to drink local water, and how their families experience daily life.

Bad Science, Good Science, Popular Science

One way to look at official support for lay involvement is to view it as simply "good politics," whereby the government provides a formal mechanism for citizen participation in such areas as Environmental Impact Statements and Recombinant DNA Advisory Panels. However, public participation was limited in these cases to minor roles on panels that already had an official agenda (Jasanoff 1986; Krimsky 1984).

But as we observe in popular epidemiology, lay involvement is not merely "good politics." It is also "good science," since it changes the nature of scientific inquiry. This involves four elements addressed throughout:

1. Lay involvement identifies the many cases of "bad science," e.g., poor studies, secret investigations, failure to inform local health officials.
2. Lay involvement points out that "normal science" has drawbacks, e.g., opposing lay participation in health surveys, demanding standards of proof that may be unobtainable or inappropriate, being slow to accept new concepts of toxic causality.
3. The combination of the above two points leads to a general public distrust of official science, thus pushing laypeople to seek alternate routes of information and analysis.
4. Popular epidemiology yields valuable data that often would be unavailable to scientists. If scientists and government fail to solicit such data, and especially if they consciously oppose and devalue them, such data may be lost.

We see these four elements in many contaminated communities, but in Woburn the lay contribution to scientific endeavor has been exceptional. The Woburn case was the major impetus for the establishment of the state cancer registry (Clapp 1987). Activism also has contributed to increasing research on Woburn: the DPH and CDC are conducting a major five-year reproductive outcome study of the city, utilizing both prospective and retrospective data, and citizens have a large role in this process. The DPH is conducting a case-control study of leukemia, and an MIT study will study genetic mutations caused by trichloroethylene (TCE), to investigate their role in causing leukemia (Latowsky 1988).

Popular epidemiology can sway government opinion over time, especially when activists doggedly stick to their own work while constantly participating with official bodies. The since-retired DPH Commissioner who took office in 1988, Deborah Prothrow-Stith, asked for a more official relationship with FACE. Upon visiting Woburn, the commissioner said she was "struck with how epidemiology is dependent on the role the public plays in bringing these things to light" (Mades 1988). A DPH official in 1990, going over the chronology of events, noted that the FACE/Harvard School of Public Health study found positive associations between well water and adverse reproductive outcomes, a position the DPH avoided for the preceding six years since the study was published (Kruger 1990). Other officials now view the study as an important source of research questions and methods that informs the ongoing official studies, as well as a prompt to government action. Indeed, they expect that the new studies may show evidence of adverse health effects (Condon 1991; Knorr 1991).

Popular epidemiologists also provide continuity in the scientific process. As a leading activist stated: "We have been the institutional memory of studies in Woburn. We have seen agency heads come and go. We have seen project directors come and go. Our role has been to bring those efforts together and to help the researchers investigate what was going on all throughout the area" (Latowsky 1990). To understand the significance of that position, we may observe that as late as 1990 an EPA Remedial Project Manager for the Woburn site could hear a question, "Is the leukemia cluster a cause for urgency of cleanup?" and respond that "Our investigation is not concerned with the cluster of leukemia. It's really irrelevant. We're on a schedule based on our regulations" (Newman 1990). While EPA does not consider its responsibilities to include cluster detection (Newman 1991), all elements are related for residents.

Ozonoff (1988) sums up the Woburn impact:

> In hazardous waste, three names come up—Love Canal, Times Beach, and Woburn. Woburn stands far and above them all in the amount of scientific knowledge produced. All over the country, Woburn has put its stamp on the science of hazardous waste studies.

Of particular value is the discovery of a TCE syndrome involving three major body systems—immune, cardiovascular, and neurological—which is increasingly emerging in other TCE sites.

How Do We Know If Lay Investigations Provide Correct Knowledge?

It is obviously necessary to evaluate the correctness of findings that result from popular epidemiology. Such knowledge is not "folk" knowledge with an antiscientific basis. In most cases, popular epidemiology findings are the result of scientific studies involving trained professionals, even if they begin as "lay mapping" of disease clusters without attention to base rates or controls. Indeed, lay-involved surveys are sometimes well-crafted research with defendable data. Lay-people may initiate action and even direct the formulation of hypotheses, but they work *with* scientists, not in place of them. Thus, the end results can be judged by the same criteria as any study. However, since all scientific judgments involve social factors, there are no simple algorithms for ascertaining truth. Scientific inquiry is always full of controversy; what is different here is that laypeople are entering that controversy.

Public health officials worry that some communities might exaggerate the risks of a hazard, or be wrong about the effects of a substance. Yet if this occurs, it must be seen in context: community fears are too often brushed aside and data has been withheld. Given the increasing cases (or at least recognition of those cases) of technological disasters, drug side effects, and scientific fraud, public sentiment has become more critical of science. In response, lay claims may be erroneous. But this is the price paid for past failures and problems, and is a countervailing force in democratic participation (Piller 1991). Exaggerated fears may be understood as signs of the need to expand public health protection, rather than justifications to oppose lay involvement. Even if a community makes incorrect conclusions, their data base may still remain useful for different analyses. As mentioned before, the DPH disagrees with the Harvard/FACE conclusions, yet they are now testing those same relationships in their own study.

Even if they do not exaggerate claims, lay investigators may pursue specific inquiries with their own agenda in mind. For example, they may emphasize certain health data and minimize other reports. This may stem from the salience of certain hazards or diseases, the population affected (especially children), and the dynamics between residents and corporate and governmental actors. Citizens' efforts are typically more avowedly political and media-oriented, since their lack of power compels them to mobilize mostly in public rather than scientific venues.

Conclusion

Limitations of This Study

Despite the triangulation of various forms of data, this study is based on a single site. The laypeople who played such a major role may be a unique group in their abilities and tenacity, leading them to be at the forefront of challenges to official science. As well, the Woburn activists are one of the few groups to find such high-quality scientific collaboration. To the extent that these activists and their scientific colleagues are unique, parts of the Woburn experience may not be widely generalizable to other toxic waste sites. The literature does tell us that some parts of the Woburn experience are indeed common to other sites, especially the challenge to, and distrust of, scientific and governmental elites. Those parts that are perhaps less generalizable are the successful application of health survey techniques, and eventual acceptance of activists' role by public health agencies. Lastly, both the study of the existence of disease clusters and the causal role of toxic substances are in their early stages of development. The intense conflicts over these issues may make it difficult to judge the validity of research methods, results, and interpretation.

Causes and Implications of Popular Epidemiology

Popular epidemiology stems from the legacy of health activism, growing public recognition of problems in science and technology, and the democratic upsurge regarding science policy. This chapter has pointed to the difficulties faced by communities due to differing conceptions of risk, lack of resources, poor access to information, and unresponsive government. In popular epidemiology, as in other health-related movements, activism by those affected is necessary to make progress in health care and health policy. In this process a powerful reciprocal relationship exists between the social movement and new views of science. The striking awareness of new scientific knowledge, coupled with government and professional resistance to that knowledge, leads people to form social movement organizations to pursue their claims-making. In turn, the further development of social movement organizations leads to further challenges to scientific canons. The socially constructed approach of popular epidemiology is thus a result of both a social movement and a new scientific paradigm, with each continually reinforcing the other.

Dramatically increasing attention to environmental degradation may make it easier for many to accept causal linkages previously considered too novel. Further, this expanding attention and its related social movements may lead to the identification of more disease clusters. This then could lead to the reevaluation of problems of low base rates in light of how other sciences (e.g., physics, paleontology) conduct research on low base-rate phenomena. As well, growing numbers of similar cases containing small sample sizes and/or low base-rate phenomena may allow for more generalizability. These increasing cases also produce more anomalies, allowing for a paradigm shift.

Causal explanations from outside of science also play a role. Legal definitions of causality, developed in an expanding toxic tort repertoire, are initially determined by judicial interpretation of scientific testimony. Once constructed, they can take on a life of their own, directing public health agencies and scientists to adhere to scientific/ legal definitions that may or may not accord completely with basic science. At the least, they set standards by which scientific investigations will be applied to social life (e.g., court-ordered guidelines on claims for disease caused by asbestos, nuclear testing, DES).

Lay and professional approaches to knowledge and action on environmental health risks are structurally divergent, much as Freidson (1970) conceives of the inherent differences and conflicts between patient and physician. Yet just as modern efforts from both medicine and its clientele seek an alternate model, so too does popular epidemiology offer a new path. Popular epidemiology offers a bridge between the two perspectives, a bridge largely engineered and constructed by lay activists, yet one with the potential to bring citizens and scientists together.

NOTES

1. The *interview* included questions such as: "What has the progress of the Woburn situation taught you about the nature of environmental hazards?", "Did you know what you were getting at when you filed the suit?", "Do you expect you or your family will have any future health problems as a result of the Woburn pollution from Wells G and H?" A complete interview guide is available from the author.

2. The researchers conducted extensive analyses to demonstrate that the data were not biased. They found no differences when they compared baseline rates of adverse health effects for West Woburn (never exposed to Wells G and H water) and East Woburn (at a period prior to the opening of the wells). They examined transiency rates to test whether they were

related to exposure and found them to be alike in both sectors. Other tests ruled out various biases potentially attributable to the volunteer interviewers (Lagakos et al. 1984).

REFERENCES

Anderson, Henry A. 1985. "Evolution of Environmental Epidemiologic Risk Assessment." *Environmental Health Perspectives* 62:389–92.

Aronowitz, Stanley, 1988. *Science as Power: Discourse and Ideology in Modern Society*. Minneapolis: University of Minnesota Press.

Berman, Daniel. 1977. "Why Work Kills: A Brief History of Occupational Health and Safety in the United States." *International Journal of Health Services* 7:63–87.

Blocker, T. Jean, and Douglas Lee Eckberg. 1989. "Environmental Issues as Women's Issues: General Concerns and Local Hazards." *Social Science Quarterly* 70:586–93.

Brown, Phil. 1987. "Popular Epidemiology: Community Response to Toxic Waste-Induced Disease in Woburn, Massachusetts." *Science, Technology, and Human Values* 12:78–85.

———, and Edwin J. Mikkelsen. 1990. *No Safe Place: Toxic Waste, Leukemia, and Community Action*. Berkeley: University of California Press.

Clapp, Richard. 1987. Personal Interview, March 14.

Commonwealth of Massachusetts. 1987. "Cancer Case Reporting and Surveillance in Massachusetts." Senate Committee on Post Audit and Oversight. Boston, Massachusetts (September).

Condon, Suzanne. 1991. Personal Interview, August 7.

Couto, Richard A. 1985. "Failing Health and New Prescriptions: Community-Based Approaches to Environmental Risks." Pp. 53–70 in *Current Health Policy Issues and Alternatives: An Applied Social Science Perspective*, edited by Carole E. Hill. Athens: University of Georgia Press.

Cutler, John J., Gerald S. Parker, Sharon Rosen, Brad Prenney, Richard Healy, and Glyn G. Caldwell. 1986. "Childhood Leukemia in Woburn, Massachusetts." *Public Health Reports* 101:201–5.

Dickson, David. 1984. *The New Politics of Science*. New York: Pantheon.

DiPerna, Paula. 1985. *Cluster Mystery: Epidemic and the Children of Woburn, Mass.* St. Louis: Mosby.

Edelstein, Michael. 1988. *Contaminated Communities: The Social and Psychological Impacts of Residential Toxic Exposure*. Boulder: Westview Press.

Epstein, Samuel S., Lester O. Brown, and Carl Pope. 1982. *Hazardous Waste in America*. San Francisco: Sierra Club Books.

Fisher, Sue. 1986. *In the Patient's Best Interests*. New Brunswick, NJ: Rutgers University Press.

Freidson, Elliott. 1970. *Profession of Medicine*. New York: Dodd, Mead.

Freudenberg, Nicholas. 1984a. *Not in Our Backyards: Community Action for Health and the Environment*. New York: Monthly Review.

———. 1984b. "Citizen Action for Environmental Health: Report on a Survey of Community Organizations." *American Journal of Public Health* 74:444–48.

Frumkin, Howard, and Warren Kantrowitz. 1987. "Cancer Clusters in the Workplace: An Approach to Investigation." *Journal of Occupational Medicine* 29:949–52.

Gibbs, Lois Marie. 1982. "Community Response to an Emergency Situation: Psychological Destruction and the Love Canal." Paper presented at the American Psychological Association (August).

Goggin, Malcolm L. 1986. "Introduction. Governing Science and Technology Democratically: A Conceptual Framework." Pp. 3–31 in *Governing Science and Technology in a Democracy*, edited by Malcolm L. Goggin. Knoxville: University of Tennessee Press.

Greenberg, Michael, and Daniel Wartenberg. 1991. "Communicating to an Alarmed Community about Cancer Clusters: A Fifty State Study." *Journal of Community Health* 16:71–82.

Gusfield, Joseph R. 1981. *The Culture of Public Problems*. Chicago: University of Chicago Press.

Gute, David. 1991. Personal Interview, August 14.

Hill, Austin Bradford. 1987. "The Environment and Disease: Association or Causation." Pp. 15–20 in *Evolution of Epidemiologic Ideas*, edited by Kenneth Rothman. Chestnut Hill, MA: Epidemiology Resources, Inc.

Jasanoff, Sheila. 1986. "The Misrule of Law at OSHA." Pp. 155–78 in *The Language of Risk: Conflicting Perspectives on Occupational Health*, edited by Dorothy Nelkin. Beverly Hills: Sage.

Kennedy, Dan. 1988. "EPA to Say Pollutants Caused Leukemia." *Woburn Daily Times*, May 9.

Kleinman, Arthur. 1980. *Patients and Healers in the Context of Culture*. Berkeley: University of California Press.

Knorr, Robert. 1991. Personal Interview, August 7.

Krauss, Celene. 1989. "Community Struggles and the Shaping of Democratic Consciousness." *Sociological Forum* 4:227–39.

Krimsky, Sheldon. 1984. "Beyond Technocracy: New Routes for Citizen Involvement in Social Risk Assessment." Pp. 43–61 in *Citizen Participation in Science Policy*, edited by James C. Peterson. Amherst: University of Massachusetts Press.

Krimsky, Sheldon, and Alonzo Plough. 1988. *Environmental Hazards: Communicating Risks as a Social Process*. Boston: Auburn House.

Kruger, Elaine. 1990. "Environmental Exposure Assessment and Occurrence of Adverse Reproductive Outcomes in Woburn, Mass." Presentation at "Investigations in the Aberjona River Watershed," Woburn, Massachusetts (May 18).

Lagakos, Steven. 1987. Personal Interview, April 6.

———, Barbara J. Wessen, and Marvin Zelen. 1984. "An Analysis of Contaminated Well Water and Health Effects in Woburn, Massachusetts." *Journal of the American Statistical Association* 81:583–96.

Latour, Bruno. 1987. *Science in Action: How to Follow Scientists and Engineers through Society*. Cambridge: Harvard University Press.

Latowsky, Gretchen. 1988. Personal Interview, May 26, 1988.

———. 1990. "FACE's Role in the Community." Presentation at "Investigations in the Aberjona River Watershed," Woburn, Massachusetts (May 18).

Levine, Adeline Gordon. 1982. *Love Canal: Science, Politics, and People*. Lexington, MA: D. C. Heath.

Levy, Barry S., David Kriebel, Peter Gann, Jay Himmelstein, and Glenn Pransky. 1986. "Improving the Conduct of Environmental Epidemiology Studies." Worcester, Massachusetts: University of Massachusetts Medical School, Department of Family and Community Medicine, Occupational Health Program.

Lilienfeld, Abraham. 1980. *Foundations of Epidemiology*, second edition. New York: Oxford University Press.

Mades, Nancy. 1988. "Commissioner Wants FACE-DPH Pact." *Woburn Daily Times*, April 7.

Nash, June, and Max Kirsch. 1986. "Polychlorinated Biphenyls in the Electrical Machinery Industry: An Ethnological Study of Community Action and Corporate Responsibility." *Social Science and Medicine* 23:131–38.

Nelkin, Dorothy (Ed.). 1985. *The Language of Risk: Conflicting Perspectives in Occupational Health*. Beverly Hills: Sage.

Neuffer, Elizabeth. 1988. "Court Orders New Hearings in Woburn Pollution Case." *Boston Globe*, December 8.

Newman, Barbara. 1990. "Investigations and Remediation at the Wells G and H Site." Presentation at "Investigations in the Aberjona River Watershed," Woburn, Massachusetts (May 18).

————. 1991. Personal Interview, September 6.

Ozonoff, David. 1988. Presentation at "Examining Woburn's Health" (April 24).

————, and Leslie I. Boden. 1987. "Truth and Consequences: Health Agency Responses to Environmental Health Problems." *Science, Technology, and Human Values* 12:70–77.

Paigen, Beverly. 1982. "Controversy at Love Canal." *Hastings Center Report* 12(3):29–37.

Parker, Gerald, and Sharon Rosen. 1981. "Woburn: Cancer Incidence and Environmental Hazards." Massachusetts Department of Public Health (January 23).

Piller, Charles. 1991. *The Fail-Safe Society: Community Defiance and the End of American Technological Optimism*. New York: Basic Books.

Rodriguez-Trias, Helen. 1984. "The Women's Health Movement: Women Take Power." Pp. 107–26 in *Reforming Medicine: Lessons of the Last Quarter Century*, edited by Victor Sidel and Ruth Sidel. New York: Pantheon.

Roth, Julius. 1963. *Timetables*. Indianapolis: Bobbs-Merrill.

Rothman, Barbara Katz. 1986. *The Tentative Pregnancy: Prenatal Diagnosis and the Future of Motherhood*. New York: Viking.

Rubin, James H. 1987. "Justices Limit Right of Citizens to Sue on Water Pollution Violations." *New York Times*, December 2.

Schlictmann, Jan. 1987. Personal Interview, May 12.

Scott, Wilbur J. 1988. "Competing Paradigms in the Assessment of Latent Disorders: The Case of Agent Orange." *Social Problems* 35:145–61.

Smith, Barbara Ellen. 1981. "Black Lung: The Social Production of Disease." *International Journal of Health Services* 11:343–59.

Stimson, Gerry V., and Barbara Webb. 1975. *Going to See the Doctor*. London: Routledge.

Strauss, Anselm, Leonard Schatzman, Rue Bucher, Danuta Ehrlich, and Melvin Sabshin. 1964. *Psychiatric Ideologies and Institutions*. New York: Free Press.

Upton, Arthur C., Theodore Kneip, and Paolo Toniolo, 1989. "Public Health Aspects of Toxic Chemical Disposal Sites." *Annual Review of Public Health* 10:1–25.

Waitzkin, Howard, 1983. *The Second Sickness: Contradictions of Capitalist Health Care*. New York: Macmillan.

————. 1989. "A Critique of Medical Discourse: Ideology, Social Control, and the Processing of Social Context in Medical Encounters." *Journal of Health and Social Behavior* 30:220–39.

Zelen, Marvin. 1987. Personal Interview, July 1.

Zola, Irving K. 1973. "Pathways to the Doctor—From Person to Patient." *Social Science and Medicine* 7:677–89.

QUESTIONS

1. How does popular epidemiology differ from traditional epidemiology? Be specific.

2. What roles are professional medical people likely to play in popular epidemiology movements?

Chapter Twenty-Two

Environmental Movements and Expert Knowledge
Evidence for a New Populism

Stephen R. Couch and Steve Kroll-Smith

Common to most social movements is a populist appeal to rights and entitlements based upon the idea of citizenship (Waltzer 1991; Seligman 1992). A rhetoric of moral entreaty fashions appeals to freedom of speech, thought, and faith, the right to own property, the right to economic welfare, the right to clean environments, and so on. In social movements moral understandings of right and wrong, good and bad, proper and improper are created, affirmed, and changed (Gusfield 1963).[1]

Like their counterparts in the feminist, labor, and civil rights movements, environmental movements typically appeal to issues of justice and rights to make their claims. At the end of the nineteenth century and the beginning of the twentieth, for example, people organized in response to a perceived need to protect and conserve species and habitats (Schnaiberg 1980; Nash 1989). Their moral appeal was based on accepting a transition from liberalism's natural rights philosophy to a "rights of nature" ethic (Nash 1989, p. 7). More recently, appeals to environmental justice and the more provocative charge of environmental racism direct attention to unequal distributions of risks (Bullard 1992).

Evidence reveals contemporary environmental movements organizing around more than a populist appeal to moral or ethical rights. Specifically, people believing they are endangered by the production, use, and disposal of toxic materials are fusing a moral appeal to safe environments with a popular appropriation of expert knowledge. This is designed to make a particularly persuasive claim on institutions to change or modify their behaviors and policies (Kroll-Smith, Couch, and Floyd 1995). These movements are doing more than relying on a rhetoric of civil rights or environmental justice, as important as these appeals are; they are also arming themselves with the lingual resources of toxicology, risk assessment, biomedicine, environmental impact inventories, nuclear engineering, and other instruments of reason. We find this development interesting for what it represents about the activities of the movement groups. It is also interesting for what it represents about the complex exchange between citizens, expert knowledge, and expert systems in the waning years of the twentieth century.

If Locke could write in the seventeenth century that the "rights of man" would be assured by joining the ordinary person to "instrumental rationality," by the nineteenth century ordinary people were effectively separated from technical ways of knowing the world. From the early twentieth century to the present, appeals to human rights were increasingly dissociated from rationality and its instruments (Touraine 1995). Expert knowledge was the province of the professions, licensed and protected by the state (Giddens 1990).

Expert systems emerged, mysterious and complicated, almost magical, artfully manipulating weights and measures, microscopes, slide-rules, tests of all sorts—in short, the instruments of rational knowledge. Social movements relied on ethical and moral, not scientific, appeals to lobby for change. If an expert opinion was needed, the best a person or group could do was hire an expert to represent their interests. Sociologists wrote about "symbolic polities" (Gusfield 1963, p. 180) and "rhetorics of transcendence" (Stewart, Smith, and Denton 1984, p. 121). Ordinary citizens could certainly appeal to scientific ways of knowing to assist them in constructing a rhetorical message; but they were not themselves claiming to know something new and legitimate based on their use of scientific knowledge or on their own experience or insight. Separating citizens from instrumental rationality insured that modernity would succeed, as Alain Touraine writes, in separating the "world of nature, which is governed by the laws discovered and used by rational thought, and the world of the Subject" (1995, p. 57).

But nature and the Subject collided. First, in the 1970's it was with the Love Canal, the dioxin contamination of Times Beach, Missouri, the nuclear accident at Three Mile Island, and the underground mine fire in Centralia, Pennsylvania; and the collisions continued into the 80's with the Bhopal chemical disaster, the Chernobyl nuclear fire, the eleven million gallons of crude oil spilled in Prince William Sound by the Exxon Valdez; and into the 90's with the decommissioning of seventeen high and low level radioactive depositories owned by the U.S. Department of Energy, coupled with the massive problems of global environmental change. These signal events, of course, occurred and are occurring in a tandem with hundreds, if not thousands, of more local, less newsworthy, events.

A plethora of environmental disasters, and threats of countless more, some literally global in scope, are changing the relationships of ordinary people to experts and expert knowledge. Citizens now know that environmental dangers require technical solutions, but they are increasingly distrustful of the experts. In response to this gap between knowing a problem requiring a rational, technical solution and a decreasing readiness to trust in the good intentions or know-how of experts, groups and organizations are unhinging the languages of expertise from expert systems. They are, then, taking these attributes of expertise into their communal worlds. They tinker with them sufficiently to make rational sense of their miseries and appeal to significant institutional others to change, based on a rhetoric of communal, or what we might call, moral rationality. It is one thing for a young woman cradling an infant to appeal to a government hearing board in the language of home, hearth, and children; it is quite another when this woman clutches her child to her breast and talks in the complicated language of toxicology, discussing the neurotoxic effects of boron or dichlorotetrafluoroethane on her child's cognitive development.

Contemporary environmental grassroots mobilization is joining populism with instrumental rationality. Appeals fashioned from personal and communal experiences linked to local claims to know something new about physical and chemical processes are changing patterns of political participation, institutional decision making, and, not surprisingly, the very methodology of science itself. This admittedly bold thesis will direct the remarks to follow.

Plan of Work

The recent work of Anthony Giddens (1990; 1991; 1994) and Ulrich Beck (1992; 1995a; 1995b), and to a lesser extent Kai Erikson (1991) is assembled as groundwork for understanding why grassroots groups are increasingly active in producing their own lay or citizen science. The strength of Beck and Giddens, in particular, is their provocative conceptualization of large-scale social changes, including changes in expert knowledge and expert systems, and the linkages between these changes and humanly-induced environmental problems.

A weakness in their approach, however, is a marked lack of interest in how these social changes might actually be occurring. Scott Lash, among others (see Irwin 1995), observes Giddens' and Beck's studied interest in experts and their knowledge

> with relative neglect of the grassroots. [They concentrate] on the formal and institutional at the expense of the increasing proportion of social, cultural and political interaction in our increasingly disorganized capitalist world that is going on outside of institutions. (1994, p. 200)

Preferring to theorize without systematic case studies, they invite others to examine the fit between their provocative ideas and situated, grounded relationships between people, problems, and science. Contemporary environmental movements, we will argue, are creating a new more populist structure within which expert knowledge is being created, organized, and transmitted. Careful inquiry into three movements reveals citizens responding to toxins in their local environments. Their responses involve removing scientific, technical, and medical forms of expertise (and sometimes one or two experts) from traditional institutions and locating them in local, communal fields of action. In combining scientific knowledge with populism, anti-toxics responses are examples of a new structural arrangement between science and people. This arrangement both specifies and extends the theoretical arguments of Giddens and Beck.

Two cases are taken from the United States, where social movements have traditionally been an especially important part of the political landscape. The first case is the Love Canal and the anti-toxics movement; it is constructed from secondary sources and primary literature from environmental organizations. The second case is the environmental illness movement; it is reconstructed from a recent book by the second author. A third case is found in Great Britain to illustrate the parallel development of grassroots mobilization and the popular appropriation of scientific expertise out of the United States. This case, labor mobilization in response to the use of 2,4,5-T, is drawn from secondary sources. Each of the three cases illustrates a fusion of traditional populist appeals with scientific reasoning accomplished at the local level by movement remembers.

A final discussion aggregates observations made from the cases to suggest that as ordinary people fashion scientific accounts of their troubles, they often challenge traditional methodological assumptions, constructing an alternative epistemology based on immediate, practical concerns. And, importantly, there is evidence that institutions are willing to learn from a practical epistemology exercised by people who are not licensed to speak the languages of instrumental rationality. If a new pattern of institutional learning is emerging in response to ordinary people joining languages of expertise to moral appeals for change, the new populism we are describing is evidence of a new form of political participation and simultaneously of a new science.

Technological Hazards, Citizens, and Science

Ulrich Beck's "risk society" (1992), Anthony Giddens' "risk culture" (1991), and Kai Erikson's more colorful phrase, "a new species of trouble" (1991; 1994) highlight the increasing incapacity of science to reach a consensus on the scope and severity of invisible biospheric dangers that trouble, if not disable, an increasing number of people. Relying on images from the Old Testament, Erikson sees environmental and technological dangers expressing "a malevolence that the authors of Revelations would have found difficult to believe" (1991, p. 15). Amidst escalating danger, Giddens writes, "risk becomes fundamental to the way both lay actors and technical specialists organise the social world" (1991, p. 3). Beck concurs: "(I)n the risk society . . . unknown and unintended consequences come to be a dominant force in history . . ." (Beck 1992, p. 22).

The source of Beck's "unintended consequences" is, not surprisingly, technology's intervention into physical environments. Instrumental rationality and natural processes are interacting to produce catastrophic risks that have outrun the ability of our institutions to control them. Ecological devastation, in other words, is produced (at least in part) by scientific knowledge and measurement procedures.

Ironically, however, the failures of science and technology that put each of us at risk are also a principal resource for their further development. "(S)cience and technology," Giddens writes, "are the only means of bringing their own damage into view" (1994, p. 208). Indeed, for Beck, the greater the threat, the greater is our dependence on science (1995a, p. 92; see also 1995b, p. 112).

Complicating this picture is a growing skepticism among lay people who question the validity of expert knowledge. After all, if routine actions by expert systems are creating dangers these systems can neither predict nor control, but nevertheless appear to benefit from, the public would be foolish to be anything but wary. Science, it would appear, "is the Sorcerer's Apprentice, having itself succumbed to the doubt it let loose" (Beck 1995b, p. 119). From the vantage point of the lay person, institutional science is experienced as increasingly incoherent. It creates problems it cannot solve while simultaneously extending its control over everyday life.

While a crisis of trust threatens the institution of science, however, it does not threaten the development or continued utility of scientific knowledge. Importantly, as people are becoming skeptical of experts they are also becoming increasingly aware of their dependence on expert knowledge to address the sources of their misery (Giddens 1990, p. 150; Beck 1992, Ch. 8). Beck asks, "Are we witnessing the last days of the lay person, the individual dependent upon the technical knowledge of others?" (1995a, p. 55). Instrumental or scientific rationality is moving from its privileged locations in universities and corporations to lay communities where citizens are engaging in the collection and analysis of data that bears directly on their particular, local circumstances. Thus, it is not the demise of science we are witnessing, a claim made by many postmodernists, but a strikingly new phase in its development.

But what specifically does this new phase look like? How is it organized? Giddens talks vaguely about "open spaces for public dialogue" between citizens with expertise and experts (1994, p. 17), while Beck writes of the hope that "the culturally manufactured perceptibility of hazards [will] restore to us the competence to judge for ourselves" (1995b, p. 184). Spaces and spheres are vague images, useful when more concrete behaviors, situations, and events are left unanalyzed. A careful look at the concrete events and life circumstances of people who are struggling to interpret untoward changes in their bodies and the local environments they inhabit, however, can dispense with such figurative expressions. Moreover, by locating the good, albeit abstract, ideas of Beck and Giddens in local histories, we encourage them to recognize a nascent and novel form of populism, one that joins traditional ethical and moral appeals to claims to know something valid and rational about how the world works.

The three cases we are about to discuss describe in some detail how citizen movements are appropriating scientific and medical knowledge, moving them into their communal organizations and, importantly, changing a few of their key assumptions about how the world is to be known. Common to each of these movements are people who find the authoritative voices of science and medicine unable to make sense of their bodies and environments. Importantly, they are doing more than questioning the uses of expert knowledge. Indeed, often, they become experts themselves. In so doing, they alter a few of the foundational assumptions of classical rational thought. Each case documents one or more of the following ideas to be taken up and discussed further in the conclusion. First, as people find medicine and science incoherent explanations of their local troubles, they are learning to trust the validity of their immediate, sensory, somatic experiences. In each of these movements, people are insisting that subjective experience be reintroduced as a valid criterion of judgment in science and thus collapsing the subjective/objective distinction dear to classic rational thought.

A second and related point documented in the cases is the possibility that local or contextual knowledge is necessary to a scientifically valid way of knowing. Local knowledge, in the forms of past experiences, daily routines, perceived relationships (between, for example, animal behaviors and herbicide spraying) is routinely dismissed as anecdotal by scientists and medical researchers. But an increasing number of environmental movements are constructing a popular version of research that acknowledges the importance of contextual knowledge. While their view of the world might be more local than that of a research team from a distant university, their idea of relationships is arguably more complicated, including as it does a confluence of variables typically ignored by conventional science.

A third point made by the cases is really a consequence of the first two: namely, accumulated knowledge about one local setting may not always be valid in another setting. If institutional science rests on the assumption that knowledge is cumulative and, in principal, universal, environmental movements counter with the argument that good science might also rest on particular, situated forms of knowing.

Finally, and perhaps not surprisingly, evidence in each of the three cases suggests that the appropriation of rational modes of inquiry are becoming key organizing strategies for environmental movements. The perceived needs for a citizen-based science is contributing to movement goals, used as a recruitment strategy, and, importantly, fashioning the very language of grievance.

Love Canal and the Citizens Clearinghouse for Hazardous Waste

In the summer of 1978, a housewife in upstate New York learned that the school her son attended was on top of a chemical waste dump. Concerned that her son's health problems were connected with the dump, she was rebuffed in her attempts to have him transferred to another school. Summoning all her courage, this shy woman went door-to-door, asking her neighbors if their families were experiencing any health problems. Armed with her "housewife data," Lois Gibbs began a battle with the New York State Department of Health that would result in the eventual relocation of most of the Love Canal community. Gibbs would go on to found the Citizens Clearinghouse for Hazardous Waste (CCHW), a national organization providing assistance to community groups dealing with waste problems. Her Love Canal experience would shape her approach to dealing with hazardous waste problems in communities throughout the United States.

The Love Canal story has taken on almost iconic proportions within parts of the environmental community. The story has been told elsewhere and need not be recounted here in detail (see Levine 1982). But for our purposes, Gibbs's experience with experts and expert knowledge at Love Canal provides an excellent empirical example of what was discussed above. Over time, Gibbs came to distrust the expert system and the knowledge produced by that system, but without losing complete faith in the utility of scientific knowledge. But where scientific knowledge contradicted her experience and the experience of others living in the community, experience was the knowledge she chose to trust.

Once problems at Love Canal were discovered, the New York Department of Health undertook a health study. The results of this study showed a high incidence of miscarriages and birth defects among residents living close to the Canal. Based on these results, at a meeting in Albany with some Love Canal residents on August 2, 1978, the State Health Commissioner recommended that all pregnant women and children under the age of two who lived near the Canal should relocate. This infuriated Lois Gibbs and others; Gibbs shouted at the Commissioner, "If the dump will hurt pregnant women and children under two, what, for God's sake, is it going to do to the rest of us?" (Gibbs 1982, p. 30).

Instead of reluctantly accepting the findings and recommendations of professional science, Gibbs objected. What science found ran counter to her experience. In talking with another Health Department physician at this meeting, Gibbs made her feelings abundantly clear:

> He [the physician] walked up and down, up and down, insisting that he couldn't find any problem. There just wasn't that much abnormality. I told him I thought he was dead wrong. . . . I found sick people all around the canal. 'You can't stand there and tell me there's no problem at Love Canal!' According to his survey, he didn't see any. I kept telling him the survey must not have been conducted properly. I told him about the five crib deaths, that most of these women had been breast-feeding. (Gibbs 1982, pp. 30–31)

Over the next few years, additional health studies were done by the government, but residents put little faith in them. In fact, with the assistance of independent scientific advisors, the Love Canal Homeowners Association undertook their own health study, finding that "many families in the area were affected with an abnormally high rate of illnesses" (Gibbs 1982, p. 5).

That health study was prompted by a pattern that emerged out of the initial "housewife" data collected by Gibbs. Oddly enough, the illnesses she discovered did not cluster in areas nearest to the canal, but were spread out in an initially confusing pattern. The key to the pattern was discovered by Gibbs, with the help of locally situated knowledge about underground stream beds, or swales, that ran through the area:

> I knew of one swale, an old stream bed that went behind my house. I drew that swale on the map. Later, I drew a swale that Art Tracy, Mary Richwalter, and some of the other old-time residents had told me about. Actually, my neighbors drew the line for the swales. I was surprised: the illnesses clustered along the swales. (Gibbs 1982, pp. 66–67)

Gibbs called Dr. Beverly Paigen, a scientist who took up the Love Canal residents' cause, and told her what she had found.

> She suggested we do a full survey of illnesses along the swales and swampy areas. She also told me to have people tell us exactly what illnesses they had and whether they had been to the doctor for diagnosis. The following day, we started our own medical survey. (Gibbs 1982, p. 67)

In the end, the swale theory, while controversial, gained significant scientific respect. However, it did not lead to the relocation of the remaining Love Canal residents. In the end, on October 1, 1980, President Carter signed a bill evacuating all remaining Love Canal residents because of the mental anguish caused by living at the contaminated site (Gibbs 1982, p. 5).

Following Love Canal, Gibbs did not return to her homemaker role; rather, she moved to Washington and started the Citizens Clearinghouse for Hazardous Waste. Since its founding in 1982, CCHW has been contacted by over 7,000 local groups trying to deal with contamination issues. Shaped by Gibbs's experiences at Love

Canal, which gave her a strong belief in the importance of local initiative and response, CCHW adamantly maintains an emphasis on grassroots organizing, rather than top-down directing.

> Our mission is to assist people in building strong, community-based organizations that can fight against corporate polluters and unresponsive government agencies. We provide organizing, training and technical assistance over the phone, by visiting sites, and through our 60+ manuals and handbooks. (Citizens Clearinghouse 1996, p. 3)

Many of those manuals and handbooks deal with scientific issues and give a glimpse into the approach of this movement organization to scientific knowledge and its uses. Produced in language easily understood by the layperson, titles include "Environmental Testing," "Hazardous Waste," "Solid Waste Incineration," "Medical Waste," and "Common Questions About Health Effects." There is a manual, complete with questionnaires, on how a community group can do its own health survey. There are also "fact packs" of news clips on scientific subjects, including "Cancer Clusters," "Dioxin Toxicity," "Lead Toxicity," and "Burning Hazardous Waste in Cement Kilns." The emphasis is on "do-it-yourself" science wherever possible, and on helping community residents understand the science behind their problems and the potential technical solutions of them.

Throughout CCHW's handbooks, manuals and newsletters, one can see the partial institutionalization of an approach to science that grew from Gibbs's Love Canal experience. The approach begins with distrust, not of science itself, but of its institution and expert system. Part of this distrust is based on questioning the independence of science and the disinterested, unbiased scientist:

> We have "independent" studies on which we are supposed to rely. The majority of those studies are funded by the chemical companies. (Zeff, Love, and Stults, 1989, p. 13)

> Many "experts" have bad reputations with environmental groups because they act more like "hired guns" than scientists and professionals. (Zeff, Love, and Stults 1989, p. 17)

> Beware of "hidden agendas." If this expert is approaching you for something other than just the money, WHAT IS IT? Do they want to exploit your issue, grab your publicity? FIND OUT, or else you might get burned. (Collette and Gibbs 1985, p. 6)

> MYTH: "This guy's the expert, so the recommendations must be right."
> REALITY: Experts do not come with any guarantees, no 90-day warranties. They are as human as you and I and as prone to error, fatigue, burn-out, time constraints and screw-ups. (Collette and Gibbs 1985, p. 13)

Part of the distrust also comes from people's experiences running counter to scientific knowledge. CCHW publications are full of testimonials of people's health problems that stumped the experts, eventually to be traced to contamination; for example:

> When Eric started school, instead of things getting better, like everybody told me he would, "he'd outgrow his asthma," he just got worse. We went through a year of really severe abdominal pains and the doctors just didn't know what was going on.... They finally said it was an epileptic stomach.... One night he had to have emergency surgery. They thought it was appendicitis, but it wasn't. So they ... took out Eric's gall bladder.

They decided that's what it was. A six year old with a gall bladder problem!' Eventually, a potential link with the Stringfellow Acid Pits was investigated, but the state, despite "health effects that included an increase in cancer, urinary tract infections, respiratory problems, ear infections, heart "problems" ... considered this "no significant health impact." (Zeff, Love, and Stults 1989, pp. 8,10)

One result of this distrust is to hire experts only when necessary, and to be careful of how they are used. In a manual entitled "Experts: A User's Guide," the authors urge community groups to first make certain they really need to hire an expert:

> There are many ways your group can get the same information without spending money on an expert. There are lots of groups across the country (see "Resource" list) who can give you information on the toxicity of chemicals, groundwater flow, toxicity levels and standards, health surveys and many other technical topics. (Collette and Gibbs 1985, p. 1)

In the end, if a group decides to hire an expert, the authors council the group to make certain the group maintains control:

> NEVER hire an expert to speak for your group! Experts should be used to support what you say, but never be the spokesperson. (Collette and Gibbs 1985, p. 2)

An important emphasis in this document and throughout CCHW publications, is the importance and legitimacy of local knowledge, based both on direct experience and local culture:

> There are two types of experts ... community experts and professional experts. Community experts are people like you. You live in the neighborhood and know more about the community than anyone on the outside. Chances are good that you or your neighbors know, for example, who dumped what, how it was dumped, when it was dumped, what the land looked like before it was dumped, which way the water flows and which way the wind blows ... If you live in a farm community, isn't it common sense that people who've worked the land all their lives will have opinions about the ground that are at least as valid as some "hired gun" expert from some far-away university? (Collette and Gibbs 1985, p. 5)

Sometimes a group is fortunate enough to find an expert within their neighborhood, one who can be trusted and who can combine expert and local knowledge:

> If you're lucky, when you go door-to-door in your community, you may find that there are genuine, professional experts living right in your community who can help you. Now there's a powerful combination—a neighbor who also has the credentials. (Collette and Gibbs 1985, p. 5)

But an emphasis of CCHW, as of the anti-toxics movement as a whole, is on empowering communities by giving residents the scientific information to make their own decisions about their future. CCHW publications represent one approach. Another is through conferences and workshops, such as one recently run by the CCHW Science Director Stephen Lester in Tifton, Georgia, on "how to evaluate soil testing results and what do the numbers mean in terms of health effects" (Citizens Clearinghouse 1996, p. 18).

A third approach is through recognizing and publicizing "citizen experts," people who themselves gained expert knowledge and used it to help their community. A recent CCHW newsletter tells of an award presented by an Illinois group to an eighth-grader whose science fair project "helped show local citizens the extent of contamination caused by [a local] incinerator" (Citizens Clearinghouse 1996, p. 18). Also, as its lead article, this issue of the newsletter published an extended testimonial of a woman who trained herself as an expert in order to learn about the effects of carbon incineration:

> Ten months ago, if someone had told me that I—*hardly* a scientific sort—would one day know the difference between a benzene and a chlorobenzene or, when prompted, launch into an impassioned discourse on the ideal residence time in an afterburner, I would have suggested politely that this person was not playing with a full deck. But ten months ago, I had no idea that I would soon be thrust into an unexpected and intense crash course on the joys of living next to a carbon reactivation unit. (Cocek 1996, p. 4)

Her motivation to become a citizen expert grew out of dissatisfaction with information given by an expert employed by the incinerator company, who

> spoke glowingly about the virtues of their proposed . . . unit, but when we asked our first question, . . . he simply turned to us with condescension and said, "I'd like to answer that but, unfortunately, I think this is all a little too technical for y'all to understand." As I looked at him looking so earnestly smug, three words came to my mind: "You wanna bet?" From that point on, my "Hazardous Wastes and You" education began in full force. (Cocek 1996, p. 5)

Overall, the approach of CCHW—indeed, of the entire anti-toxics movement—to science can be seen as a breaking down of boundaries—boundaries between experts and non-experts, or between expert, local, and experiential knowledge, or between science and politics. Experts and expert knowledge are being unhinged from the legitimized scientific institution and moved into the domain of a loose network of social movement organizations which use science, in combination with other types of knowledge, to better understand their predicaments.

In fact, the movement is attempting to use science not only as a way to understand chemicals and their effects, but as an organizing tool. This was clearly evident at a recent workshop on dioxin in Baton Rouge, Louisiana, in which over 550 people attended

> one of the most inspiring, diverse and unified environmental gatherings in many years. . . . Along with new scientific findings and international political developments regarding dioxin, panels addressed the primary importance of environmental justice, strategies for allying with chemical workers, the need to challenge corporate power and the development of economic as well as technological alternatives. (Tokar 1996)

A new book on dioxin (Gibbs and Citizens Clearinghouse 1995) is explicit in its goal to use science to enhance grassroots organization.

> The first nine chapters of the book . . . provide the scientific and medical facts about dioxin and how it is harming the health of the American people. Much of the information in these chapters comes from the EPA's 1994 dioxin reassessment . . . [The second]

nine chapters provide a step by step guide on how to put together a grassroots campaign to stop dioxin exposure. In addition to providing information on specific strategies, the book explains why organizing to stop dioxin exposure is, in reality, rebuilding democracy. (Brody and Froelich 1996, p. 28)

An excerpt from the book reinforces this point:

Our campaigns must not be only about the danger of dioxin, but also the dangers of a society where money buys power. To create the equality and justice of a true democracy, our organizing must restore the people's inalienable right to govern and protect themselves. (Gibbs and Citizens Clearinghouse 1995, as quoted in Brody and Froelich 1996, p. 29)

So we see here an example of the new populism—an appeal to moral values of justice and democracy, but basing that appeal on the definition of problems based on the development and use of expert knowledge. As with corporate science, the boundaries of the scientific institution have become permeated, and scientific knowledge (and some scientists) have become captured by non-scientific organizations. But unlike the corporate/science relationship, we here see the explicit use of science for political organization. In addition, through the validation of local contextual and experiential knowledge, we see science's content and methods themselves altered in ways that may portend, in Beck's words, "a new phase in its development" (Beck 1995b, p. 112).

Multiple Chemical Sensitivity, Citizens, and Medicine

People who identify themselves as multiply chemically sensitive argue that their bodies negatively react to a broad range of synthetic chemicals at doses far lower than those affecting most, less sensitive people. The symptoms of this hypersensitivity are not viewed as ordinary manifestations of toxicity or allergy, but are assumed to be triggered by trace amounts of chemicals found in air, food, and water, as well as such everyday substances as particle board, carpeting and adhesives, toxic and fragranced cleaning products, detergents, cosmetics, and so on (Rest 1992; Hileman 1991). Moreover, and complicating the problem of multiple chemical sensitivity (MCS), after the initial chemical insult(s) that cause the disorder, the body becomes intolerant to a seemingly endless range of unrelated chemical compounds (Hileman 1991; Kroll-Smith and Ladd 1993).

A striking feature of this nascent public health problem is its distribution across a wide spectrum of demographic groups, including "industrial workers, office workers, housewives, and children" (Ashford and Miller 1991, p. 5). The Environmental Research Foundation (1987) and the National Academy of Sciences (1987) suggest that between 15–20% of the U.S. population may have an increased allergic sensitivity to chemicals commonly found in the environment which place them at increased risk of contracting a debilitating illness. While the disabling symptoms of MCS can range from mild to severe, many people are forced to make major lifestyle changes in their work, home, and social environments to cope with their reactions to chemicals.

Despite the alarming estimates of the number of people with MCS and the recognition of the problem by the National Academy of Sciences, professional medical societies are seemingly unanimous in their opinion that it is not a legitimate physical disorder. Local medical boards reportedly threaten to censure physicians who diagnose people with MCS (Hileman 1991, pp. 27–28). Several national medical societies, including the American Academy of Allergy and Immunology (1989) and the American College of Physicians (1990), have published position papers denying the reality of MCS as a physical disorder and criticizing physicians who treat the patient "as if" the disease existed. A report in the *Annals of Internal Medicine* labeled the population of people claiming to suffer from MCS a "cult" (Kahn and Letz 1989, p. 105; see also Rest 1992).

Frustrated by the unwillingness or inability of medical experts to acknowledge their somatic misery and recommend treatment, the multiply chemically sensitive are rejecting physicians while embracing the abstract language of biomedicine. Three national citizen organizations—the Chemical Injury Information Network (CIIN), Human Ecology Action League (HEAL), and the National Center for Environmental Health Strategies (NCEHS)—are organized around the acquisition and use of biomedical knowledge.

The CIIN, with offices in six states, boasts local chapters in all fifty states and several foreign countries. Its monthly newsletter, *Our Toxic Times*, regularly includes technical articles on chemicals and disease written by non-experts. Its November 1995 issue includes an essay by Cindy Duehring, a young woman with a high school education, titled, "Porphyrinogenic Chemicals & Metabolism: MCS Patients at Risk." The HEAL has chapters in all fifty states, with multiple chapters in many states. Its quarterly newsletter, *The Human Ecologist*, provides an array of useful information to the chemically reactive, often pointing out the need for patients to take responsibility for knowing biomedical "facts" about their illness that might be unknown to their attending physicians. The NCEHS counted 4,000 subscribers to its newsletter, the *Delicate Balance*. In addition to these three national newsletters, we have found sixteen other MCS newsletters circulating throughout the U.S. A regular feature of these more local newsletters is a biomedical account of MCS.

Interview data reveal *in situ* how and why biomedical knowledge is appropriated by ordinary people to fashion accounts of themselves as environmentally ill. As part of a larger study by the second author of this chapter, 137 people who self-identify as MCS were interviewed on a broad range of topics (Kroll-Smith 1997). A strikingly common pattern emerges in these data. People report their bodies changing in seemingly random ways to nothing particularly unusual. The debilitating nature of these somatic changes forces people to monitor their bodies in relationship to their social and physical surroundings (recall from chapter 4 Jack's inventory of what Kroll-Smith should and should not wear to the interview). Monitoring reveals an association between certain specific environments or consumer products and the troubling symptoms.

For the chemically reactive, knowledge about their sickness originates with embodied experiences that are typically tracked, classified, and arranged into meaningful clusters. It is a practical epistemology insofar as it joins experiences to practices and is mindful of the results. A former intensive care nurse describes what she would say to a person who wanted to understand this stage in the process of becoming multiply chemically sensitive:

> I would . . . ask the person to take a notebook and for one day write down the chemicals found in every item they eat or drink that day (as listed in the ingredients). Ask them to list every item in their home that is scented, every cleaning product and the chemicals they contain. Chemicals they encounter in stores, gas stations, day cares, etc. What pesticides they used and how often they used them . . . Then ask them to look up the exposure limits of each and the accumulative effect."

People reorganize their thinking about their bodies and environments by a meticulous, detailed process of assigning somatic responses to ordinary chemical products and routine places. In the language of statistics, associations quickly become correlations as their trial and error methods yield a more manageable, but also more circumscribed, world. Illustrated in their accounts is the obvious fact that truth for the environmentally ill is not being sought outside a deliberately rational practice, though that practice originates, some might say heretically, with human experience.

A former advertising executive, for example, developed a sophisticated classification scheme. She listed fifty-two separate "Chemicals and Irritants" on the left side of the page and fifty-two "Reactions" on the right side. She listed several reactions more than once. Consider an excerpt from her taxonomy of environmental agents and somatic responses:

Chemicals and Irritants	Reaction
Polyester	Throat tightens
Clothing dyes	Skin crawls
Newsprint	Body vibrates
Computers	Difficulty concentrating
Magic markers	Naso-pharyngeal passages irritated

A retired Army Colonel developed a rating system to catalogue his varied somatic reactions to environments and his body's capacity to withstand the "insults":

> 0 to 5 is my "insult scale." I figure my body can be only so insulted before it breaks down . . . Say I've been exposed to two 4s and one 5 in the morning and I am invited to someone's house that evening for dinner. I will usually decline because I will likely get sick if I am exposed to anymore insults.

Among the twelve 5s on the "insult scale" are "traffic jams," "visits to the Veterans Administration office," "shopping malls," and "filling my car with gas."

Expressed in these excerpts is the complicated use of local, contextual knowledge to identify and manage the sources of their miseries. But the multiply chemically sensitive are also contesting epistemic assumptions of the medical profession in its own language by defining the sources of their disorder in acute or chronic exposures to chemical agents in dose magnitudes below those officially recognized as dangerous. Of the 137 interviews conducted with MCS respondents, nine attributed the origin of their illness to chemical exposures officially recognized as toxic. The remaining 128 respondents narrated etiology stories that located the origin of their troubles in routine, sub-clinical exposures to chemical agents.

The body's increasing intolerance to ordinary, putatively benign, places and mundane consumer products is a key feature of this illness and baffles most physicians. "We don't dismiss these people, they are truly ill," admits a prominent allergist and medical researcher who speaks for the majority of practicing physicians,

> but batteries of chemical tests can't pinpoint any specific sensitivity. Some are definitely allergic and we all agree that they are suffering, but we simply don't understand the cause of the disease as determined by medical diagnosis. (Selner, quoted in *Chemecology* 1991, p. 2)

Another sympathetic but discouraging assessment concludes that

> there is no laboratory test that can diagnose MCS, no fixed constellation of signs and symptoms, and no single pathogen to isolate and transmit through a cell line. . . . Even worse, some chemicals are neurotoxic and may produce symptoms that resemble anxiety attacks or mood disorders. (Needleman 1991, p. 33)

"I know my body is more sensitive than standard medical tests reveal," observes a former accountant with MCS. When asked how he knew this, he criticized the logic of

> double-blind testing with placebo controls that expects us to take a supposedly benign pill when in fact everything is made of chemicals . . . Why wouldn't I react negatively to a foreign chemical even though they call it a placebo?

Finally, and complicating an already heretical biomedical theory, is the assumption of MCS that each chemical irritant may trigger a different constellation of symptoms in each person and every system in the body can be adversely affected. Thus, combinations of body systems and symptoms interact geometrically, creating, at least theoretically, a seemingly endless configuration of somatic miseries (Ashford and Miller 1991; Kroll-Smith and Ladd 1993). In response to exposure to fresh paint, one respondent reports losing his balance and becoming disoriented, while another claims she becomes nauseous and tired; both are likely to manifest different symptoms when exposed to different chemical agents. The somatic troubles of these two respondents and dozens of others we interviewed challenge the biomedical assumption that each disease is caused by a specific aversive agent affecting an identifiable body system (Kroll-Smith and Ladd 1993; Freund and McGuire 1991).

Confusing and confounding the once protected distance between expert and layperson, people with MCS are appropriating the language of biomedicine and modifying several of its basic assumptions to make medical sense of their miseries. Subordinating clinical medicine to their own interests does not, of course, mean that the medical reasoning of the chemically reactive is necessarily valid. Nor can we as sociologists find one medical theory more worthy than another. What sociologists can know, however, is why a person with MCS believes his or her biomedical account is sound. When asked about this problem of validity, people with MCS are apt to invoke a simple, but compelling, pragmatist version of truth: it's true if it works.

MCS is a popular medical theory that gives a measure of control over sick bodies, helping people to reduce their symptoms, while providing a rational explanation for how bodies adapt or fail to adapt to chemical places and things. In short, it works.

Citizens, Science, and Herbicide

During the 1980s, the herbicide 2,4,5-T was a source of controversy between British regulatory authorities and British farmers. Produced since the 1940s, 2,4,5-T has acquired a notorious reputation from its use as a defoliant in the Vietnam War. Pictures of children screaming in fear and agony after being sprayed with the chemical agent are now among the symbolic images of that war. And American pilots and ground troops who claimed to be permanently disabled from exposure to the chemical sued the federal government years after the war alleging wrongful misrepresentation of its hazardous properties. 2,4,5-T is suspected of causing, among other things, chloracne, birth defects, spontaneous abortions, and several types of cancers.

In spite of its less than wholesome reputation, 2,4,5-T was commonly used in Britain. Homeowners used it to control weeds in their gardens. Railway employees sprayed it on tracks to kill troublesome vegetation. Farmers and forestry workers regularly used it to protect crops and clear underbrush.

The position of the British government on the use of the herbicide was, at least at the start of the conflict, simple and straightforward. Its Advisory Committee on Pesticide concluded a review of the scientific literature on 2,4,5-T, proclaiming that it "can safely be used in the UK in the recommended way and for the recommended purposes" (Irwin 1995, p. 17). The Committee's conclusion, however, is not as interesting for our immediate purposes as the mode of reasoning it used to investigate the herbicide.

Defining the public as "concerned" but not "informed" (Irwin 1995, p. 112), the Committee of scientists and regulators could approach its task as if it were operating in a laboratory, free from outside noises and interference. Deliberating on the fate of the herbicide with the solemnity of authorities who must weigh evidence and act in the best interests of naive subjects, the personal experiences of these subjects could, in good conscience, be ignored. The Committee claimed that "its own knowledge and experience is backed up by a valuable body of medical and scientific expertise within and beyond the machinery of government" (Irwin 1995, p. 112). A model consisting of the objective experiences of a committee of experts examining the results of professional, laboratory experiments can safely ignore the subjective experiences of a naive public. Indeed, it would violate accepted research protocol to do otherwise.

The position of the National Union of Agricultural and Allied Workers (NUAAW), however, approached the herbicide problem from a quite different epistemic vantage point. For these ordinary citizens, laboratory studies and literature surveys were inadequate sources of knowledge about their personal experiences with 2,4,5-T. Statistical rates of chloracne or miscarriages, for example, are unable to account for why a neighbor is inflicted with painful skin rashes after using the herbicide or a woman whose husband carries chemical residue home on his clothes spontaneously aborts what appeared to be a healthy fetus. The frequency of these personal, non-statistical, experiences prompted the NUAAW to sponsor a survey among its membership to discern what lay people know about how to apply the herbicide and how to avoid physical contact with it or its gaseous vapors. Additional questions included frequency of use and symptoms after use. A total of forty questions were asked.

In addition, the survey included fourteen case studies of people who suffered medical problems after being exposed to the herbicide directly or indirectly. In developing its case studies, the union relied on medical records, family illness histories and employment records documenting contact with the herbicide. The intention "was to establish the level of exposure involved and the scale of alleged effects" (Irwin 1995, p. 20) for each case.

In its report to Britain's Minister of Agriculture, Fisheries, and Food, the NUAAW concluded that the personal experiences of people who use 2,4,5-T could not be replicated in the laboratory. "It is the NUAAW's conviction, distilled from the experience of thousands of members working in forests and on farms, that the conditions envisaged by members of the (Advisory Committee) are impossible to reproduce in the field" (Irwin 1995, pp. 17–18). In other words, experimental data generated in laboratories cannot reliably predict the seemingly endless personal experiences of people who come into contact with 2,4,5-T. "In their evidence to the (Advisory Committee)," Irwin writes, "the farmworkers discuss what they consider to be the 'realities' of pesticide use" (1995, p. 19). A good example of the argument from reality—that ultimately knowledge is based on concrete experiences—is the dispute over the word "recommended."

A point strongly made by the Advisory Committee was the importance of users following the handling procedures recommended by the manufacturer and the regulatory agencies (Irwin 1995, p. 112). If these procedures were followed, the Committee reasoned, risks from using the herbicide would be negligible. One, perhaps unintended, consequence of this style of reasoning was to relieve industry and government from liability in the event an injury. More importantly for our purposes, however, was the response of the lay public to the idea that government regulators could foresee all the particularities, unusual circumstances or chance occurrences typically encountered in a normal work environment.

A farmworker and member of the union notes, "They (the 'experts') may know the risks of 2,4,5-T. They may handle the stuff properly. They tell us we'll be alright if we use the spray normally. But have they any idea what 'normally' means in the fields?" (Irwin 1995, p. 112). Normal or routine from the vantage point of the worker in the field (and, indeed, for most of us who work in today's complex world) is best captured in Murphy's infamous Law. Perhaps the most reliable approach to recommended use instructions is to base an application protocol on the assumption that what can go wrong probably will. Such an approach, however, would require considerably more imagination than is typically found in one or two paragraphs on the back of a container. And, furthermore, it raises the indelicate political and moral question of whether the risks can be safely managed, or even adequately identified.

The controversy over the use of the word "recommended" is a symptom of a more complicated conflict between two quite different ways of knowing, each claiming to be based on instrumental, rational reasoning. Blurring the traditional distinction between lay person and expert, farm and forestry workers are claiming to know something about the use of an industrial chemical that should be the basis for government regulation of the substance. The basis of their knowledge claims are subjective, human experiences. For the scientific committee, on the other hand, valid and reliable knowledge in the modern world is the product of experimental science and pointedly not subjective experiences. People, of course, can trust their senses as sources of knowing; it is simply that when they do so they cannot claim to know something scientific or medical about themselves or their world. Sensory knowledge, in other words, should not be a basis for government or corporate decision-making.

For farm and forestry workers, scientific knowledge is incoherent, unable to account for their personal experiences. For scientists, lay knowledge is unreliable insofar as it is based on experiences. The conflict, it seems, is between a local, concrete knowledge based on systematic observation and analysis and a remote, abstract knowledge, based on experimental design and statistical coefficients. Such local, practical knowledge has always existed, of course; what is new is its public role in challenging the validity and reliability claims of institutional science. A final point worth considering in this case is the NUAAW's effective use of what it called a "balance of probabilities" argument.

The NUAAW did more than attack the premises of Advisory Committee reports. It constructed its own standard of validity to oppose the tests of significance and cluster measures relied upon by the Committee's scientists. Based on its survey and case study data base, British farm and forestry workers (the NUAAW) proposed a principle for assessing the particular dangers associated with the use of a potentially risky substance. The concept, "balance of probabilities," was introduced as a reasonable approach to the inherent difficulties in foreseeing all of the ways in which 2,4,5-T would be used in mundane, uncontrolled settings.

The balance of probabilities test asks the British government to consider how the herbicide is used in local settings and to make a reasonable determination whether or not it is possible to control for all of the contingencies and unexpected, chance events in applying the chemical. Not surprisingly, the NUAAW had already made up its mind on this issue: "In our view the decision (to ban the herbicide) has to be made on the balance of probabilities . . . where lives are at stake a responsible body cannot wait, as was the case with asbestos, until there is a sufficiently impressive death toll" (Irwin 1995, p. 21).

Importantly for our purposes, the union coined its test of validity based on several indices: direct observation and experience with 2,4,5-T, local knowledge of the difficulties in predicting how the chemical is used and systematic data collection using both survey and case study methods. Its appeal to the British government for the right to work in safe environments is reasoned from a rational assessment of personal experiences, mundane knowledge and their own socio-medical study.

Strongly suggested in the 2,4,5-T case is the primacy of mundane human experiences in building rational knowledge movements. Is it reasonable for a person who watches as a co-worker is diagnosed with cancer to question whether the workplace might be the source of the cancer? It is sensible to become more concerned about the workplace when additional unusual illnesses appear among co-workers? Is it sound practice to begin exploring what science and medicine are reporting about the chemicals (or other substances) people typically use in the workplace? Finally, and most importantly, is a person exercising good judgment when, in spite of official assurances that the workplace is safe, he or she, in cooperation with fellow employees, design and conduct a systematic inquiry into their experiences with suspect materials?

Answers to these questions are simple and straightforward from a modern science perspective. To wit, it is unreasonable to generalize beyond personal experience; it is unsound to assume a pattern exists in the absence of rigorous statistical tests; and, finally, it is impractical, if not dangerous, for lay people to think they can conduct valid and reliable inquiries into disease or behavioral patterns, toxicology, and so on. In Beck's words, "Experience—understood as the individual's sensory understanding of the world—is the orphan child of the scientized world" (1995a, p. 15).

Conclusion

The three cases discussed above show how non-scientific movements and organizations are appropriating and using scientific and medical knowledge, dislodging this knowledge from its institutional structures and relocating it within different organizational spheres. This process illustrates how contemporary environmental movements are organizing around both a populist appeal to moral or civil rights, and what they see as valid scientific truths. However, the scientific truths are based on knowledge which does not conform to all of the standards of truth held by the institution of science. Rather, it is altered in some significant ways.

As illustrated above, the "new populism's" use of science views science based solely on experimental methods as inadequate. This is consistent with the views of Beck, who argues that experimental science has taught us little about modern hazards; people have learned about them from experience, from a "crash course" (Beck 1995a, p. 32). These experiences point out the inadequacy of theory based on experimentation, in that theory refuses to acknowledge the validity of the experiences (Beck 1995a, p. 56). People now insist that experience be reintroduced as a valid criterion of judgment in science. Beck writes:

> Science has risen to its present dominating image of technical power and objectivity by virtue of its repression of experience, so to speak. . . . Experience, which was once the main authority and judge of truth, has become the quintessence of the subjective, a relic, a source of illusions that attack the understanding and make a fool of it. In this view, it is not science but rather the subject and subjectivity that are wrong. (Beck 1995a, p. 15)

The truth of the 2,4,5-T controversy or the toxics and MCS movements begins with the subject and that subject's subjective experience—specifically, with ordinary people who experience untoward changes in their bodies in relationship to perceived changes in their environments. If modernity "actively eliminates the idea of the Subject" (Touraine 1995, p. 27), replacing the unique and personal qualities of the knower by what is objectively known, it might be said that the Subject is returning in the form of a sick body. Knowing that begins with the sensations of the body, rekindles an epistemology that is anything but Cartesian—namely, I feel, therefore I think. When bodies resist being the objects of biomedical theory and somatic experiences become a source of objective knowledge, it is not difficult to imagine a fault line in the foundation of modern rationality.

The above cases also illustrate the breakup of the institutional association of science with various jurisdictions, and the arising of new public forms of science. "The equation of scientific rationality with jurisdictional assignments (certain methods, theories, or divisions of labor) begins to break up, and new, publicly oriented forms of expert scientific action arise" (Beck 1995a, p. 53). Giddens writes of the "democratizing of democracy," a process which creates "a public arena in which controversial issues—in principle—can be resolved, or at least handled, through dialogue rather than through pre-established forms of power" (Giddens 1994, p. 16).

All three of our case studies illustrate a broadening of the forum for dialogue and debate, and of the debaters. In addition, there are other, sometimes more formal, efforts at building structures within which experts and lay people can meet and discuss scientific issues and controversies (see, for example, Irwin 1995; Kindler 1996). At the same time, our cases caution us to be realistic concerning the extent to which scientific "truths" will emerge from such processes, which are inevitably political. The 2,4,5-T case involved a union fighting for its members. Multiple chemically sensitive people not only combine rational knowledge with experience to explain their maladies; they work to achieve fragrance free zones and chemical free public spaces. The anti-toxics movement explicitly uses scientific conclusions about dioxin as an organizing strategy to unite a wide variety of social movement organizations in a campaign to eliminate dioxin's use and to rebuild America. These are far cries from public arenas in which scientific questions are discussed with the goal of coming to a better understanding of how the world works.

There is some irony here. It has been pointed out that democratic institutions have developed in tandem with modern science and have provided the framework within which scientific ideas can be openly shared and debated. But as it develops more fully, democracy is breaking down institutional barriers to participation in the scientific debate. It may indeed lead to the "democratizing of democracy." But this leveling also destroys whatever protection the scientific enterprise had from the operation of political-economic systems and may mean that science becomes even more caught up in extra-scientific struggles than it has been to date. Students of science have known for some time that a truly disinterested, objective science is impossible. And yet this should not blind us to the real consequences of the professionalization and institutionalization of knowledge, one of which is to insulate knowledge production somewhat from the vicissitudes of larger societal struggles.

Of course, much has been made of corporations' use of science for its own purposes, and the effects of this on science. Now perhaps it's the turn of certain social movements.

REFERENCES

American Academy of Allergy and Immunology, Executive Committee. 1989. "Clinical Ecology." *Journal of Allergy and Clinical Immunology* 78: 269–271.

American College of Physicians. 1990. "Clinical Ecology." *Internal Medicine* 3: 168–178.

Ashford, Nicholas, and Claudia S. Miller. 1991. *Chemical Exposures.* New York: Van Nostrand Reinhold.

Beck, Ulrich. 1992. *Risk Society.* London: Sage.

———. 1995a. *Ecological Enlightenment.* New Jersey: Humanities.

———. 1995b. *Ecological Politics in an Age of Risk.* Cambridge: Polity Press.

Brody, Charlotte, and Veronica Froelich. 1996. "Dying from Dioxin: The Book, the Tour, the Nationwide Campaign." *Everyone's Backyard* 13(4): 28–30.

Bullard, Robert, ed. 1992. *Confronting Environmental Racism.* Boston: Southend Press.

Chemecology. 1991. "Editorial: What Is Making These People Sick?" *Chemical Manufacturers Association* 20: 1–2.

Citizens Clearinghouse for Hazardous Waste. 1996. *Everyone's Backyard,* 14(1).

Cocek, Christina. 1996. "A Lantern in the Window:Notes from a New and Hopeful Activist." *Everyone's Backyard* 14(1): 4–6.

Collette, Will, and Lois Marie Gibbs. 1985. *Experts: A User's Guide.* Arlington, VA: Citizens Clearinghouse for Hazardous Waste.

Duehring, Cindy. 1995. "Porphyrinogenic Chemicals & Metabolism: MCS Patients at Risk." *Our Toxic Times* v.6 (11): 1–3.

Environmental Research Foundation. 1987. "Hazardous Waste News" 53. Princeton, NJ.

Erikson, Kai. 1991. "A New Species of Trouble." Pp. 11—29 in *Communities at Risk,* edited by Stephen R. Couch and J. Stephen Kroll-Smith. New York: Peter Lang.

———. 1994. A New Species of Trouble. New York:Norton.

Freund, Peter E. S., and Meredith E. McGuire. 1991. *Health, Illness and the Social Body: A Critical Sociology.* Englewood Cliffs, NJ: Prentice-Hall.

Gibbs, Lois. 1982. *The Love Canal: My Story.* Albany: SUNY Press.

Gibbs, Lois, and Citizens Clearinghouse for Hazardous Waste. 1995. *Dying from Dioxin: A Citizens Guide to Reclaiming Our Health and Rebuilding Democracy.* Boston: South End Press.

Giddens, Anthony. 1990. *The Consequences of Modernity.* Stanford, CA: Stanford University Press.

———. 1991. *Modernity and Self-Identity.* Stanford, CA: Stanford University Press.

———. 1994. *Beyond Left and Right: The Future of Radical Politics.* Stanford, CA: Stanford University Press.

Gusfield, Joseph R. 1963. *Symbolic Crusade.* Urbana, IL: University of Illinois Press.

Hileman, Betty. 1991. "Chemical Sensitivity: Experts Agree on Research Protocol." *Chemical and Engineering News.* April 1. Washington, DC: American Chemical Society.

Irwin, Alan. 1995. *Citizen Science.* London: Routledge.

Kahn, Ephraim, and Gideon Letz. 1989. "Clinical Ecology: Environmental Medicine or Unsubstantiated Theory." *Annals of Internal Medicine* 111: 104–106.

Kindler, Jeffrey D. 1996. *Evaluating a Community Discussion in a Technological Disaster: Using Antheil's Comprehensive Evaluation Model as a Conceptual Guide.* Ph.D. Dissertation, Temple University.

Kroll-Smith, Steve. 1997.*Bodies in Protest.* New York: New York University Press.

Kroll-Smith Steve, Stephen R. Couch, and Hugh Floyd. 1995. "Non-Experts, Experts, and Expert Knowledge." Paper presented at the Mid-South Sociology Meeting, Mobile, Alabama.

Kroll-Smith, Steve, and Anthony Ladd. 1993. "Environmental Illness and Biomedicine: Anomalies, Exemplars, and the Politics of the Body." *Sociological Spectrum* 13: 7–33.

Lash, Scott. 1994. "Expert Systems or Situated Interpretation? Culture and Institutions in Disorganized Capitalism." Pp. 198–215 in *Reflexive Modernization*, edited by Ulrich Beck, Anthony Giddens, and Scott Lash. Stanford, CA: Stanford University Press.

Lawson, Lynn. 1993. *Staying Well in a Toxic World.* Chicago: Noble Press.

Levine, Adeline. 1982.*Love Canal:Science, Politics and People.* Lexington, MA: Lexington Books.

Nash, Roderick. 1989. *The Rights of Nature.* Madison: University of Wisconsin Press.

National Academy of Sciences. 1987. "Workshop on Health Risks from Exposure to Common Indoor Household Products in Allergic or Chemically Diseased Persons." (Unpublished report) Washington, DC.

Needleman, Herbert L. 1991. "Multiple Chemical Sensitivity." *Chemical & Engineering News*, June 24: 32–33.

Rest, Kathleen. 1992. "Advancing the Understanding of Multiple Chemical Sensitivity: An Overview and Recommendations from an AOEC Workshop." *Proceedings of the AOEC Workshop on Multiple Chemical Sensitivity* 8: 1–14.

Schnaiberg, Allan. 1980. *The Environment.* New York: Oxford University Press.

Seligman, Adam B. 1992. *The Idea of Civil Society.* Princeton: Princeton University Press.

Stewart, Charles, Crain Smith, and Robert E. Denton, Jr. 1984. *Persuasion and Social Movements.* Prospects Heights, IL: Waveland Press.

Tokar, Brian. 1996. "Campaigning against Dioxin." Paper presented at Loyola University, New Orleans, LA, March.

Touraine, Alain. 1995. *Critique of Modernity.* London: Blackwell.

Waltzer, Michael. 1991. "The Idea of Civil Society." *Dissent* Spring: 293–304.

Zeff, Robbin Lee, Marsha Love, and Karen Stults. 1989. *Empowering Ourselves: Women and Toxics Organizing.* Arlington, VA: Citizens Clearinghouse for Hazardous Waste.

QUESTIONS

1. Regarding the idea of new populism, why is it populism and why is it new?

2. What criticisms did farm and forestry workers raise about scientific studies of the health risk of the herbicide 2,4,5-T?

Part 8

Setting the Environmental Health Agenda

Agendas refer to those issues, items, concerns, and goals that are slated for substantial effort and attention. Individuals set personal agendas, for example, deciding that over the next two years they will save up enough money for a down payment on a house, lose 30 pounds, and read at least one novel a month. Social collectivities also set agendas: religious denominations raise money to build new churches, corporations seek ways to increase their profit margins, and community associations work to reduce crime in residential neighborhoods. Research agendas guide scientific work at both the level of the individual scientist, who must decide whether she is going to study health systems in the United States or rural development in Zimbabwe, as well as funding agencies, which must decide whether they should target scarce resources for research on vitamin deficiencies, global warming, male contraception, or chronic lead poisoning. Concerns we share as citizens of a particular community, state, or society achieve expression through public and policy agendas. The public agenda is composed of those issues that are currently receiving considerable media coverage and that are being discussed and debated by ordinary citizens. As these words are being written school violence dominates the public agenda, due to the massacre at Columbine High School in Littleton, Colorado. Policy agendas refer to possible changes in legislation or regulation currently being considered by local, state, or federal governmental bodies.

Determining what items are going to be on an agenda is referred to as agenda setting. Setting agendas typically involves conflict and contention, as pursuing a particular goal or trying to change some aspect of the world entails costs. This can happen even at the personal level: individuals are often conflicted intra-personally about difficult-to-reconcile goals, such as pleasing one's parents and pursuing a low-paying but satisfying career option. Setting public and policy agendas are activities particularly likely to be rife with strife. Public and policy agendas are directed at a broad cross-section of society, at least some segments of which will view the associated costs as unacceptable. To understand the setting of public and policy agendas we have to examine the actions and interests of an array of support and opposition groups, the resources available to them, and the historical context within which they operate.

Throughout this book we have encountered readings that have argued the need for increased public, government, medical, scientific, community, and corporate attention to the issue of environment-health linkages. These readings have focused on particular aspects of the agenda-setting process: disputes about how to measure levels of toxic substances and best determine their potential health impacts, questions about how to formulate public policies to address potential health threats, citizens' efforts to mobilize for public and official consideration of their concerns and lay knowledge, opposition to these activities by vested interests, and efforts to redefine the boundaries of scientific and medical disciplines. This final set of readings demonstrates how these various aspects work together in efforts to redefine public, policy, and scientific agendas.

Wilbur Scott's contribution, chapter 23 "Competing Paradigms in the Assessment of Latent Disorders: The Case of Agent Orange," overviews efforts to bring health problems experienced by Vietnam veterans exposed to this herbicide defoliant to public and official attention. Media coverage and efforts by organizations representing Vietnam veterans played crucial roles in placing Agent Orange on the public agenda. This placement was opposed by groups concerned about potential costs, including the companies who manufactured the chemical components of Agent Orange (2,4-D and 2,4,5-T), the Veterans Administration which was charged with treating service-related health problems, and the U.S. military which had directed the spraying of millions of gallons of the defoliant on Vietnamese forests. The time-lag between exposure and the onset of symptoms created a sphere of uncertainty, allowing conflicting interpretations of whether or not Agent Orange truly was "the cause" of the veterans' health problems. The waxing and waning interest in Agent Orange illustrates that it is challenging to keep an issue on public and policy agendas long enough to receive compensation for past harms.

Chapter 24, Michael Reich's "Environmental Politics and Science: The Case of PBB Contamination in Michigan," recounts efforts to bring public and official attention to widespread contamination of the human food chain with the synthetic fire retardant polybrominated biphenyl (PBB) in the state of Michigan in the mid-1970s. Following the accidental mixing of PBB into dairy cattle food, individual farmers struggled on their own to deal with the health problems of herds that had unknowingly ingested the substance. It took a year for the contaminant to be discovered, but initially the scope of the problem was underestimated. Michigan agricultural interests, concerned about the economic impacts of a consumer boycott of Michigan dairy and beef products, sought to minimize the problem. Mobilization and demonstrations by affected farmers, and health investigations by oppositional professionals, were important in gaining broader media and public attention to the problem. In cases such as these where key policy issues revolve around such measurement disputes as "safe" exposure levels, politically adept scientists play important roles in setting policy agendas.

In chapter 25 on "Environmental Health Research: Setting an Agenda by Spinning Our Wheels or Climbing the Mountain?" John Eyles seeks to provide clearer direction for research on environment-health linkages. The agenda-setting issues in this piece are more narrowly focused on the scientific community, though Eyles' suggestions do have important policy implications. Scientific knowledge about the connection between workplace and community environmental hazards and health impacts is divided into three categories: linkages for which we have overwhelming evidence, linkages for which we have plausible evidence, and linkages for which we have equivocal evidence. It is important that plausible evidence of environment-health linkages be kept on the public agenda, so that government policies to protect against these hazards do not appear arbitrary (a claim likely to be made by those whose interests are harmed by the policies). Traditionally, scientists have sought to overcome problems of equivocal evidence with calls for further research, with particular attention directed toward more precise measurements of the phenomenon under investigation. Eyles questions whether this strategy is an effective one to combat the uncertainty characteristic of many environment-health disputes. Like Reich, Eyles implies that conflicts over environmental hazards have an inherent political dimension, and therefore can never be resolved solely through the application of science. Eyles seeks to advance the research agenda on environment-health linkages by suggesting ways five social scientific theories (symbolic interactionism, structuration, structuralism/ post-structuralism, functionalism, and ecologism) could be used to generate pertinent research topics. Application of these theories might shed important light on various aspects of the agenda-setting process, including interpretations of environmental hazards, mobilization efforts, and the structure of oppositional forces.

Competing Paradigms in the Assessment
of Latent Disorders
The Case of Agent Orange

Wilbur J. Scott

Nearly 220,000 Vietnam veterans have requested physical examinations from the Veterans' Administration (VA) to confirm suspicions that their health problems are caused by exposure to the herbicide Agent Orange (VA, 1986:6). They allege that exposure to the herbicide some 10 to 20 years ago has produced health abnormalities among them and their children. The amount and type of exposure is virtually impossible to document and the symptoms in question include many conditions that are not uniquely associated with exposure to a herbicide. Further, the scientific evidence concerning the effects of exposure among humans is mixed. Exposure may have no significant long-term effects or it may contribute to the growing list of so-called latent disorders: environmentally produced, degenerative diseases in which there is a substantial time lag between exposure and consequence. These circumstances make the determination of cause and effect extremely difficult.

If all Agent Orange claims were worthy of payment, compensation could reach several billion dollars. Vietnam veterans' organizations have accused the VA of negligence in investigating the claims. VA representatives have suggested that opportunistic veterans are pushing the issue to qualify undeservedly for disability ratings and compensation payments. Subsequent litigation includes a landmark class action suit against The Dow Chemical Company and five other manufacturers of Agent Orange and a third party suit against the federal government filed by Dow Chemical. An out-of-court settlement which requires the manufacturers to establish a $180 million fund for compensating veterans has produced grumbling and dissatisfaction on all sides.

While many central questions concerning what happened and who now is responsible remain essentially unanswered, a sociologial account of the case provides insight into how it happened. Disputes about toxic substances and latent disorders bring into play the institutions charged with identifying disease as a social problem: science, medicine and health care systems and, with recent increases in malpractice suits and personal injury litigation, the courts. The subject underscores two schools of thought concerning disease and social problems (Pfohl, 1985; Schneider, 1985; Spector and Kitsuse, 1977; 1987). One approach presumes that a problem exists when acts and conditions become severe or aggravated. In this view, if exposure to asbestos, radon, or dioxin causes cancer, the evidence will eventually accumulate and invite discovery. In contrast, a constructionist approach argues that participating activists and organized interests create social problems by advancing competing claims and versions of evidence. Accordingly, phenomena become recognized as problems because they have been sponsored successfully rather than because they are inherently troublesome. The latter view fuels the inquiry presented here.

This approach to disease goes beyond Parsons' (1951) formulation of the functions and normative expectations embodied in the "sick role" and fits neatly with Freidson's (1970) articulation of how medical practitioners create sickness as an official role. Following their lead, medical sociologists have found it useful to distinguish the terms "disease," "illness," and "sickness" (Susser, 1973). Disease refers to a condition in which biological functioning of the body is impaired. The perception by the individual that he or she does not feel well, that something is wrong, is known as illness. Finally, sickness occurs when appropriate authorities formally confirm that an individual has a disease and should feel ill.

These distinctions invite the recognition that sickness and its consequences may differ from the person's biological state. While sickness often has some biological basis, not all diseased people may legitimately act sick and some people who are permitted to act sick actually are not diseased. In disputes about whether or not someone is sick—and, if so, what is the extent or source of the illness—the assumptions, values, and interests of several institutions come into play. These include science, medicine and health care systems, and the courts. All of these institutions become involved in political struggles that involve disease. They may give shape and meaning to the process of determining the patient's physical condition as well.

Science, medicine, and the courts each recognizes a distinction between disease and sickness. Each has a language and a decision-theory to address the likelihood of reaching erroneous conclusions about the patient's condition. Each of the three domains uses its own guidelines for drawing conclusions, considers the consequences of incorrect decisions, and establishes justifications for its way of coping with the uncertainty which typically pervades latent disorders.

In this chapter, I synthesize materials from newspapers, magazines, and books to identify the protagonists, explore the claims-making activity, and review the varieties of evidence put forth in the case of Agent Orange. The focus of the chapter initially is on the telling of the story and then on making sociological sense of it. The evidence, I believe, shows that the Agent Orange controversy peaks and wanes on the coat-tails of a changing array of interest groups and interested persons who advance and oppose the claim that the herbicide has harmed our troops. The battle over Agent Orange has been fought with varying degrees of success in several institutional domains. The uncertainty in the case, coupled with questions of enormous liability and responsibility, allows us to see the politics of facts and certification in an unusually clear light.

"Only We Can Prevent Forests"

Ideally suited for guerilla warfare, the dense jungles and forests of South Vietnam allowed Communist troops to fight when they had the advantage and hide when they did not. Therefore, as early as 1962, the American military toyed with the idea of defoliating combat areas by aerial spraying and purchased several specially prepared herbicides. Chemical companies mixed the ingredients according to military specifications and packaged the different herbicides in color-coded fifty-five gallon drums which gave them their names—Agents White, Blue, Purple, Pink, Green and Orange (ACSH, 1981; CSA, 1981; Lacey and Lacey, 1982). Each of the six herbicides contained compounds that were, or were suspected to be, poisonous. The U.S. Air Force assigned the spraying mission to the 309th Air Commando Squadron and designated the program Operation Hades. It subsequently was given the more neutral designation of Operation Ranch Hand. The squadron, whose members coined the motto, "Only We Can Prevent Forests," sprayed large tracts of terrain from C-123 transport planes equipped with storage tanks and high-pressure nozzles. Other military units sprayed much smaller areas using helicopters, trucks, and backpacks to transport the herbicide.

The most commonly used herbicide was Agent Orange, a 50–50 mix of 2,4-D (n-butyl-2,4-dicholorophenoxy-acetate) and 2,4,5-T (butyl-2,4,5-trichlorophenoxy-acetate; CSA, 1981; Davis and Simpson, 1983; Galston, 1979). United States armed forces sprayed approximately 11 million gallons of Agent Orange between 1965 and 1971. About 90 percent of the spraying took place in jungle and forest regions to deprive enemy troops of cover; about 8 percent was devoted to destroying croplands under their control and the remaining 2 percent was used to clear U.S. base camp perimeters and communication lines.

Although the U.S. Army originally developed 2,4,5-T during the Second World War for chemical and biological warfare, the military apparently considered it safe to use as a herbicide. An Army training manual used in Vietnam described Agent Orange as "relatively nontoxic" to humans and animals (Department of the Army, 1969). Further, farmers and ranchers, the forestry industry, and government agencies had used phenoxy herbicides containing 2,4,5-T in the United States since the 1950s. In 1969, however, a scientific study reported that laboratory animals exposed to trace amounts of dioxin, an unintended byproduct of the manufacture of 2,4,5-T, developed cancer and had offspring with birth defects. At about the same time, Vietnamese officials received reports of birth defects and other health problems in rural areas sprayed with Agent Orange. In response to a growing public concern that dioxin might pose a significant hazard to human health, the military in 1969 restricted the spraying program in Vietnam and then officially discontinued it in 1970. (Some units failed to receive word of the ban and continued spraying until 1971.) On the homefront, the U.S. Department of Agriculture placed restrictions on the use of 2,4,5-T in products or areas where direct human contact was likely.

In 1973, Dr. Matthew Meselson of Harvard University developed a technique for detecting dioxin at extremely low levels. His discovery allowed scientists to question a large body of research which failed to find evidence of 2,4,5-T's harmfulness. They soon concluded that dioxin was very toxic; doses of three or four parts per billion killed some laboratory animals (ACSH, 1981; CSA, 1981; Tschirley, 1986). In 1974, the National Academy of Sciences dispatched a team to Vietnam to assess the defoliation program's effects on humans and the environment. On-going military activity made it difficult to collect valid, systematic data. The team did not find enough evidence to draw conclusions about the effects on humans but recommended a study of the long-term consequences (NAS, 1974).

The Lull before the Storm

The military withdrew the last American combat troops from Vietnam in 1973 and left behind a small number of advisors. North Vietnam took Saigon in 1975, ending the American military presence. With this event, Vietnam veterans temporarily disappeared from public view. They lacked a national organization dedicated to their interests and, further, had no effective advocate in Congress. The American Legion, Veterans of Foreign Wars (VFW), and Disabled American Veterans (DAV), who traditionally have provided veterans with fraternal support and political clout, attracted few Vietnam veterans. These powerful organizations lobbied Congress during the 1970s for pensions and expanded health care in VA hospitals, programs benefitting older veterans comprising the bulk of their memberships. In 1978, there were only 11 Vietnam-era veterans serving in Congress; in contrast, 92 Second World War veterans—almost 20 percent of the combined House and Senate—were in Congress by 1948, two years after the end of that war (Bonior et al., 1984:136). In 1977, President Jimmy Carter appointed Max Cleland, a Vietnam veteran and triple amputee, as Director of the VA. Cleland was an outsider to the political machinery of veterans' affairs: his three predecessors were Second World War veterans and former commanders of either the American Legion or VFW.

Vietnam veterans had pressing, immediate needs. Like veterans of all previous wars (Schuetz, 1980; Wecter, 1980), Vietnam veterans experienced problems of readjustment upon returning home. However, there were significant differences. Unlike soldiers in the First and Second World Wars, who served for the duration of the war, most soldiers in Vietnam spent one year there and then returned to civilian life. Hence, most Vietnam veterans entered and exited the combat zone individually as the war dragged on. The one-year rotation system deprived them of the primary ties which ordinarily bind combat troops (Moskos, 1976). Since the war lacked popular support, re-entry often took place in a divisive atmosphere. When the war ended, there was no triumphant return en masse, and there were no victory parades (MacPherson, 1984:54–64).

Universal conscription during the Second World War also produced a veteran population drawn from the mainstream of American life. In contrast, those who served in Vietnam were drawn from "the ragged fringes of the Great American Dream" (Caputo, 1977:26). They were on the average 19 years of age (compared to an average age of 26 among Second World War soldiers), and they were poorer and less educated than the men of draft age who avoided active duty. Finally, the Vietnam War had a greater wounded-to-killed ratio than any previous war, producing a large number of traumatized and disabled veterans (Kuramoto, 1980:298).

In 1971, Senator Alan Cranston (California) introduced legislation to provide specialized health care and readjustment counseling for Vietnam veterans. Although the cost of the program for the recommended five-year period amounted to less than one percent of the VA budget for those years, the American Legion and VFW opposed the legislation, fearing that it would divert funding from other programs favored by them (Bonior et al., 1984:131–33). The Senate passed the legislation, but the House allowed it to die. Senator Cranston reintroduced the bill in 1973 and 1975, but it suffered the same fate. In 1977, Senator Cranston, by then chair of the Senate Committee on Veterans' Affairs, endorsed a House bill giving congressional veterans' committees a voice in choosing sites for VA hospitals in exchange for support in the House for readjustment counseling. The House bill failed, taking with it the proposed programs for Vietnam veterans. Attempts to make the benefits of the Vietnam GI Bill comparable to those of the Second World War GI Bill encountered the same resistance.

Almost five years after the last combat unit returned from Vietnam, legislation to assist Vietnam veterans, opposed by the American Legion and VFW, had yet to clear Congress. The DAV did provide about $250,000 for a study, the Forgotten Warrior Project (Wilson, 1977). The report revealed significant problems in adjusting to civilian life among a half of the veterans surveyed. However, the VA, the agency responsible for providing assistance, had conducted no study of their special needs. This pattern of neglect set the stage for a flurry of activity by Vietnam veterans in which Agent Orange played a key role.

Agent Orange Gets White-Hot

The VA received the first claims contending illness and disability due to herbicides in 1977. By then a scattering of Vietnam veterans had collected circumstantial evidence linking their health problems to exposure to Agent Orange in Vietnam (Wilcox, 1983). Among them was Air Force veteran Charlie Owens. Shortly after his retirement in 1977, he complained of stomach pain. His doctors diagnosed it as cancer. Owens attributed his cancer to the Agent Orange sprayed by his unit in Vietnam. He expressed this concern to Maude DeVictor, a counselor in Chicago's VA regional office, contending that the spray often was "so thick it looked just like Los Angeles smog" (Kurtis, 1983:43). Within a month after the diagnosis, Owens died of stomach cancer.

DeVictor had not heard of Agent Orange; she asked other veterans if they recalled having been exposed to Agent Orange and requested information from the Department of Defense (DOD) about its possible harmful consequences. The DOD sent her reprints of scientific studies concluding that there was no evidence of the herbicide's harmfulness to humans. She compiled files on about two dozen veterans who reported exposure to the herbicide. DeVictor turned over the files to Bill Kurtis, a reporter for WBBM-TV, Chicago's CBS affiliate.

Sensing a big story, Kurtis began work on a documentary about Agent Orange (Kurtis, 1983:42–46). He interviewed veterans who were themselves sick or who had children with birth defects. He also taped interviews with researchers, including Dr. Meselson, who testified that dioxin was a suspected carcinogen and teratogen. Since the documentary would be broadcast to a local audience late at night, Kurtis sought to increase its exposure. In the week before its airing, he provided each senator and representative from Illinois with a transcript and videotape of the documentary and, on the day before the showing, held a private viewing for reporters from local newspapers.

WBBM aired the documentary, "Agent Orange: Vietnam's Deadly Fog," at 10:30 p.m. on March 23, 1978. The story attracted national attention. Hundreds of inquiries from veterans flooded into the VA office in Chicago. Representative Abner Mikva of Chicago arranged for a subsequent screening of the documentary before the House Committee on Veterans' Affairs. He invited other interested parties, including administrators from the VA, the Environmental Protection Agency (EPA), and the chemical companies who manufactured Agent Orange. In one dramatic swoop, Agent Orange went from the private rumblings of a handful of veterans to the center of national attention.

In April, 1978, Representative Ralph Metcalfe (Illinois) asked the General Accounting Office (GAO), the official research arm of Congress, to elicit from DOD a review of herbicide use in Vietnam. DOD took the position that only the approximately 1,200 participants in Operation Ranch Hand were at risk but nonetheless started compiling a computerized record of the spraying missions (called the HERBS tapes). Simultaneously, the EPA announced it would revoke the registration of (and hence preclude the further domestic use of) products containing 2,4,5-T unless manufacturers could document its safety for humans. The EPA cited studies of laboratory animals and of workers exposed to dioxin in industrial settings and accidents. The agency encouraged interested parties, including the general public, to submit information concerning exposure to dioxin and health problems. Residents of Alsea, Oregon reported a curious increase in miscarriages after a nearby forest was sprayed with Silvex, a herbicide containing 2,4,5-T. The EPA ordered an epidemiological study.

In 1978, the VA circulated a memorandum describing procedures for dealing with Agent Orange-related inquiries. The memorandum stated that the effects of exposure to Agent Orange were short-term and fully reversible and urged the staff to reassure worried veterans (Wilcox, 1983: appendix A). In December, 1978, the VA also organized the Steering Committee on Health Related Effects of Herbicides. The committee noted that the VA considers an injury or illness as service-connected only if it arises during military service. Since the symptoms did not occur during active duty or within a year of discharge, the committee ruled that the ailments were not service-connected. The committee also insisted on scientific evidence based on human subjects before admitting a link between exposure to Agent Orange and harmful consequences. These rulings meant that veterans could not receive medical treatment from the VA for Agent-Orange related conditions and that they could not qualify for compensation.

If the CBS documentary thrust Agent Orange into the national limelight, a lawsuit initiated in 1978 kept it there. Paul Reutershan, a Vietnam veteran dying of stomach cancer, filed a claim with the VA that year contending that his exposure to Agent Orange while flying helicopters in Vietnam caused the cancer. The documentary reinforced his belief in the herbicide's harmfulness. The VA acknowledged that he had cancer but denied that it was caused by Agent Orange.

Reutershan explored the possibility of legal action. Judicial relief, however, is not easy to obtain in cases like these. The doctrine of sovereign immunity and its related principle of intra-military immunity, the Feres doctrine, prohibit veterans from suing the government for damages incurred during military service (*Feres v. United States*, 1950; Lacey and Lacey, 1982:153–80; Schuck, 1986: 58–69). Further, congressional legislation prohibits veterans from suing the VA over the disposition of a case (Phillips, 1980). Therefore, Reutershan filed a $10 million personal injury suit in a New York state court, naming Dow Chemical and two other manufacturers of Agent Orange as defendants (*Reutershan v. The Dow Chemical Company*, 1978). Shortly before he died in December, 1978, Reutershan founded Agent Orange Victims International (AOVI) and pledged its handful of members to carry on the suit after his death.

In the fall of 1978, two other organizations devoted to the needs of Vietnam veterans emerged. Vietnam Veterans of America was founded in Washington, DC as a national organization to advocate their interests, and the 11 Vietnam-era Veterans in Congress formed an official caucus, Vietnam Veterans in Congress. Fledgling organizations already in existence, such as Citizen Soldier and Vietnam Veterans Against the War, also organized task forces on Agent Orange.

The Issue Broadens

In January of 1979, a month after Reutershan's death, AOVI hired Victor Yannacone, an accomplished attorney in workman's compensation disputes and a passionate foe of the chemical industry, as their "legal field commander" (Schuck, 1986:44). Yannacone refiled Reutershan's complaint as a class-action suit in federal court on behalf of all Vietnam veterans and their children who might have been damaged by the herbicide (*Reutershan et al. v. The Dow Chemical Company et al.*, 1979). The suit named all six manufacturers of Agent Orange—Dow Chemical, Hercules, Northwest Industries, Diamond Shamrock, Monsanto, and North American Phillips—as defendants. Yannacone estimated the damages at $4 billion.

Recent trends in the legal system favored the Agent Orange suit (Schuck, 1986:26–34). First, a series of decisions gradually extended a manufacturer's liability to all individuals who use or are exposed to its product. Previously, the obligation to produce and market nondefective goods—the duty of care—applied only to the purchaser of a product. In addition, strict liability without fault became the standard for assessing a manufacturer's culpability whereas, in the past, the plaintiff was required to show that the manufacturer failed to exercise reasonable precaution. The courts also no longer required the plaintiff to prove that a specific defendant made the product which caused the damage so long as some reasonable method existed for allocating damages.[1] Finally, the amounts which courts and juries awarded in compensation for damages skyrocketed. Awards in excess of one million dollars became commonplace.

Agent Orange quickly proved to be no ordinary product liability case. From the moment it was filed, the Reutershan suit attracted attention and notoriety in the media. Attorneys on both sides felt that this worked in favor of the veterans. Even after the CBS documentary and its coverage, very few veterans—and an even smaller percentage of the American public—knew about Agent Orange. Press coverage of the suit developed awareness and encouraged veterans who thought they might be afflicted to initiate claims and join the suit. It also created sympathy for the veterans and suspicion about the chemical companies.

Early in 1979, Senator Charles Percy (Illinois) requested a second GAO inquiry into herbicide use in Vietnam. This investigation included an initial assessment of the HERBS tapes (USGAO, 1979). The report acknowledged that the Air Force clearly had used herbicides without proper precautions. Spraying from aircraft often took place close enough to troops that wind drift may have exposed them to the herbicide, and ground troops frequently entered contaminated areas shortly after spraying. Further, it stated that troops used empty herbicide drums for constructing latrines and even home-made hibachis. The report concluded that as many as 40,000 Army, Marine, and Air Force personnel experienced significant exposure to the herbicide; it also provided a listing of units whose personnel it considered officially at risk.

The GAO report legitimated the concerns of worried veterans. It could no longer be argued that defoliation took place under carefully controlled conditions in areas remote from American troops. The remaining question concerned the consequences of exposure. Agent Orange was now a point of conflict in three arenas: the VA, Congress and civilian regulatory agencies, and the courts.

The VA, the nation's single largest health care system, caters primarily to 12 million veterans of the Second World War whose average age in 1979 exceeded 60 years and who have chronic ailments associated with aging. At age 65, veterans are eligible to receive all their medical care from the VA free of charge. Only 30 percent of patients treated by the VA have service-connected disabilities and, of all medical treatment provided by the VA, only 16 percent is service-connected. In contrast, the average Vietnam veteran is 30 years younger. They receive acute treatment and physical and occupational rehabilitation from injuries associated with warfare. The largest single diagnosis in VA facilities, alcohol abuse, is an especially acute problem among Vietnam veterans with combat experience. The Agent Orange claims involve latent disorders which did not fit this profile. VA Director Cleland admitted that many in the VA simply viewed the Agent Orange claims "as a big hoax, like witchcraft" (Bonior et al., 1984:167).

The politics of veterans' compensation belies the aura of administrative rationality that surrounds it. In time of war, soldiers incur risk of injury and death for which they cannot be repaid easily. Compensation and other benefits—which are funded by Congress and executed by the VA—stem from a recognition of this altruistic contribution and stand as visible symbols of collective gratitude or even guilt. Two principles serve as guidelines for compensating veterans (House Committee, 1956). First, the highest priority should go to programs which extend medical care for service-connected injuries or illnesses, ease re-entry into civilian life, and provide adjustments for career advantages lost while in the service. Second, when there is doubt about a claim, the veteran should be given the benefit of the doubt.

In June 1979, the House Committee on Interstate and Foreign Commerce held the first congressional hearings on Agent Orange and its effects on Vietnam veterans (House of Representatives, 1979). Veterans with cancer or who had children with birth defects testified before the committee. In response to this hearing and the growing political presence of Vietnam veterans, Congress enacted P.L. 96–151 commissioning the VA to conduct an epidemiological study of veterans exposed to Agent Orange. Significantly, Congress also ratified a bill—first introduced by Senator Cranston in 1971 and defeated on five separate occasions—that established readjustment counseling for Vietnam veterans.

The VA, however, stood firm on its position of requiring scientific evidence based on human subjects. By this time, about 750 veterans had submitted claims to the VA and about 5,000 had requested physical examinations or treatment for ailments related to Agent Orange. No conclusive evidence linked Agent Orange with disease and birth defects. Evidence consisted of personal accounts and claims by veterans, circumstantial evidence from industrial accidents, and inconclusive scientific studies. Nonetheless, to require scientific evidence as a precondition for treatment and compensation ran contrary to the spirit of VA guidelines and placed an extensive burden of proof upon Agent Orange claimants.

In March 1979, WBBM broadcast a second documentary update of the Agent Orange story. This documentary stressed the presence of dioxin in many commonly used compounds and hence the exposure of many civilians in the United States to this toxin. Kurtis (1983) again arranged for a private showing of the documentary for members of Congress. In June, the Senate requested a long-term study of the effects of exposure to dioxin by the Department of Health, Education, and Welfare. The EPA and various industry groups and universities also had studies of dioxin under way.

For years, agricultural and industry researchers had conducted research on Agent Orange's active ingredients, 2,4,5-T and 2,4-D. They knew that the ingredients killed plants by mimicking natural plant-growth hormones. The impostures caused rapid and uncontrolled cell growth; the plants literally grew themselves to death. But the early studies by and large did not ask if the active ingredients could produce unscheduled and abnormal cell growth in animals or humans. During the early 1970s, other researchers developed an effective methodology for assessing dioxin—an unintentional by-product in Agent Orange—and, by all accounts, everyone involved soon considered dioxin to be exceedingly toxic to laboratory animals (for summaries, see ACSH, 1981; CSA, 1981; Davis and Simpson 1983; Galston, 1979).

In part because establishing toxicity is complicated, answers to questions about dioxin's effects on humans lacked experimental evidence and sparked debate (cf., Epstein, 1983; Tschirley, 1986). Epidemiologists trace about 25 percent of all cancers to humanly produced environmental and occupational exposures and the rest to voluntary consumption of foods, drugs, and stimulants, and to naturally occurring carcinogens (Katzman 1986). While common foodstuffs such as peanut butter or well-done steak may contain naturally occurring carcinogens in levels of parts per thousand, regulators try to protect the public from exposures to *synthetic* chemicals measured in parts per million. Exposure to dosages of dioxin as minute as parts per billion have proved toxic to some laboratory animals (ACSH, 1981; CSA, 1981; Tschirley, 1986).

Humans have been exposed to dioxin in several well-documented industrial accidents or oversights. The incidents date back to the exposure in 1949 of about 250 workers in the Monsanto chemical plant in Nitro, West Virginia. The most spectacular accident occurred in Seveso, Italy in 1976, where a plant explosion exposed 37,000 people to tricholorophenol (Whiteside, 1979). In these and other incidents, the most common initial symptom was chloracne, a blistering of the skin (Taylor, 1979). The condition often subsided in a few days but in some cases lasted indefinitely. Other symptoms included nausea, dizziness, and numbing of the fingers and toes. Time-lags between initial exposure and serious disease made it difficult to establish long-term effects. For example, follow-up studies of Swedish workers exposed during the 1950s to compounds containing dioxin revealed higher than expected incidences of lymphomas and soft-tissue sarcomas. However, time-lags of several years made it almost impossible to eliminate other potential causes of these diseases. Further, the diseases allegedly caused by dioxin may be caused by factors other than exposure to dioxin and, with the exception of chloracne, exposure to dioxin apparently does not produce one distinctive disease.

Some claims of illness lacked credibility because they were not consistent with prevailing medical opinion. Medical researchers simply did not understand how dioxin worked, and some of the symptoms attributed to dioxin suggested unfamiliar disease processes. This was especially true of claims by veterans that exposure to dioxin may have produced birth defects in their children: medical science thought mothers had to be exposed to the toxin during pregnancy.[2] Even this question could not be assessed in studies of the Seveso victims. The incidence of stillborn births and birth defects was lower among women who had been exposed than for matched controls because of therapeutic abortions available by special arrangement after the incident.

Amidst this confusion, several states took the initiative in addressing exposure to dioxin among veterans and civilians. New York and New Jersey instructed state commissions to hold hearings and conduct investigations. In late 1978 and early 1979, the Love Canal controversy—in which chemicals from buried metal drums leaked into the ground water in Buffalo, New York—received extensive media scrutiny in both states. Investigators discovered in 1979 that dioxin was one of the contaminants seeping from the drums. Both states also have large numbers of Vietnam veterans.

By 1979, the slow and arduous task of moving the Agent Orange case to trial was under way. A Multi-District Litigation Panel assigned the case to Judge George Pratt of the Federal District Court in Uniondale, New York. From the outset, the case bogged down in a stupefying array of legal entanglement. Product liability cases ordinarily fall under state tort law; applying varied state laws in federal court would be complicated and burdensome. Nevertheless, the chemical companies argued that the product liability laws of each state in which individual plaintiffs reside should apply. The plaintiffs maintained that federal common law should govern the case in order to simplify matters. Either way a decision would be controversial.

Meanwhile, the chemical companies filed third-party complaints against the government for indemnity and sought immunity as war contractors. They argued that the government had not simply purchased herbicides from them, but also had specified their content, supervised their production, and controlled their application in Vietnam. Further, they noted that government regulations forbade them from selling the same mixtures in civilian markets. In November 1979, Judge Pratt accepted the case as a federal question and refused to accept the chemical companies motion to dismiss (*In re "Agent Orange,"* 1979). The companies immediately filed an appeal.

One other event took place in November 1979. Acting on a request by a small group of Vietnam veterans, twenty-six senators co-sponsored a bill which set aside land next to the Lincoln Memorial for a Vietnam veterans' memorial. It had been a big year for Vietnam veterans.

Into the '80s: The Issue Wanes

The CBS documentaries and the lawsuit propelled Agent Orange into the limelight at a time when neither veterans nor the public knew much about it. These events became vehicles for disseminating information and heightening awareness, both about the herbicide and its potential harm to veterans and about dioxin and its potential danger to civilians. In April 1980, WBBM broadcast the third installment of its series on the issue, "Agent Orange: The View from Vietnam" (Kurtis, 1983: 78–79). It verified the destruction of forests and croplands and reported Vietnamese cases of cancers and birth defects which Vietnamese health officials attributed to Agent Orange. Again, Congress received a private screening of the documentary. Although the installment did not evoke the dramatic responses of its two predecessors, it did help sustain congressional interest in the issue. A number of congressional committees held special hearings about Agent Orange, dioxin, their effects on veterans and civilians, and the formulation of policies to deal with them (House of Representatives, 1980). In June 1980, Congress passed the bill which set aside two acres for a Vietnam veterans' memorial. Situated prominently next to the Lincoln Memorial on some of the most coveted real estate in Washington, DC, the memorial was an apt indication of how much the political power of Vietnam veterans had increased in the past two years.

Although the lawsuit continued to attract inquiries from worried veterans, it suffered a major setback in November 1980. The Second Court of Appeals ruled that substantial federal interest was lacking in the Agent Orange case (*In re "Agent*

Orange," 1980). The veterans sought a higher ruling on the appeal. Meanwhile, Judge Pratt again certified the case as a class action and denied the chemical companies motion to dismiss as war contractors.

In 1981, Congress continued its response to initiatives on behalf of Vietnam veterans. That same year, however, President Ronald Reagan placed a hiring freeze on the readjustment counseling program established in 1979 and placed restrictions on other expenditures in the program. A subsequent budget message recommended getting rid of all outreach programs for Vietnam veterans, but in passing the 1979 bill, Congress specifically had exempted the programs for Vietnam veterans from this kind of budget trimming. Vietnam Veterans of American and Vietnam Veterans in Congress both filed lawsuits to overturn the illegal freeze and restrictions. Congress not only overrode President Reagan's proposals, but also passed P.L. 91–72 which extended G.I. bill benefits for job training, contained a small business loan package for Vietnam veterans, and directed the VA to provide Vietnam veterans with medical care for any condition which might possibly result from exposure to Agent Orange. While the measure was under consideration, the VA lobbied vigorously against it. President Reagan expressed his disapproval of the bill on budgetary grounds before reluctantly signing it into law (Bonior et al., 1984:76–78, 152–54).

In June, the Supreme Court sought the opinion of the U.S. attorney general on the question of federal interest in the Agent Orange lawsuit. After consultation with the VA and DOD, the attorney general reported no special federal interest in the case. The Supreme Court then declined to rule on the appeal, thereby allowing the decision of the appeals court to stand (*In re "Agent Orange,"* 1981). Yannacone refiled the suit in federal court, leaving two questions yet unanswered: which state's law should apply and which choice-of-law rules would that state's court apply? These issues were not resolved until February 1984.

Meanwhile, newspapers in St. Louis revealed that streets and stable areas in Times Beach, Missouri had been contaminated mistakenly with waste oil containing dioxin (Kurtis, 1983; *Science News*, 1983; *Time*, 1984). For years a local contractor had purchased waste oil from a nearby chemical plant to use in tarring dirt roads and arenas. Reports of peculiar deaths among farm animals and birds dated back to 1971, when the same contractor tarred an arena and stable areas in Moscow Mills, Missouri. In 1983, the EPA acted quickly and decisively. It evacuated and relocated the town's population at a cost of about $33 million; it fenced off the town and declared it a disaster area for the next 20 years.

The actions of the EPA stand in sharp contrast to those of the VA. Students of formal organizations use the concept of the "lazy monopoly" to explain the inertia of some large organizations in the face of demands for change (Hirschman, 1970; Seidler, 1979). The concept is also useful in explaining the actions of the VA in the Agent Orange controversy. The lazy monopoly has control over a service or product. Its executives are slow to improve the quality of the product, change their policies, or alter the structure of the organization. They prefer instead to lose clients or personnel who criticize the organization. The absence of competition encourages recalcitrance in the lazy monopoly; it invites unsatisfied clients to renounce their membership in the organization and obtain the product or service elsewhere.

By 1983, more than 100,000 veterans had requested physicals for complaints related to Agent Orange. While this would be a large number of cases for most hospital systems, the numbers were small within the VA's overall caseload. However, the inquiries represented the wrong kind of patient for the medical care VA hospitals typically provide. Caring for these veterans would require a substantial cost and effort. Nevertheless, P.L. 97–72 required the VA to provide treatment for Agent Orange-related conditions. On the face of it, the VA was compliant. It began a newsletter, *Agent Orange Review,* to summarize current information and procedures for receiving treatment. It had already established a registry from which to compile a listing of veterans who inquired about the herbicide and its effects. (In 1982, the registry was found to have 89,000 names but no addresses.) However, the VA admitted no link between Agent Orange and disease, with the exception of chloracne. It defined a condition which may have been produced by Agent Orange as one that was not congenital, not a result of an injury, or did not have a specific or well-established cause (VA, 1983).

Noting that the VA opposed the legislation requiring it to provide health care, many veterans sought comprehensive testing and, when necessary, treatment from private physicians—services which they paid for out-of-pocket or through private health insurance. Veterans who were sick and unable to work also found that they could file claims with the Social Security Administration (SSA). Compensation by the SSA for a disabling condition did not require that the claimant prove how the condition came into being, only that it existed. This process siphoned troublesome clients away from the VA. It stemmed the influx of claims whose authenticity was debatable and whose medical profiles were not synchronized with the structure of services the VA typically provided.

Ironically, the pensions and old-age health care which take up the majority of the VA budget and medical activity, and which the American Legion and VFW previously supported at the expense of programs for Vietnam veterans, provide assistance and care for conditions which are not service-connected. As noted earlier, these organizations did not absorb significant numbers of Vietnam veterans and consequently did not serve as their advocates. In fact, they opposed legislation designed to assist Vietnam veterans. With the Agent Orange controversy, Vietnam Veterans of America emerged as a strong national organization. These developments meant that the VFW and American Legion, organizations with aging memberships, were in danger of losing an entire generation of veterans who very shortly would be a necessary source of members. It was time to adjust to changing political realities. In 1982, the VFW adopted a resolution advocating compensation for Agent Orange-related conditions. In 1983, the American Legion followed suit.

Scientists continued to chip away at the problem. Researchers proposed explanations of how dioxin and similar toxic substances cause disease in laboratory animals (Matasumura, 1983) and possibly humans. Other studies reported higher rates of soft-tissue sarcomas and lymphomas among Swedish workers exposed to dioxin in the workplace (Sterling and Arundel, 1986) and among Kansas farmers who used herbicides containing 2,4,5-T (Hoar et al., 1986) than for matched controls. Finally, Ranch Hand participants (Lathrop et al., 1984) and residents of Times Beach, Missouri

(Hoffman et al., 1986) showed physiological abnormalities not found among matched samples of unexposed persons, although the clinical significance of these abnormalities was not clear.

As the Agent Orange lawsuit dragged into its fourth year, Judge Pratt resigned from the case. It was reassigned to Judge Jack Weinstein. Weinstein, who enjoyed a reputation as an innovative judge and precocious professor of law at Columbia University, quickly reconceptualized the case. Taking the initiative, he ruled one by one on the lingering legal issues and placed pressure on the lawyers to proceed directly to trial or to settle out of court. Of the two outcomes, Weinstein's preference was for settlement. Schuck (1986:143) states: "To try such a case, (Weinstein) felt, would consume an almost unthinkable amount of time . . . , money, talent, and social energy, and with the inevitable appeals and possible retrials, the outcome might be uncertain for many years to come. . . . Finally, a negotiated settlement offered the prospect of everybody obtaining something rather than one side losing everything."

However, even a dynamic judge of Weinstein's caliber could not easily induce a settlement. Agent Orange had become the basis of a highly visible, emotionally charged, symbolic suit; accomplished, aggressive attorneys were gearing up for the long-awaited day in court. Through a series of carefully crafted legal and tactical maneuvers, Weinstein skewed the options and incentives towards settlement. On May 7, 1984, nine years to the day after the last American troops were evacuated from Saigon and only hours before the trial was to begin, Weinstein extracted a cease-fire from lawyers representing the two sides. The manufacturers of Agent Orange, who already had spent $100 million on legal fees in the case, agreed to establish a $180 million fund from which veterans harmed by dioxin exposure would be compensated. No payments have yet been made since aspects of the settlement are still under appeal. When compensation begins, the court estimates that the maximum payment will be $3,400 to survivors of deceased veterans who were exposed to Agent Orange and $12,800 ($1,280 per year for 10 years) to veterans exposed to the herbicide who are now 100% disabled (Kubley et al., 1985:85–86). According to the agreement, establishing the fund did not affirm that exposure to Agent Orange caused or contributed to disease among veterans and did not represent an admission of liability by the chemical companies (*In re "Agent Orange,"* 1984).

The terms of the settlement are (Schuck, 1986:165): (1) Defendants would pay $180 million plus interest. . . . (2) The settlement fund would advance moneys to pay class notice and settlement administration expenses. (3) No other distribution of settlement funds would be made, until appeals from a final settlement order had been exhausted. (4) Defendants could obtain reverse indemnification for veteran opt-out claims upheld by state courts up to $10 million until January 1, 1999. (5) The class definition would be interpreted to include service people whose injuries had not yet been manifested. (6) Plaintiffs could retain defendants' documents for one year. (7) All parties reserved all rights to sue the United States. (8) Defendants denied all liability. (9) Defendants reserved the right to reject the settlement if a "substantial" number of class members opted out. (10) Any class member who opted out would have an opportunity to opt back in. (11) Unclaimed funds would revert to the defendants after twenty-five years. (12) The settlement agreement was subject to

Rule 23(e), a "fairness" hearing. (13) Although after-born claimants were not included in the class and could therefore sue, the distribution plan would make special arrangements to address their needs. (14) The court would retain jurisdiction until the settlement fund was exhausted.

Veterans reacted to this settlement with varying proportions of disappointment and realism. Particularly galling was the lack of answers to questions about dioxin's harmfulness or the manufacturer's culpability. As Schuck (1986:256) observes:

> The veterans expected the law to provide the essential elements of simple justice . . . with comprehensible narrative evidence, a clear determination of guilt or innocence, and a swift, straightforward remedy. Their lawyers knew, however, that this view of adjudication was chemical, . . . that many of the complex-technical aspects of the Agent Orange case contradicted their clients' intuitive notions of justice and even defied common sense. The intricacies (of the case) simply had no counterpart, no coherent meaning, in the veterans' experiences. . . . For them, the law had become mystifying, alien, and unjust.

Discussion

The constructionist perspective rests on the assumption that facts and their interpretations are established in a process studded with entrenched practices and interests. Latent disorders—diseases in which there is a substantial time lag between exposure and consequence—are especially useful in illustrating the politics of facts and certification. The uncertainty accompanying them creates special personal, organizational, and political problems (Couch and Kroll-Smith, 1985:566–70).

To begin with, the time lag between exposure and consequence strains the methodologies of science, medicine, and law to the point where it is difficult to establish evidence. Simply finding and identifying the victims may be a formidable task. Even in dramatic instances of disease, rival explanations may be posed with no way to judge between them. The growing controversy often becomes a convenient and symbolic medium for venting other antagonisms. Organizations, communities, and government entities may become embroiled in disputes over blame and responsibility. Interest groups may form as advocates of different kinds of perspectives and evidence. When this occurs, contradictory conclusions may be obtained by asking the same question in different domains, and conflicts may emerge over whose version will be accepted as true. In the protracted period of suffering and haggling, frustration and delusion may become a prominent feature of the victims' response.

The case of Agent Orange encompasses all of these considerations and provides an ideal historical incident for examining the socio-political processes in which disease, illness, and sickness are invoked, sustained, or challenged. The term "disease" refers to a disruptive biological condition, "illness" to the perception by the individual that he or she is not well, and "sickness" to the recognition by appropriate authorities that the individual indeed has a disease. Evidence of disease, illness, and sickness is assessed in three institutional domains—science, medicine, and law— each with its own decision rules for certifying them.

Scientific research is guided by the null hypothesis, i.e., the proposition that the effect under investigation is *not* produced by the suspected cause. The null hypothesis here is that dioxin does not produce cancers in exposed subjects or birth defects in their offspring. By focusing on the null hypothesis, scientific research favors type II (failing to reject a false null hypotheses) over type I (rejecting a true null hypothesis) errors (Hays, 1973). If an error is going to be made, scientific investigators would rather accept a false null hypothesis than a false alternate hypothesis. Hence, it is a greater sin in scientific studies of phenoxy herbicides to conclude that dioxin harms health, when it actually does not, than to conclude that dioxin is harmless to humans when it actually is.

Medical diagnoses often carry the opposite bias. Two types of error may result here also. Well persons may be diagnosed as sick, or diseased ones may be considered healthy. In general physicians are socialized to believe that it is less of a sin to diagnose a well person as sick than to fail to detect disease in a diseased person (Scheff, 1972). Two considerations mediate this posture. One, medical knowledge retains a firm base in current scientific consensus. Physicians who act otherwise run the risk of being labelled "quacks" or of being sued. Two, the tendency to suspect harm until doubt is removed may be reversed in organizational contexts where directives endorse a particular position regarding etiology, diagnosis and/or treatment, or when diagnosing sickness may imply organizational culpability for treatment and compensation (Nathanson and Becker, 1981). Although treatment and compensation take place in separate administrative channels in the VA, the cause of an injury or disease takes on a significance in this setting beyond that usually required for diagnosis and treatment. In this situation, the patient's needs and interests may become divorced from those of the organization. Certainly it is necessary to ask, "Can the organization afford to define people who report having illnesses—and hence who may be diseased—as sick?"

The legal system provides yet a third set of criteria for establishing cause and effect in personal injury cases. Rules allow for the introduction and evaluation of several kinds of evidence including, but not limited to, scientific and medical evidence (Levine and Howe, 1985). The basic principle in a civil suit is "preponderance of the evidence." A verdict may be reached when the evidence simply is tipped in either direction (Kaye, 1983). The judicial regulation of hazards recognizes two relevant standards of liability: strict liability and negligence (Page, 1983). Strict liability assigns legal responsibility whether or not the defendant was careless. Hence the decision rule under strict liability is: if, in the judgment of the court, the probability that the defendant's activity caused the plaintiff's injury is greater than 50 percent, find for the plaintiff; otherwise, find for the defendant. Therefore, if a case is wrongfully decided, it is as likely to favor an undeserving plaintiff as an undeserving defendant.

In the Agent Orange controversy, a small number of Vietnam veterans complained of illnesses of unknown and, ultimately, disputed origin. Civilian medical authorities diagnosed cancers and other disabling or life-threatening conditions. Some veterans sought assistance from the VA to confirm suspicions that the exposure to Agent

Orange in Vietnam caused the diseases. This scenario seemed unlikely to VA physicians and costly to VA administrators. Hence, the inquiries had no effective representation in the institutional arena in which the illnesses of veterans are certified as sicknesses. The issue remained obscure until covered in a television documentary. The WBBM-TV news team operated under different organizational and methodological constraints than those of VA officials. The documentary blurred the distinction between illness and sickness by portraying the veterans' stories sympathetically and by revealing the uncertainty in the medical community outside the VA about Agent Orange's safety.

The documentary and the lawsuit vaulted Agent Orange into the national limelight. The lawsuit also moved the question of whether Agent Orange causes disease into an arena where standards of evidence permit circumstantial as well as scientific evidence and where the preponderance principle provides for definitive decision-making despite uncertainty. However, circumstances peculiar to this case undermined the certification of sickness as adjudicated by law. A class-action suit in a mass toxic tort case raised problems of unprecedented scope and complexity. Previous lawsuits involving latent disorders, such as the asbestos and DES cases, provided no helpful guidelines for evaluating in court the inconclusive scientific evidence and contradictory medical-opinion of the Agent Orange case. In those previous court cases, authorities were able to document levels of exposure and to detect a distinctive latent disorder as a consequence of exposure. This was not possible in the case of Agent Orange. After much delay and a considerable amount of legal sparring, the two sides settled out-of-court in 1984.

The controversy over Agent Orange also became a vehicle for organizing and securing other benefits for Vietnam veterans. When the VA reacted slowly and reluctantly, Vietnam veterans were angered by their inability to receive treatment for troubling and serious health problems. They interpreted the VA's hesitancy as a violation of the compensation guidelines and as part of an overall pattern of indifference to them in American society. Vietnam Veterans of America established itself as a national organization during the early stages of the controversy and remains the largest organization solely for Vietnam veterans. Congress approved legislation to help Vietnam veterans adjust to civilian life only after the Agent Orange controversy aroused Vietnam veterans and their sympathizers and sparked political activity among them. Congressional mandates also require the VA to provide medical care for veterans who claim to have been exposed to Agent Orange and to compensate veterans should definitive scientific evidence ever link Agent Orange with disease.

Though the issue has waned, the controversy over Agent Orange is not over. In 1987, a study of mortality records conducted by the VA found higher rates of death due to lung cancer and non-Hodgkin's lymphomas than expected among Marine units who were exposed to Agent Orange (Boffey, 1987). The VA withheld the findings from the public for several months to assess their relevance; Vietnam Veterans of America (VVA) accused them of a cover-up. The bumper sticker, "Sprayed and Betrayed," while not selling briskly, still has a market.

NOTES

1. In a landmark case involving diethylstilbestrol (DES), the plaintiff could not show which company produced the actual pill which harmed her. However, the court required the manufacturers to pay damages in proportion to their individual shares of the DES market (*Sindell v. Abbott Laboratories.* 1980).

2. Subsequent research has explored the possibility that toxins to which males are exposed prior to reproductive activity are linked to birth defects. For example, Hales et al. (1986:423) now have reported that a toxin introduced into male mice accrued in the reproductive tract and subsequently was transmitted to female mice during mating, thereby "altering progeny outcome."

REFERENCES

American Council on Science and Health (ACSH). 1981. *The Health Effects of Herbicide 2,4,5-T.* New York: American Council on Science and Health.

Boffey, Phillip. 1987. "Lack of military data halts Agent Orange study." *New York Times* September 1:A1,C5.

Bonior, David, Stephen Champlin, and Timothy Kolly. 1984. *The Vietnam Veteran: A History of Neglect.* New York: Praeger.

Caputo, Philip. 1977. *A Rumor of War.* New York: Ballentine Books.

Couch, Stephen, and J. Stephen Kroll-Smith. 1985. "The chronic technical disaster: Toward a social scientific perspective." *Social Science Quarterly* 66:564–75.

Council on Scientific Affairs (CSA). 1981. *The Health Effects of "Agent Orange" and Polychlorinated Dioxin Contaminants.* Chicago: American Medical Association.

Davis, Miriam, and Michael Simpson. 1983. "Agent Orange: Veterans' complaints and studies of health effects." Issue Brief Number IB83043. Washington, DC: The Library of Congress.

Department of the Army. 1969. *Employment of Riot Control Agents, Flame, Smoke, Antiplant Agents, and Personnel Detectors in Counterguerilla Operations.* Training CircularTC 3–16. Washington, DC: Department of the Army.

Epstein, Samuel. 1983. "Agent Orange diseases: Problems of causality, burdens of proof, and restitution." *Trial* November: 91–138.

Freidson, Eliot. 1970. *Profession of Medicine: A Study of the Sociology of Applied Knowledge.* New York: Dodd, Mead and Company.

Galston, Arthur. 1979. "Herbicides: A mixed blessing." *BioScience* 29:85–90.

Hales, Barbara, Susan Smith, and Bernard Robaire. 1986. "Cyclophosphamide in the seminal fluid of treated males." *Toxicology and Applied Pharmocology* 84:423–30.

Hays, William. 1973. *Statistics for the Social Sciences.* 2d ed. New York: Holt, Rinehart and Winston.

Hirschman, Albert. 1970. *Exit, Voice, and Loyalty: Responses to Decline in Firms, Organizations, and States.* Cambridge, MA: Harvard University Press.

Hoar, Sheila, Aaron Blair, Frederick Holmes, Cathy Boysen, Robert Robel, Robert Hoover, and Joseph Fraumeni. 1986. "Agricultural herbicide use and risk of lymphoma and soft-tissue sarcoma." *Journal of the American Medical Association* 256:1141–47.

Hoffman, Richard, Paul Stehr-Green, Karen Webb, Gregory Evans, Alan Knutsen, Wayne Schramm, Jeff Staake, Bruce Gibson, and Karen Steinberg. 1986. "Health effects of long-term exposure to 2,3,7,8-tetracholorodibenzo-p-dioxin." *Journal of the American Medical Association* 255:2031–38.

House Committee on Veterans' Affairs. 1956. *Veterans' Benefits in the United States: A Report to the President by the President's Commission on Veterans' Pension.* Washington, DC:House Commission Print No. 235, 84th Congress, 2nd Session.

House of Representatives. 1979. *Involuntary Exposure to Agent Orange and Other Toxic Spraying: Hearings before the Subcommittee on Oversight and Investigation of the Committee on Interstate and Foreign Commerce,* 96th Congress, 1st Session, June 25–27.

————. 1980. *Agent Orange: Exposure of Vietnam Veterans, Hearings before the Sub-committee on Oversight and Investigation of the Committee on Inter-state and Foreign Commerce,* 96th Congress, 2d Session, September 25.

Katzman, Marvin. 1986. "Chemical catastrophes and the courts." *Public Interest* 86:91–105.

Kaye, David. 1983. "Statistical significance and the burden of persuasion." *Law and Contemporary Problems* 46:13–23.

Kubley, Craig, David Addlestone, Richard O'Dell, Keith Snyder, and Barton Stichman. 1985. *The Viet Vet Survival Guide.* New York: Random House.

Kuramoto, Frank. 1980. "Federal mental health programs for the Vietnam veteran." Pp. 293–304 in Charles Figley and Seymour Leventman (eds.), *Strangers at Home: Vietnam Veterans Since the War.* New York: Praeger.

Kurtis, Bill. 1983. *Bill Kurtis on Assignment.* Chicago: Rand-McNally.

Lacey, Pamela, and Vincent Lacey. 1982. "Agent Orange: Government responsibility for the military use of phenoxy herbicides." *Journal of Legal Medicine* 3:137–80.

Lathrop, George, William Wolf, Richard Albanese, and Patricia Moynahan. 1984. *Ranch Hand II. An Epidemiologic Investigation of Health Effects in Air Force Personnel Following Exposure to Herbicides.* San Antonio, TX:Brooks Air Force Base.

Levine, Murray, and Barbara Howe. 1985. "The penetration of social science into legal culture." *Law and Policy* 7:173–98.

MacPherson, Myra. 1984. *Long Time Passing: Vietnam and the Haunted Generation.* Chicago: Signet Books.

Matasumura, Fumio. 1983. "Biochemical aspects of action mechanisms of 2,3,7,8-tetracholoro-dibenzo-p-dioxin (TCDD) and related chemicals in animals." *Pharmacology and Therapeutics* 19:195–209.

Moskos, Charles, Jr. 1976. "The military." Pp. 55–77 in Alex Inkeles, James Coleman, and Neil Smelser (eds), *Annual Review of Sociology.* Volume 2. Palo Alto, CA: Annual Reviews Inc.

Nathanson, Constance, and Marshall Becker. 1981. "Professional norms, personal attitudes, and medical practice: The case of abortion." *Journal of Health and Social Behavior* 22:198–211.

National Academy of Sciences (NAS). 1974. *Report of the Committee on the Effects of Herbicides in South Vietnam; Part A, Summary and Conclusions.* Washington, DC: National Academy of Sciences.

Page, Talbot. 1983. "On the meaning of the preponderance test in judicial regulation of chemical hazard." *Law and Contemporary Problems* 46:267–83.

Parsons, Talcott. 1951. *The Social System.* New York: The Free Press.

Pfohl, Stephen. 1985. "Toward a sociological deconstruction of social problems." *Social Problems* 32:228–31.

Phillips, Dean. 1980. "Subjecting the Veterans Administration to court review." Pp. 325–41 in Charles Figley and Seymour Leventman (eds.), *Strangers at Home: Vietnam Veterans Since the War.* New York: Praeger.

Scheff, Thomas. 1972. "Decision rules and types of error, and their consequences in medical diagnosis." Pp. 309–23 in Eliot Freidson and Judith Lorber (eds.), *Medical Men and their Work.* Chicago:Aldine.

Schneider, Joseph. 1985. "Social problems theory: The constructionist view." Pp. 209–29 in Ralph Turner and James Short (eds.), *Annual Review of Sociology*. Palo Alto, CA: Annual Reviews Inc.

Schuetz, Alfred. 1980. "The homecomer." Pp. 115–22 in Charles Figley and Seymour Leventman (eds.), *Strangers at Home: Vietnam Veterans Since the War*. New York: Praeger.

Schuck, Peter. 1986. *Agent Orange on Trial: Mass Toxic Torts Disasters in the Courts*. Cambridge, MA: Harvard University Press.

Science News. 1983. "Dioxin digest." *Science News* September 3:156–57.

Seidler, John. 1979. "Priest resignations in a lazy monopoly." *American Sociological Review* 44: 763–83.

Spector, Malcolm, and John Kitsuse. 1977. *Constructing Social Problems*. Menlo Park, CA: Cummings.

———. 1987. *Constructing Social Problems*. Hawthorne, NY: Aldine.

Sterling, Theodor, and Anthony Arundel. 1986. "Health effects of phenoxy herbicides." *Scandinavian Journal of Work and Environmental Health* 12:161–73.

Stoscheck, Christa, and Lloyd King Jr. 1986. "Role of epidermal growth factor in carcinogensis." *Cancer Research* 46:1030–37.

Susser, Mervyn. 1973. *Causal Thinking in the Health Sciences*. New York: Oxford University Press.

Taylor, James. 1979. "Environmental chloracne: Update and overview." *Annuals of the New York Academy of Science* 320:295–307.

Time. 1984. "Hazards of a toxic wasteland: Learning to cope with high-tech risks." *Time* December 17:32–34.

Tschirley, Frank. 1986. "Dioxin." *Scientific American* 254:29–35.

United States General Accounting Office (USGAO). 1979. *U.S. Ground Troops in South Vietnam Were in Areas Sprayed with Herbicide Orange*. FPCD-80-23. Washington, DC: USGAO.

Veterans Administration (VA). 1983. "Health care services under public law 97–72." *Agent Orange Review* 2:1–4.

———. 1986. "A summary of VA Agent Orange activities." *Agent Orange Fact Sheet* 1–6.

Wecter, Dixon. 1980. "When Johnny comes marching home: Veterans' 'benefits' in historical perspective." Pp. 257–66 in *Strangers at Home: Vietnam Veterans Since the War*. New York: Praeger.

Whiteside, Thomas. 1979. *The Pendulum and the Toxic Cloud*. New Haven: Yale University Press.

Wilcox, Fred. 1983. *Waiting for an Army to Die*. New York: Vintage Books.

Wilson, John. 1977. *Identity, Ideology, and Crises: The Vietnam Veteran in Transition*. A partial and preliminary report on the Forgotten Warrior Project. Funded under a grant from the Disabled American Veterans.

Cases Cited

Feres v. The United States of America, 340 U.S. 135, 1950.

In re *"Agent Orange" Product Liability Litigation* (MDL No. 381), 475 F. Supp. 928, (E.D.N.Y) 1979. 635 F. 2d 987, (2d Cir.) 1980. 454 U.S. 1128, 1981. 597 F. Supp. 740, (E.D.N.Y) 1984, (Weinstein, C.J.).

Reutershan v. The Dow Chemical Company, No. 78-CV-14365, (N.Y. Sup. Ct., N.Y. County), filed July 20, 1978.

Reutershan et al. v. The Dow Chemical Co. et al. and The United States of America,

No. 78-CV-4253, (S.D.N.Y.), filed January 9, 1979. Memorandum Supporting Plaintiffs' Amended Verified Complaint, filed July 15, 1979.

QUESTIONS

1. How did the media transform the problem of Agent Orange into a public issue?
2. Why are latent health disorders easier to turn into political conflicts than more visible, manifest disorders?

Chapter Twenty-Four

Environmental Politics and Science
The Case of PBB Contamination in Michigan

Michael R. Reich

Introduction

Contamination of the environment by toxic chemicals raises difficult problems of illness, safety levels, and disposal. These problems often contain large areas of scientific uncertainty and affect conflicting interests in society. When that happens, toxic problems are resolved not simply as technical matters but become complex public issues and political controversies.[1] The combination of scientific uncertainty and conflicting interests makes decisions about policy depend on value judgments and political bargaining, as well as scientific information. Another reason for political controversy is that dissatisfied groups seek to expand the scope of a public issue into the political realm to get what they want.[2]

The sources of uncertainty in environmental contamination are many. Environmental exposures are usually low-level and long-term, and can interact with other factors. The chemicals often accumulate and persist in the environment and in human tissue, creating a continued presence in the food chain and a continued internal toxic burden for people. The increased risk of cancer and of reproductive hazards due to environmental pollutants exist against a background of those same risks associated with other causes, making epidemiological studies extraordinarily difficult and indecisive.

Epidemiology has other uncertainties as well. Studies may be inadequate in design or measurement. The population under study may be too small to detect an infrequent but meaningful effect that is occurring. The population may be studied at an inappropriate interval after exposure, so that the effects have already occurred and disappeared or have not yet occurred. The study may involve a subpopulation not susceptible to the exposure. The institutional context of epidemiology also produces uncertainties. Those problems can result from lack of manpower, lack of resources, or lack of understanding of epidemiology's purposes and uses.[3]

These uncertainties of scientific information pose a dilemma for regulation. Steven Jellinek, a former official in the US Environmental Protection Agency (EPA), wrote that the regulator must make decisions about chemicals "in the midst of pervasive uncertainty." Since the regulator does not have "the luxury of putting off decisions

TABLE 24.1
Agricultural Cases of Chemical Contamination Reported in 1976

State	Year	Animal	Estimated value	Contaminant
Arkansas	1969	poultry	$ 4,000,000	Heptachlor
Missouri	1969	cattle	50,000	Dieldrin
Oregon	1969	cattle	250,000	Dieldrin
New York	1970	poultry	500,000	Dieldrin
New Mexico	1970	swine	63,000	Mercury
North Carolina	1971	swine	10,000	Dieldrin
New York	1971	poultry	1,000,000	PCB
Southeast US	1971	poultry	2,500,000	PCB
California	1971	poultry	50,000	PCB
Georgia	1971	poultry	2,500	Dieldrin
Maine	1972	poultry	3,000,000	PCB
Minnesota	1972	poultry	336,000	PCB
Maine	1972	poultry	150,000	Dieldrin
Missouri	1972	poultry	78,000	Dieldrin
California	1972	poultry	90,000	Dieldrin
North Carolina	1973	poultry	2,000,000	Chlordane
North Carolina	1973	poultry	88,000	Dieldrin
Louisiana	1973	poultry	20,000	Dieldrin
California	1973	lambs	25,000	HCB
Missouri	1973	poultry	567,000	PCB
Louisiana	1973	cattle	400,000	HCB
Mississippi	1974	poultry	6,900,000	Dieldrin
Michigan	1974	cattle & poultry	75,000,000* (approx.)	PBB

SOURCE: See reference no. 6.

* In 1979, the estimated cost of PBB contamination in Michigan was $215,000,000.

until certainty arrives," there exists an "inevitability of being wrong" sometimes.[4] Yet regulators prefer to present their decisions as if based on certainty. For public legitimacy, they prefer their decisions to appear grounded on scientific fact and to mask the margins of error. Scientific uncertainty thus tends to be overwhelmed by organizational demands to maintain routine procedures and to protect policy spaces. Once an agency becomes committed to a position, the uncertainty tends to fade away and the definition of the problem tends to resist change.

Some toxic victims and their allies contest the bureaucratic definitions of illness, safety, and disposal. They use social conflict to expose uncertainties and values of existing policies, and to provide pressures and incentives for changes in policy. In such controversies, scientists are not detached technical specialists but are active political participants. Cases of toxic contamination thus raise problems of politics in science and of science in politics.

In 1973 and 1974, the state of Michigan suffered one of the worst chemical disasters in United States history. By the end of 1975, as a result of contamination by poly-brominated biphenyls (PBB), about 28,900 cattle, 5,920 pigs, and 1.5 million chickens had been destroyed; buried around the state were 865 tons of contaminated animal feed, 17,940 lbs of cheese, 2,630 lbs of butter, 34,000 lbs of dry milk products, and nearly 5 million eggs.[5] No one knows the total cost of cleaning up the statewide contamination, but estimates reached the hundreds of millions of dollars (see Table 24.1).[6] In addition, even five years after the contamination, about 97 per cent of Michigan's residents showed measurable levels of the chemical and the most highly exposed groups showed little significant decline in PBB levels.[7] The possible long-term health consequences remain a point of concern and uncertainty.

This chapter explores the interaction of science and politics in the Michigan case of PBB contamination. It first reviews the discovery and the causes of the contamination. It then examines how and why the problems of illness, safety, and disposal became public issues and political controversies. The final section analyzes how public policies for PBB contamination were affected by bureaucratic processes, political conflict, and involved scientists.

A Private Trouble

For about one year, from the time the contamination occurred in May 1973 until the contaminant was identified in April 1974, the problem was considered a private trouble, the difficulty of a single dairy farmer. That farmer, Rick Halbert in southwest Michigan, first noticed health and production problems in his herd of 400 dairy cows in September 1973. Halbert checked for the usual infectious diseases, but the symptoms did not fit. His veterinarian then examined the cows, but he too could not diagnose the unfamiliar illness.

Both men suspected something wrong with the feed, possibly in a recent order of high-protein feed pellets supplied by Farm Bureau Services, Michigan's largest feed distributor and a subsidiary of the state's most important farmer organization, the Michigan Farm Bureau. Although the company denied any problems, Halbert decided to run an experiment and gave 12 of his own calves a diet of only the pellets. Within six weeks, five of the calves had died.

From the experiment with calves and from observation of the herd, Halbert and his veterinarian noted two phases in the cows' symptoms. The first phase included decreased appetite and milk production, and increased urination and tearing. In the second phase, cows developed hematomas and abscesses, abnormal hoof growth (becoming long and curling upward), matted hair that eventually fell out, and severe reproductive abnormalities.[8]

At Halbert's urging, the State Department of Agriculture repeated the feed tests, but the state used mice not cows. In two trials, all treated mice died. Even then, the president of Farm Bureau Services insisted that the feed was pure and healthy. The company veterinarian, Dr. James McKean, explained to Halbert that the mice died because they had eaten "cattle feed" not "mice food." Halbert considered that nonsense.[9]

In early 1974, however, the feed company began to accept that a problem existed in its high-protein feed pellets. The company hired research institutes to conduct chemical analyses of the feed and to perform a feeding trial on calves. The company, however, did not report the feed problem to public officials, and the company veterinarian misrepresented experimental results and other information to Halbert on several occasions.

Halbert, meanwhile, continued to seek help from state and federal scientists to identify the poison in the feed. Many responded that they could not study the problem of a single farmer. But others agreed to help.

In March 1974, a toxicologist analyzing the feed with gas-liquid chromatography accidentally left the machine on for the unusually long time of eight hours, producing an unexpected series of peaks and indicating the presence of an unidentified chemical. One month later, the feed was analyzed by low-resolution mass spectrometry. Halbert passed the results to a scientist in the US Department of Agriculture, who immediately recognized the compound as one he had worked with: polybrominated biphenyls, a flame retardant produced by Michigan Chemical Corporation. This company, Halbert learned, sold magnesium oxide to Farm Bureau Services, which added the substance to dairy feed, to increase cows' milk and buttermilk production. Apparently, there had been a mix-up.

In April 1974, when first informed of the mix-up, Michigan Chemical denied that its PBB product, Firemaster BP-6, could have been confused for its magnesium oxide (MgO) product, Nutrimaster. Company officials explained that the two products were stored and manufactured in separate buildings, and were totally different in consistency and color: BP-6 was chunky and amber, and MgO was granular and whitish. In addition, the company reportedly packaged the chemicals in color-coded bags, one bright red (BP-6) and the other royal blue (MgO).

On April 30, however, an inspector for the federal Food and Drug Administration (FDA) discovered a half-used bag of PBB in a Michigan feed mill, thereby linking Michigan Chemical with the sick cows. But the bag was Firemaster FF-1, from an experimental batch of PBB in which the chemical was ground into powder and mixed with an anti-caking agent (calcium polysilicate). The special processing transformed BP-6 into a substance remarkably similar to magnesium oxide in both consistency and color. Moreover, the discovered bag was *not* color-coded, and its label did not list ingredients or manufacturer. The plain brown bag showed only the trade name, Firemaster FF-1, stenciled across the top. Once the bag was opened, even those meager markings became nearly impossible to read.

Several factors at Michigan Chemical thus contributed to causing the mix-up. In spring 1973, Michigan Chemical ran out of color-coded bags and used plain brown 50-pound bags for both PBB and MgO. Neither product was clearly or adequately marked. Also, according to an internal company memorandum, the storage of the experimental FF-1 was "*very* poor," with broken bags in some warehouse areas.[10] Additional confusion occurred because Michigan Chemical used three different commercial names for its MgO product.

Problems also existed at the feed company, Farm Bureau Services, as detailed in sworn court statements of several employees. The men employed to mix the feeds had little job training, and one employee could not read well enough to recognize the word "Nutrimaster." Some employees, however, could read perfectly well and did report to a supervisor the appearance of a new trade name in the warehouse: "Firemaster." The supervisor told them it was just another name for MgO and to keep adding it as required.[11] Estimates of the amount of PBB introduced into the dairy feed, as a result of the mix-up, ranged from 500 to 1,000 pounds, although it could have been more.

Michigan Chemical first manufactured PBB in 1970 as a fire retardant for molded plastic parts, such as the cases of televisions, typewriters, and business machines. Firemaster BP-6 contained mostly hexabromobiphenyl (about 60 per cent), and several additional isomers. Production of Firemaster BP-6 rose rapidly, from 20,000 pounds in 1970 to 2.2 million pounds in 1972, to 4.8 million pounds in 1974. For Michigan Chemical and the company's owner, Northwest Industries, PBB became a successful product.

In the early 1970s, Michigan Chemical recognized some possible health problems with PBB. A private firm tested the acute toxicity of Michigan Chemical's PBB product and, in 1970, concluded that Firemaster BP-6 was non-toxic for ingestion or dermal application, not a skin or eye irritant, and not highly toxic when inhaled.[12] But in late 1971, Michigan Chemical prepared for workers a one-page health and safety statement on BP-6 that recommended against prolonged exposure and noted that the BP-6 probably accumulates in fatty tissue and the liver, "which certainly is undesirable and possibly could be dangerous." The statement warned against allowing BP-6 to contaminate any food or feed.[13]

Other private companies considered the probable chronic toxicity of PBB too risky for production. In the early 1970s, two of America's largest chemical companies, Dow and DuPont, separately performed animal laboratory tests on similar PBB compounds. Both companies decided not to manufacture a PBB product, because the tests showed detrimental toxic and environmental effects—evidence of liver damage, bioaccumulation, and high probabilities of carcinogenicity and teratogenicity—and because of knowledge about the high toxicity in humans of a related compound, polychlorinated biphenyls (PCB). The companies publicly announced their decisions and their research in 1972.[14,15]

The PBB mix-up in Michigan thus created a major case of contamination with a relatively unknown but probably toxic chemical. For about nine months, while the problem was considered Halbert's private trouble, Michigan farmers and consumers unwittingly ate dairy and other farm products contaminated by PBB.

A Public Issue

With the contaminant identified as PBB, state and federal agencies officially recognized the problem as not just Halbert's personal plight. Once the contaminant and the problem had a name, the problem was transformed from a "private trouble" into a "public issue."[16] Private and public institutions then began taking more effective steps to deal with the contamination.

But public officials were slow to understand the scope of the problem. Both federal and state governments initially underestimated the extent of the contamination. In May 1974, state agriculture officials repeatedly announced that PBB affected only a "very few farms." Officials publicly defined the problem as an agricultural problem affecting a small group of farmers.

Official policy also did not fully recognize the complexity of the contamination problem. In confronting PBB contamination in May 1974, government officials tried to deal with three questions: *illness*—Which people and animals are "sick"?; *safety*—What levels of PBB contamination are "safe"?; and *disposal*—How should contaminated goods be disposed? These three problems became the issue to be addressed, understood, and resolved. But as Michigan Health Director, Dr. Maurice S. Reizen, later recalled: "No one . . . immediately recognized the full extent of the problem, its urgency, or its real and potential impact on the economy, the health and the lives of Michigan people."[17]

In the spring of 1974, few people in Michigan or elsewhere had heard of PBB. According to one state official, Dr. Kenneth Wilcox, the Michigan Department of Public Health started with "absolutely zero knowledge" about the chemical.* Even federal agencies lacked information on PBB. The FDA, for instance, found few scientific studies to help determine what level, if any, might be safe in human food. The main experimental study available compared the capacities of PBB and PCB to increase the activity of the liver's microsomal enzymes, which are involved in metabolizing toxins and drugs. It showed PBB on a molar basis to be five times stronger than PCB.[18]

Nonetheless, on May 10, 1974, about two weeks after identification of the contamination as PBB, the FDA set an "action" level of 1.0 part per million for milk and milk products (on a fat basis). An "action" level is not a "tolerance," which requires formal and public procedures, but is a temporary administrative guideline that can be informally and quickly set by the FDA. It requires no public participation, can be based on incomplete scientific data, and can stand for years before a tolerance level is set. It provides for relatively quick regulatory action, despite scientific uncertainty.

FDA scientists chose the action level for PBB solely on measurement capability, officials explained, since the Food, Drug, and Cosmetic Act requires in cases of "avoidable" contamination that the guideline be set at the lowest detectable level. The scientists also regarded 1.0 ppm in milk and milk products as probably safe, since the PCB "temporary tolerance" in the US then was 2.5 ppm, and the PBB molecule is heavier than the PCB molecule.[19] About one month later, the US Department of Agriculture set the same level for meat.

Using the federal guideline for food safety, the State Department of Agriculture began to quarantine dairy farms in Michigan, attempting to keep additional PBB-tainted products off the market and to contain the contamination. By the end of May 1974, the state had quarantined 30 farms; these farms had all bought feed from Farm Bureau Services.

PBB poisoning in cattle and other animals was initially defined by the level of contamination. State and federal agencies used the FDA action level for human food to indicate the health of animals. In spring 1974, very little scientific literature existed about PBB effects in cattle. But officials reasoned that if humans could safely consume milk and meat contaminated up to 1.0 ppm of PBB, then the animals that produced the milk and meat would be considered healthy. That definition of illness in animals gradually became a focus of controversy, as farmers questioned its accuracy.

As with animal illness, the main controversy over illness in people centered on the

*Interview with Dr. Kenneth Wilcox, Bureau of Disease Control and Laboratory Services, Michigan Department of Public Health, August 22, 1978.

definition of PBB poisoning: Who was sick due to PBB contamination? Again, no scientific reports existed about the health effects of PBB contamination for people, but many studies had been published about human poisoning due to a related chemical, PCB. The high toxicity of PCB in humans suggested that PBB might also cause human health problems.

The State Department of Health first performed a screening of 211 farm people. The Department concluded in July 1974 that although PBB could be detected in the farmers' blood, and some people showed medical disorders, the results "have not revealed a medical syndrome, or group of symptoms, which can be related to PBB."[20] The State Health Department next designed a "short-term" epidemiological study of 300 persons, divided into a PBB-exposed group and a non-exposed control group. But the results of that study were not made public until spring 1975, when they became a focus of controversy, as discussed below.

The FDA also conducted a survey of health problems in all quarantined farm families in May and June 1974. That survey similarly found various human health problems, but the federal agency did not analyze or publicize the data. The survey and its results remained unknown to the subjects and to the public until revealed in 1977 by a Congressional investigation.[21]

In the matter of disposal, the state initially had no idea what to do with PBB-contaminated animals. It therefore instituted quarantines until it could decide on an appropriate policy. In early July 1974, nearly two months after the first quarantine, the governor signed a law to open a disposal site for killing and burying contaminated animals. The law allowed the state to condemn animals, to approve disposal facilities, and to use a civil suit to recover costs from the responsible companies. But state officials decided *not* to condemn any animals and *not* to order disposal, because they did not want to open the possibility for farmers to file suit against the state or for the state to be held financially responsible. The state thus quarantined farms and monitored the disposal operation, while the farmers and the companies decided on their own whether to destroy the animals. Farmers, in turn, felt enormous pressure to dispose of their animals. As Halbert put it: "Farmers were simply shut off from their markets and stuck with useless animals—after months of going backwards in double time, most farmers who were faced with this impossible ruinous situation agreed to have their animals destroyed."[22]

Another problem related to disposal arose in July 1974. People living near the disposal site, in the sparsely populated area of Kalkaska County in north-central Michigan, opposed the plan. Two days after the site opened, the county's commissioners filed a lawsuit to stop the disposal. The state had never informed local officials about the disposal plan. The trial delayed the start of disposal for another two months. But even after that time local opposition persisted to the burial of PBB-contaminated animals.

In November 1974, the FDA lowered the PBB action level for food from 1.0 to 0.3 ppm. Publicly, the agency explained that the reduction resulted solely from improved analytical capability, as required by law.

But other factors also influenced the decision. In the summer of 1974, some farmers began to complain about animals contaminated below 1.0 ppm that suffered from symptoms similar to PBB toxicity, with serious economic consequences. Then in the fall, two FDA veterinarians, Drs. Richard H. Teske and D. J. Wagstaff, defined in an unpublished study a chronic syndrome in cows of PBB poisoning that included higher rates of mortality, reproductive problems, and congenital abnormalities.[23] The study confirmed the farmers' observations about health disorders in cattle contaminated below the FDA action level of 1.0 ppm and provided an additional stimulus for change in policy. A third important factor was an experimental study independently initiated by a Michigan scientist, Dr. Thomas Corbett, showing that mice fed high doses of PBB developed gastrointestinal bleeding, enlarged livers, and birth defects—including cleft palate and brain defects.[24] After obtaining his results, Corbett urged state officials in September and October 1974 to lower the PBB action level.

By fall 1974, public officials recognized a broader contamination problem than initially understood. The contamination resulted not only from the initial mix-up of PBB and MgO, which affected herds like Halbert's at a high level, but also from widespread secondary contamination (see Figure 24.1).[25] Machinery that mixed feed passed PBB to other animal feeds and feed additives that did not directly include MgO. And animals unsuited for human consumption were slaughtered, processed, and added to feeds. These two feedback cycles helped produce a low level of chronic contamination throughout Michigan's farm animals—not just cattle—and throughout Michigan's food chain.

The new action level, by defining a much larger group of cattle as unfit for production or consumption, expanded the problem of disposal. By November 1974, the state had quarantined about 10,000 cattle, and about 9,000 had been killed and buried at Kalkaska. Under the new action level, the number of cattle for disposal rose rapidly. The State Department of Natural Resources had approved the Kalkaska site for the burial of 13,000 animals, but burial continued far beyond that number. The Department agreed to continue disposal there because no other site was available and because it seemed safer than burying on individual farms. In 1976, when the Department of Natural Resources closed the area, the Kalkaska site held about 30,000 animals.[26] While local residents argued that the burial site was not safe with that many decaying PBB carcasses, officials from the Department of Natural Resources contended the site posed no threat to public health.

The new action level also affected policies on compensation for damages. Farm Bureau Services and Michigan Chemical had begun in summer 1974 to settle out of court with farmers whose herds were quarantined by the 1.0 ppm level. The new action level greatly expanded the number of claimants against the companies, and raised the stakes for everyone involved.

A Political Controversy

Even with the new federal level in November 1974, some farmers complained about cows measuring below 0.3 ppm in milk and meat—"low-level" animals—that were sick and unproductive. The demand to lower the action level arose again from farmers but also from politicians and the press as the problem became a political

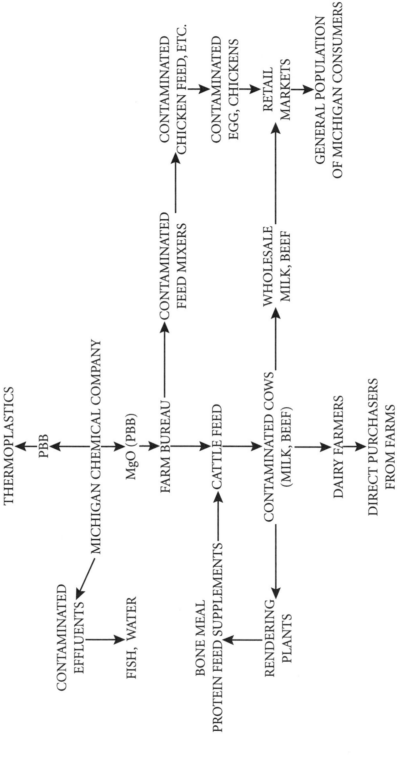

FIGURE 24.1

The Pathway of Polybrominated Biphenyl (PBB) Environmental Contamination in Michigan

SOURCE: Selikoff IJ, Anderson HA (reference no. 25).

438

controversy. These groups redefined the issues from an agricultural problem involving some farmers to a public health problem involving all Michigan consumers. And that new definition compelled changes in policy.

In early 1975, the Michigan press published prominent reports of human illness allegedly due to PBB. By March, state and federal officials had become concerned about public "panic" in Michigan over PBB.

Partly as a response, the FDA sent six veterinarians to Michigan in March to survey the health status of state dairy herds, especially to study health problems of low-level cattle (under 0.3 ppm). Their report concluded that no significant differences in health problems could be found between 16 "PBB-exposed" cattle herds and 15 "non-exposed" herds.[27] But that survey had various defects. Most glaring was that only two control herds were negative for PBB. Data on other herds were alleged to be "not available"—even one year after the survey's completion. Feed used on both exposed and control farms was contaminated with PBB. And one farm in the control group was later quarantined.

One FDA scientist recommended a more competent epidemiological survey of cattle, with follow-up studies, to separate two groups: cattle with low levels of PBB that were survivors of earlier high-level feed contamination and showed clinical symptoms; and cattle with low levels of PBB that were later exposed to low-level feed contamination only and showed no clinical symptoms.[28] The head of the veterinary team rejected the suggestion and the epidemiological approach.

Also in March 1975, the Michigan legislature passed a resolution urging the Agriculture Commission to hold a hearing on the removal of all food products showing a detectable level of PBB. To review the proposed reduction, the State Department of Agriculture held a public hearing in late May. Scientists from state and federal agencies testified that a lower level was not necessary, while one Michigan doctor (Dr. Walter Meester) and several farmers called for reducing the levels to protect public health. Not surprisingly, the Michigan Agriculture Commission rejected the proposed reduction and upheld the FDA action level.

At that hearing, the Michigan Department of Public Health presented the results of its short-term health survey. The study found no significant difference in health problems between the supposedly exposed and control groups, and therefore concluded that no disorders could be associated with PBB.[29] But a respected clinical toxicologist in Michigan, Dr. Walter Meester, criticized the state's survey, particularly because 70 per cent of the subjects in the control group had detectable PBB blood levels. Meester concluded that the state's study was "poorly planned, does not conform to the standards of adequate scientific, medical and epidemiological evaluation, was incomplete, possibly biased, and does not support the conclusions reached and publicized in the lay press."[30]

In fall 1975, farmers increased their public complaints and protests about the state's handling of the PBB contamination and about persistent animal health problems on farms contaminated at low levels. In November, farmer Al Green dramatically protested the state's refusal to help farmers with cattle contaminated by low levels of PBB. In an act that gained national media attention, he and other farmers with low-level cattle shot and buried the farmer's herd of 112 cows and calves.

Farmers moved their complaints to the political arena in early 1976. In March 1976, a committee from the state legislature held several public hearings in farm districts to collect first-hand accounts of problems associated with PBB contamination. At these meetings gathered farmers with cattle contaminated at low levels. As the farmers spoke publicly, they discovered common troubles. The farmers especially criticized bureaucrats who blamed herd health problems not on low levels of PBB but on "poor management"—that is, on the farmers themselves. Two weeks after the last hearing, farmers organized a march on the state capital, and dumped the carcasses of PBB-contaminated cows on the steps of the state capitol building to confront the politicians with the PBB problem.

Just before the farmers marched on Lansing in March 1976, Michigan's governor appointed an advisory panel of scientists to review all technical data on PBB. In late May, that blue-ribbon panel delivered its report and surprised the governor and his bureaucracy by unanimously recommending a reduction in the PBB guideline to the minimum detectable limits—because of the compound's similarity to PCB, its accumulation in tissue, and the high probability that it causes cancer and birth defects.[31] Subsequent research found neoplastic nodules in livers of four of five rats 10 months after a single dose of 1 gm/kg body weight, supporting early suspicions that PBB is carcinogenic.[32]

The State Agriculture Department, in June 1976, held another public hearing to review the FDA action level and the blue-ribbon report. At the hearing, FDA official Dr. Albert C. Kolbye, Jr. testified that available toxicological data suggested that 0.3 ppm was probably safe, and that the agency was no longer required by the Food, Drug, and Cosmetic Act to lower the action level, because the agency considered the contamination had changed from "avoidable" to "unavoidable." The Michigan Agriculture Commission followed the FDA and refused again to alter the PBB action level.

One study that the expert panel and the public hearing did not review was an FDA experimental feeding of PBB to beagle dogs that showed immunological effects. Although the study was begun in November 1975, and the animals sacrificed in January 1976, analysis of the data did not occur until October 1977, after a Congressional committee publicly questioned FDA officials about the experiment and its findings.[33]

When Michigan's executive branch refused to lower the PBB level, the state legislature began to debate a bill to lower the federal level within Michigan by state law. About one year later, in August 1977, the state legislature passed a law that reduced the state's PBB action level from 300 parts per billion (ppb) to 20 ppb in cattle fat, and required a test of each cow sent to slaughter. The Michigan Department of Agriculture and Michigan Farm Bureau had desperately opposed the bill, but public concern and political action about the public health consequences of PBB contamination forced approval of the law—four years after the contamination of Michigan began.

A key group active in pushing for a new PBB policy was farmers with herds contaminated at low levels. In August 1976, these farmers formed the PBB Action Committee, to bring their complaints about state policy to the politicians, to the press, and to the public. The group, stressing that low levels of PBB could damage animal and human health, helped redefine the PBB problem as a public health hazard to Michigan consumers. Many group members also had filed damage suits against the companies responsible for the PBB contamination.

The discovery of widespread PBB contamination of Michigan residents also helped redefine the PBB problem as a public health hazard. In August 1976, the Department of Public Health found that 22 of 26 samples of human breast milk from the general population in Michigan showed the presence of PBB, while none of 10 samples from women outside Michigan contained even a trace of the chemical. In announcing this news to the public, the health director expressed uncertainty about the health consequences of the finding but nonetheless did not recommend that women discontinue breast feeding.[34] Later, a more scientific study detected PBB in the breast milk of 95 per cent of nursing mothers tested in Michigan's lower peninsula.[35]

Additional evidence about the human health consequences of PBB became an important argument to lower the state's PBB action level. In November 1976, an eminent clinical epidemiologist from New York, Dr. Irving Selikoff, arrived in Michigan with a 35-member medical team to examine farm families exposed to PBB. In early January 1977, Selikoff stated in a preliminary report that real health problems existed among Michigan dairy farmers and that those problems could be related to PBB exposure.[36] Michigan Health Director Reizen responded: "This is the first time that I can recall we have knowledge about what PBB's effect is on people."[37]

The formal reports from Selikoff's study supported the early evaluation. A comparison of the Michigan farmers with farmers in Wisconsin found significantly more musculo-skeletal problems in the Michigan group, especially joint disorders (pain, swelling, and crepitation) and neurological symptoms (more tiredness, fatigue, headaches, dizziness, and unusually long sleep hours).[38] Two indicators of liver function, SGOT and SGPT, were also significantly higher in Michigan men than in Michigan women and than in the entire Wisconsin group. These increases did not correlate with alcohol consumption, thereby supporting PBB as probable cause.[39] Also, Michigan farmers showed immunological abnormalities (including decreased number of circulating lymphocytes and altered responses to tests of functional integrity of these cells) when compared with Wisconsin farmers and New York City residents.[40]

Selikoff's findings influenced public views and public policy in Michigan, despite a conflicting study by state and federal health officials. The government study found no dose-response relationship. It reported the most symptoms among farmers with low-level PBB cattle contamination but not enough for compensation. The report concluded that bias in selection factors and other non-PBB factors produced the appearance of a higher number of symptoms in the low-level PBB farmers when compared with the control group.[41]

The legal questions of animal health problems were settled through litigation and negotiation. While farmers with animals contaminated above 0.3 ppm were slowly receiving compensation from the insurance companies of Michigan Chemical and Farm Bureau Services, farmers with low-level contaminated animals continued with no payment whatsoever. Many of these low-level-contamination farmers filed civil damage suits against the companies involved. The first case went to trial in February

1977, and became the longest and most expensive litigation in Michigan history. In October 1978, after 16 months in court, 63 witnesses, and 25,000 pages of court transcript, the judge ruled against the farmer and for the companies. Simply stated, the judge decided that the farmer's lawyers had not proved any damage to the health of the dairy cows or any reduction in milk production due to low-level PBB contamination.[42] Other low-level farmers subsequently settled out of court for small fractions of their losses.

The question of human health problems due to PBB continued to be debated. In October 1977, lawyers for 246 persons filed a civil damage suit against the companies for PBB syndrome, including symptoms of fatigue, loss of balance, impotence, and aching joints and muscles. But this suit never went to trial. In 1979, as part of an out-of-court settlement, the plaintiffs withdrew their court claims of present (but not future) human health damages. In 1980, a representative of the State Department of Health wrote: "Although people have complained of health problems . . . it has not been proven scientifically that PBB is directly responsible." That person recognized, however, that no one has repeated Selikoff's survey of the low-level farmers,[43] suggesting that his findings have not been disproven. Many questions about health problems thus remain unresolved.

The question of disposal was raised again by the state's new action level in October 1977, since the state needed another burial site. Rural residents who lived near the newly selected site at Mio in Oscoda County formed a local chapter of the PBB Action Committee to oppose in court and in demonstrations the state's policy on disposal. Even though the cattle for burial at Mio contained extremely small amounts of PBB (estimated at a total of two ounces in 3,500 animals), opposition persisted. Under court order, some animals were buried at Mio.[44] Then in 1979, local opposition forced the state to ship 1,500 PBB-contaminated carcasses to a burial ground for radioactive wastes in Death Valley, Nevada.[45]

Another problem of disposal arose in the late 1970s, as some farmers discovered low-level PBB contamination of their lands and buildings. The presence of contaminated animals from 1973 to 1976 had created a diffuse and long-lasting pollution of soils and dusts on some farms. Even new animals brought to the farms became contaminated by low levels of PBB. The farmers once again confronted problems of whether the milk from their animals was safe, whether their animals were healthy, how to clean up their farms, and who would pay for damages.

The controversies over illness, safety, and disposal thus combined to make PBB a major political issue in Michigan. The organized group of farmers and the public sense of crisis pushed the state legislature to pass measures to remove the last PBB-contaminated cattle from farms and to provide loans to suffering farmers. Then, in the fall election of 1978 for governor, PBB contamination again became intensely political, when Democratic challenger William B. Fitzgerald made Republican Governor Milliken's alleged mishandling of the PBB disaster into a central issue of the campaign. In October 1978, when the judge in the low-level-contamination herd trial ruled against the farmer, he also suggested that the state government had not mishandled the contamination problem, although that was not an issue in the case. The judicial decision just before the election greatly boosted the incumbent governor's campaign, and two weeks later he was reelected. PBB then faded away as a political controversy.

Science and Politics

For about five years, with some ups and downs, Michigan people lived PBB politics.[44,46] During that period, the politics of PBB remained intimately interwoven with the science of PBB. That mixing of science and politics resulted from the complex nature of the issue, from the uncertain character of the science, and from the high stakes of the interests. These traits affect various technical controversies, not just chemical disasters like PBB contamination in Michigan. Such technical controversies can be better understood by analyzing three patterns in the interaction of science and politics: the power of government agencies to define the problem; the ability of political conflict to produce changes in policy for technical issues; and the role of scientists as political participants in controversies.

The Persistence of Bureaucratic Definitions

The definition of a problem tends to become frozen in the position of a bureaucratic agency, and thereby to resist change. Organizational theorists Cyert and March wrote that organizations seek to avoid uncertainty by following routine procedures and do not predict problems but respond to feedback. Organizations thus tend to "move from one crisis to another" while relying on standard operating procedures to make decisions. The structures and strategies of public and private institutions thus can contribute to inadequate understanding and inappropriate responses to new problems.[47]

In the PBB case, Halbert confronted the powerful inertia of public and private organizations, when state agencies and the feed company insisted that his cattle problem represented a private and not a collective trouble. Halbert struggled against the narrow definition of his problem, a definition that served as a form of blaming-the-victim.[48] Later, farmers with cattle contaminated at low levels struggled against similar efforts to blame the farmers for their contaminated cattle's health problems. As with other social problems, the PBB case reminds us that to influence policy individuals and groups must transform a private trouble into a public issue and often into a political controversy.[49]

In each area of illness, safety, and disposal, government officials publicly presented decisions as correct, even when surrounded by various uncertainties. The Michigan Department of Agriculture, for example, adamantly supported the 0.3 ppm level as safe and persistently defined low-level cattle as healthy. The Department's defense of the 0.3 level arose not only from its assessment of scientific evidence but also from a structural conflict of interest. The Department was supposed to protect both the common consumer and the agricultural industry, but the Department leaned more toward protecting the industry, through close ties to the Michigan Farm Bureau.

That structural conflict of interest represents a common problem of government agencies that must meet the interests of a specific constituency and the interests of a diffuse public. Federal agencies susceptible to this kind of structural conflict of interest include the Food and Drug Administration[50] and the Federal Trade Commission.[51] The situation within the Michigan Department of Agriculture parallels that of the Atomic Energy Commission which suffered from a conflict between promoting and regulating nuclear power and ultimately led to its reorganization into two separate agencies.[52] Such conflict can create a situation of "regulatory capture" in which the interests of the regulated industry come to dominate the actions of the regulatory agency.

The Michigan case also demonstrates how uncertain figures become "golden numbers," as Robert Socolow observed in another environmental conflict. He wrote: "A number that may once have been an effusion of a tentative model evolves into an immutable constraint.... Apparently, the need to have precision in the rules of the game is so desperate that the administrators seize on numbers (in fact, get legislators to write them into laws) and then carefully forget where they come from."[53] An organization becomes committed to a number or a policy not only for institutional interests but also for broader social and political reasons. Protecting the number or the policy comes to represent protecting organizational integrity.

Michigan health officials also became committed to a definition of PBB contamination as an agricultural problem and not a health problem, as affecting cattle and not people. Officials recognized persons with ill health but refused to admit a connection to PBB. The State Department of Health continued to defend its position that no "PBB syndrome" existed, unable to admit publicly the uncertainties and problems of the Department's initial epidemiological survey. That position resulted in a policy that lasted three years, a policy of no medical care or assistance to farm families which had consumed large quantities of PBB-contaminated products.

The Michigan case thus illustrates two basic structural problems for public health departments: the conflicts around organizational boundaries, and the tensions between public and private sectors.[54] Chemical contamination, like many technical problems with broad social and political consequences, does not fit easily within the bailiwick of a single agency. Health problems commonly overlap with questions of agriculture, environment, labor, schools, and other areas, thereby creating complex conflicts for health departments around responsibility, coordination, and communication with other government agencies. Such conflicts around organizational boundaries represent typical problems for institutions in a complex social environment. The structural dichotomy between public and private sectors is a long-standing tension for public health departments in the United States. Private medicine has traditionally opposed public provision of medical services, except under extraordinary circumstances. While health departments are responsible for the epidemiologic investigation of environmental problems (about which private medicine has little interest), the agencies lack mechanisms to provide victims of chemical contamination with health services (which would cross the interests of private medicine). Deciding what role public health departments take toward victims of toxic contamination will remain a source of continuing tension.

Recently, on August 12, 1982, the US House of Representatives Subcommittee on

Investigations and Oversight, Committee on Science and Technology, held hearings on the problem of compensating victims of toxic contamination and included a panel of witnesses on Michigan's PBB contamination. As several experts testified, bureaucratic difficulties in assisting victims of toxic contamination result in part from inadequacies of the legal system, especially tort law. The hearings discussed two bills to provide compensation to victims of toxic contamination (HR 9616 in 95th Congress and HR 5074 in 96th Congress). Those bills propose an administrative system of compensation to avoid problems created by litigation and to deal better with the uncertainty in scientific evidence on toxic contamination. Among other ideas for victim compensation discussed at the hearing were a national insurance system funded out of general tax revenues and a modified national insurance system specifically for victims of toxic contamination which would include the possibility of subrogation for provable losses (the fund could sue the responsible parties to recover its payments). While the hearings did not focus on a particular legislative proposal and did not reach a precise recommendation, the session demonstrated persistent public and political concern that the victims of toxic contamination are not receiving adequate redress from either public administration or private litigation. That concern has been heightened by recent petitions for bankruptcy filed by the Manville and UNR Corporations to halt litigation by victims of asbestos poisoning.

On the problem of disposal, state officials defined burial as the correct scientific solution, and they decided on a site without consulting local politicians or local residents. Officials sought to portray burial as a narrow technical solution, to give legitimacy to their choice of policy. But that emphasis on rapid burial had an important political dimension: to remove contaminated cattle from the public sphere as fast as possible. As Dorothy Nelkin noted about technical decisions in general, the definition of inherently political problems as technical reflects a greater value on efficiency than on democracy. "Yet technical planning limits public choice and threatens the widely held assumption that people should be able to influence decisions that affect their lives."[55] The efforts to bypass the social and political aspects of burial delayed disposal and increased the burden on farmers and, ironically, exacerbated the social and political problems of disposal. People refused to consider burial as simply a technical problem and refused to let public officials ignore the social and political dimensions.

Social Conflict and Changes in Policy

One commentator on environmental politics criticized the occurrence of conflict in "technical" decisions and proclaimed the need to control and reduce conflict.[56] The PBB case illustrates the difficulties of avoiding conflict in toxic contamination and also demonstrates important consequences of conflict in exposing uncertainties, redefining problems, and making policy makers more accountable. Conflict is especially important in changing persistent bureaucratic definitions and procedures. As Michel Crozier explained, a "bureaucratic system will resist change as long as it can; it will move only when serious dysfunctions develop and no alternatives exist."[57]

In the Michigan case, the main source of social conflict was farmers who owned cattle contaminated by low levels of PBB but who were not compensated for damages. Those farmers constituted a relatively concentrated group who believed they suffered health problems and unfair costs and therefore organized around the issue of PBB contamination. Political controversy created by those farmers and their allies helped expose uncertainties in the definition of illness in cattle and people and uncertainties in the levels of safety. In their protests, farmers used dramatic images of death and disease—the carcasses of cattle—to get public attention, to put pressure on politicians, and to challenge the authority of official statements. As other relatively powerless groups have done,[58] the farmers learned how to use "institutional disruptions" to create power that could force changes in public policy and force concessions from public and private institutions. They also learned how to create an organization (the PBB Action Committee), how to find allies in politics and in the press, and how to use both organization and allies to work for their goals.

A public issue becomes a political controversy as affected groups seek allies but also as political entrepreneurs seek issues. The structure of political competition in a society thus influences the form and the timing of the political controversy. In the Michigan case, some Democrats used the PBB issue to criticize the Republican governor and to advance individual political careers. But their actions also provided assistance to farmers who were spurned by their traditional political patrons in the Republican party, the Michigan Department of Agriculture, and the Michigan Farm Bureau. That PBB became more of a political controversy during elections was no accident. For elections provided an opportunity for low-level farmers to gain concessions from more powerful groups in society, as Piven and Cloward concluded generally for disadvantaged groups.[58]

Conflicts over disposal involved typical siting controversies: the "not-in-my-backyard" phenomenon, and the "if it's so safe, put it in the city" response. Those conflicts raised uncertainties about the safety of underground water supplies (in burial) and the problem of air pollution (for incineration). Protests in court and in direct action compelled politicians and bureaucrats to explain, justify, and improve the technical decisions. One political solution was to export remaining carcasses out of Michigan, to export the problem to someone else's (already contaminated) backyard in Nevada.

Conflict served to make policy makers more accountable by raising the political stakes. Conflict tended to expose organizational biases, such as the connection between the State Department of Agriculture and the Michigan Farm Bureau. In response to growing controversy, both federal and state politicians organized various public hearings, some weighted to support one side and one position, others designed to hear both sides and do nothing. Policy makers spoke at all hearings and sometimes were subjected to sharp questioning, to make them at least explain publicly the basis of decisions about the contamination.

Scientists in Controversies

The popular image of the scientist is the rational problem solver, somehow not affected by emotions or values. Former EPA official Jellinek, for example, set forth that conventional notion: "When confronted with a great deal of uncertainty, [scientists] avoid drawing conclusions, and instead call for additional careful study and research."[4] Government officials need that image of the rational scientist, especially when officials are seeking a source of legitimacy for a particular policy.

The PBB case shows, however, that actual scientists respond in various ways to uncertainty, not always upholding professional ideals and scientific principles, and sometimes submitting to organizational demands and private interests. The case shows how scientists can become deeply involved in public issues and political controversies. As James B. Conant put it: "The notion that a scientist is a cool, impartial, detached individual is, of course, absurd. The vehemence of conviction, the pride of authorship burn as fiercely among scientists as among any creative workers." Conant also warned that "this emotional attachment to one's own point of view is particularly insidious in science because it is so easy for the proponent of a project to clothe his convictions in technical language."[59]

In the first nine months, when the contamination was considered Halbert's private trouble, Dr. McKean, the veterinarian for Farm Bureau Services, who worked as a company scientist, showed some of the same conflicts as a company doctor.[60] The vet followed company policy and did not report the feed problem to public officials. And in his contacts with Halbert, the vet tended to protect his company against potential claims more than help the farmer with animal problems. The company vet thereby critically contributed to the long delay in public recognition of the feed contamination. That company scientist clearly acted with a greater commitment to private organization than to professional ideals or social ethics.

Government scientists also worked under organizational constraints. In the first nine months, these scientists were limited for organizational and budgetary reasons from doing research on a "private" problem. The state laboratory that performed feed tests on mice, for example, was not designed for research, lacking adequate personnel, equipment and funding. These scientists also lacked full information about Halbert's feed problem, information held by the company which could have hastened efforts to solve the puzzle. Some individual scientists stretched rules and budgets to study Halbert's problem, but most government scientists stuck by routines in organizations not designed to identify or research chemical contamination.

Once the problem became a public issue, scientists in state and federal agencies assumed major roles in managing the contamination problem. Their analyses of the extent of the problem became limited by organizational tendencies to underestimate the problem. State bureaucracies sought to maintain total control over scientific information, and the governor's office only began to seek outside advice in early 1976. Federal scientists, such as those in the FDA, saw themselves as professionals, but they also worked under organizational constraints. The study on chronic PBB poisoning in cattle in fall 1974 remained unpublished; and the memorandum critical of the FDA herd health survey of March 1975 did not result in further epidemiological efforts. Some scientists thus registered dissent within the organization but without pushing for a change in policy.

Nongovernment scientists who became active in the public issue often came from universities. Some Michigan scientists (Meester and Corbett) pursued epidemiological and experimental research on their own initiative, criticizing government positions, raising health issues, and providing legitimacy to complaints of farmers. Other scientists performed studies that supported positions of the private companies or public agencies involved, often with research funds from those institutions. Outside scientists with the greatest public impact were those like Selikoff, who maintained a strong scientific reputation, a prominent public presence, and an independent organizational base outside Michigan.

These comments suggest several conclusions about scientists in conflicts over chemical contamination and other technical controversies:

- Scientists figure on all sides of the issues;
- Opposing groups all seek to mobilize their own "legitimate" scientists and scientific data;
- The resolution of conflicts depends not only on the scientific information but also on the mobilization of scientists as well as supporters;
- In a polarized situation, few scientists and few scientific studies are perceived as "neutral"; even if not intentionally instrumental, scientists and studies become identified and used more by one side than another.

In the Michigan case, as in other cases of technical controversy, the more politically adept scientists had the greater impact on policy. It is misleading to view scientists as "disinterested and rational" and politicians as somehow the opposite. As a former staff member of the Presidential Science Advisory Council in the 1960s wrote: "The Science Advisor's most potent weapon was and is a reputation for clear, disinterested technical knowledge. If he does not use or understand the political process, he will be bypassed and ineffective, regardless of the quality of his technical advice."[61] The public scientist thus needs legitimacy from a neutral reputation but gains effectiveness from political action. The effective scientist in the area of public policy thus must know about the limits of scientific uncertainty, the demand for certainty by bureaucratic organizations, and the consequences of social conflict—and how to use those constraints. In sum, the scientist must know how to be a political actor.

ACKNOWLEDGMENTS

The initial research for this article was supported by a grant from the Ford Foundation. The article benefited greatly from comments by the column's editor, Barbara G. Rosenkrantz, and two anonymous reviewers, by Rose H. Goldman, Marc J. Roberts, Rashid Shaikh, Robert B. Leflar, Robert Quinn, David Morell, Fred Fry, Steven M. Soble, Les Boden, and George Weeks, and by discussion at a seminar of the Interdisciplinary Programs in Health (IPH) at the Harvard School of Public Health. Penny Kohn provided secretarial assistance in preparation of the manuscript. In preparing the final article, the author was supported as an IPH Research Fellow by Grant #CR807809 from the US Environmental Protection Agency.

NOTE

On November 18, 1982, the U.S. Environmental Protection Agency, the State of Michigan, and Velsicol Chemical Corporation announced a consent judgment of $38.5 million to settle clean-up costs associated with PBB and other chemical contamination from Michigan Chemical Corporation (purchased by Velsicol in 1970). The agreement included $13.5 million to the State and $500,000 to the EPA, to reimburse public expenses for clean-up at the Gratiot County Landfill, which was used by Michigan Chemical, and to settle a $120 million suit filed in 1978 by the State to recover costs of investigating and managing the PBB disaster. According to Velsicol officials, the company had already set aside the other $24.5 million, for clean-up of Michigan Chemical's factory in St. Louis, Michigan, and of toxic waste sites used by the company, and for materials and services in the State's clean-up of the Gratiot County Landfill. State officials considered the agreement a success for accelerating clean-up at Michigan's worst toxic waste dump, for avoiding more years of litigation, and for obtaining reimbursement at one estimate of the costs of PBB to state government—thereby helping contain recent conflicts in Michigan about toxic waste dumps and PBB.

REFERENCES

1. Reich M R: Toxic politics: a comparative study of public and private responses to chemical disasters in the United States, Italy, and Japan. PhD dissertation, Yale University, 1981.

2. Schattschneider E E: The Semi-Sovereign People, A Realist's View of Democracy in America. New York: Holt, Rinehart and Winston, 1960.

3. MacMahon B: Strengths and limitations of epidemiology. *In:* The National Research Council in 1979: Current Issues and Studies. Washington, DC: National Academy of Sciences, 1979.

4. Jellinek S D: On the inevitability of being wrong. Technology Review, August/September 1980.

5. Polybrominated Biphenyl (PBB). Lansing: Michigan Department of Agriculture, December 1975.

6. US Senate, Committee on Agriculture and Forestry: Report Loans for Agricultural Producers Suffering Losses Due to Chemical Contamination. Washington, DC: US Govt Printing Office, March 4, 1976.

7. Wolff M S, Anderson H A, Selikoff I J: Human tissue burdens of halogenated aromatic chemicals in Michigan. JAMA 1982. 247:2112–2116.

8. Jackson T F. Halbert F L: A toxic syndrome associated with the feeding of polybrominated biphenyl-contaminated protein concentrate to dairy cattle. J Amer Vet Med Assoc 1974; 165:437–439.

9. Halbert F, Halbert S: Bitter Harvest, The Investigation of the PBB Contamination: A Personal Story. Grand Rapids: Eerdmans Publishing, 1978.

10. Michigan Chemical Corp.: BP-6 Plant Housekeeping Inspection, Inter-Office Memorandum. Saint Louis, MI: Michigan Chemical Corp., 1973.

11. Courter P, Lehnert D: PBB: answers taking shape. Michigan Farmer, November 1, 1975.

12. Hilltop Research: Acute Toxicity and Irritation Studies on Firemaster BP-6. Miamiville, OH: Hilltop Research, 1970.

13. Michigan Chemical Corp.: Firemaster BP-6, Health and Safety Chicago: Michigan Chemical Corp., 1971.

14. Aftosmis J G, Culik R, Lee K P, et al: Toxicology of brominated biphenyls, I. oral toxicity and embryotoxicity. Toxicol Appl Pharmacol 1972; 22:316.

15. Norris J M, Ehrmantraut J W, Gibbons C L, et al: Toxicological and environmental factors in selection of decabromobiphenyl oxide as a fire retardant chemical. In: Polymeric Materials for Unusual Service Conditions, Proceedings of the Applied Polymer Symposia No. 22, New York, 1973.

16. Mills C W: The Sociological Imagination. New York: Oxford University Press, 1959.

17. Brody J E: Illnesses linked to chemical PBB. The New York Times, January 5, 1977.

18. Farber T M, Baker A: Microsomal enzyme induction by hexabomobiphenyl. Toxicol Appl Pharmacol 1974; 29:102.

19. Kolbye A C: Testimony in Hearings before the US Senate Subcommittee on Science, Technology, and Space, Committee on Commerce, Science, and Transportation, March 31, 1977. Washington, DC: US Govt Printing Office, 1977.

20. Michigan Department of Public Health: Press release. Lansing Michigan Department of Public Health, July 25, 1974.

21. Cordle F: Testimony in Hearings before the US House of Representatives Subcommittee on Oversight and Investigation. Committee on Interstate and Foreign Commerce, August 2 and 3, 1977. Washington, DC: US Govt Printing Office, 1977.

22. Halbert F: Testimony in Hearings before the US Senate Sub-committee on Science, Technology, and Space, Committee on Commerce, Science, and Transportation, March 28–30, 1977 Washington, DC: US Govt Printing Office, 1977.

23. Teske R H, Wagstaff D J: Review of the current status of polybrominated biphenyl (PBB) toxicosis in dairy cattle in Michigan. (Memorandum) Washington, DC: FDA, October 25, 1974.

24. Corbett T H, Beaudoin A R, Cornell R G, et al: Toxicity of polybrominated biphenyls (Firemaster BP-6) in rodents. Environ Res 1975; 10:390–395.

25. Selikoff I J, Anderson H A: A Survey of the General Population of Michigan for Health Effects of Polybrominated Biphenyl Exposure. Report to the Michigan Department of Public Health. September 30, 1979. New York: Environmental Sciences Laboratory, Mt. Sinai School of Medicine.

26. Clark E: Kalkaska burial site. (Memorandum) Lansing: Michigan State House Democratic Staff, September 28, 1976.

27. Mercer H D, Teske R H, Condon R J, et al: Herd health status of animals exposed to polybrominated biphenyl (PBB). J Toxicol Environ Health 1976; 2:235–249.

28. Kolbye A C: Low level polybrominated biphenyl contamination in dairy herds. (Memorandum) Washington, DC: FDA, June 25, 1975.

29. Michigan Department of Public Health: The Short-Term Effects of PBB on Health. Lansing: Michigan Department of Public Health, 1975.

30. Meester W D: Critique on the Michigan Department of Public Health Study. Grand Rapids, 1975.

31. PBB Scientific Advisory Panel: Report to William G. Milliken, Governor, State of Michigan, on PBB. Lansing, May 24, 1976.

32. Kimbrough R D, Burse V W, Liddle J A, et al: Toxicity of brominated biphenyls. (Letter to Editor) Lancet 1977; 2:602–603.

33. Appendix C. Hearings before the US House Subcommittee on Oversight and Investigations, Committee on Interstate and Foreign Commerce, August 2 and 3, 1977. Washington DC: US Govt Printing Office, 1977.

34. Michigan Department of Public Health: Press release. Lansing: Michigan Department of Public Health, August 19, 1976.

35. Brilliant L B, Wilcox K, Van Amburg G, *et al*: Breast-milk monitoring to measure Michigan's contamination with polybrominated biphenyls. Lancet 1978; 2:643–646.

36. Selikoff I J: PBB Health Survey of Michigan Residents, Initial Report of Findings. New York: Environmental Sciences Laboratory, Mt. Sinai School of Medicine, January 4, 1977.

37. Transcript of Governor Milliken's News Conference; Joint News Conference Held with House Speaker Bobby Crim, Dr. Maurice Reizen, and Dr. Irving Selikoff, on PBB. Lansing: Michigan Executive Office, January 4, 1977.

38. Anderson H A, Lilis R, Selikof I J, *et al*: Unanticipated prevalence of symptoms among dairy farmers in Michigan and Wisconsin. Environ Health Perspect 1978; 23:217–226.

39. Anderson H A, Holstein E C, Daum S M, *et al*: Liver function tests among Michigan and Wisconsin farmers. Environ Health Perspect 1978; 23:333–340.

40. Bekesi J G, Holland J R, Anderson H A, *et al*: Lymphocyte function tests among Michigan dairy farmers exposed to polybrominated biphenyls. Science 1978; 199:1207–1209.

41. Budd M L, Hayner N S, Humphrey H E B, *et al*: Polybrominated biphenyl exposure— Michigan. Morbid Mortal W Rep 1978; 27:115–116.

42. *Tacoma v. Michigan Chemical Corp, et al*, No. 2933, Cir. Ct., Wexford, MI, unreported opinion, October 26, 1978.

43. Climo M: Office of Communication Services, Michigan Department of Public Health. Letter to M. R. Reich, May 8, 1980.

44. Egginton J: The Poisoning of Michigan. New York: Norton, 1980.

45. Nevada gets first of PBB cattle. Lansing: The State Journal, January 23, 1980.

46. Chen E: PBB: An American Tragedy. Englewood Cliffs, NJ: Prentice-Hall, 1979.

47. Cyert R M, March J G: A Behavioral Theory of the Firm. Englewood Cliffs, NJ: Prentice-Hall, 1963.

48. Ryan W: Blaming the Victim (rev. ed.). New York: Vintage Books, 1976.

49. Cobb R W, Elder C D: Participation in American Politics. The Dynamics of Agenda-Building. Baltimore: Johns Hopkins University Press, 1972.

50. Turner J S: The Chemical Feast. New York: Grossman Publishers, 1970.

51. Cox E, Fellmeth R, Schultz J: The Nader Report on the Federal Trade Commission. New York: Grove Press, 1970, c1969.

52. Primack J, von Hippel F: Advice and Dissent, Scientists in the Political Arena. New York: Basic Books, 1974.

53. Socolow R H: Failures of discourse. *In:* Feiveson H A, Sinden F W, Socolow R H (eds): Boundaries of Analysis, An Inquiry into the Tocks Island Dam Controversy. Cambridge, MA: Ballinger, 1976.

54. Pickett G: The future of health departments: the governmental presence. Ann Rev Public Health 1980; 1:297–321.

55. Nelkin D: Science, technology, and political conflict: analyzing the issues. *In:* Nelkin D (ed): Controversy; Politics of Technical Decisions. Beverly Hills: Sage Publications, 1979.

56. Gladwin T H: The Management of Environmental Conflict: A Survey of Research Approaches and Priorities. New York: New York: New York University Faculty of Business Administration Working Paper, 1978.

57. Crozier M: The Bureaucratic Phenomenon. Chicago: University of Chicago Press, 1964.

58. Piven F F, Cloward R A: Poor People's Movements, Why They Succeed, How They Fail. New York: Pantheon, 1977.

59. Conant J B: Modern Science and Modern Man. New York: Doubleday Anchor Books, 1953, c1952.

60. Hardy H L: Beryllium disease: a clinical perspective. Environ Res 1980; 21:1–9.
61. Robinson D Z: Politics in the science advisory process. Technology in Society 1980; 2:153–163.

QUESTIONS

1. What do the actions of Michigan dairy farmer Rick Halbert suggest about the role "ordinary people" might play in getting an issue on public and policy agendas?

2. What role did the oppositional professional Irving Selikoff play in the political controversy over PBB contamination?

Chapter Twenty-Five

Environmental Health Research

Setting an Agenda by Spinning Our Wheels or Climbing the Mountain?

John Eyles

What Is Environmental Health?

Last (1987, p. 131) defines environmental health as "the aspect of public health concerned with all the factors, circumstances, and conditions in the environment or surroundings of humans that can exert an influence on human health and well-being." He adds: "One sharply defined aspect of the environment is the workplace, where many kinds of exposures to hazards have been identified." While this definition helps frame the content of environmental health, it begs the issue of public health. Last (1987, p. 1) answers this question himself by extending the purposes of medicine identified by Sigerist (1951; 1961), namely the preservation of health, restoration of health when sickness occurs, relief of suffering and the promotion of good health. Winslow (1920) furthers the argument by characterizing public health practice as the science and art of preventing disease, prolonging life, and promoting health and well-being through organized community effort for the sanitation of the environment, control of communicable infections, organization of medical and nursing services for the early diagnosis and prevention of disease, education of the individual in personal health, and development of the social machinery to assure everyone a standard of living adequate for the maintenance or improvement of health.

In these early thoughts, we see the kernel of what constitutes environmental health. It is about the prevention of disease and the promotion of health in environments or geographically defined populations. It is a field that is action oriented. It is an applied science, not particularly concerned with theory. Indeed, as Green and Ottoson (1994, p. 381) remark, "ecology and environment are convenient terms for everyday discussion, but they are concepts mainly of theoretical value to the health scientist. They represent bodies of scientific knowledge. They suggest ways of organizing reality and ideas so as to keep sight of the whole while trying to analyse and solve problems that represent only parts of the whole. Effective community health practice requires the concentration of efforts and actions on specific parts of the environment." Their view of theory is not positive. As we shall see, concepts of theoretical value determine approaches, questions and, to some extent, answers. But we may thus assert that the debate on what constitutes health is also of little practical value to practitioners in environmental or public health. Whether health is seen as functional capacity or as human potential, or as a resource for living (see Aggleton, 1988) serves merely as a backdrop against which practice occurs or is measured. It is the preservation, restoration and promotion of that which society conventionally sees as important, i.e. health, that drives the environmental health agenda which may be seen as not only disease prevention and health promotion, but also as the monitoring of particular environments for adverse human health effects and outcomes. It is this agenda that may lead to a spinning of our wheels and rather narrow development of our discipline.

The Content of Environmental Health

This argument may be advanced by briefly examining, through examples, the content of what we may term traditional environmental health. Setting the environmental health agenda in terms of disease prevention and health promotion means emphasizing actions to monitor and then alter the circumstances that lead to disease or good health. Further, we must recognize the significance of individual life-style behaviours and options, as well as the community or population effects of threats to the environment. Thus, while hazards may impact on individuals, their effects may be more wide-ranging. In this discussion, we do not want to replay the history of public health. We are all familiar with the story of public health and its part in the successful improvement of sanitary and living conditions (Rosen, 1958; Hamlin, 1991). We are further aware of the importance of nutrition and food quality in improving and maintaining human health (Howe, 1971; McKeown, 1976). We will select three examples to illustrate the content, namely food quality and safety, occupational hazards and urban environments. These examples help demonstrate the continued importance of "traditional" applied environmental health. After the examples, there will be a brief commentary on method and evidence to drive the argument forward . . . up the mountain.

Many traditional environmental health issues seem to have continued salience in countries of the South. As much as one-third of the world's population goes hungry most of the time. For them, food quantity is as important as quality. There are many nutritional deficiency diseases, including protein-energy malnutrition graphically documented by Learmonth (1978), vitamin deficiencies leading potentially to rickets (vitamin D), pellagra (vitamin B_3), xerophthalmia (vitamin A) and other problems, e.g. goitre (iodine) and anaemia (iron). These nutritional diseases are diseases of impoverished environments, modifiable with economic and political will. We should remember McKeown's (1976) estimate that some 50% of the decline in premature mortality in England and Wales was accomplished by improved nutrition.

In countries of the North, other environmental concerns dominate, e.g. over the use of pesticides in the production of food. Chlorinated hydrocarbons (including DDT) were the first and best known effective insecticides. But their mode of action is not precisely known. DDT is permanently stored in fat and concentrates in various food chains including breast milk (Rogan *et al.*, 1980). Its long-term effects on human health are not known and despite its beneficial role in malaria control, it itself is strictly controlled. Phenol-based herbicides such as 4,6 dinitro-ortho-cresyl can cause acute poisoning while the phenoxyaliphatic acids (2,4,5-T and 2,4-D) can also have an adverse effect on human health because of the presence of dioxins. Pesticides or herbicides are problematic because they are environmentally persistent, and accumulate in water and food chains. While such accumulations potentially affect many communities, especially in rural areas, other food concerns affect few but worry many. One of the latest is the food-poisoning outbreak of Escherichia coli 0157:H7 in hamburger meat in the Pacific Northwest which led to one death, and cases of diarrhea and renal failure. The outbreak led to a review of meat processing phases and recommendation of thorough cooking and the avoidance of raw or pink hamburger (Griffin and Tauxe, 1991; Powell, 1994).

The second example concerns occupational environments and their hazards, identified by Last (1987) as a key component in environmental health. Occupational health and safety has seen much progress within the late nineteenth century, primarily through labour laws, and regulation and compensation settlements. Further, through careful epidemiologic research, a great deal is known about the effects of dusts, gases, fumes, chemicals and noises of the workplace. For example, in 1895 bladder cancer was first described among a small group of workers engaged in the manufacture of magenta from aniline, which was incorrectly diagnosed as the most likely agent. During the 1914–1918 war, the manufacture of dyestuffs increased in Britain as German supplies were halted. The health problem was not, however, revealed by national occupational mortality statistics because the population-at-risk was small and included in other groups. (Classifications may hinder the search for associations and causations.) Tumours did develop though at younger ages than for non-occupational cancers, having an incubation period of 15–20 years and an individual risk of one in 5–10. Indeed, for one small group distilling β-naphthylamine the risk was one in one. In fact, in 1948 this product along with x-naphthylamine and benzidine were found to be responsible for producing the cancer. But the cancer was also found in rubber goods workers who were used as a "control group" and an analysis of death certificates found cable workers had an excess of the disease, rubber being used as insulation on cables. While the manufacture of β-naphthylamine was halted in Britain in 1952 and screening introduced, cases of bladder cancer still occur in rubber and cable workers. The causes are still not fully known (Eyles and Woods, 1983; also Morris, 1975).

More generally, Table 25.1 shows the major occupational hazards and those workers affected. In other words, we know a great deal about occupational hazards. While some work environments will continue to be inherently dangerous, prevention and legislation are the likely courses of action to ensure the protection and restoration of health, and the maintenance of healthy workplaces. Further, we should note that in food safety and occupational environments, the unit of analysis is the individual or work-site so that the conventional models, approaches and designs of epidemiology and environmental health are likely to discern significant associations between expo-

TABLE 25.1
Major Occupational Hazards and the Estimated Numbers of Workers Exposed in the United States

Potential dangers	Diseases that may result	Workers exposed (U.S.)
Asbestos	White-lung disease (asbestosis); cancer of lungs and lining of lungs; cancer of other organs	Miners; millers; textile, insulation and shipyard workers—estimated 1.6 million exposed
Lead	Kidney disease, anemia; central nervous system damage; sterility; birth defects	Metal grinders; lead-smelter workers; lead storage-battery workers—estimated 835,000 exposed
Arsenic	Lung cancer, skin cancer, liver cancer	Smelter, chemical, oil refinery workers; pesticide makers and sprayers—estimated 660,000 exposed
Benzene	Leukemia; aplastic anemia	Petrochemical and oil-refinery workers; dye users; distillers; chemists; painters; shoemakers—estimated 600,000 exposed
Cotton dust	Brown-lung disease (byssinosis); chronic bronchitis; emphysema	Textile worker—estimated 600,000 exposed
Coal dust	Black-lung disease	Coal miners—estimated 208,000 exposed
Coke-oven emissions	Cancer of lungs, kidneys	Coke-oven workers—estimated 30,000 exposed
Radiation (x-rays)	Cancer of thyroid, skin, breast, lungs and bone; leukemia; reproductive effects (spontaneous abortion, genetic damage)	Medical technicians, uranium miners; nuclear power and atomic workers
Vinyl chloride	Cancer of liver, brain, lung	Plastic-industry workers—estimated 10,000 directly exposed

SOURCE: From Occupational Safety and Health Administration; Nuclear Regulatory Commission, U.S. Departments of Energy, Interior, and Health and Human Services; Health and Welfare Canada, Green and Ottoson 1994, 468.

sure and outcome. This is not the case with the final example which considers urban and residential environments.

Housing and neighbourhoods surrounding our homes are usually seen as private matters. But given their importance in personal worth and status, social well-being and as commodities of stored wealth, these environments have become the most significant element of the traditional public health agendas. It was always this way (ignoring the emphasis on individual risk factors of much personal health promotion) with attention to sanitation, crowding, zoning, undesirable neighbours and facilities, and the consequences of the actions of others on this environment. For example, air pollution from industrial and automobile exhausts, home heating, incineration, open fires and dumps, crop spraying, road dust and chemicals affect all residential environments, albeit with differential impact. Table 25.2 identifies the seven major air pollutants in the US and their health effects. These effects are particularly bad where temperature inversions are easily created as in Los Angeles and Las Vegas. In addition, groups of cancer-causing pollutants have been identified. Of about 700 atmospheric contaminants, 47 have been labelled as recognized carcinogens (adequate evidence), 42 as suspected carcinogens (limited evidence), 22 as cancer promoters and 128 as mutagens. While emitted in small quantities, their potential adverse health effects are seen as severe (Green and Ottoson, 1994), although for individuals they may not add significantly to the burden of risk. It becomes difficult to sort out the effects of these pollutants alongside all other risk factors: the evidence is not clear because the situation is complex. There are similar difficulties with evidence with more concentrated "pollutants" in urban environments, namely toxic waste and leukemia risk (Brown and Mikkelson, 1990; Neutra *et al.*, 1991) and electro-magnetic fields and cancer risk (Savitz *et al.*, 1990; Jauchem and Merritt, 1991; Theriault *et al.*, 1994).

While by their nature examples are selective, a similar set of conclusions could be drawn from other environmental health issues. What comes over is a fragmented field, committed to action-oriented research and applying the highest epidemiologic standards to its empirical findings. Indeed, traditional environmental health is very much geared towards evidence-based action. In some instances, the evidence is so overwhelming and based on decades-old notions that saying more is merely embarrassing, viz. malnutrition, insanitary conditions, application of disease-prevention controls, especially in the South. In other instances where plausible biological pathways have been identified, e.g. occupational hazards, noise, air pollutants, there seems

TABLE 25.2

The Seven Air Pollutants For Which U.S. Standards Exist, Their Main Sources, and Their Health Effects

Pollutant	Description	Main sources	Health effects
Ozone	Main component of smog, formed in air when sunlight "cooks" hydrocarbons (like gasoline vapors) and nitrogen oxides from automobiles	Not directly emitted but formed from emissions of automobiles, etc.	Irritation of eyes, nose, throat; impairment of normal lung function
Carbon monoxide	By-product of combustion	Cars and trucks	Weakens heart contractions; reduces oxygen available to body; affects mental function, visual acuity and alertness
Particles	Soot, dust, smoke, fumes, ash, mists, sprays, aerosols, etc.	Power plants, factories, incinerators, open burning, construction, road dust	Respiratory and lung damage, in some cases cancer; hastens death
Sulfur dioxide	By-product of burning coal and oil, and of some industrial processes; reacts in air to form sulfuric acid which can return to earth as "acid rain"	Power plants, factories, space heating boilers	Increases acute and chronic respiratory disease; hastens death
Nitrogen dioxide	By-product of combustion; is "cooked" in the air with hydrocarbons to form ozone (smog); on its own, gives smog its yellow-brown color	Cars, trucks, power plants, factories	Pulmonary swelling; may aggravate chronic bronchitis and emphysema
Hydrocarbons	Incompletely burned and evaporated petroleum products; are "cooked" with nitrogen dioxide to form ozone (smog)	Cars, truck, power plants, space heating boilers, vapors from gasoline stations	Negligible
Lead	A chemical element	Leaded gasoline	Brain and kidney damage; emotional disorders; death

SOURCE: Adapted from U.S. Environmental Protection Agency, Green and Ottoson 1994, 514.

little further point in documenting more suffering. The key is to ensure that evidence remains firmly on the public agenda. This may not be the case as the deterioration in working conditions in countries like the US, UK and Canada continues, especially in non-unionized shops. In further instances, the evidence is equivocal, e.g. in relation to some pesticides and to much concerning urban and residential environments. In these instances, there may be a need to develop further epidemiologic investigations, although often the numbers involved and the size of the effect mean that the evidence is likely to remain equivocal. Further, especially where there appears to be no plausible pathway between biophysical condition and toxin or pollutant, it may be that we are simply looking at the wrong target. While such environmental health may not identify diseases, it may lead to the recognition of illness—the experiences of adverse functioning (Kleinman, 1988). In other words, psychosocial impacts and perception of problems may also be evidence of some adverse effect of the environment on human health and well-being. To recognize this point requires a change in mind-set concerning the practice of science, the nature of method and the acceptability of different types of evidence. Is this possible in what has been labelled traditional environment health? The answer is both no and nor should it be. The success of such endeavour has been based on careful epidemiologic investigation of the relationships between human health and a specific environment to preserve, protect and restore that health. This is an important agenda but, for a social science, it is limited. It does not seem to progress the discipline in that similar research questions are posed in (somewhat) different environments. We appear to be spinning our wheels. To progress, it is argued, theory must be engaged. How then might this mountain be climbed?

Environmental Health—A Basis for a Research Agenda

Perhaps it is necessary to reiterate why the mountain should be climbed? Environmental health seems, despite some issues concerning the equivocal nature of evidence, to have a vigorous research and policy agenda, focusing on environmental quality and impacts on human health. Indeed, through its relations with public and community health, and its emphasis on disease prevention and health promotion, this applied mandate may be seen as broadened and strengthened through examining individual risk factors for disease and ill-health in the context of environment. The inequalities debate (Townsend and Davidson, 1982; Smith and Jacobson, 1988) and the determinants of health framework (Canada, 1974; Evans and Stoddart, 1990; Ontario, 1991) now engage these factors and the environment with attention to the environmental burden of illness and environmental risk assessments (see Cutter, 1994). Further, traditional environmental health research employs respected scientific methods, largely derived from epidemiology, and has firm links with policy through public health departments.

But should these new debates simply be grafted on to the traditional? The answer is no. While recognizing the importance and vitality of this research and policy, the new debates point to the need for different ways of seeing and examining the world.

Such grafting appears to tread water, while to begin with a different stance is to point to challenging ways of examining the relationships between environment and health. The traditional way seems to assume that most problems can be understood by more precise measurements and that those identified relationships which do not have plausibility with respect to the criteria of causal science (see Susser, 1987; Jones and Moon, 1987) are in some ways "irrational" and therefore irrelevant. Traditional ways are not only driven by a particular view of science, but also of what constitutes a problem. While this view of science may be helpful, it is limited. Its problems with equivocal evidence are indicative of that limitation. If evidence does not fit our criteria, we should challenge the criteria as well as the evidence. To do that raises issues of the nature of science and of the nature of theory in science (see Eyles and Taylor, 1994). To engage in the new debates requires a reorientation, an agenda set by theory as a wide-ranging system of ideas for helping to understand and explain centrally important issues of social life (see Ritzer, 1992). As Stinchcombe (1968, p. 3) so succinctly puts it, "theory ought to create the capacity to invent explanations."

Directions for Environmental Health Research

The centrally important issues for this discussion are of course environment and health, conjoined as environmental health. Definition of the root-terms—environment and health—is problematic. They are core-values and life concerns, identified as such in the quality of life literature (Eyles, 1994). With respect to health, which is often used coterminously with quality of life and well-being, there seems to be a move away from seeing it in terms of potential—the complete physical, social and mental well-being—to the still positive, and contingent resource for living. In other words, health is that state that allows for effective functioning in society in the light of particular physical, demographic and social attributes. Its unit of analysis or attribution is the individual.

"Environment" conversely has been broadened in its scope. That breadth takes two forms. The first sees environment as a synonym of nature and emphasizes the relationships between environment and society (e.g. Simmons, 1993; Dickens, 1992). These relations are often expressed as world-views which orient action and behaviour, and inform on the nature of existence, truth, beauty, etc. In this way, world-views act as ideologies which, although partial, provide existential orientation for the members of a society (see Therborn, 1980). With respect to environment, Lynch (1981) suggests three normative theories, namely cosmic, in which there is a magical or mystical relation between environment and the gods to ensure order and harmony in the cosmos; machine, in which there are interdependent repairable parts the stability of which is ensured by their predictability; and the organic, which rejects the standardization of the machine and argues for a dynamic, self-regulatory entity the health of which is determined by a balance between diverse elements. This last view—the organic (or organismic)—enables the introduction of the second form—the extension of world-views (and environments) to encompass the natural world, viewed as "ecosystem" through which human beings must be understood by their relationships with (especially impact on) the natural world. While there is a questioning of the shape of these relationships and the meaning of "nature" (see Schrader-Frechette and McCoy, 1993), ecosystem is a powerful metaphor for environment. We may see its power in Dunlap and Catton's (1980) distinction between the dominant western world-view and the new ecological paradigm. In the former, people are seen as fundamentally distinct from and possess dominance over all other things on earth. Further, people can choose their goals and learn whatever is necessary to achieve them in a world of vast resources, providing unlimited opportunities for ceaseless progress. In the latter paradigm, while people are influenced by social and cultural forces, they are also affected by the biophysical environment and relations with other species in competition for space, food, and water. They are also shaped by physical laws, especially regarding the creation and transportation of energy, which cannot be overridden. The ecological paradigm (and organic world-view) are powerful ideas which shape the environmental agenda.

These ideas or world-views about environment are collective and relational, concerning humans as a species and the environment as a (fragile) resource. Health, on the other hand, is an individualized phenomenon, albeit still a resource. But these differences in scale mean that part of the environmental health agenda is to work out how environment and health are forged in environmental health. Until that time, a working definition of environmental health is the health and well-being of human populations in specific environments (physical, social, and societal). It is, however, recognized that the relationships between environment and health are multi-directional, and that just as environments affect human health so too do issues of human health and well-being affect environments. Given this working definition, how might we understand the effects of environments on human health and well-being? What systems of ideas are helpful for that understanding? In other words, what does the rest of this suggested environmental health agenda look like?

From that definition, the primary focus insofar as disentanglement is possible, is human health with environments forming the contexts within which disease, illness, and health issues are played out. To examine these relations and to understand how significant parts of social reality are formed and operate, five integrated building blocks are examined, namely:

(i) how individuals and groups talk about and perceive health and illness in specific locales and environments;
(ii) how individuals and communities find their place or identities in environments to negotiate everyday practices concerning health and illness;
(iii) how through practices, rationalization and routinization, structures are created that possess the power to affect ideas and actions about health and illness in particular environments;
(iv) how structures become integrated as systems which shape the nature of actions and the interrelations between structural elements in particular environments; and
(v) how the ecosystems (modified by human actions) in particular functions to impact on health, illness, and well-being.

Although not perfect fits, all these agenda items are related to particular theories, which in turn help pose questions, suggest perspective and shape approach to content issues. Respectively, these theories are symbolic interactionism, structuration, structuralism and post structuralism, functionalism and ecologism. This is not an exhaustive list. Indeed, it is possible to argue that Marxism and feminism are useful ways of examining structure and power, just as cultural theories can help exploration of language, perception, agency and negotiation. All these (and close variants, e.g. symbolic interactionism and sociological phenomenology) presuppose particular research and policy issues, asked and answered (through objectives and methods) in particular ways. These items will now be briefly discussed in turn, recognizing that different environmental health researchers may adopt different theoretical stances. Theories are not about truth, they are about explanation and must be judged by their utility. As Stinchcombe (1968, p. 4) observes "Constructing theories of social phenomena is done best by those who have a variety of theoretical strategies to try out ... the crucial question to ask of a strategy is not whether it is true, but whether it is sometimes useful." Debate should focus on utility, with truth measured by its usefulness rather than its definitions.

Language structures everything. We cannot think, perceive or act without the use of language. Through it, reality is objectified. In other words, experience is stabilized into discrete, identifiable objects in terms of material objects, ideas and roles. Language thus has symbolic force with symbols being used to represent whatever people agree they should mean (Charon, 1985). These shared meanings enable people to carry out distinctively human actions and interactions. For symbolic interactionism, interaction is the key process in understanding how the ability to think and perceive is developed and expressed. Interaction then produces perceptions and meanings which orient individuals for action. These perceptions and meanings are not codified for one time. They can be altered through interpretive schemes through which individuals face new situations and evidence (see Blumer, 1969).

This interactionist approach has been used to help understand how people construct risk through events, meanings and interactions, and how these perceptions of risk concerning health, children, security, and property lead to uncertainty and anxiety (see Eyles *et al.*, 1993a). Further, Schrader-Frechette (1991) has argued that two very different conceptions of reality and perceptions of risk clash over many environmental concerns ranging from carcinogenic pesticides to loss of global ozone. Experts regard lay beliefs and concepts as distrustful, biased, and irrational. The public, in demanding zero tolerance for pollutants like organochlorines, is unrealistic and, in failing to recognize the safety advances in the environmental field, is irrational. Further, Hadden (1991) has documented the differences between lay and expert perceptions of risks from hazardous waste, while others (e.g. Dunwoody and Neuwirth, 1989; Laird, 1989) have noted the importance of, and problems with, communicating risk information. But it does seem worthwhile utilizing the interactionist perspective to explore the perceptions and language of risk, anxiety and uncertainty with respect to environmental health. While there is a large technical literature on which to draw, the connections between risk perceptions and core-values are only just being made (e.g. Edelstein, 1988; Tonn *et al.*, 1990; Eyles *et al.*,

1993b). It may be that the causes of risk perception can be investigated in core-values through the generation of narrative and text, with a central research question being "how do people talk about risk?"

The second item, related to the first, engages agency and negotiation. Utilizing elements of structuration theory (Giddens, 1984), agency may be seen as the capabilities of people and their related behaviours. The individual is active, knowledgeable, and reasoning (see Dear and Moos, 1994). Individuals reflexively monitor their actions and can articulate why certain actions were carried out, although many motives for action operate unconsciously through taken-for-granted rationalizations or through the rules of language. (Our first research item is to try and reveal some of this "practical consciousness.") But agents do not operate atomistically. The importance of context is recognized through "duality of structure"—the mutual dependency of structure, i.e. rules and resources, organized as properties of social systems, and agency, i.e. the continuous flow of conduct. And also by the recognition that in interaction, relations between people or communities may be asymmetrical. It thus seems appropriate to examine the flow of human conduct as a negotiation process in which what things mean for the self and others is determined, and which is then transformed into feasible actions.

These theoretical ideas are relevant to help understand how individuals (and communities) determine their identities (through their capabilities and actions), and view themselves as healthy or sick, desirable or contaminated and so forth. For example, Eyles and Donovan (1990) explored how most individuals in three working-class, urban environments defined themselves as healthy despite significant chronic clinical illnesses, and adverse social and cultural circumstances, particularly sexism and racism. To be healthy is identified with effective functioning which is in turn linked to personal worth, hence the need to be healthy despite these burdens (see also Litva and Eyles, 1994b). Ill-health is negotiated away—out of the practices of everyday life. Indeed, such negotiation illustrates agency par excellence, as it "concerns events of which an individual is a perpetrator." Whatever happened would not have happened if that individual had not intervened (Giddens, 1984, p. 9).

This idea of negotiation in the practices of everyday life is furthered by Bourdieu's (1984) conception of field which provides the strategies through which individuals and communities seek to safeguard and improve their positions. Fields are arenas of struggle through which the asymmetries of negotiation can be recognized. Thus, field specifies position as a network of relations. It is possible to see the positions of the industrial company-public health group and the residents' group, as reported by Brown (1992) as fields, negotiating the ways of knowing and acting about environmental degradation and potential leukemia risk in Woburn, Mass. Further, the capabilities of groups and communities in the negotiation process may determine the location of what are regarded as undesirable facilities such as mental health care homes with zones of intolerance equating directly with economic well-being (see Taylor, 1988) or municipal solid or hazardous waste sites in which age, education, life cycle stage and socio-economic status are good predictors of willingness to accept such a facility (see Eyles et al., 1992 for a review). Agency and negotiation are key in understanding public facility location, providing a different perspective than the traditional focus of distance-decay effects and location-allocation models (see Joseph and Phillips, 1984). Thus, a key research question is "how do communities live with (negotiate) themselves when stigmatized as contaminated in some way?"

The third agenda-item, related to the second, concerns the creation of structures through rationalization and routinization. Indeed, it is possible to follow Giddens (1984), arguing that structures simply give form and shape to social life. They are not themselves frameworks (see Held and Thompson, 1989). In this light, structures enable as well as constrain human activity and thought. But more theoretical insight may be obtained from structuralism and poststructuralism, which assert that a structure is not a directly visible reality. It is a level of reality, the functioning of which constitutes the underlying logic of the system (see Godelier, 1972). The purpose of structuralism is, therefore, to provide objective, universal knowledge which is not provided in particular individual, historically given subjectivities (Anderson *et al.*, 1986). This knowledge may be achieved by deconstructing texts or phenonema of interest to discover asymmetries of power, the determination of actions and so forth (see, for example, Barthes, 1970; Eagleton, 1983; Dews, 1989). Such structural analyses have been undertaken in health care (e.g. Navarro 1978; Gough, 1979; Doyal, 1979) to examine the economic class structures that underlie the British and American systems. Further, Smith's (1981) study of black lung in Appalachia may be used to reveal how structural interests redefined the disease to ensure their continual relevance and dominance. Parallels may be seen in Jones and Moon's (1987) development of radical epidemiology in which structural context is of central importance. The environmental health arena awaits further such analyses.

But further insight comes from poststructuralism, particularly through the work of Foucault (1965; 1975) who has directly examined the asylum and the isolation of the mad and the birth of the clinic through medical practice. Foucault searches not only for "real" structures, but through his archaeology of knowledge (Foucault, 1966) that which determines what is possible and what can be said in a particular discourse (see Sheridan, 1980). Thus power is not only a matter of structural interests but also of the domination of knowledge. Institutions thus use the techniques and technologies of knowledge to exert control over people (Foucault, 1969) not in a conscious way because history is seen as lurching from one knowledge-based system of dominance to another. This structuring of knowledge—which is always contested—is a powerful tool for understanding paradigm shifts in environmental health (e.g. why from machine to ecologic?) and the nature of environmental health policy formulation. For that latter task, some formal policy analytic tools to identify stakeholders, interests and values that rest on knowledge-claims (see Sahatier and Jenkins-Smith, 1993) are also required. Thus, a key research question is why does "environment" appear to dominate "health" in policy discourse?

The fourth, related to the third item, concerns the integration of structures as practices into systems. Parsons (1937; 1951), in his development of functionalism, argues that social systems build from norms and values which, together with other actors, make up part of the environment. If interaction leads to gratification then there is reinforcement and the action is repeated. Repeated actions lead to certain expectations and responses so that rules or values develop. These values become institutionalized in a network of roles. This describes the subsystem of action encompassing the personality, cultural, biological, and social system. In Parsons' scheme, these connect to higher and lower order systems. Craib (1984) has correctly seen this as a theory for a filing system (see also Rocher, 1974). Further, functionalism, which thus involves identifying the parts of a social system is seen as descriptive rather than explanatory. It is also seen as individualistic rather than collectivist, emphasizing motivations and capabilities rather than order, conflict and context (Lockwood, 1967; Burger, 1977). But Parsons' idea on functional prerequisites for system survival and development remains useful.

There are four such prerequisites, namely adaptation (each system must adapt to its environment), goal attainment (each must have a means of mobilizing resources to achieve goals), integration (each must maintain internal coordination and develop ways of dealing with problems), and pattern maintenance (each system must maintain itself as nearly as possible in a state of equilibrium through pattern-variables of particularly, affectivity, instrumentality and specificity). Such ideas seem worthy of attention in exploring an important content area in environmental health research, namely healthy cities and healthy communities. The healthy cities movement (see Ashton, 1992) is predicated on reorienting social actions so that cities work better (integration) with due regard to the environment (adaptation) through mutual aid (pattern maintenance) and resource mobilization (goal attainment). Not only healthy cities can be usefully examined in this light, but so too may the determinants of health and achieving health for all research and policy arenas. These are based on system integration in which consensus rather than conflict are emphasized. Thus, examination of these environmental and public health fields may be couched in functionalist terms and the implications of such theorizing can be drawn out.

But there are variants in functionalism. Thus, Gouldner (1971) emphasizes that system integration is a matter of degree with dependence of parts [even at the level of health, economy, environment (see Hancock, 1990)] varying from complete to parallel. Merton (1968) too notes that unity and equilibrium are matters of degree, dependent upon the processes that bring them about. Merton in fact enriched functionalism by adding such ideas as dysfunction (an institution can have negative consequences for some parts of the system, e.g. noise in the workplace and its effect on human health), nonfunction (consequences apparently irrelevant to the system under consideration such as a "survival" from earlier times or a distant place, e.g. the apparent dismissal of folk ideas—medical plants, environmental practices—in modern medical care), manifest and latent functions (respectively, anticipated and unanticipated consequences of actions, e.g. the corporate definition of black lung with the unanticipated consequence of helping bind workers into unions), and anomie (the disjuncture between values and people's capacities to act in accordance with them, e.g. the limitations of healthy public policy such as non-smoking environments). While the examples may seem unrelated and forced, they point to the significance of using theories to get different views of particular environmental health subject-matter. But they all relate to a key research question: "how and why are environmental health policies made in particular ways?"

The final agenda item, related to the fourth, concerns the examination of a particular system—the environment or ecosystem—and its relations to human health and well-being. In some ways, this may be seen as a special case of system integration and analysis. Traditionally, this item has been examined by epidemiologists and medical geographers looking at the relations between environment and health. Important studies have examined the physiological and psychosocial effects of exposures to toxic events (e.g. Davidson *et al.*, 1986) or waste disposal facilities (e.g. Hertzmann *et al.*, 1987; Elliott *et al.*, 1993). Further, likely outcomes and their effects on such relationships can be quantified in risk assessments (see O'Brien, 1986). Such research is extremely important, but is perhaps more in keeping with the traditional research agenda in environmental health. Such research should also be critically appraised in terms of its implicit theoretical stances: a preliminary assessment suggests that much is based on functionalism (see Litva and Eyles, 1994a).

But if we recognize the importance of the definitional debate around environment and health, there is a qualitative difference in the theoretical stance as well as research content of "new" environmental health. Based on the work of the human ecologists and neo-ecologists (see Theodorson, 1961), ecologism has gone through several guises to connect "environment" and "health." Park (1936) in classical ecology attempted to apply systematically the basic theoretical scheme of plant and animal ecology to the study of human communities, through such concepts as natural area, invasion, succession, dominance, equilibrium, etc. Human society was seen as organized at the biotic level (that of existence, a subsocial, non-thoughtful level of struggle and competition) and the cultural (that of community, based on communication and consensus). Later work (e.g. Hawley, 1944) dismissed the distinction regarding all human relations as social. The main task of human ecology is thus seen to be the analysis of community structure, in terms of the division of labour. Community structure is seen as the organization of sustenance activities, which may perhaps be extended to way of life to connect with traditional geographic concerns established by the work of Vidal de la Blache (1913), Fleure (1919), and Geddes (see Freeman, 1965) on society-environment relations.

Today, ecosystem is the key construct. As Rolston (1992, p. 142) puts it "the ecosystem is the community of life; in it the fauna and flora, the species, have entwined destinies. Ecosystems generate and support life, keep selection processes high, enrich situated fitness, evolve diverse and complex kind." Ecosystem returns attention to plants and animals as well as their environing conditions. Rolston (1992, p. 142) continues "the plants and animals within an ecosystem have needs, but their interplay can seem simply a matter of distinction and abundance, birth rates and death rates, population densities, parasitism and predation, dispersion, checks and balances, stochastic processes." Ecosystems are selective systems. They are increasingly regarded as fragile systems too, requiring protection and human intervention. Indeed, there is concern over ecosystem health (Rapport, 1989; Costanza *et al.*, 1992) and ecosystem stress (Rapport *et al.*, 1985). The theory underpinning such research may be viewed as primarily organismic, emphasizing interdependence for survival, growth, and maturation. Schrader-Frechette and McCoy (1993) have highlighted the problematic, often normative, nature of such theory in the non-human world.

Ehrenfeld (1992) suggests that non-equilibrium theories, emphasizing the inverse relationship between diversity and stability, may be more appropriate.

What, though, is the relevance of such research to the environmental health agenda with its emphasis on human health and well-being? In some ways, the answer to this question brings us back to the definitional debate because its importance lies in the need to bring together human health and ecosystem issues. Thus, an extremely important part of the environmental health agenda is to engage the ecologically and biologically driven ecosystem health agenda with the impacts of such concerns on human well-being. The insertion of humans in the ecosystem approach has as yet been largely schematic (see Allen *et al.*, n.d.). But it may be achieved both quantitatively and critically. Quantitatively there is a growing body of literature on the human costs of environmental actions and the environmental costs of human actions (e.g. Pearce, 1992). This balancing of costs and actions drives much policy debate and formulation, as evidenced by initiatives such as restoration of wetlands, the clean-up of Hamilton harbour and the barring of certain pesticides. Critically, there is a need to examine the implications of the apparent anthropomorphism of the environment and the possibilities of bringing together human health and ecosystem health in more than mere assertion, although potential elitism of the ecological "ideology" has been identified (see Luke, 1991). There remains a need to assess policy critically to discover what drives the agenda and with what implications (e.g. Harrison, 1991; Jasanoff, 1991). In this, we can see close parallels with the research agenda of "system integration." More specifically in relating ecosystem and human well-being concerns, we can ask "are people willing to accept adverse health effects for current environmental life-style options?"

Conclusions

None of the topics or questions suggested in this environmental health agenda is new. In fact, most have long and/or encouraging track records, e.g. risk perception, environmental health economics, environmental policy analysis, the relations between environmental quality, and human health, and so forth. Further, all the agenda items are inextricably linked together and it is likely that different theoretical stances can be used to shed light on different agenda items. That should be expected. But what this chapter is at pains to do is to put theory (theories) front and foremost in our exploration of environmental health questions. Theories help us understand and explain the operation of the social world, why its parts are related in the ways they are and so forth. Theory informs empirical investigation, suggesting research directions and particular places to look for answers. No one theory answers or explains everything, hence the apparent potpourri of suggestions in this chapter, although there is a structure to the suggested building blocks (and theories) from the linguistic and perceptual to the systemic.

Table 25.3 summarizes the discussion in the chapter. In its suggestions for research, it must again be emphasized that much effort in many disciplines has already been focused on these topics. Further, recall the chapter established the antecedents of

TABLE 25.3
Environmental Health Research Agenda

Building block	One useful theory	Illustrative example	One suggested research topic
(i) Language and perception	Symbolic interactionism	Risk, anxiety, uncertainty	Language and perception of environmental risk
(ii) Agency and negotiation	Structuration	Lay beliefs and actions in everyday life	Living in a stigmatized community
(iii) Structure and power	Structuralism and poststructuralism	Conceptual and material domination	"Environment" and "health"
(iv) System integration	Functionalism	Functioning environments	Consensus and conflict in environmental health policy
(v) Ecosystem and well-being	Ecologism	Environmental quality definitional questions	Connecting ecosystem and human well-being

present-day environmental health research, these being seen as largely applied and empirical. This work has not been dismissed or rejected. It may be found under agenda item (v), although there remain questions about how "environment," "health" and "theory" are treated in traditional environmental health research. Theory must be seen as providing a reorientation of these traditional environmental health issues, enabling different perspectives on these themes. Engaging theory enhances the agenda by locating environmental health firmly in the social as well as health and natural sciences. This engagement is of course only one element for disciplinary progress, others being social relevance and a commitment to social justice. All this may appear to be climbing a mountain but it is not—such engagement is central to the role of science and the scientist. It does not distance science from society. The traditional linkages through thorough empirical investigation to policy are not broken. Our stance may become more critical. In any event, the nature of and reasons for particular stances (and policy positions) will become clearer than if theory was not engaged. Environmental health will be socially relevant if it spins its wheel. By climbing the mountain, it has the potential to contribute as it has always done and as a constructive critic. Let's climb!

REFERENCES

Aggleton, P. (1988) *Health*. Routledge, London.
Allen, T. *et al.* (n.d.) *The Ecosystem Approach*. International Joint Commission, Windsor.
Anderson, R. J. *et al.* (1986) *Philosophy and the Human Sciences*. Croom Helm, London.
Ashton, J. (ed) (1992) *Healthy Cities*. Open UP, Milton Keynes.
Barthes, R. (1970) *Elements of Semiology*. Beacon, Boston.
Blumer, H. (1969) *Symbolic Interaction*. Prentice-Hall, Englewoods Cliffs.
Bourdieu, P. (1984) *Distinction*. Harvard UP, Cambridge.
Brown, P. (1992) Popular epidemiology and toxic waste contamination. *Journal of Health and Social Behavior* 33, 267–281.
Brown, P. and Mikkelson E. (1990) *No Safe Place*. Univ. California Press, Berkeley.
Burger, T. (1977) Talcott Parsons, the problem of order in society and the program of analytic sociology. *American Journal of Sociology* 83, 320–334.

Canada (1974) A new perspective on the health of Canadians, Government Printer, Ottawa.

Charon, J. (1985) *Symbolic Interaction*. Prentice-Hall, Englewood Cliffs.

Costanza, R. *et al.* (1992) *Ecosystem Health*. Island Press, Washington.

Craib, I. (1984) *Modern Social Theory*. Wheatsheaf Books, Brighton.

Cutter, S. (ed) (1994) *Environmental Risks and Hazards*. Prentice-Hall, Englewood Cliffs.

Davidson, L. M. *et al.* (1986) Toxic exposure and chronic stress at Three Mile Island. In *Advances in Environmental Psychology*, eds. A. H. Labovits *et al.*, Vol 6. Erlbaum, Hillsdale.

Dear, M. J. and Moos, A. (1994) Structuration theory in urban analysis. In *Marginalized Places and Populations*, eds D. Wilson and J. O. Huff. Greenwood Press, Westport.

Dews, P. (1989) *Logics of Disintegration*. Verso, London.

Dickens, P. (1992) *Society and Nature*. Temple, UP, Philadelphia.

Dorn, M. and Laws, G. (1994) Social theory, body politics and medical geography, *Professional Geographer* 46, 106–110.

Doyal, L. (1979) *The Political Economy of Health*. Pluto Press, London.

Dunlap, W. and Catton, R. (1980) A new ecological paradigm for post-exuberant sociology. *American Behavioral Scientist* 24, 15–47.

Dunwoody, S. and Neuwirth, K. (1989) Coming to terms with the impact of communication on scientific and technological risk judgements. In *Science as Symbol*, eds L. Wilkins and P. Patterson Greenwood Press, Westport.

Eagleton, T. (1983) *Literary Theory*. University of Minnesota Press, Minneapolis.

Edelstein, M. (1988) *Contaminated Communities*. Westview, Boulder.

Ehrenfeld, D. (1992) Ecosystem health and ecological theories. In eds Costanza *et al.* op. cit.

Elliott, S. J. *et al.* (1993) Modelling psychosocial effects of exposure to solid waste facilities. *Social Science and Medicine* 37, 791–805.

Evans, R. and Stoddart, G. (1990) Producing health consuming health care. *Social Science and Medicine* 31, 1347–1363.

Eyles, J. (1994) Social indicators, social justice and social well-being. McMaster University, Centre for Health Economics and Policy Analysis, Working Paper 94–1.

Eyles, J. and Donovan, J. (1990) *The Social Effects of Health Policy*. Avebury Press, Aldershot.

Eyles, J. and Taylor, S. M. (1994) Environmental health research: the meaning of measurement and the measurement of meaning, paper presented at Environmental Epidemiology Symposium, Hamilton, Ontario.

Eyles, J. and Woods, K. J. (1983) *The Social Geography of Medicine and Health*. St. Martin's Press, New York.

Eyles, J. *et al.* (1992) Risk and anxiety in a rural population. In *La geografia medica e gli ecosistemi*, eds C. Palagiano *et al.* Rux, Perugia.

———. (1993a) The social construction of risk in a rural community. *Risk Analysis* 13, 281–290.

———. (1993b) Worrying about waste. *Social Science and Medicine* 37, 805–812.

Fleure, H. J. (1919) Human regions. *Scottish Geographical Magazine* 35, 94–105.

Foucault, M. (1965) *Madness and Civilization*. Vintage, New York.

———. (1966) *The Order of Things*. Vintage, New York.

———. (1969) *The Archaeology of Knowledge*. Harper, New York.

———. (1975) *The Birth of the Clinic*. Vintage, New York.

Freeman, T. W. (1965) *One Hundred Years of Geography*. Duckworth, London.

Giddens, A. (1984) *The Constitution of Society*. Polity Press, Cambridge.

Godelier, M. (1972) *Rationality and Irrationality in Economics*. New Left Books, London.

Gough, I. (1979) *The Political Economy of the Welfare State*. Macmillan, London.

Gouldner, A. (1971) *The Coming Crisis of Western Sociology*. Basic Books, New York.

Green, L. and Ottoson, J. (1994) *Community Health*. Mosby, St. Louis.

Griffin, P. M. and Tauxe, R. V. (1991) The epidemiology of infections caused by E. coli 0157: H7. *Epidemiological Review* 13, 60–71.

Hadden, S. G. (1991) Public perception of hazardous waste. *Risk Analysis* 11, 47–57.

Hamlin, C. (1991) *The Science of Impurity*. Univ. California Press, Berkeley.

Hancock, T. (1990) From public health in the 1980s to healthy Toronto 2000. In *Local Healthy Public Policy*, eds A. Evers *et al.* Westview Press, Boulder.

Harrison, K. (1991) Between science and politics. *Policy Sciences* 24, 367–388.

Hawley, A. N. (1944) *Ecology and human ecology*. Social Forces 22, 398–405.

Held, D. and Thompson. J. B. (eds) (1989) *Social Theory of Modern Societies*. Cambridge UP, Cambridge.

Hertzmann, C. *et al.* (1987) Upper Ottawa Street landfill site health study. *Environmental Health Perspectives* 75, 173–195.

Howe, C. M. (1971) *Man, Environment and Disease in Britain*. David and Charles, Newton Abbott.

Jasanoff, S. (1991) Cross-national differences in policy implementation. *Evaluation Review* 15, 103–119.

Jauchem, J. and Merritt, J. (1991) The epidemiology of exposure to electromagnetic fields. *Journal of Clinical Epidemiology* 44, 895–906.

Jones, K. and Moon, G. (1987) *Health, Disease and Society*. RKP, London.

———. (1993) Medical geography—taking space seriously. *Progress in Human Geography* 17, 515–524.

Joseph, A. and Phillips, D. (1984) *Accessibility and Utilization*. Harper and Row, London.

Kearns, R. A. (1993) Place and health. *Professional Geographer* 45, 139–147.

Kleinman, A. (1988) *The Illness Narratives*. Basic Books, New York.

Laird, F. N. (1989) The decline of deference. *Risk Analysis* 9, 543–550.

Last, J. (1987) *Public Health and Human Ecology*. Appleton and Lange, East Norwalk.

Learmonth, A. (1978) *Patterns of disease and hunger*. David and Charles, Newton Abbott.

Litva, A. and Eyles, J. (1994a) Coming out: exposing ourselves to social theory in medical geography. Paper presented at the Sixth International Symposium on Medical Geography, Vancouver.

———. (1994b) Health or healthy? *Social Science and Medicine* 39, 1083–1091.

Lockwood, D. (1967) Some remarks of "The Social System." In *System Change and Conflict*, eds W. J. Demereth and R. O. Peterson. Free Press, New York.

Luke, T. (1991) Community and ecology. *Telos* 88, 69–79.

Lynch, K. (1981) *Good City Form*. MIT Press, Cambridge.

Mayer, J. D. and Meade, M. S. (1994) A reformed medical geography reconsidered. *Professional Geographer* 46, 103–106.

McGlashan, N. (1972) *Medical Geography*. Methuen, London.

McKeown, T. (1976) *The Role of Medicine*. Blackwell, Oxford.

Merton, R. (1968) *Social Theory and Social Structure*. Basic Books, New York.

Morris, J. N. (1975) *The Uses of Epidemiology*. Churchill Livingstone, Edinburgh.

Navarro, V. (1978)*Class Struggle, the State and Medicine*. Martin Robertson, Oxford.

Neutra. R. *et al.* (1991) Hypotheses to explain the higher symptom rates observed around hazardous waste sites. *Environmental Health Perspectives* 94, 31–38.

O'Brien, B. (1986) *What are my Chances, Doctor?* Office of Health Economics, London.

Ontario (1991) *Nurturing Health*, Premier's Council on Health Strategy, Toronto.

Park, R. (1936) Human ecology. *American Journal of Sociology* 42, 1–15.

Parsons, T. (1937) *The Structure of Social Action*. McGraw-Hill, New York.

————. (1951) *The Social System*. Free Press, New York.

Pearce, D. (1992) *Economic Values and the Natural World*. Earthspan, London.

Powell, D. (1994) Risk communication for health professionals. In Risk communication: papers and proceedings, ed. Environmental Health Program. McMaster University Environmental Health Program, Working Paper Series 1, Hamilton, Ontario.

Rapport, D. (1989) What constitutes ecosystem health? *Perspectives in Biology and Medicine*, 33, 120–132.

Rapport, D. *et al.* (1985) Ecosystem behaviour under stress. *American Naturalist* 125, 617–640.

Ritzer, G. (1992) *Contemporary Sociological Theory*. McGraw-Hill, New York.

Rocher, G. (1974) *Talcott Parsons and American Sociology*. Nelson, London.

Rogun W. J. *et al.* (1980) Pollutants in breast milk. *New England Journal of Medicine* 302, 1450–1453.

Rolston, H. (1992) Challenge in environmental ethics. *In The Environment in Question,* eds D. E. Cooper and J. A. Palmer. Routledge, London.

Rosen, G. A. (1958) *A History of Public Health*. MD Publications, New York.

Sahatier, P. and Jenkins-Smith, H. (eds) (1993) *Policy Change and Learning*. Westview Press, Boulder.

Savitz, D. A., John, E. M. and Kleckner, R. C. (1990) Magnetic field exposure from electrical appliances and childhood cancer, *American Journal of Epidemiology* 131, 763–771.

Schrader-Frechette, K. (1991) *Risk and Rationality*. Univ. California Press, Berkeley.

Schrader-Frechette, K. and McCoy, E. (1993) *Method in Ecology*. Cambridge UP, Cambridge.

Sheridan, A. (1980) *Michel Foucault*. Tavistock, London.

Sigerist, H. E. (1951) *A History of Medicine*. Vol. 1. Oxford UP, New York.

————. (1961) *A History of Medicine*. Vol. 2, Oxford UP, New York.

Simmons, I. G. (1993) *Interpreting Nature,* Routledge, London.

Smith, A. and Jacobson, B. (1988) *The Nation's Health*. Kings Fund, London.

Smith, B. E. 1981. "Black Lung: the Social Production of Disease." *International Journal of Health Services* 11:343–59.

Stinchcombe, A. (1968) *Constructing Social Theories*. Harcourt, Brace and World, New York.

Susser, M. (1987) *Epidemiology, Health and Society*. Oxford UP, New York.

Taylor, S. M. (1988) Community reactions to deinstitutionalization. In *Location and Stigma,* eds C. J. Smith and J. A. Giggs. Unwin Hyman, Boston.

Theodorson, G. A. (ed) (1961) *Studies in Human Ecology*. Harper and Row, New York.

Therborn, G. (1980) *The Power of Ideology and the Ideology of Power*. Verso, London.

Theriault, G. *et al.* (1994) Cancer risks associated with occupational exposure to magnetic fields. *American Journal of Epidemiology* 139, 550–555.

Tonn, B. E. *et al.* (1990) Knowledge-based representations of risk beliefs. *Risk Analysis* 10, 169–184.

Townsend, P. and Davidson, N. (1982) *Inequalities in Health*. Penguin-Harmondsworth.

Vidal da la Blache, P. (1931) Des caractères distincifs de al gégraphie *Annales de Géographie* 22, 289–299.

Winslow, C. A. (1920) The untilled fields of public health. *Science* 51, 23–28.

QUESTIONS

1. What types of issues are on the environmental health agendas in countries of the South, and in countries of the North?

2. In what ways has the concept "environment" been broadened in scope over the past several decades?

3. How does Eyles see symbolic interactionism contributing to the environmental health research agenda?

Contributors

David Allen is Professor and Chair of the Department of Sociology at the University of New Orleans.

Thomas A. Arcury is Senior Research Associate at the Center for Urban and Regional Studies, University of North Carolina.

Colin K. Austin is Senior Research Associate at the Center for Urban and Regional Studies, University of North Carolina.

Martha Balshem is Associate Professor at the University of Portland.

Barbara Berney works at the Massachusetts Public Health Association in Newton, Massachusetts.

Phil Brown is a Professor of Sociology, Brown University.

Lawrence Busch is a University Distinguished Professor at the Department of Sociology, Michigan State University.

Stella M. Čapek is an Associate Professor of Sociology/Anthropology, Hendrix College.

Barry I. Castleman is an environmental scientist.

Stephen R. Couch is a Professor of Sociology and the Director of the Center for Environment and Community at Pennsylvania State University.

Veena Das is a professor at the School of Economics, New Delhi University, in New Delhi, India.

Dawn Downey is a Research Associate at the Department of Epidemiology and Public Health, University of Newcastle upon Tyne.

Elaine Draper is a Visiting Scholar at the Institute for the Study of Social Change, University of California at Berkeley.

John Eyles is a Professor in the Environmental Health Program, McMaster University, Hamilton, Ontario, Canada.

H. Hugh Floyd is a Professor of Sociology at Stamford University.

Valerie J. Gunter is an Associate Professor of Sociology at the University of New Orleans.

Eve Hudson is a Research Associate at the Department of Epidemiology and Public Health, University of Newcastle upon Tyne.

Judith Kirwan Kelley is an Assistant Professor of Sociology in the Criminal Justice Department at Curry College.

Sheldon Krimsky is a Professor of Urban and Environmental Policy at Tufts University.

Steve Kroll-Smith is a Research Professor of Sociology at the University of New Orleans.

Lynn Lawson is a freelance writer and editor of the *Canary News*.

Gerald E. Markowitz is a Professor of History at John Jay College and CUNY Graduate Center.

Suzanne Moffatt is a Lecturer in the Department of Epidemiology and Public Health, University of Newcastle upon Tyne.

Peter Phillimore is a Senior Lecturer of Social Policy at the University of Newcastle upon Tyne.

Sara A. Quandt is an Associate Professor of Public Health Sciences, at Wake Forest University School of Medicine.

Michael R. Reich is a Professor at the Harvard School of Public Health.

David Rosner is a Professor in the Program in the History of Public Health and Medicine, Columbia School of Public Health.

Rosa M. Saavedra works at the Southern and Appalachian Leadership Training Program, Highlander Center. She was a community organizer at the North Carolina Farmworkers' Project while co-authoring "Farmworker and Farmer Perceptions of Farmworker Chemical Exposure in North Carolina."

Wilbur J. Scott is a Professor and Chair of the Department of Sociology, University of Oklahoma.

Janet Siskind is an Associate Professor of Anthropology living in New York City.

Sandra Steingraber works at the Women's Community Cancer Project in Cambridge, Massachusetts.

Keiko Tanaka is a Lecturer, Department of Sociology, University of Canterbury.

Patricia Widener is an MA candidate and Research Assistant at the Department of Sociology, University of New Orleans.

Steve Wing is an Associate Professor of Epidemiology at the School of Public Health, University of North Carolina.

Grace E. Ziem is an occupational health physician.

Index

Agency for Toxic Substances and Disease Registry (ATSDR), 11, 49, 52, 53, 72, 238, 243, 247
Agent Orange, 270–271, 274–275, 406, 409–429
air pollution, 215, 221, 224, 227, 229
asthma, 13, 14, 222

Beck, Ulrich, 17, 23, 386–388, 394, 401
Bhopal, India, 216, 270–286, 385
black lung disease, 9, 10
bovine spongiform encephalopathy (BSE), 12
breast cancer, 20, 290, 292, 296

Canadian Food and Drug Directorate, 110
cancer registries, 287, 290–293, 297, 306, 377
Carson, Rachel, 19, 20, 21, 23, 293, 295, 296–298, 340
Carter, Jimmy, 168
Center For Disease Control (CDC), 243, 249–250, 259, 262–263, 267, 368–369, 371–372, 377
chlorofluorocarbons (CFCs), 96–97, 100, 301–302
Citizens Clearing House for Hazardous Waste (CCHW), 389–393
Clean Air Act, 246, 247
Clean Water Act(s), 105, 247
Commoner, Barry, 1, 5
contamination: chronic, 18; toxic, 8

Department of Defense (DOD), 413, 421
Department of National Health and Welfare (Canada), 114
dichlorodiphenyl trichloroethane (DDT), 98–101, 103, 301, 339, 454
dioxin, 352–353
diseases, contested, 4

endometriosis, 343, 345–363
environmental carcinogen, 290
environmental endocrine hypothesis (EEH), 93, 95–107
environmental epidemiology, 215, 217–218
environmental illness, 48, 72, 288
environmental justice, 357
Environmental Protection Agency (EPA), 2, 4, 13, 14, 20, 47, 65, 77, 101–102, 104–105, 176, 247–248, 252, 261, 339, 353, 371–372, 376, 414, 420, 430

environmental risk factors, 218
epidemiology, 10, 29–45, 373
Erikson, Kai, 386–387
ethylene dibromide (EDB), 99
erucic acid, 94, 109, 114

Food and Drug Administration (FDA), 433, 435–438, 447

General Accounting Office (GAO), 101–102, 264, 414, 416
genetic monitoring, 202–204
genetic screening, 196–197, 199–202
genetic testing, 194, 196–197, 200
Giddens, Anthony, 16, 24, 386–388, 463
Grinder's disease (see also silicosis), 150–157
Gulf War Syndrome, 14, 15

knowledge, experiential, 367, 394, 399
knowledge, expert, 84, 380, 385, 385–387
knowledge, folk, 378
knowledge, lay, 366, 376, 380
knowledge, local, 85, 388, 396, 400
knowledge, New Age, 83
knowledge, problems of, 4, 22
knowledge, uncertain, 4

lead epidemiology, 236, 251
Lead Paint Poisoning Prevention Act, 240, 246–247
lead poisoning, 15, 235–269, 457
leukemia, 364, 368, 370
lung cancer, 295

Maximum Allowable Concentrations (MACs), 121–122
McKibben, Bill, 2, 5
methyl isocyanate (MIC), 271, 273, 276–278, 282–284
Mine Safety and Health Administration (MSHA) 163
Montreal Protocol (1987), 96
multiple chemical sensitivity, 11, 28, 48, 72–89, 288, 344, 394–398, 401–402

National Academy of Sciences, 15, 395
National Cancer Act, 293

National Cancer Institute, 303
National Institute for Occupational Safety and
    Health (NIOSH), 120, 125, 138, 144, 163, 166–
    172, 247–248
National Institutes of Health, 96
National Research Council, 14, 48, 77
Natural Resources Defense Council, 96

Occupational Safety and Health Act, 124, 166,
    170
Occupational Safety and Health Administration
    (OSHA), 76, 94, 102, 120, 125, 127, 130, 136–
    138, 140, 163, 166–167, 169–172, 198, 202, 248

polybrominated biphenyls (PBBs), 406, 430–452
polychlorinated biphenyls (PCBs), 100, 102, 301,
    348, 354, 436
polyvinyl chlorides (PVCs), 348, 354
popular epidemiology, 18, 219, 343, 364–383
postnatural environment, 16, 93
postnatural world, 2, 3, 4, 5, 16
pesticides, 2, 4; aerial spraying of, 77

radiation: exposure, 38; ionizing, 39; releases, 49
Reagan, Ronald, 168

silicosis (*see also* Grinder's disease), 143, 148, 162–174

tetrachloroethylene (PCE), 368
threshold limit values (TLVs), 94, 120–141, 166–167,
    197–198
Toxic Substances Control Act, 124
toxic tort, 129, 270–272; litigation overview, 171
toxins, 4, 19; environmental, 10
trichloroethylene (TCE), 368, 377–378
2,4,5-trichlorophenoxyacetic acid (2,4,5-T), 303, 386,
    398–401, 406, 411–412, 414, 417, 455

U.S. Department of Health, 76
U.S. Housing and Urban Development (HUD),
    246, 250, 261

Veterans' Administration (VA), 409, 412–413, 415–
    417, 420, 424–425

Worker Protection Standard (WPS), 176–177, 189